Breeding Farm Animals: An Integrated Approach

Breeding Farm Animals: An Integrated Approach

Editor: Roger Greer

CALLISTO REFERENCE

www.callistoreference.com

Callisto Reference,
118-35 Queens Blvd., Suite 400,
Forest Hills, NY 11375, USA

Visit us on the World Wide Web at:
www.callistoreference.com

ISBN: 978-1-64116-047-6 (Hardback)

Cataloging-in-Publication Data

Breeding farm animals : an integrated approach / edited by Roger Greer.
   p. cm.
Includes bibliographical references and index.
ISBN 978-1-64116-047-6
1. Livestock--Breeding. 2. Animal breeding. I. Greer, Roger.
SF105 .B74 2019
636.082--dc23

Table of Contents

Preface

This book was inspired by the evolution of our times; to answer the curiosity of inquisitive minds. Many developments have occurred across the globe in the recent past which has transformed the progress in the field.

The study of techniques and methods of breeding farm animals fall under the field of animal husbandry. This field is concerned with breeding animals and managing livestock in an agricultural setting for the purpose of obtaining meat, milk, eggs, leather, etc. Practices of animal rearing are integral to the science of cultivation. The principles of animal breeding are developed from the study of population genetics, molecular genetics, statistics, etc. Artificial insemination, selective breeding, intensive animal farming are some of the popular techniques of this field. This book provides information about certain key concepts and practices related to animal breeding while also discussing new methods and systems that have come into place owing to the technological advancement in the past decades. This book also presents researches and studies that have transformed this discipline and aided its advancement. It aims to equip students and experts with the advanced topics and upcoming concepts in this area. Those who are looking to develop a thorough understanding of breeding farm animals will be immensely benefited by the extensive content offered in this book.

This book was developed from a mere concept to drafts to chapters and finally compiled together as a complete text to benefit the readers across all nations. To ensure the quality of the content we instilled two significant steps in our procedure. The first was to appoint an editorial team that would verify the data and statistics provided in the book and also select the most appropriate and valuable contributions from the plentiful contributions we received from authors worldwide. The next step was to appoint an expert of the topic as the Editor-in-Chief, who would head the project and finally make the necessary amendments and modifications to make the text reader-friendly. I was then commissioned to examine all the material to present the topics in the most comprehensible and productive format.

I would like to take this opportunity to thank all the contributing authors who were supportive enough to contribute their time and knowledge to this project. I also wish to convey my regards to my family who have been extremely supportive during the entire project.

Editor

Dietary carnosic acid, selenized yeast, selenate and fish oil affected the concentration of fatty acids, tocopherols, cholesterol and aldehydes in the brains of lambs

Agnieszka J. Rozbicka-Wieczorek, Katarzyna A. Krajewska-Bienias, and Marian Czauderna

The Kielanowski Institute of Animal Physiology and Nutrition,
Polish Academy of Sciences, 05-110 Jabłonna, Poland

Correspondence to: Marian Czauderna (m.czauderna@ifzz.pl)

Abstract. The function of the brain is to exert centralized control over the other internal organs and tissues of the body. Thus, the objective of our studies was to evaluate changes in the concentration of fatty acids (FAs), cholesterol (CHOL), cholest-4-en-3-one (CHOL-4-3), tocopherols, malondialdehyde (MDA) and fatty aldehydes in the brains of lambs fed supplemented diets. Thirty male Corriedale lambs with a body weight of 30.5 ± 2.6 kg were allotted to five groups of six lambs and housed individually. After the preliminary period, for 35 days the animals were fed a diet containing 3 % rapeseed oil (RO) (the RO diet), a diet enriched with 2 % RO and 1 % fish oil (FO) (the FO diet) or the diets with combined addition of 2 % RO, 1 % FO, 0.1 % carnosic acid (CA) (the CA diet) and 0.35 ppm Se as the selenized yeast (SeY) (the CASeY diet) or selenate (SeVI) (the CASeVI diet). The CASeVI diet most efficiently increased the accumulation of FAs (including unsaturated FAs), CHOL-4-3 and fatty aldehydes in the lamb brain. This diet most effectively decreased the concentration of CHOL and MDA in the brain. The CASeY diet showed a different impact on the level of FAs, CHOL, CHOL-4-3, tocopherols, MDA and fatty aldehydes in the brain as compared with the CASeVI diet. The CA diet reduced the concentration of CHOL-4-3, the sums of fatty aldehydes, FAs, atherogenic- and thrombogenic-saturated FAs in the brain compared with the CASeVI diet; the CA diet most effectively increased the value of polyunsaturated FA (PUFA) peroxidation index in the brain. The RO diet most efficiently increased the concentration of CHOL and values of the ratios of saturated FAs to PUFAs and long-chain n-6PUFAs to long-chain n-3PUFAs in the brain.

The current studies provide new useful information for nutritionists carrying out further investigations aimed at improving farm-animal health, growth performance, reproductive system and the nutritional quality of feed for ruminants.

1 Introduction

Reactive oxygen agents are known to elicit oxidative damage of cholesterol (CHOL), its metabolites (like cholest-4-en-3-one) and polyunsaturated fatty acids (PUFAs), especially long-chain PUFAs (LPUFAs) (Xiao et al., 2009; Orth and Bellosta, 2012; Sultana et al., 2013). Cholest-4-en-3-one (CHOL-4-3) and other CHOL metabolites inhibit body weight gain and body fat content (Suzuki et al., 1998). The brain is sensitive to oxidative damage since it contains high contents of CHOL, its metabolites and oxidizable LPUFAs

(like C22:6n-3 (DHA) and C20:4n-6 (ArA)) (Patterson et al., 2012; Rosa et al., 2013) and is poorly endowed with protective antioxidant enzymes (Stadelmann-Ingrand et al., 2004). Oxidative damage of PUFAs leads to the formation of aldehydic end-products, including malondialdehyde (MDA), 4-hydroxy-2,3-nonenal (4-HNE) or 4-hydroxy-2,3-alkenals (4-HAKs) of different chain length; 4-HNE and 4-HAKs are active molecules in physiological or pathological conditions (Sultana et al., 2013). Aldehydes may be produced during decomposition of hydroperoxides of fatty acids (FAs) (Albert et al., 2013). Aldehydes can affect several cell functions, including signal transduction, gene expression, cell proliferation and more generally the response of the target cells (Stadelmann-Ingrand et al., 2004). Fatty aldehydes are naturally occurring species in tissues (Dannenberger et al., 2006; Chen at al., 2011); for example, during the synthesis of sphingosine by the brain, palmitoyl CoA is reduced by NADPH to palmitic aldehyde, which is then reincorporated into dihydrosphingosine (Sultana et al., 2013).

The susceptibility of tissues to a radical attack is a function of the balance between the magnitude of the oxidative stress and its own antioxidant potential. Studies have shown that adequate levels of Se, carnosic acid (CA) or tocopherols in diets protect against the oxidative degradation of CHOL, CHOL metabolites (like CHOL-4-3), unsaturated FAs (UFAs) and accumulation of carbonyl moieties on proteins produced by oxidative stress (Han et al., 2009; Morán et al., 2012). CA can modify microbiota, resulting in changes of the ruminal bacterial metabolism (Jordan et al., 2013). Similarly, Se compounds are used as nutritional sources of Se. The chemical form of dietary Se affects a synthesis yield of Se complexes and Se proteins in ruminal microorganisms (Mainville et al., 2009). Se derived from dietary selenized yeast (SeY) is more efficiently accumulated in the animal's body than inorganic forms of Se (e.g. selenate) (Navarro-Alarcon and Cabrera-Vique, 2008). The principal physiological roles of Se proteins are to maintain the appropriate metabolism of ArA and low levels of free radicals within cells, thus decreasing oxidative stress and peroxidative damage of lipids and CHOL in tissues (Tapiero et al., 2003; Schweizer et al., 2005; Yu et al., 2008).

Considering the above, we hypothesized that selenate (SeVI) or especially SeY added to the diet with CA and fish oil (FO) would reduce levels of fatty aldehydes, MDA, CHOL and CHOL-4-3 (the neuro-protector; Xiao et al., 2009) in the ruminal brain. Importantly, the brain comprises the body's central nervous system (Martin et al., 2004; Ogata et al., 2011).

Thus, the first aim of our study was to explore effects of FO added to the diet with rapeseed oil (RO) on contents of FAs, CHOL, CHOL-4-3, tocopherols and MDA (the marker of lipid peroxidation) and fatty aldehydes in the lamb brain. We expect that 1 % FO (rich in LPUFA) added to the diet with 2 % RO will be stimulated the oxidative stress in the brain compared with the diet with 3 % RO. RO (rich in C18:2n-6 (LA)) was added to the diet because RO stimulated the synthesis of health-promoting conjugated LA isomers in ruminants (Lee and Jenkins, 2011); moreover, LA is the precursor of ArA, which plays important physiological functions in the ruminant brain (Patterson et al., 2012; Rosa et al., 2013).

The second aim of our study was to investigate effects of different chemical forms of Se (as SeY or SeVI) added to the diet with CA, FO and RO on levels of FAs, tocopherols, MDA, CHOL, CHOL-4-3 and fatty aldehydes in the brain. We expect that SeY and SeVI added to the diet will be re-

Table 1. Chemical composition of the concentrate-hay diet with vitamins and mineral mixture (the basal diet) and odourless fish oil (FO)[a] fed to lambs.

Specification[b]	Meadow hay[d]	Concentrate[c]		
		Barley meal	Soybean meal	Wheat starch
Dry matter (%)	88.4	87.6	89.7	87.3
Crude protein (%)	9.50	9.94	41.8	0.90
Crude fibre (%)	27.3	2.87	4.34	–
Crude fat (%)	3.40	2.50	2.25	0.09
Ash (%)	4.85	1.84	6.16	0.12
Neutral detergent fibre (%)	59.2	18.0	18.8	–
Acid detergent fibre (%)	32.1	4.61	6.44	–
Acid detergent lignin (%)	4.47	1.14	1.49	–

[a] The iodine value of FO: 50–65 g 100 g^{-1} FO; the acid value of FO: 20 mg KOH g^{-1} FO;
[b] % in dry matter (DM);
[c] The main fatty acids in concentrate (µg g^{-1}): C14:0 104, C16:0 3189, C18:0 1425, c9C18:1 774, C18:2n-6 29163, C18:3n-3 1014; the gross energy (MJ per kg of dry matter (DM)): barley meal: 16.3, soybean meal: 17.8, wheat starch: 16.7;
[d] The gross energy: 17.1 MJ per kg of DM.

duced oxidative stress in the brain compared with the diet with CA, FO and RO.

2   Material and methods

2.1   Animals and experimental design

Thirty male Corriedale lambs were carefully selected from a large herd of lambs ($\sim$ 110 male lambs). Thus, selected lambs have similar body weight (23.3 $\pm$ 2.1 kg) and age (82–90 days). During a 3-week preliminary period the lambs were given free access to the basal diet (the standard concentrate-hay diet with vitamins and mineral premix) (Table 1); water was offered ad libitum. The basal diet (BD) consists of the following components: meadow hay ($\sim$ 36 %), a mixture of soybean meal ($\sim$ 36 %), barley meal ($\sim$ 16.5 %), wheat starch ($\sim$ 9 %) and mineral–vitamin mixture (20 g kg^{-1} BD). The basal diet contains crude protein 120 g, crude fibre 12 g, and 11 MJ metabolizable energy in 1 kg dry mater. The basal diet was enriched in 3 % rapeseed oil (RO) (the RO diet) or 2 % RO and 1 % odourless FO (the FO diet) (Table 2).

After a 3-week preliminary period, 30 lambs with an average body weight (BW) of 30.5 $\pm$ 2.6 kg at the beginning of the experiment were individually penned and divided into 5 groups of 6 lambs (Table 2). The animals were distributed into five groups according to the initial weights of lambs, so that the average initial body weights of lambs between the groups were similar. This study was conducted under the authority of the Third Local Commission of Animal Experiment Ethics at the University of Life Sciences, 02-786 Warsaw, Poland.

In the first experiment (Table 2), for the next 35 days following the 3-week preliminary period the lambs were fed the RO diet (the non-supplemented diet; group RO) or the FO diet (group FO).

Table 2. The experimental scheme and the composition of the experimental diets, the body weight (BW) of lambs and brain masses[a].

Group	Additives added to the basal diet	The body weights of lambs		Body weight gain	Brain mass[g] (g)
		$BW_{initial}$[b] (kg)	BW^c (kg)	$(BWG)^d$ (%)	
RO[e]	3 % RO (RO diet)	30.7 ± 3.2	36.3 ± 3.4	18.5 ± 1.3[a]	115 ± 8[a]
FO[f]	2 % RO, 1 % FO (FO diet)	30.6 ± 2.4	37.7 ± 2.1	23.4 ± 1.1[b]	123 ± 10[a]
CA[f]	2 % RO, 1 % FO and 0.1 % CA (CA diet)	30.6 ± 2.6	37.2 ± 2.3	21.5 ± 1.1[a]	112 ± 9[a]
CASeY[f]	2 % RO, 1 % FO, 0.1 % CA and 0.35 ppm Se as SeY (CASeY diet)	30.3 ± 2.7	36.8 ± 2.7	21.6 ± 1.5[a]	113 ± 8[a]
CASeVI[f]	2 % RO, 1 % FO, 0.1 % CA and 0.35 ppm Se as SeVI (CASeVI diet)	30.3 ± 3.0	38.5 ± 3.1	26.8 ± 1.6[b]	119 ± 8[a]

[a] Results are expressed as means ± standard deviations (SDs). Mean values in columns having the different superscripts are significantly different at
[a,b] $P < 0.05$. Statistical analyses were carried out between groups RO and FO and between groups CA, CASeY and CASeVI;
[b] The average initial body weight (kg) of lambs after the 3-week preliminary period;
[c] The average body weight ($BW_{35\,days}$; kg) of lambs after 35 days of the experimental period;
[d] The relative body weight gain of lambs after 35 days of the experimental period; BWG,% = [($BW_{35\,days}$ − $BW_{initial}$) × 100 %]/$BW_{initial}$;
[e] For the 3-week preliminary period the lambs were fed the basal diet enriched in 3 % RO;
[f] For the 3-week preliminary period the lambs were fed the basal diet enriched in 2 % RO and 1 % FO;
[g] The concentration of Se in the brains of lambs fed the CASeY and CASeVI diets were higher than in the brains of lambs fed the diets without addition of SeY or SeVI (i.e. Se concentrations of the lamb brains per unit dry mass: 0.43 ± 0.01 and 0.302 ± 0.007 ppm, respectively).

In second experiment (Table 2), after the 3-week preliminary period, for 35 days the lambs were fed the FO diet enriched in 0.1 % CA (the CA diet; group CA), the FO diet enriched in 0.1 % CA and 0.35 ppm Se as SeY (the CASeY diet; group CASeY) or the FO diet enriched in 0.1 % CA and 0.35 ppm Se as SeVI (the CASeVI diet; group CASeVI).

All diets were formulated to be isoenergetic and isonitrogenous. All diets were adjusted weekly and supplied as two equal meals at 07:30 and 16:00 each day to ensure free access to dosed feed. The diet allowance was changed weekly according to body weights of lambs. Animals completely ate served portion meals. All lambs were fed the same mass of freshly prepared diets with the appropriate additives (appropriate amounts of RO, FO, CA, SeY or SeVI were daily added to the concentrate and then vigorously mixed; Table 2). The average daily diet intake was 1.08 kg per lamb. Fresh drinking water was always available. The lambs were slaughtered at the end of the 35-day experiment (i.e. at 07:00–08:00). After 12 h of starving, lambs were made unconscious by the intramuscular ration of xylazine (2–4 mg 10 kg^{-1} BW). The whole brains were removed, weighed (Table 2), homogenized and frozen. All brains were stored in sealed tubes at −32 °C until chromatographic analyses; each brain sample was analysed separately.

2.2　Chemicals

HPLC-grade acetonitrile, methanol and GC-99 %-grade *n*-hexane were purchased from Lab-Scan (Dublin, Ire-

land); other reagents were of analytical grade (POCh, Poland). Methyl ester standards of a conjugated linoleic acid (CLA) isomer mixture and all other methyl ester standards of fatty acids, fatty aldehyde standards, cholest-4-en-3-one, 5α-cholestane (the internal standard), 2,4-dinitrophenylhydrazine, 2,6-di-*tert*-butyl-*p*-cresol, sodium selenate (SeVI), 1,1,3,3-tetramethoxy propane and 25 % BF_3 in methanol were provided by Sigma-Aldrich (St Louis, MO, USA). Chloroform, dichloromethane (DCM), methanol, tocopherol standards, cholesterol, KOH, NaOH, Na_2SO_4 and concentrated HCl were purchased from POCh (Gliwice, Poland). All other chemicals were of analytical grade and organic solvents were of HPLC grade. Carnosic acid (CA) was purchased from Hunan Geneham Biomedical Technology Ltd. (Changsha Road, Changsha, Hunan, China). Rapeseed oil (RO) and odourless fish oil (FO) were supplied by company AGROSOL (Pacanów, Poland). RO comprised the following main fatty acids (μg g^{-1} RO): C14:0 56, C16:0 13091, *c9*C16:1 33, C18:0 5490, *c9*C18:1 85859, *c12*C18:1 786, LA 282394, C18:3*n*-3 (αLNA) 74, C20:0 194, *c11*C20:1 108, C22:0 430 and *c15*C24:1 61. FO included the following main fatty acids (μg g^{-1}): C12:0 82, C14:0 12345, *c9*C14:1 215, C15:0 477, C16:0 56947, *c7*C16:1 318, *c9*C16:1 420, $\sum$C16:2 15586, C17:0 493, *c9*C17:1 193, C18:0 9452, *c6*C18:1 188, *c7*C18:1 842, *c9*C18:1 290592, *c12*C18:1 15834, *c14*C18:1 159, LA 114512, αLNA 20968, *c11*C20:1 24206, *c7c9c12c15*C18:4 473, *c11c14*C20:2 2270, *c8c11c14*C20:3

258, ArA 304, *c8c11c14c17*C20:4 607, C22:0 139, *c13*C22:1 11036, *c11*C22:1 1704, C20:5*n*-3 (EPA) 6792, *c13c16*C22:2 95, *c7c10c13c16*C22:4 144, *c15*C24:1 397, C22:5*n*-3 (DPA) 1560 and DHA 26570. The meadow hay contained the following main fatty acids (μg g^{-1} FO): C8:0 83, C12:0 142, C14:0 239, *c9*C15:1 131, C16:0 4034, *c9*C16:1 184, C18:0 459, *c9*C18:1 1266, *c12*C18:1 72, LA 13100, αLNA 4178, C20:0 58, *c11*C20:1 74, C22:0 101, C24:0 69, *c15*C24:1 71.

The vitamin and mineral mixture was purchased from POLFAMIX OK (Grodzisk Mazowiecki, Poland); 1 kg of vitamin and mineral mixture comprised 285 g calcium, 16 g phosphorus, 56 g sodium, 42 mg cobalt as carbonate, 10 mg iodine as iodate, 1 g iron as sulfate, 6 mg Se as selenite, 0.5 g copper as sulfate, 5.8 g manganese as sulfate, 7.5 g zinc as sulfate; vitamins included A (500 000 IU kg^{-1}), D3 (125 000 IU kg^{-1}), and E as α-tocopherol (25 000 IU kg^{-1}).

The selenized yeast (Se-*Saccharomyces cerevisiae*) was donated by Sel-Plex (Alltech In., USA). About 83 % of the total Se content of selenized yeast (SeY) represents Se in the form of Se methionine (Se-Met) incorporated into the proteins of *Saccharomyces cerevisiae* (Czauderna et al., 2009a); the chemical composition of SeY was presented in a previous publication (Czauderna et al., 2009a).

Water used for the preparation of mobile phases and chemical reagents was prepared using an Elix™ water purification system (Millipore, Canada). The mobile phases were filtered through a 0.45 µm membrane filter (Millipore) and then degassed for 2–3 min in a vacuum with ultrasonication prior to use.

2.3   Saponification and extraction of fatty aldehydes and CHOL-4-3 from brain hydrolysates

Fatty aldehydes in homogenized brain samples ($\sim$ 50 mg) were hydrolysed with a mixture of 2 mL of 2 M KOH in water and 2 mL of 1 M KOH in methanol. Next, 50 µL of the internal standard (IS) solution (17 mg mL^{-1} nonadecanoic acid in chloroform) was added to the obtained mixture. The resulting mixture was flushed with argon (Ar) for $\sim$ 4 min. The vial was then sealed and the mixture vortexed and heated under Ar at 95 °C for 10 min, cooled for 10 min at room temperature, and sonicated for 10 min. The resulting mixture was protected from the light and stored in the sealed vial under Ar at $\sim$ 22 °C overnight. Next, 3 mL of water was added to the hydrolysate and the solution was again vortexed. The obtained solution was acidified with 4 M HCl to $\sim$ pH 2, and free fatty acids were extracted four times with each 3 mL of DCM. Extraction was repeated four times with each 3 mL of *n*-hexane. The upper *n*-hexane layer was combined with the DCM layer, and next the resulting organic phase was dried with $\sim$ 0.1 g of Na$_2$SO$_4$. The organic solvents were removed under a stream of Ar at room temperature. Afterwards the residue was re-dissolved in 0.5 mL of *n*-hexane, and then 1 µL of the resulting solution was injected onto the GC column. The total fatty aldehyde profile in a 1 µL sample at a

split ratio of 10 : 1 was determined using the column temperature gradient programme. The column was operated at 90 °C for 10 min, then the temperature programmed at 7 °C min^{-1} to 200 °C, held for 21 min, programmed at 10 °C min^{-1} to 235 °C, held for 25 min. Helium as the carrier gas operated at a constant pressure (223.4 kPa) and flow rate of 1 mL min^{-1}. Injector and MS detector temperatures were maintained at 200 and 240 °C, respectively.

CHOL-4-3 in brain samples was determined after saponification according to Czau-derna et al. (2013). Free CHOL-4-3 in processed samples was then quantified using capillary gas chromatography coupled to a quadrupole mass selective detector (Czauderna et al., 2013).

2.4   Saponification, extraction and preparation of fatty acid methyl esters

Fatty acids (FAs) in homogenized brain samples were saponified according to Czauderna et al. (2009a) followed by gentle methylations. The base- and acid-catalysed methylations were introduced for preparation of methyl esters of fatty acids (FAMEs) in brain samples (Czauderna et al., 2009a). FAMEs were then quantified using a capillary gas chromatography according to Czauderna et al. (2009a).

2.5   Gas chromatographic equipment

The analyses of fatty aldehydes, FAMEs and CHOL-4-3 in brain samples were performed on a SHIMADZU GC-MS-QP2010 Plus EI equipped with a BPX70 fused silica capillary column (120 m $\times$ 0.25 mm i.d. $\times$ 0.25 µm film thickness; SHIM-POL, a quadrupole mass selective detector (Model 5973N) and an injection port. Fatty aldehydes and FAME identifications were validated based on electron impact ionization spectra of fatty aldehydes and FAMEs and compared with authentic fatty aldehydes, CHOL-4-3, and FAME standards and the NIST 2007 reference mass spectra library.

2.6   Saponification, extraction and analyses of tocopherols, CHOL, its metabolite and MDA in the brain

Tocopherols and CHOL in homogenized brain samples were quantified using reversed-phase liquid chromatography according to Czauderna et al. (2009b). The MDA concentration in brain samples was determined after saponification followed by derivatization according to Czauderna et al. (2011). Chromatographic analyses of tocopherols, CHOL and derivatized MDA in brain samples were performed using an ultra-fast liquid chromatography system and a photodiode array detector (SHIMADZU, Japan) (Czauderna et al., 2011).

Table 3. The concentrations ($mg\,g^{-1}$) of fatty acids, indexes of A^{SFA} ($_{index}A^{SFA}$) and T^{SFA} ($_{index}T^{SFA}$), and the concentration ratios of ΣSFAs to ΣFAs, ΣPUFAs or ΣLPUFAs in the brains of lambs[a].

Fatty acids	The first experiment		The second experiment		
	RO	FO	CA	CASeY	CASeVI
C14:0	0.125^a	0.093^b	0.097^a	0.118^a	0.123^a
C16:0	3.29^a	2.90^a	2.88^a	3.09^{ab}	3.42^b
C18:0	$1\,3.05^a$	2.83^a	2.80^a	3.00^{ab}	3.33^b
ΣSFAs[b]	6.72^a	6.03^b	5.97^a	6.48^{ab}	7.10^b
ΣFAs[c]	15.6^a	14.3^a	13.5^a	15.4^{ab}	16.3^b
A^{SFA}[d]	3.42^A	2.99^B	2.98^a	3.21^{ab}	3.55^b
T^{SFA}[e]	6.46^a	5.82^b	5.78^a	6.21^{ab}	6.88^b
$_{index}A^{SFA}$[f]	0.467^a	0.438^b	0.470^a	0.448^a	0.465^a
$_{index}T^{SFA}$[g]	0.334^a	0.296^b	0.298^a	0.325^b	0.310^{ab}
$A^{SFA}/\Sigma n$-3PUFAs	$1,890^a$	1.600^b	1.593^a	1.783^b	$1,699^b$
$T^{SFA}/\Sigma n$-3PUFAs	3.589^a	3.112^b	3.091^a	3.450^b	3.292^b
ΣSFAs$/\Sigma$PUFAs	1.71^a	1.59^b	1.63^a	1.68^a	1.68^a
ΣSFAs$/\Sigma$LPUFAs	1.77^a	1.66^b	1.72^a	1.75^a	1.76^a

[a] Mean values in rows having the different superscripts are significantly different at [a,b] $P < 0.05$;
Statistical analyses were carried out between groups RO and FO and between groups CA, CASeY and CASeVI;
[b] The sum of C8:0, C10:0, C12:0, C14:0, C16:0, C18:0, C20:0, C22:0 and C24:0;
[c] The sum of all assayed fatty acids;
[d] The atherogenic SFAs = C12:0, C14:0 and C16:0;
[e] The thrombogenic SFAs = C14:0, C16:0 and C18:0;
[f] The atherogenic index = (C12:0 + 4 × C14:0 + C16:0) / (MUFAs + n-6 PUFAs + n-3PUFAs) (Morán et al., 2013);
[g] The thrombogenic index = (C14:0 + C16:0 + C18:0) / 0.5 × MUFAs + 0.5 × n-6PUFAs + 3 × n-3PUFAs + n-3PUFAs / n-6PUFAs) (Morán et al., 2013).

2.7 Statistical analyses

Statistical analyses were performed using the Statistica software package (StatSoft, Version 10, 2010). Statistical analyses of the effects of dietary additives (FO, CA, SeY and SeVI) on the concentration of FAs, tocopherols, CHOL, CHOL-4-3, MDA and fatty aldehydes in brain samples were conducted using the non-parametric Mann–Whitney U test. The results in Table 2 are presented as mean values with the standard deviations (SDs) of six examined lambs ($n = 6$); the results in Tables 3–5 are presented as the means of the individually analysed samples ($n = 6$). Mean values in the columns or rows having the different superscripts are significantly different at [a,b] $P < 0.05$ and [A,B] $P < 0.01$.

3 Results and discussion

The brain is the major control network for the body's functions and abilities, and it enables automatic operation of vital organs; the brain centres control the reproductive system (Wainwright, 2002; Martin et al., 2004). The hypothalamus is a portion of the brain that controls body temperature, hunger, important aspects of parenting and attachment behaviours, thirst, fatigue, sleep, and circadian rhythms; the hypothalamus controls reproductive and immune systems by regulating the secretory activities of the pituitary gland (the hypo-thalamic–pituitary–gonadal axis) (Martin et al., 2004; Vadakkadath Meethal and Atwood, 2005). LPUFAs (like DHA and ArA) are essential components for lamb brain development and physiological functions (Wainwright, 2002; Patterson et al., 2012; Rosa et al., 2013). Considering the above, we study the effects of dietary FO, CA and antioxidants (SeY and SeVI) on oxidative stress and the contents of physiologically important compounds (like LPUFA, tocopherols, fatty aldehydes, cholesterol and its metabolite) in the brain.

3.1 Effects of the experimental diets on concentrations of fatty acids in the brain

In the current study, neither macroscopic lesions nor pathological changes were found in the brain as well as other internal organs of lambs fed the experimental diets enriched in FO, CA (a phenolic diterpene), SeY (rich in Se-Met) or SeVI. In fact, diets containing up to 2 mg Se per kg would not be toxic for ruminants (McDowell et al., 2005; Weiss and Hogan, 2005; Mainville et al., 2009; Krajewska et al., 2012; Eun et al., 2013; Netto et al., 2014). Our current study also indicated that the experimental diets enriched in 0.35 ppm Se as SeY or SeVI resulted in a relatively small increase ($P < 0.05$) in the concentration of Se in the liver (65–70 %), heart (48–50 %), blood (26–49 %), longissimus dorsi muscle (30–

Table 4. The concentrations (mg g^{-1}) of fatty acids, values of Δ9-desaturase index and ratios of $\sum$PUFAs, $\sum$LPUFAs and $\sum n$-3LPUFAs to $\sum$SFAs or $\sum$FAs in the brains of lambs[a].

Fatty acids	The first experiment		The second experiment		
	RO	FO	CA	CASeY	CASeVI
Δ9-desaturase index[b]	0.538[a]	0.530[a]	0.494[a]	0.534[b]	0.513[ab]
c9C18:1	3.56[a]	3.30[b]	2.80[a]	3.68[b]	3.67[b]
$t11$C18:1	2 0.027[a]	0.020[b]	0.021[a]	0.020[a]	0.022[a]
ΣMUFAs	1 4.91[a]	4.50[a]	3.84[a]	5.03[b]	5.01[b]
LA	142[a]	163[a]	192[a]	147[a]	181[a]
αLNA	0.0017[A]	0.0003[B]	0.0013[A]	0.0015[A]	0.00002[B]
C20:4n-6	1.20[a]	1.03[a]	1.03[a]	1.16[a]	1.22[a]
DPA	0.104[a]	0.121[a]	0.120[a]	0.143[a]	0.121[a]
DHA	1.70[a]	1.74[a]	1.75[ab]	1.64[a]	1.97[b]
ΣPUFAs	3.94[a]	3.80[a]	3.66[a]	3.85[ab]	4.17[b]
Σn-6PUFAs	1.41[a]	1.24[a]	1.27[a]	1.38[a]	1.45[a]
Σn-3PUFAs	1.81[a]	1.87[a]	1.87[ab]	1.80[a]	2.09[b]
ΣLPUFAs	3.79[a]	3.63[a]	3.47[a]	3.71[a]	3.99[a]
Σn-3LPUFAs	1.81[a]	1.87[a]	1.87[a]	1.78[a]	2.09[a]
Σn-6LPUFAs	1.98[a]	1.77[b]	1.60[a]	1.93[b]	1.90[b]
Σn-6/Σn-3[c]	0.788[a]	0.668[b]	0.681[a]	0.792[b]	0.704[ab]
Σn-6L/Σn-3L[d]	1.097[a]	0.947[b]	0.855[a]	1.084[b]	0.909[a]
ΣPUFAs / ΣFAs	0.253[a]	0.267[b]	0.274[a]	0.257[b]	0.262[ab]
ΣLPUFAs / ΣFAs	0.244[a]	0.255[b]	0.259[a]	0.247[b]	0.250[ab]
Σn-3LPUFAs / ΣSFAs	0.269[a]	0.310[b]	0.317[a]	0.280[b]	0.301[ab]
Σn-3LPUFAs / ΣFAs	0.116[a]	0.132[b]	0.141[a]	0.120[b]	0.132[ab]

[a] Mean values in rows having the different superscripts are significantly different at [a,b] $P < 0.05$ and [A,B] $P < 0.01$. Statistical analyses were carried out between groups RO and FO and between groups CA, CASeY and CASeVI;
[b] Δ9-desaturase index $= c$9C18:1/(c9C18:1 + C18:0);
[c] The ratio of Σn-6PUFAs to Σn-3PUFAs;
[d] The ratio of Σn-6LPUFAs to Σn-3LPUFAs.

45 %) and biceps femoris muscle (28–36 %) compared with the RO diet; in contrast, the FO or CA diets revealed negligible ($P > 0.05$) impact on the concentration of Se in these tissues in comparison with the RO diet (Ruszczyńska et al., 2016).

As can be seen from results (Table 2), all diets fed to lambs resulted in negligible changes in the brain mass of lambs. On the other hand, our study demonstrated that the supplements added to the diet affect the concentration of SFAs (Table 3) and UFAs (Table 4) in the lamb brain. Our study was especially focused on the concentrations of LPUFAs in the brain. LPUFAs are essential components of the lamb brain (Wainwright, 2002; Patterson et al., 2012; Rosa et al., 2013) and their concentrations remain fairly constant, or change little in the brains of lambs fed the supplemented diets (Table 2). It has been found (Table 4) that the brains of all examined lambs are characterized by the high concentration of LPUFAs ($\sim$ 25 % of ΣFAs), especially DHA (11–13 % of ΣFAs). In fact, lipids in central nervous tissues of mammals are characterized by a high concentration of DHA, which is especially enriched in amino-phospholipids, serine phosphoglycerides, and ethanol-amine (Schönfeld and Reiser, 2013).

DHA and other n-3PUFAs occupy the sn-2 position of phospholipids, and it has been suggested that the brain has a molecular species requirement for its function (Schönfeld and Reiser, 2013). Indeed, our study showed that DHA in the brain constitutes about 95 % of Σn-3PUFA in the lamb brain (Table 4).

3.1.1 The first experiment

The results summarized in Table 2 documented that the FO diet increased the body weight gain (BWG) of lambs compared with the RO diet. The FO diet decreased the concentration sum of saturated FAs (ΣSFAs), including atherogenic (A^{SFA}) and thrombogenic (T^{SFA}) SFAs in the brain compared with the RO diet (Table 3). As a consequence, the FO diet decreased values of the concentration ratios of ΣSFAs to the sum of PUFAs (ΣPUFAs) and the sum of LPUFAs (ΣLPUFAs) in the brain compared with the RO diet. The present results documented that the diet enriched in FO reduced the brain concentration of C14:0, which is considered to be 4 times more atherogenic than the other SFAs (like C12:0 or C16:0) (Pikul et al., 2008). The FO diet can prevent atherosclerosis (i.e. the deposition of atheromas) and throm-

Table 5. The concentrations of cholesterol, CHOL-4-3, tocopherols (T), MDA, fatty aldehydes and values of PUFA peroxidation index[a] in the brains of lambs[b].

Specification	The first experiment		The second experiment		
	RO	FO	CA	CASeY	CASeVI
CHOL, $\mu g\,g^{-1}$	467[a]	410[a]	407[a]	397[a]	380[a]
CHOL-4-3 $\mu g\,g^{-1}$	68[a]	61[a]	31[a]	101[b]	110[b]
δT, $\mu g\,g^{-1}$	2.1[a]	2.9[a]	1.2[a]	1.3[a]	1.40[a]
γT, $\mu g\,g^{-1}$	2.7[a]	2.6[a]	1.5[a]	1.8[ab]	2.2[b]
αT, $\mu g\,g^{-1}$	4.3[a]	4.1[a]	3.7[a]	3.8[a]	3.2[a]
αTAc, $\mu g\,g^{-1}$	2.3[a]	2.2[a]	2.1[a]	2.9[b]	3.2[b]
$\Sigma\alpha$T,[c] $\mu g\,g^{-1}$	6.6[a]	6.3[a]	5.7[a]	6.7[a]	6.4[a]
ΣTs,[d] $\mu g\,g^{-1}$	11.4[a]	11.8[a]	8.5[a]	9.1[a]	10.1[a]
MDA, $ng\,g^{-1}$	8.93[a]	9.63[a]	10.23[a]	9.81[ab]	8.66[b]
MDA$_{index}$, ng/mg	2.27[a]	2.54[b]	2.79[a]	2.55[a]	2.07[b]
Fatty aldehydes, $\mu g\,g^{-1}$					
AL-C10:0	96[a]	70[b]	56[a]	70[a]	74[a]
AL-C12:0	154[a]	116[b]	107[a]	101[a]	125[a]
AL-C14:0	625[a]	562[a]	494[a]	662[b]	815[c]
AL-C16:0	49[a]	49[a]	39[a]	55[b]	67[b]
AL-C18:0	984[a]	911[a]	894[a]	978[a]	1396[b]
AL-$c9$C18:1	480[a]	397[a]	330[a]	474[b]	541[b]
AL-$c11$C18:1	431[a]	342[a]	288[a]	395[b]	469[b]
ΣALs[e]	2820[a]	2446[a]	2206[a]	2735[a]	3487[b]

[a] PUFA peroxidation index: MDA$_{index}$ $(ng\,mg^{-1})$ = [MDA $(ng\,g^{-1})$]/[ΣPUFAs $(mg\,g^{-1})$].
ΣPUFAs – the concentration sum of PUFAs in the brain;
[b] Mean values in rows having the different superscripts are significantly different at [a,b] $P < 0.05$.
Statistical analyses were carried out between groups RO and FO and between groups CA, CASeY and CASeVI;
[c] The concentration sum of α-tocopherol (αT) and α-tocopheryl acetate (αTAc);
[d] The concentration sum of δT, γT, αT and αTAc;
[e] The concentration sum of all assayed fatty aldehydes.

bosis in the brain (i.e. the formation or presence of a blood clot in cerebral vessels; Ogata et al., 2011).

A better evaluation of the functional effects of dietary supplements on atherosclerosis and thrombosis may be given by the indexes of atherogenicity ($_{index}A^{SFA}$) and thrombogenicity ($_{index}T^{SFA}$) (Pikul et al., 2008; Morán et al., 2013). These indexes were calculated using concentrations of A^{SFA} and T^{SFA} as well as n-3PUFAs, n-6PUFAs and monounsaturated FAs (MUFAs) (Table 3); $_{index}A^{SFA}$ is defined as the relationship between the pro-atherogenic SFAs (A^{SFA}) and anti-atherogenic FAs (MUFAs, n-6PUFAs and n-3PUFAs). $_{index}T^{SFA}$ is defined as the relationship between the pro-thrombogenetic SFAs (T^{SFA}), the anti-thrombogenetic fatty acids (MUFAs, n-6PUFAs and n-3PUFAs) and the ratio of n-3PUFAs to n-6PUFAs; a high n-6PUFAs over n-3PUFA ratio is thrombogenic (Pikul et al., 2008; Garaffo et al., 2011). As expected, our study indicated that the FO diet also decreased $_{index}A^{SFA}$ and $_{index}T^{SFA}$ as well as the concentration ratios of A^{SFA} and T^{SFA} to Σn-3PUFA in the brain compared with the RO diet (Table 3). The A^{SFA}/Σn-3PUFA and T^{SFA}/Σn-3PUFA ratios are the relationship between A^{SFA} and T^{SFA}

as well as n-3PUFAs; n-3PUFAs play of critical roles in preventing atherosclerosis and thrombosis (Pikul et al., 2008; Garaffo et al., 2011). Thus, current study and other investigations (Pikul et al., 2008; Garaffo et al., 2011) indicated that, in contrast to the RO diet, the FO diet (rich in n-3LPUFA) reduced the risk of the pathological changes in cerebral vessels (i.e. changes resulting in athero-thrombotic stroke or hypertension) (Ogata et al., 2011); so, the concentrations of A^{SFA} and T^{SFA} as well as the values of $_{index}A^{SFA}$ and $_{index}T^{SFA}$ (Table 3) are the powerful predictors of a risk for arteriosclerotic cerebral vessel changes (e.g. athero-thrombotic stroke).

The diet containing only RO (rich in LA) stimulated the desaturation and elongation of n-6PUFAs (like LA) to n-6LPUFAs (i.e. precursors of eicosanoids involved in various pathological processes involving inflammatory conditions such as atherosclerosis, obesity, and Alzheimer's disease). Therefore, the concentration of LA was numerically lower ($P = 0.068$) in the brains of lambs fed the RO diet than in the brains of lambs fed the FO diet (Table 4). Consequently, the RO diet increased the concentrations of pro-inflammatory precursors (n-6LPUFAs) and the $\sum n$-6PUFAs / $\sum n$-3PUFAs and $\sum n$-6 LPUFAs / $\sum n$-3LPUFA ratios in the brain. Therefore, we argued that the RO diet increased the risk of diseases in the ruminant brain as the high concentrations of A^{SFA}, T^{SFA} and the high n-6PUFAs over n-3PUFA ratio (Tables 3 and 4) are hypercholesterolemic, atherogenic and thrombogenic (Pikul et al., 2008; Garaffo et al., 2011; Orth and Bellosta, 2012). On the other hand, the FO diet reduced the risk of atherosclerosis, thrombosis and reduced inflammation in the lamb brain as this diet decreased the Σn-6PUFAs / Σn-3PUFA ratio, the concentrations of A^{SFA}, T^{SFA} and increased the ratios of ΣPUFAs / ΣFAs, ΣLPUFAs / ΣFAs, Σn-3LPUFAs / ΣSFAs and Σn-3LPUFAs / ΣFAs (Table 4) compared with the RO diet. The addition of FO to the diet reduced the concentration of $\sum n$-6LPUFAs and the ratio of $\sum n$-6LPUFAs to $\sum n$-3LPUFAs in the brain compared with the RO diet. Therefore, we argued that dietary FO, rich in n-3PUFAs, decreased the capacity of the desaturation and elongation of n-6PUFAs (e.g. LA) to their n-6 long-chain anabolites (i.e. n-6LPUFAs) (Table 4). In fact, n-3PUFAs usually have higher affinity than n-6PUFAs for the elongation and desaturation enzymes in mammal tissues (Patterson et al., 2012).

3.1.2 The second experiment

It has been found that the CASeVI diet increased the body weight gain (BWG) of lambs compared with the CA and CASeY diets (Table 2). The diets enriched in FO and CA, irrespective of the presence of SeY or SeVI, affected the concentrations of fatty acids in the lamb brains (Tables 3 and 4). Indeed, concentrations of fatty acids in the brain are effectively altered by the concentrations of FAs and antioxidants in a diet, and with life stage, increasing with development and decreasing with aging (Niedźwiedzka et al., 2006; Uauy and Dangour, 2013). As can be seen from the results in Table 3, the CASeVI diet increased the concentration of $\sum$SFAs, including A^{SFA} and T^{SFA}, in the brain compared with the CA diet. On the other hand, the CASeVI diet only numerically increased ($P > 0.05$) the concentrations of $\sum$SFAs as well as C16:0, C18:0, A^{SFA} and T^{SFA} in the brain as compared with the CASeY diet. The CASeY diet revealed only numerical effects (numerically increased) on the concentrations of these fatty acids in the brain compared with the CA diet. Concomitantly, the CASeY and CASeVI diets increased the values of the $A^{SFA} / \Sigma n$-3PUFA and $T^{SFA} / \Sigma n$-3PUFA ratios in the brain compared with the CA diet. Our study showed that the CASeY diet increased the value of $_{index}T^{SFA}$ in the brain compared with the CA diet.

Thus, the present data are in agreement with our previous investigations in which diets enriched in SeVI stimulated the accumulation of $\sum$SFAs, including A^{SFA} and T^{SFA}, in subcutaneous fat tissue and blood plasma of lambs (Niedźwiedzka et al., 2008). Considering the above facts, we argued that the diet with SeY and especially SeVI (groups

CASeY and CASeVI) increased the risk of the pathological changes in cerebral vessels (Garaffo et al., 2011; Ogata et al., 2011). Chronic ingestion of SFAs results in blood-brain barrier dysfunction and significant delivery into the brains of plasma proteins. Moreover, it is well known that among the saturated fatty acids, C12:0, C16:0 and especially C14:0 are recognized as health risk factors (resulting in atherothrombotic stroke, hypertension or coronary diseases) (Ogata et al., 2011; Morán at al., 2013; Schönfeld and Reiser, 2013).

The current study showed that the CASeY diet more efficiently increased the capacity of $\Delta 9$-desaturation (the $\Delta 9$-desaturase index) in the brain than the CA diet. Moreover, the concentration of $c9$C18:1 is higher in the brains of lambs fed the CASeY and CASeVI diets compared with lambs fed the CA diet (Table 4). Indeed, the CA diet decreased the concentration of $c9$C18:1 in the brain, resulting in a drop of the concentration of $\sum$MUFAs in the brain compared with the CASeY and CASeVI diets.

The present study showed that the CASeY and CASeVI diets stimulated the accumulation of $c9$C18:1 (the product of $\Delta 9$-desaturation of C18:0), while the CASeVI diet also increased the concentration of $\sum$PUFAs in the brain compared with the CA and CASeY diets (Table 4). Thus, current investigations are consistent with our previous studies in which dietary SeVI increased the capacity of $\Delta 9$-desaturation in adipose tissues and decreased the yield of fatty acid catabolism in adipose tissues of rats (Czauderna et al., 2003). Thus, the CASeVI diet increased the concentration of $\sum$FAs (including $\sum$SFAs, $\sum$MUFAs and $\sum$PUFAs) in the brain compared with the CA diet (Tables 3 and 4).

We documented that the diets enriched in SeY or SeVI stimulated the capacity of desaturation and elongation enzymes of n-6PUFAs to n-6LPUFAs in the brain compared with the CA diet (Table 4). Therefore, the CASeY and CASeVI diets more efficiently increased the accumulation of n-6LPUFA in the brain than the CA diet. Moreover, the diet including SeY (rich in Se-Met) increased the ratios of $\sum n$-6PUFAs / $\sum n$-3PUFAs and $\sum n$-6LPUFAs / $\sum n$-3LPUFAs in the brain, while it reduced the ratios of $\sum n$-3LPUFAs / $\sum$SFAs and $\sum n$-3 LPUFAs / $\sum$FAs in the brain compared with the CA diet. Therefore, we suggest that SeVI and especially SeY added to the diet modify an impact of n-3PUFAs derived from FO on the capacity of the desaturation and elongation enzymes of n-6PUFAs to n-6LPUFAs in the brain. We argued that SeVI and especially SeY added to the diet increased the affinity of n-6PUFAs for the elongation and desaturation enzymes in the brain, whereas it reduced the affinity of n-3PUFA for these enzymes in the brain compared with the CA diet.

Considering the above facts, we suggested that the CA diet more efficiently improved brain functions as the concentration of Σn-6LPUFAs (long-chain anabolites of LA) and the $A^{SFA} / \sum n$-3PUFAs and $T^{SFA} / \sum n$-3PUFA ratios are lower in the brain of lambs fed the CA diet than in the brains of lambs fed the CASeY and CASeVI diets. In fact, cyclooxygenases and lipoxygenases can convert n-6LPUFAs (like ArA) to the 2-series of prostaglandins, the 2-series of thromboxanes, and the 4-series of leukotrienes; these eicosanoids are involved in various pathological processes involving inflammatory conditions such as atherosclerosis, obesity, and Alzheimer's disease (Wainwright, 2002; Azad et al., 2011).

3.2 Effects of the experimental diets on concentrations of CHOL, CHOL-4-3, tocopherols and aldehydes in the brain

CHOL is vital to normal physiological functions of the brain (Orth and Bellosta, 2012); CHOL in the brain is primarily derived by de novo synthesis. Excess CHOL in the brain can lead to many signalling events via cholesterol metabolites, antioxidant processes and pro-inflammatory mediators. Numerous studies showed that alternation in CHOL biosynthesis and its degradation influences higher-order brain functions (Orth and Bellosta, 2012). Considering the above fact, we study effects of the diets enriched in FO, CA or Se (as SeY or SeVI) on the concentration of CHOL and its metabolite (CHOL-4-3) in the lamb brain.

3.2.1 The first experiment

The effects of the FO diet on the concentrations of CHOL, CHOL-4-3, tocopherols and MDA in the brains of lambs were rather small compared with the RO diet, but some numerical effects of the FO diet on the concentrations of CHOL, CHOL-4-3 and MDA in the brain were observed (Table 5). In comparison with the RO diet, the FO diet numerically decreased the concentrations of CHOL (the statistical tendency: $P = 0.087$) and CHOL-4-3 ($P = 0.15$) and numerically increased the concentration of MDA (the statistical tendency: $P = 0.097$) in the brain. Interestingly, the diet enriched in FO (rich in LPUFA) increased the value of PUFA peroxidation index (MDA_{index}) in the brain compared with the RO diet. So, we argued that a better evaluation of the functional effects of dietary supplements on the capacity of lipid peroxidation may be given by MDA_{index}. Indeed, this index was calculated using concentration of MDA (the product of PUFA peroxidation) and PUFAs (i.e. the substrates which react with radicals). The yield of MDA formation depends upon the concentration of PUFAs as well as the chemical form of an antioxidant (Czauderna et al., 2011). Based on the value of MDA_{index}, we argued that the FO diet increased the yield of PUFA peroxidation in the brain compared with the RO diet.

Fatty aldehydes are present in considerable quantity in the lipid fractions of muscles and the brain and may be intermediates in lipid metabolism (Stadelmann-Ingrand et al., 2004; Dannenberger et al., 2006). Thus, we investigate the effect of the RO and FO diets on the abundance of fatty aldehydes in the brain (Table 5). Our study documented that the FO diet reduced the concentrations of AL-C10:0 and AL-C12:0 in the brain compared with the RO diet. The effects of the FO diet on the concentrations of longer chain fatty aldehydes (AL-C14:0, AL-C16:0 and AL-C18:0) and unsaturated fatty aldehydes (AL-$c9$C18:1 and AL-$c11$C18:1) in the brain were rather small compared with the RO diet, but numerical effects of the FO diet on the concentrations of these aldehydes were observed (Table 5); the FO diet numerically decreased ($P > 0.05$) the concentrations of these aldehydes in the brain compared with the RO diet. Therefore, we suggest that the diet enriched in FO (rich in LPUFA) stimulated the catabolism yield of fatty aldehydes (intermediates in lipid metabolism) in the brain compared with the RO diet. Concomitantly, the FO diet more effectively increased PUFA peroxidation (i.e. MDA_{index}) in the brain than the diet with only RO (poor in LPUFA).

3.2.2 The second experiment

It has been found that SeY or especially SeVI added to the diet stimulated CHOL metabolism to CHOL-4-3 compared with the CA diet. As a consequence, the concentration of CHOL-4-3 is higher in the brains of lambs fed the CASeY and CASeVI diets than the CA diet (Table 5). Concomitantly, our study shows that the high concentration of CHOL-4-3 in the brains of lambs fed the CASeVI diet implicates the higher increase of the BWG of lambs (Table 2) than the CA and CASeY diets.

Animal studies documented the importance of dietary tocopherols to the healthy functioning of the brain, including protection against lipid peroxidation, DNA damage and neuron loss (Lu et al., 2015; Morris et al., 2015). Therefore, we examine the impact of the diets enriched in Se (as SeY and SeVI) on the concentrations of tocopherols in the brain (Table 5). Concerning tocopherols, no significant differences were observed between the CA, CASeY and CASeVI diets in the concentrations of δ-tocopherol (δT), α-tocopherol (αT) and the sums of αT and α-tocopheryl acetate ($\Sigma\alpha$T) and all assayed tocopherols (ΣTs) in the brain. On the contrary, the CASeY and CASeVI diets resulted in increasing the concentration of α-tocopheryl acetate (αTAc) in the brain compared with the CA diet. Moreover, feeding the CASeVI diet resulted in an increase in the concentration of γ-tocopherol (γT) in the brain compared with the CA diet. Unlike αT, γT is a potent defender against disease-provoking species in the mammal body known as reactive nitrogen oxides; γT has been found to decrease inflammation, promote factors that guard against certain cancers, and activate genes involved in protecting against Alzheimer's disease (Lu et al., 2015; Morris et al., 2015). γT is even superior than αT in the protection of mammal tissues from damage by some specific free radicals (e.g. peroxynitrite or NO_x), due to a stable nitro-adduct formed in its unsubstituted position on chromanol ring (Lu et al., 2015).

In the present dietary study, the CASeVI diet decreased the concentration of MDA in the brain compared with the

CA diet. Moreover, the diet containing SeVI more efficiently reduced the value of MDA_{index} in the brain than the CA and CASeY diets. These results reinforce our suggestions that SeVI is more efficiently utilized for biosynthesis of antioxidative proteins containing Se cysteine (like glutathione peroxidases) than dietary SeY (rich in Se-Met) (Tapiero et al., 2003; Navarro-Alarcon and Cabrera-Vique, 2008). Based on the above observations, we suggest that the diet enriched in SeVI improved brain functions as decreased oxidative stress in brain tissues as well as increased the concentration of γT, which has excellent antiinflammatory, antineoplastic and natriuretic functions (Lu et al., 2015).

Interestingly, the concentration of MDA and the value of MDA_{index} in the brains of lambs from group CASeY is similar to the concentration of MDA and the value of MDA_{index} in the brains of lambs fed the FO diet (i.e. the diet without antioxidant). Therefore, we suggest that simultaneous dosage of SeY and CA (the CASeY diet) has negligible impact on the yield of PUFA peroxidation as well as MDA_{index} in the brain compared with the FO diet.

We also investigate effects of the diets enriched in FO, CA without or with Se treatments (as SeY or SeVI) on the abundance of fatty aldehydes in the lamb brain (Table 5). Our results indicated that the diet enriched in the CASeY and CASeVI diets increased the concentrations of AL-C14:0, AL-C16:0 and unsaturated fatty aldehydes (i.e. AL-$c9$C18:1 and AL-$c11$C18:1) in the brain compared with the CA diet. Moreover, the diet with SeVI more efficiently stimulated the accumulation of AL-C14:0, AL-C18:0 and the concentration sum of all assayed fatty aldehydes (ΣALs) in the brain compared with the CA and CASeY diets.

Considering the above results, we suggest that the diet enriched in SeVI reduced the catabolism yield of fatty aldehydes (intermediates in lipid metabolism) in the brain compared with the CA and SeY diets. Indeed, aldehydes are produced during decomposition of hydroperoxides of lipid PUFAs following a reaction of peroxidation; MDA is the end product of lipid PUFA peroxidation (Czauderna et al., 2011; Albert et al., 2013). SeVI added to the diet is the suitable substrate for the biosynthesis Se cysteine containing enzymes, which play important roles in detoxification of reactive oxygen species and reactive nitrogen species, and antioxidant defence mechanisms in the mammal body (Tapiero et al., 2003; Tinggi, 2003; Navarro-Alarcon and Cabrera-Vique, 2008). On the other hand, SeY (rich in Se-Met) added to the diet stimulated the biosynthesis of mainly Se-Met containing proteins, which are not considered anti-oxidative Se enzymes (Navarro-Alarcon and Cabrera-Vique, 2008).

Detailed analyses of the concentrations of ΣFAs (Table 3) and ΣALs (Table 5) in the brains of lambs fed all examined diets (the first and second experiment) documented that there is a positive correlation ($r = 0.941$) between with the concentrations of ΣFAs and ΣALs in the brain. Therefore, this confirms our suggestion that the biosynthesis yield of fatty aldehydes positively correlated with the concentration of fatty acids in the brain. In fact, fatty aldehydes from fatty acid precursors can occur using carboxylic acid reductase (i.e. a versatile enzyme for conversion of fatty acids into fatty aldehydes) (Kunjapur and Prather, 2015).

4 Conclusions

The main novelty of our studies was to investigate the influence of the organic and inorganic form of Se added to the diet including CA and FO on the concentration of FAs, CHOL, its metabolite, tocopherols, MDA and fatty aldehydes in the lamb brain. The CASeY diet showed different impact on the concentrations of FAs, CHOL, CHOL-4-3, tocopherols, MDA and fatty aldehydes in the brain compared with the CA and CASeVI diets. For example, the CASeVI diet reduced the oxidative stress in the brain and increased the concentrations of γT and ΣALs compared with the CA and CASeY diets.

We argued that our studies provide useful information for nutritionists carrying out further investigations aimed at improving farm-animal health, the growth performance, the reproductive system (by regulating the secretory activities of the pituitary gland) and the nutritional quality of feed. Indeed, physiologically, the function of the brain is to exert centralized control over the other internal organs and tissues of ruminants.

Acknowledgements. This study was in part supported by the National Science Centre (NCN): grant no. 2013/09/B/NZ9/00291 and by the statutory funds from the Kielanowski Institute of Animal Physiology and Nutrition, PAS, Jabłonna, Poland (project no. II.1; 2015).

Edited by: M. Mielenz

References

Albert, B. B., Cameron-Smith, D., Hofman, P. L., and Cutfield, W. S.: Oxidation of marine omega-3 supplements and human health, Bio. Med. Research International, 2013, 464921, doi:10.1155/2013/464921, 2013.

Azad, N., Rasoolijazi, H., Joghataie, M. T., and Soleimani, S.: Neuroprotective effects of carnosic acid in an experimental model of alzheimer's disease in rats, Cell J., 13, 39–44, 2011.

Chen, C. T., Liu, Z., and Bazinet, R.: Rapid de-esterification and loss of eicosapentaenoic acid from rat brain phospholipids: an intracerebroventricular study, J. Neurochem., 116, 363–373, 2011.

Czauderna, M., Kowalczyk, J., Wąsowska, I., Niedźwiedzka, K. M., and Pastuszewska, B.: The effects of selenium and conjugated linoleic acid (CLA) isomers on fatty acid composition, CLA isomer content in tissues, and growth of rats, J. Anim. Feed Sci., 12, 865–881, 2003.

Czauderna, M., Kowalczyk, J., Niedźwiedzka, K. M., Leng, L., and Cobanova, K.: Dietary selenized yeast and CLA isomer mixture affect fatty- and amino acid concentrations in the femoral muscles and liver of rats, J. Anim. Feed Sci., 18, 348–361, 2009a.

Czauderna, M., Kowalczyk, J., and Niedźwiedzka, K. M.: Simple HPLC analysis of tocopherols and cholesterol from specimens of animal origin, Chem. Anal.-Warsaw, 54, 203–214, 2009b.

Czauderna, M., Kowalczyk, J., and Marounek, M.: The simple and sensitive measurement of malondialdehyde in selected specimens of biological origin and some feed by reversed phase high performance liquid chromatography, J. Chromatogr., 879, 2251–2258, 2011.

Czauderna, M., Marounek, M., Duskova, D., and Kowalczyk, J.: The sensitive and simple measurement of underivatized cholesterol and its oxygen derivatives in biological materials by capillary gas-chromatography coupled to a mass-selective detector, Acta Chromatograph., 25, 655–667, 2013.

Dannenberger, D., Lorenz, S., Nuernberg, G., Scollan, N., Ender, K., and Nuernberg, K.: Analysis of fatty aldehyde composition, including 12-methyltridecanal, in plasmalogens from longissimus muscle of concentrate- and pasture-fed bulls, J. Agr. Food Chem., 54, 182–188, 2006.

Eun, J.-S., Davis, T. Z., Vera, J. M., Miller, D. N., Panter, K. E., and ZoBell, D. R.: Addition of high concentration of inorganic selenium in orchardgrass (Dactylis glomerata L.) hay diet does not interfere with microbial fermentation in mixed ruminal microorganisms in continuous cultures, Prof. Anim. Sci., 29, 39–45, 2013.

Garaffo, M. A., Vassallo-Agius, R., Nengas, Y., Lembo, E., Rando, R., Maisano, R., Dugo, G., and Giuffrida, D.: Fatty acids profile, atherogenic (IA) and thrombogenic (IT) health lipid indices, of raw roe of blue fin tuna (Thunnus Thynnus L.) and their salted product "Bottarga", Food Nutr. Sci., 2, 736–743, 2011.

Han, F., Chen, D., Yu, B., and Luo, W.: Effects of different selenium sources and on serum biochemical parameters and tissue selenium retention in rats, Front. Agric. China, 3, 221–225, 2009.

Jordan, M. J., Lax, V., Rota, M. C., Loran S., and Sotomayor, J. A.: Effect of the phenological stage on the chemical composition, and antimicrobial and antioxidant properties of Rosmarinus officinalis L. essential oil and its polyphenolic extract, Ind. Crop. Prod., 48, 144–152, 2013.

Krajewska, K. A., Rozbicka-Wieczorek, A. J., Kowalczyk, J., and Czauderna, M.: Dietary linseed oil and selenate affect the concentration of fatty acids and selenium in the spleen, pancreas, and kidneys of lambs, J. Anim. Feed Sci., 21, 285–301, 2012.

Kunjapur, A. M. and Prather, K. L. J.: Microbial engineering for aldehyde synthesis. Minireview, Appl. Environ. Microb., 81, 1892–1901, 2015.

Lee, Y.-J. and Jenkins, T. C.: Identification of enriched conjugated linoleic acid isomers in cultures of ruminal microorganisms after dosing with 1-^{13}C-linoleic acid, J. Microbiol., 49, 622–627, 2011.

Lu, D., Yang, Y., Li, Y., and Sun, C.: Analysis of tocopherols and tocotrienols in pharmaceuticals and foods: a critical review, Curr. Pharma. Anal., 11, 66–78, 2015.

Mainville, A. M., Odongo, N. E., Bettger, W. J., McBride, B. W., and Osborne, V. R.: Selenium uptake by ruminal microorganisms from organic and inorganic sources in dairy cows, Can. J. Anim. Sci., 89, 105–110, 2009.

Martin, G. B., Milton, J. T. B., Davidson, R. H., Banchero-Hunzicker, G. E., Lindsay D. R., and Blache, D.: Natural methods for increasing reproductive efficiency in small ruminants, Anim. Reprod. Sci., 82–83, 231–246, 2004.

McDowell, L. R., Davis, P. A., Cristaldi, L. A., Wilkinson, N. S., Buergelt, C. D., and Van Alstyne, R.: Toxicity of selenium: fear or precaution?, Feedstuffs, 77, 12–13, 2005.

Morán, L., Andres, S., Bodas, R., Benavides, J., Prieto, N., Perez, V., and Giraldez, F. J.: Antioxidants included in the diet of fattening lambs: effects on immune response, stress, welfare and distal gut microbiota, Anim. Feed Sci. Tech., 173, 177–185, 2012.

Morán, L., Giráldez, F. J., Panseri, S., Aldai, N., Jordán, M. J., Chiesa, L. M., and Andrés, S.: Effect of dietary carnosic acid on the fatty acid profile and flavour stability of meat from fattening lambs, Food Chem., 138, 2407–2414, 2013.

Morris, M. C., Schneider, J. A., Li, H., Tangney, C. C., Nag, S., Bennett, D. A., Honer, W. G., and Barnes, L. L.: Brain tocopherols related to Alzheimer's disease neuropathology in humans, Alzheimers Dement., 11, 32–39, 2015.

Navarro-Alarcon, M. and Cabrera-Vique, C.: Selenium in food and the human body: A review, Sci. Total Environ., 400, 115–141, 2008.

Netto, A. S., Zanetti, M. A., Correa, L. B., Del Claro, G. R., Salles, M. S. V., and Vilela, F. G.: Effects of dietary selenium, sulphur and copper levels on selenium concentration in the serum and liver of lamb, Asian Australas. J. Anim. Sci., 27, 1082–1087, 2014.

Niedźwiedzka, K. M., Wąsowska, I., Czauderna, M., Kowalczyk, J., and Pastuszewska, B.: Influence of dietary conjugated linoleic acid isomers and Se on fatty acids profile in blood plasma and some tissues of rats, J. Anim. Feed Sci., 15, 471–489, 2006.

Niedźwiedzka, M. K., Kowalczyk, J., and Czauderna, M.: Influence of selenate and linseed oil on fatty-acid and amino-acid profiles in the liver, muscles, fat tissues and blood plasma of sheep, J. Anim. Feed Sci., 17, 328–343, 2008.

Ogata, J., Yamanishi, H., and Ishibashi-Ueda, H.: Review: Role of cerebral vessels in ischaemic injury of the brain, Neuropath. Appl. Neuro., 37, 40–55, 2011.

Orth, M. and Bellosta, S.: Cholesterol: Its regulation and role in central nervous system disorders, Cholesterol, 2012, 292598, doi:10.1155/2012/292598, 2012.

Patterson, E., Wall, R., Fitzgerald, G. F., Ross, R. P., and Stanton, C.: Health implications of high dietary omega-6 polyunsaturated fatty acids, J. Nutr. Metab., 2012, 539426, doi:10.1155/2012/539426, 2012.

Pikul, J., Wójtowski, J., Danków, R., Kuczyńska, B., and Łojek, J.: Fat content and fatty acids profile of colostrum and milk of primitive Konik horses (Equus caballus gmelini Ant.) during six months of lactation, J. Dairy Res., 75, 302–309, 2008.

Rosa, A., Scano, P., Incani, A., Pilla, F., Maestralec, C., Manca, M., Ligios C., and Pani, A.: Lipid profiles in brains from sheep with natural scrapie, Chem. Phys. Lipids., 175–176, 33–40, 2013.

Ruszczyńska, A., Rutkowska, D., Bulska, E., and Czauderna, M.: Effects of carnosic acid, fish oil and seleno-compounds on the level of selenium and fatty acids in lamb muscles, XLV Scientific Session of Group of Animal Nutrition KNZ PAN, Olsztyn, Poland, 21–22 June 2016, Book of Abstracts, in press, 2016.

Schönfeld, P. and Reiser, G.: Why does brain metabolism not favor burning of fatty acids to provide energy? – Reflections on disadvantages of the use of free fatty acids as fuel for brain, J. Cereb. Blood F. Met., 33, 1493–1499, 2013.

Schweizer, U., Streckfub, F., Pelt, P., Carlson, B. A., Hatfield, D. L., Köhrle, J., and Schomburg, L.: Hepatically derived selenoprotein P is a key factor to kidney but not for brain selenium supply, Biochem J., 386, 221–226, 2005.

Stadelmann-Ingrand, S., Pontcharraud, R., and Fauconneau, B.: Evidence for the reactivity of fatty aldehydes released from oxidized plasmalogens with phosphatidylethanolamine to form Schiff base adducts in rat brain homogenates, Chem. Phys. Lipids, 131, 93–105, 2004.

Sultana, R., Perluigi, M., and Butterfield, D. A.: Lipid peroxidation triggers neuro-degeneration: A redox proteomics view into the Alzheimer disease brain, Free Radical. Bio. Med., 62, 157–169, 2013.

Suzuki, K., Shimizu T., and Nakata, T.: The cholesterol metabolite cholest-4-en-3-one and its 3-oxo derivatives suppress body weight gain, body fat accumulation and serum lipid concentration in mice, Bioorg. Med. Chem. Lett., 8, 2133–2138, 1998.

Tapiero, H., Townsend, D. M., and Tew, K. D.: The antioxidant role of selenium and selenocompounds, Biomed. Pharmacother., 57, 134–144, 2003.

Tinggi, U.: Essentiality and toxicity of selenium and its status in Australia: A review, Toxicol. Lett., 137, 103–110, 2003.

Uauy, R. and Dangour, A. D.: Nutrition in brain development and aging: role of essential fatty acids, Nutr. Rev., 64, S24–33, 2006.

Vadakkadath Meethal, S. and Atwood, C. S.: The role of hypothalamic-pituitary-gonadal hormones in the normal structure and functioning of the brain, Cell. Mol. Life Sci., 62, 257–270, 2005.

Wainwright, P. E.: Dietary essential fatty acids and brain function: a developmental perspective on mechanisms, P. Nutr. Soc., 61, 61–69, 2002.

Weiss, W. P. and Hogan, J. S.: Effect of selenium source on selenium status, neutrophil function, and response to intramammary endotoxin challenge of dairy cows, J. Dairy Sci., 88, 4366–4374, 2005.

Xiao, W. H., Zheng, F. Y., Bennett, G. J., Bordet, T., and Pruss, R. M.: Olesoxime (cholest-4-en-3-one, oxime): Analgesic and neuroprotective effects in a rat model of painful peripheral neuropathy produced by the chemotherapeutic agent, paclitaxel, Pain, 147, 202–209, 2009.

Yu, L. L., Wang, R. L., Zhang, Y. Z., Kleemann, D. O., Zhu, X. P., and Jia, Z. H.: Effects of selenium supplementation on polyunsaturated fatty acid concentrations and anti-oxidant status in plasma and liver of lambs fed linseed oil or sunflower oil diets, Anim. Feed Sci. Tech., 140, 39–51, 2008.

Using microsatellite markers to analyze genetic diversity in 14 sheep types in Iran

Mohammad Taghi Vajed Ebrahimi, Mohammadreza Mohammadabadi, and Ali Esmailizadeh

Department of Animal Science, Faculty of Agriculture, Shahid Bahonar University of Kerman, Kerman, Iran

Correspondence to: Mohammadreza Mohammadabadi (mmohammadabadi@yahoo.ca)

Abstract. Investigation of genetic relationship among populations has been traditionally based on the analysis of allele frequencies at different loci. The prime objective of this research was to measure the genetic polymorphism of five microsatellite markers (McMA2, BM6444, McMA26, HSC, and OarHH35) and study genetic diversity of 14 sheep types in Iran. Genomic DNA was extracted from blood samples of 565 individuals using an optimized salting-out DNA extraction procedure. The polymerase chain reaction (PCR) was successfully performed with the specific primers. Some locus–population combinations were not at Hardy–Weinberg equilibrium ($P < 0.05$). The microsatellite analysis revealed high allelic and gene diversity in all 14 breeds. Pakistani and Arabi breeds showed the highest mean number of alleles (11.8 and 11 respectively), while the highest value for polymorphic information content was observed for the Arabi breed (0.88). A UPGMA (unweighted pair group method with arithmetic mean) dendrogram based on the Nei's standard genetic distance among studied breeds showed a separate cluster for Arabi and Pakistani breeds and another cluster for other breeds. The Shannon index (H0) for McMA2, BM6444, McMA26, HSC, and OarHH35 was 2.31, 2.17, 2.27, 2.04 and 2.18, respectively, and polymorphic information content (PIC) values were 0.88, 0.92, 0.87, 0.84, and 0.86 for McMA2, BM6444, McMA26, HSC, and OarHH35, respectively. The high degree of variability demonstrated within the studied sheep types implies that these populations are rich reservoirs of genetic diversity that must be preserved.

1 Introduction

Small ruminants, especially native breed types, play an important role in the livelihoods of a considerable part of human population in the tropics from socioeconomic aspects. Therefore, an integrated attempt in terms of management and genetic improvement to enhance production is of crucial importance (Mohammadabadi and Sattayimokhtari, 2013). Economical and biological efficiency of sheep production enterprises generally improves by increasing productivity and reproductive performance of ewes (Mohammadabadi and Sattayimokhtari, 2013). There are more than 50 million sheep in Iran, of 27 breeds and ecotypes (Khodabakhshzadeh et al., 2016) that have not defined well as distinct breeds. However, they are considered as geographically defined populations. The need to maintain and improve local genetic resources has been recognized as a priority at the world level. Biodiversity studies depicting a deep picture of the genetic variability of the available sheep breeds provide favorable opportunities

for both genetic conservation programs and enhancing production efficiency by means of controlled and well-designed crossbreeding systems exploiting breed diversities, heterosis and breed complementarity (Esmailizadeh et al., 2012).

Genetic diversity in indigenous breeds is a major concern considering the necessity of preserving what may be a precious and irreplaceable richness with regard to new productive demands (Khodabakhshzadeh et al., 2016). Conservation should be based on a deep knowledge of the genetic resources of the specific breed (Zamani et al., 2015). Therefore, it is important to try to characterize genetically indigenous breeds. Genes affecting polygenic traits and characterizing milk or meat productions are difficult to identify (Soufy et al., 2009; Shojaei et al., 2011). The maintenance of genetic diversity in livestock species requires the adequate implementation of conservation priorities and sustainable management programs, which should be based on comprehensive information regarding the structure of the populations, including sources of genetic variability among and within breeds. Genetic diversity is an essential component for population survival, evolution, genetic improvement and adaptation to changing environmental conditions (Kumar et al., 2006). Molecular methods based on molecular markers, such as random amplification of polymorphic DNA (RAPD), restriction fragment length polymorphism (RFLP) and microsatellites, are useful tools to study the genetic variations. Short tandem repeats known as microsatellites are widely used as molecular markers of choice for genetic studies. Advantages of microsatellites are a high degree of polymorphism due to existence of several alleles at each locus, their large number, distribution throughout the genome, a high level of polymorphism, neutrality with respect to selection, codominant inheritance and easy automation of analytical procedures (Canon et al., 2006; Mohammadifar et al., 2009; Mohammadabadi et al., 2010). Several studies have investigated the genetic diversity in sheep using microsatellites (Buchanan and Thue, 1998; Esmaeilkhanian and Banabazi, 2006; Bhatia and Arora, 2007; Nanekarani et al., 2010; Sun et al., 2010; Jakaria et al., 2012; Musthafa Muneeb et al., 2012; Hepsibha et al., 2013; Crispim et al., 2014), but a study of all Iranian sheep together has not been performed until now. Hence, the aim of the present study was to evaluate the genetic diversity within and between 14 sheep types in Iran (Kermani, Pakistani, Lori, Arabi, Dalagh, Baluchi, Iran-Black, Gharegol, Arman, Lori–Bakhtiari, Kermani–Pakistani, Kermani–Romanov, Lori–Bakhtiari–Romanov and Lori–Bakhtiari–Pakistani) using five microsatellite markers and to measure the distance among these breeds.

2 Materials and methods

2.1 Animals and sampling

In this study, 565 blood samples were collected from different individuals of 14 sheep types in Iran (Kermani (KER), $n = 102$ (75 females and 27 males); Pakistani (PAK), $n = 25$ (20 females and 5 males); Lori (LOR), $n = 45$ (36 females and 9 males); Arabi (ARB), $n = 47$ (34 females and 13 males); Dalagh (DAL), $n = 44$ (32 females and 12 males); Baluchi (BAL), $n = 41$ (30 females and 11 males); Iran-Black (IRB), $n = 44$ (40 females and 4 males); Gharegol (GHA), $n = 43$ (33 females and 10 males); Arman (ARM), $n = 46$ (31 females and 15 males); Lori–Bakhtiari (LRB), $n = 25$ (19 females and 6 males); Kermani–Pakistani (KER-PAK), $n = 7$ (4 females and 3 males); Kermani–Romanov (KER-ROM), $n = 48$ (28 females and 20 males); Lori–Bakhtiari–Romanov (LRB-ROM), $n = 33$ (20 females and 13 males); and Lori–Bakhtiari–Pakistani (LRB-PAK), $n = 15$ (8 females and 7 males)). Five-milliliter blood samples of both sexes were collected via the jugular vein in tubes containing EDTA for prevention of coagulation. Genomic DNA was extracted using a modified salting-out method (Abadi

et al., 2009). DNA quality definition was determined using both spectrophotometry and agarose gel (0.8 %).

2.2 Microsatellite analysis

In this study, five microsatellite markers across the sheep genome were used. Gradient PCR was used to optimize the annealing temperature for each marker. The PCR products were tested in agarose gel (0.8 %) to estimate the best annealing temperature for each primer. The studied microsatellite markers, their primer sequences, detected annealing temperature and their allele size ranges are shown in Table 1. The selected microsatellites were amplified with PCR using genomic DNA extracted from individual animals (in total 25 mL). This mixture included 2.5 mL of PCR buffer, 1 mL of MgCl$_2$, 0.5 mL of dNTP mix, 0.3 mL of Taq DNA polymerase, 16.7 mL of sterile water and 2 mL of template DNA. The reaction conditions were 94 °C for 5 min; 30 cycles of 95 °C for 30 s; annealing temperature, differing for each primer (Table 1), for 45 s and 72 °C for 30 s; and final extension at 72 °C for 5 min. A 50 bp DNA ladder was used as a standard size for sizing PCR products. The PCR products were electrophoresed at 200 V for 60 to 90 min on the 8.0 % polyacrylamide gel and visualized by staining with silver nitrate; the genotypes were scored by UVIdoc software after drying and scanning the gels.

2.3 Statistical analysis

Genotypes were assigned for each animal based on allele size data. Frequencies and number of alleles for each locus, observed and expected heterozygosity were estimated using FSTAT (version 2.9.3.2) (Goudet, 2002). The polymorphic information content (PIC) value was calculated according to Buchanan and Thue (1998). Nei's standard genetic distances (DS) among populations were computed by POPGENE (Yeh et al., 1999). This software was also used to construct the dendrogram of unweighted pair group with arithmetic mean (UPGMA).

3 Results

3.1 Allelic diversity

The PCR reactions were successfully performed with all primers. All the microsatellite loci were found to be highly polymorphic. In total, 65 alleles were detected; the HSC marker in PAK sheep breed and overall showed the highest number of alleles per locus (14 and 15 alleles respectively) while the OarHH35 marker in KER-PAK Sheep tape showed the lowest number of alleles (5 alleles) (Table 2) with a mean of 13 ± 1.22 alleles, whereas the effective number of alleles ranged from 3.26 (McMA2 marker) to 10.3 (McMA26 marker) with a mean of 6.59 ± 0.96 alleles per locus (Table 2). The number of actual and effective alleles of studied

Table 1. Characteristics of selected microsatellite in present study.

Marker (chromosome)	Primers (5'–3')	Accession numbers*	Type of repeat	Annealing temperature (°C)	Allele size range (bp)	Reference
McMA2 (13)	F: TCA CCC AAC AAT CAT GAA AC R: TTA AAT CGA GTG TGA ATG GG	AF098773	AC	52	157–201	Maddox et al. (2000)
McMA26 (18)	F: TCT CTG CTT TCC AGC CTT ATT C R: AGA GCT TTT AGG ACA GCC ACC	AF098961	GT	52.5	184–218	Maddox et al. (2000)
HSC (20)	F: CTG CCA ATG CAG AGA CAC AAG A R: GTC TGT CTC CTG TCT TGT CAT C	M90759	–	60.2	267–301	NCBI
OarHH35 (4)	F: AAT TGC ATT CAG TAT CTT TAA ACA TCT GGC R: ATG AAA ATA TAA AGA GAA TGA ACC ACA CGG	L12554	GT	59.7	87–170	Henry et al. (1993)
BM6444 (2)	F: CTC TGG GTA CAA CAC TGA GTC C R: TAG AGA GTT TCC CTG TCC ATC C	G18444	GT	57.5	75–178	Buchanan et al. (1998)

* http://www.ncbi.nlm.nih.gov:80/entrez/viewer.cgi; F: forward; R: reverse

Table 2. The number of actual alleles (Na) and effective alleles (Ne) for different combinations of locus–population and for each population.

Marker	McMA2		BM6444		McMA26		HSC		OarHH35		Mean (Ne)
Breed	Ne	Na	Ne	Na	Ne	Na	Ne	Na	Ne	Na	
ARB	8.2	11	7.7	11	10.3	12	10.1	12	7	9	8.6
ARM	7.88	10	7.1	9	7.9	11	6.3	7	7.1	9	7.2
BAL	8.7	12	8.04	11	7.3	11	6.2	8	7.07	9	7.4
DAL	8.56	12	8.01	10	7.7	11	5.5	6	5.9	8	7.1
GHA	7.73	11	7.29	10	8.04	11	6.08	8	8.6	12	7.5
IRB	8.56	12	5.28	8	8.01	10	4.36	8	7.75	11	6.8
KER	10.22	12	7.25	8	8.38	11	6	7	7.52	10	7.8
KER-PAK	3.26	6	3.76	6	4.08	7	4.9	6	4.66	5	4.13
KER-ROM	8.15	12	6.41	9	5.48	9	6.4	7	6.5	10	6.59
LOR	7.27	10	7.14	9	8.42	12	4.39	5	7.19	9	6.85
LRB	6.31	12	6.21	13	5.23	10	4.61	7	5.38	9	5.5
LRB-ROM	4.69	11	4.03	9	5.62	10	4.01	7	6.05	8	4.8
LRB-PAK	3.46	9	3.94	8	4.68	9	4.5	6	4.68	6	4.2
PAK	5.23	10	4.96	11	6.06	11	7.81	14	6.21	13	6.05
Overall	8.48	12	7.24	13	8.08	12	6.5	15	7.4	13	7.54

sheep types has been shown in Table 2. Among the sheep, the mean number of alleles ranged from 6 in KER-PAK to 11.8 in PAK sheep.

3.2 Polymorphism information content (PIC) and Shannon's information index (I)

PIC and the Shannon information index are another measure of genetic variability indicating the informativeness of the assessed loci. PIC values ranged from 0.84 (HSC marker) to 0.92 (BM6444 marker) with a mean PIC value of 0.87 (Table 3), indicating that all loci were highly polymorphic.

Table 3. Polymorphic information content (PIC), Shannon information index (I) and expected heterozygosity (H_e) in 14 sheep types.

Marker	PIC	I	H_e
McMA2	0.88	2.31	0.87
BM6444	0.92	2.17	0.86
McMA26	0.87	2.27	0.87
HSC	0.84	2.04	0.82
OarHH35	0.86	2.18	0.87
Mean	0.87	2.2	0.85

Table 4. Mean polymorphic information content (PIC), mean Shannon information index (I), mean expected heterozygosity (H_e) and mean number of actual alleles (MNa) per type across five loci.

Breed	PIC	I	H_e	MNa
ARB	0.88	2.27	0.88	11
ARM	0.87	2.09	0.86	9.2
BAL	0.86	2.15	0.86	10.2
DAL	0.84	2.07	0.85	9.4
GHA	0.87	2.17	0.86	10.4
IRB	0.84	2.05	0.84	9.8
KER	0.86	2.14	0.87	9.6
KER-PAK	0.73	1.58	0.75	6
KER-ROM	0.85	2.04	0.84	9.4
LOR	0.84	2.02	0.84	9
LRB	0.80	1.97	0.81	10.2
LRB-ROM	0.78	1.83	0.79	9
LRB-PAK	0.72	1.69	0.76	7.6
PAK	0.82	2.08	0.83	11.8
Overall	0.87	2.20	0.86	13
Mean	0.82	2.01	0.83	9.47

3.3 Heterozygosity and interbreed inbreeding estimate

The mean expected heterozygosity over all breeds in the present study was 0.85 (Table 3), so it does not likely encounter problems that result from inbreeding depression. Mean estimates of expected heterozygosity overall loci and types were 0.86 (Table 4). ARB had the highest expected heterozygosity for all the loci ($H_e = 0.88$) with an average number of actual allele of Na $= 11$, while KER-PAK had the lowest expected heterozygosity ($H_e = 0.75$) with an average number of actual allele of Na $= 6$ (Table 4).

3.4 Hardy–Weinberg equilibrium (HWE)

When the Hardy–Weinberg testing was performed for the loci, deviations from the Hardy–Weinberg equilibrium were found to be significant ($p < 0.05$) in some loci in several breeds (29 out of the 70 population–locus combinations (41.4 %) were at Hardy–Weinberg equilibrium).

3.5 Genetic differentiation

Population differentiation examined by fixation indices such as F_{IS}, F_{IT} and F_{ST} for each of the five analyzed loci across 14 sheep breeds is given in Table 5. The genetic relationships between the sheep breeds were calculated using Nei's genetic distances (D) (Nei et al., 1983). The matrix of Nei's standard genetic distances (D) among breeds is presented in Table 6 and the corresponding phylogenetic tree is presented in Fig. 1. The genetic distances between breeds ranged from 0.04 for Kermani and Kermani–Pakistani to 0.25 for Kermani and Lori–Bakhtiari–Pakistani. The phylogenetic tree (Fig. 1) provides a method of visualizing the genetic relationship be-

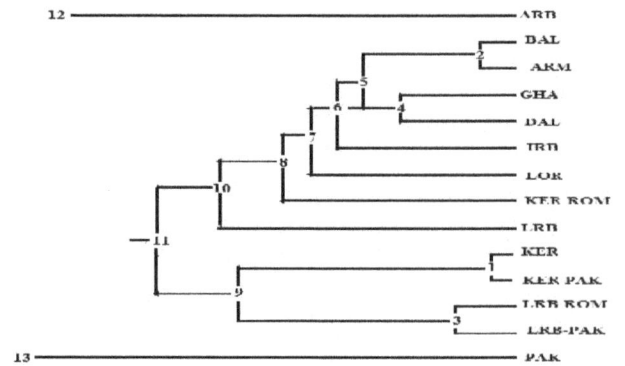

Figure 1. UPGMA phylogenetic tree based on Nei genetic distance. Location of these 14 studied breeds is shown in Fig. 2. Subspecies are indicated as Kermani (KER), Pakistani (PAK), Lori (LOR), Arabi (ARB), Dalagh (DAL), Baluchi (BAL), Iran-Black (IRB), Gharegol (GHA), Arman (ARM), Lori–Bakhtiari (LRB), Kermani–Pakistani (KER-PAK), Kermani–Romanov (KER-ROM), Lori–Bakhtiari–Romanov (LRB-ROM) and Lori–Bakhtiari–Pakistani (LRB-PAK).

Table 5. Estimators of F statistics at each locus across the 14 sheep types.

Marker	F_{IS}	F_{IT}	F_{ST}
McMA2	0.0358	0.0624	0.0276
BM6444	0.0149	0.0394	0.0248
McMA26	0.0162	0.0390	0.0232
HSC	0.1075	0.1317	0.0271
OarHH35	0.0356	0.0530	0.0181
Mean	0.042	0.0651	0.0241

tween populations, with the values in the nodes of the tree indicating the proportion of 1000 replicates of the 5 microsatellites.

4 Discussions

Genetic variation is a basic requirement for animal breeding, whereas a high genetic variation is needed for genetic improvement of domestic animals (Askari et al., 2011). The number of alleles at different marker loci serves as a measure of the genetic variability having direct impact on differentiation of breeds within a species. Since 80 % of the markers exhibited four or more alleles, the microsatellite loci screened in this study were appropriate in expressing the molecular characteristics and/or genetic variation in the population. The effective number of alleles at each locus provides information on predominant alleles. Since allelic diversity in the studied sheep breed populations was high, it can be concluded that, in these animals, genetic diversity is sufficiently high and they have a good gene pool for breeding programs. Because of the limitations in studied population number and

Table 6. Genetic distance matrices based on D_A (upper diagonal matrix) and D_{AL} (lower diagonal matrix) using 1000 bootstrap replications.

	ARB	ARM	BAL	DAL	GHA	IRB	KER	KERPAK	KERROM	LOR	LRB	LRBROM	LRBPAK	PAK
ARB	–	0.87	0.87	0.85	0.88	0.85	0.82	0.8	0.85	0.86	0.84	0.83	0.8	0.81
ARM	0.129	–	0.91	0.91	0.93	0.91	0.92	0.8	0.89	0.8	0.89	0.87	0.86	0.88
BAL	0.128	0.05	–	0.92	0.85	0.93	0.9	0.84	0.89	0.84	0.89	0.86	0.87	0.86
DAL	0.15	0.08	0.07	–	0.91	0.91	0.81	0.82	0.89	0.85	0.89	0.83	0.84	0.88
GHA	0.125	0.07	0.06	0.09	–	0.84	086	0.84	0.87	0.85	0.86	0.88	0.88	0.87
IRB	0.15	0.09	0.07	0.08	0.08	–	0.87	0.83	0.9	0.87	0.91	0.89	0.88	0.87
KER	0.18	0.08	0.09	0.1	0.14	0.13	–	0.93	0.89	0.87	0.84	0.79	0.77	0.84
KERPAK	0.21	0.21	0.17	0.18	0.16	0.17	0.04	–	0.82	0.81	0.85	0.84	0.91	0.83
KERROM	0.15	0.21	0.09	0.11	0.09	0.09	0.1	0.19	–	0.82	0.87	0.86	0.087	0.86
LOR	0.14	0.11	0.06	0.08	0.07	0.06	0.13	0.2	0.09	–	0.9	0.88	0.87	0.87
LRB	0.17	0.11	0.09	0.1	0.09	0.08	0.16	0.16	0.13	0.1	–	0.94	0.9	0.88
LRBROM	0.18	0.13	0.14	0.17	0.11	0.1	0.23	0.16	0.14	0.12	0.08	–	0.92	0.87
LRBPAK	0.21	0.14	0.13	0.17	0.11	0.12	0.25	0.08	0.13	0.13	0.09	0.06	–	0.91
PAK	0.2	0.12	0.13	0.12	0.12	0.13	0.17	0.18	0.14	0.13	0.08	0.13	0.12	–

microsatellite loci, further research with more population and loci numbers needs to be done to be able correctly evaluate the genetic relationship among the populations. However, a higher number of alleles for each locus showed that all the markers used were appropriate to analyze diversity in studied breeds. The level of variation depicted by number of alleles at each locus was similar to earlier reports on sheep breeds (Arora and Bhatia, 2006; Jakaria et al., 2012; Hepsibha et al., 2013; Crispim et al., 2014).

Among the types, the mean of PIC ranged from 0.72 in LRB-PAK to 0.88 in ARB, and Shannon information index ranged from 1.58 in KER-PAK to 2.27 in ARB. . However, PIC in the range of 0.69 to 0.92 was reported for this locus in some Iranian sheep breeds in a previous study (Esmaeilkhanian and Banabazi, 2006; Banabazi et al., 2007; Nanekarani et al., 2010). Our estimate of Shannon information index was in agreement with that reported by Nanekarani et al. (2010) on three types of Iranian sheep populations and Musthafa Muneeb et al. (2012) on Najdi sheep populations. This index estimate was higher than that of Mehraban sheep (Zamani et al., 2011) and Indian Bellary sheep (Kumar et al., 2006).

A high value of average expected heterozygosity within a breed could be attributed to the large number of alleles detected in the tested loci (Kalinwski, 2002). The high heterozygosity is attributed to the wider distribution of population and to presumably the larger flock size. It also proves that no controlled breeding is followed as males and females are allowed to graze as a large flock. Heterozygosity values were in concordance with those of Najdi sheep populations (Musthafa Muneeb et al., 2012), Indian sheep breeds (Bhatia and Arora, 2007) and five Iranian indigenous sheep populations (Esmaeilkhanian and Banabazi, 2006). The expected heterozygosity in this study was higher than that of some of loci (OarHH35) used in Hu sheep in China (Sun et al., 2010). Thus, differences of heterozygosity values obtained could be due to type of markers used, the sequence of simple sequence repeat (SSR) markers, and sampling and sample size.

In total, 58.6 % of the loci in some breeds deviated from HWE ($p < 0.05$). This can be attributed to the excess of heterozygote individuals than homozygote individuals, migration, high mutation rate in microsatellite, artificial selection in all breeds and population subdivision owing to genetic drift (Aminafshar et al., 2008). Deviation from HWE at microsatellite loci has also been reported in various studies (Esmaeilkhanian and Banabazi, 2006; Banabazi et al., 2007; Nanekarani et al., 2010; Jakaria et al., 2012; Hepsibha et al., 2013; Crispim et al., 2014). It is known that a population is considered to be in HWE only when it is able to maintain its relative allelic frequencies. This deviation from HWE can be probably due to level of crossbreeding in some populations. However, crossbreeding of local and exotic improved breeds can show a faster impact on performance than long selection schemes for the improvement of local breeds; therefore, it is more beneficial in the short term. However, as another consequence, it is one of the major threats to the existence of local genetic diversity, inducing displacement or genetic erosion. Indiscriminate crossbreeding or extensive use of exotic germplasms can lead to genetic erosion by dilution or eradication of the local genetic pool; hence, it is better to use controlled crossbreeding.

The within-breed inbreeding estimate (F_{IS}) in the investigated population was found to be low and similar to a value of 0.0024 ± 0.07 reported by Pariset et al. (2003). This indicates that the population has not suffered any inbreeding but rather has a heterozygosity excess. Further, a high level of expected heterozygosity in this study correlates with the low F_{IS} value, suggesting low selection pressure, no inbreeding and relatively more alleles.

There are three separate clusters on the dendrogram. One includes 12 breeds and another consists of Arabi and then Pakistani in a separate branch. Since D_A has been found to be more useful for obtaining of correct topology (Takezaki and Nei, 1996), we presented only the results of D_A. The lowest D_A was between Kermani and Kermani–Pakistani and

ISLAMIC REPUBLIC OF IRAN

Map Sources: ESRI, UNCS.
The boundaries and names shown and the designations used on this map do not imply official endorsement or acceptance by the United Nations. Map created in Sep 2013

Figure 2. Location of the 14 study sites in Iran. The putative subspecies are indicated as Kermani (KER), Pakistani (PAK), Lori (LOR), Arabi (ARB), Dalagh (DAL), Baluchi (BAL), Iran-Black (IRB), Gharegol (GHA), Arman (ARM), Lori–Bakhtiari (LRB), Kermani–Pakistani (KER-PAK), Kermani–Romanov (KER-ROM), Lori–Bakhtiari–Romanov (LRB-ROM) and Lori–Bakhtiari–Pakistani (LRB-PAK). Source of original map: OCHA/ReliefWeb (http://img.static.reliefweb.int/sites/reliefweb.int/files/resources/irn_ocha_1000px.png).

between Baluchi and Arman (Table 6). These distances are rational due to co-descendance of two breeds of sheep and neighboring geographic distributions. Their phenotypic similarities also agree with these distances. A possible cross-migration between Gharegol and Dalagh may occur due to the short geographic distance between the areas in which these two breeds are distributed resulted in small genetic distance between two breeds. The obtained tree revealed that the most closely related breeds were the two synthetic lines (Lori–Bakhtiari–Romanov and Lori–Bakhtiari–Pakistani). Although one local breed (Lori–Bakhtiari) participated in the creation of two analyzed breeds, clear differences can be noticed. Also, the UPGMA tree showed that the Arabi forms the most distinct breed. The rugged topography and the Zagros Mountains (see Fig. 2) limit the movement of sheep and impose reproductive isolation between Arabi and other population distribution areas. Therefore, Arabi is in a separate branch.

5 Conclusions

The breeds included in this study are of great importance to the sheep holders in Iran. This study presents an investigation of variability at the DNA level within and between some Iranian sheep breeds. The results indicated that all studied breeds exhibited considerable genetic variation, based on their high mean number of alleles and gene diversity. According to the selective standard of microsatellite loci, microsatellite loci ought to have at least four alleles per locus to be considered useful for the evaluation of genetic diversity. Based on this criterion, the five microsatellite loci used in the present study can be considered useful for the evaluation of genetic diversity within and among populations and for the selection of breeding animals from divergent groups maximizing genetic variation and consequently fitness. A future direction to our study can be studying all of the Iranian indigenous sheep breeds to better evaluate the level of inbreeding and establish appropriate conservation strategies with the aim to avoid losses of genetic diversity.

Author contributions. MM and AEK designed this project, MTVE collected blood from animals and performed experiments in the laboratory, and MM and MTVE wrote the paper.

Competing interests. The authors declare that they have no conflict of interest.

Acknowledgements. We would like to thank the sheep farmers who provided sheep blood samples for this study.

Edited by: Steffen Maak

References

Abadi, M. R. M., Askari, N., Baghizadeh, A., and Esmailizadeh, A.: A directed search around caprine candidate loci provided evidence for microsatellites linkage to growth and cashmere yield in Rayini goats, Small Ruminant Res., 81, 146–151, 2009.

Aminafshar, M., Amirinia, C., and Torshizi, R. V.: Genetic diversity in buffalo population of guilan using microsatellite marker, J. Anim. Vet. Adv., 7, 1499–1502, 2008.

Arora, R. and Bhatia, S.: Genetic diversity of Magra sheep from India using microsatellite analysis, Asian Austral. J. Anim., 19, 938–942, 2006.

Askari, N., Abadi, M. M., and Baghizadeh, A.: ISSR markers for assessing DNA polymorphism and genetic characterization of cattle, goat and sheep populations, Iran. J. Biotechnol., 9, 222–229, 2011.

Banabazi, M. H., Esmaeilkhanian, S., Miraei Ashtiani, S. R., and Moradi, S. M.: Genetic variation within and between five Iranian sheep populations using microsatellite markers, J. Sci. Technol. Agric. Nat. Resour., 10, 481–488, 2007 (Abstract in English).

Bhatia, S. and Arora, R.: Genetic diversity in Kheri-A pastoralists developed Indian sheep using microsatellite markers, Indian J. Biotechnol., 7, 108–112, 2007.

Buchanan, F. C. and Thue, T. D.: Interbreed polymorphic information content of microsatellites in cattle and sheep, Can. J. Anim. Sci., 78, 425–428, 1998.

Canon, J., Garćıa, D., Garćıa-Atance, M. A., Obexer-Ruff, G., Lenstra, J. A., Ajmone-Marsan, P., and Dunner, S.: The Econogene Consortium, Geographical partitioning of goat diversity in Europe and the Middle East, Anim. Genet., 37, 327–334, 2006.

Crispim, B. D. A., Seno, L. D. O., Egito, A. A. D. E., Junior, F. M. D. V., and Grisolia, A. B.: Application of microsatellite markers for breeding and genetic conservation of herds of Pantaneiro sheep, Electron. J. Biotechn., 17, 317–321, 2014.

Esmaeilkhanian, S. and Banabazi, M. H.: Genetic variation within and between five Iranian sheep populations using microsatellites markers, Pak. J. Biol. Sci., 13, 2488–2492, 2006.

Esmailizadeh, A. K., Nemati, M., and Mokhtari, M. S.: Fattening performance of purebred and crossbred lambs from fat-tailed Kurdi ewes mated to four Iranian native ram breeds, Trop. Anim. Health Prod., 44, 217–223, 2012.

Goudet, J.: FSTAT (version 2.9.3.2): a program to Estimate and Test gene Diversities and Fixation Indices, available at: http://www. softpedia.com/get/Science-CAD/FSTAT.shtml, 2002.

Henry, H. M., Penty, J. M., Pierson, C. A., and Crawford, A. M.: Ovine microsatellites at the OarHH35, OarHH41, OarHH44, OarHH47and OarHH64 loci, Anim. Genet., 24, 222–222, https://doi.org/10.1111/j.1365-2052.1993.tb00300.x, 1993.

Hepsibha, P., Karthickeyan, S. M. K., and Guru, V.: Microsatellite marker based assessment of genetic structure of Coimbatore breed of sheep Ovis Aries in Tamil Nadu, Indian J. Biotechnol., 13, 203–206, 2013.

Jakaria, Z., Sulandari, M. S., Subandriyo, A., and Muladno, S.: The use of microsatellite markers to study genetic diversity in Indonesian sheep, J. Indonesian Trop. Anim. Agric., 37, 1–9, 2012.

Kalinwski, S. T.: How many alleles per locus should be used to estimate genetic distances, Heredity, 88, 62–65, 2002.

Khodabakhshzadeh, R., Mohammadabadi, M. R., Esmailizadeh, A., Moradi Shahrebabak, H., Bordbar, F., and Ansari Namin, S.: Identification of point mutations in exon 2 of GDF9 gene in Kermani sheep, Pol. J. Vet. Sci., 19, 281–289, 2016.

Kumar, D., Sharma, R., Pandey, A. K., Gour, D. S., Malik, G., Ahlawat, S. P. S., and Jain, A.: Genetic diversity and bottleneck analysis of Indian Bellary sheep by microsatellite markers, Russ. J. Genet., 9, 996–1005, 2006.

Maddox, J. F., Riffkin, C. D., and Beh, K. J.: Dinucleotide repeat polymorphism at ovin MCMA1, MCMA2, MCMA5, MCMA8, MCMA9, MCMA11, MCMA14, MCMA20, MCMA24, MCMA26 loci, Anim. Genet., 31, 148–149, 2000.

Mohammadabadi, M. R. and Sattayimokhtari, R.: Estimation of (co) variance components of ewe productivity traits in Kermani sheep, Slovak. J. Anim. Sci., 46, 45–51, 2013.

Mohammadifar, A., Amirnia, S., Omrani, H., Mirzaei, H. R., and Mohammadabadi, M. R.: Analysis of genetic variation in quail population from Meybod Research Station using microsatellite markers, Pajouhesh and Sazandegi, 22, 72–79, 2009 (Abstract in English).

Mohammadabadi, M. R., Nikbakhti, M., Mirzaee, H. R., Shandi, M. A., Saghi, D. A., Romanov, M. N., and Moiseyeva, I. G.: Genetic variability in three native Iranian chicken populations of the Khorasan province based on microsatellite markers, Russ. J. Genet., 46, 572–576, 2010.

Musthafa Muneeb, M., Aljummah, R. S., and Alshaik, M. A.: Genetic diversity of Najdi sheep based on microsatellite analysis, Afr. J. Biotechnol., 1183, 14868–14876, 2012.

Nanekarani, S., Amirinia, C., Amirmozafari, N., Vaez Torshizi, R., and Gharahdaghi, A. A.: Genetic variation among pelt sheep population using microsatellite markers, Afr. J. Biotechnol., 9, 7437–7445, 2010.

Nei, M., Tajima, F., and Tateno, Y.: Accuracy of estimated phylogenetic trees from molecular data, J. Mol. Evol., 19, 153–170, 1983.

Pariset, L. M., Savarese, C., Cappuccio, I., and Valentini, A.: Use of microsatellites for genetic variation and inbreeding analysis in Sarda sheep flocks of central Italy, J. Anim. Breed. Genet., 120, 425–432, 2003.

Shojaei, M., Mohammadabadi, M. R., Asadi Foz, M., Dayani, O., Khezri, A., and Akhondi, M.: Association of growth trait and Leptin gene polymorphism in Kermani sheep, J. Cell Mol. Res., 2, 67–73, 2011.

Soufy, B., Mohammadabadi, M. R., Shojaeyan, K., Baghizadeh, A., Ferasaty, S., Askari, N., and Dayani, O.: Evaluation of Myostatin gene polymorphism in Sanjabi sheep by PCR-RFLP method, Anim. Sci. Res., 19, 81–89, 2009 (Abstract in English).

Sun, W., Chang, H., Hussein Musa, H., and Chu, M.: Study on relationship between microsatellite polymorphism and producing ability on litter size trait of Hu sheep in China, Afr. J. Biotechnol., 9, 8704–8711, 2010.

Takezaki, N. and Nei, M.: Genetic distances and reconstruction of phylogenetic trees from microsatellite DNA, Genetics, 144, 389–399, 1996.

Yeh, F. C., Yang, R., and Boyle, T.: PopGene. Version 1.31. Microsoft Window-based Freeware for Population Genetic Analysis, University of Alberta, Edmonton, AB, Canada, 1999.

Zamani, P., Akhondi, M., Mohammadabadi, M. R., Saki, A. A., Ershadi, A., Banabazi, M. H., and Abdolmohammadi, A. R.: Genetic variation of Mehraban sheep using two intersimple sequence repeat (ISSR) markers, Afr. J. Biotechnol., 10, 1812–1817, 2011.

Zamani, P., Akhondi, M., and Mohammadabadi, M. R.: Associations of inter-simple sequence repeat loci with predicted breeding values of body weight in sheep, Small Ruminant Res., 132, 123–127, 2015.

Influences of breed, sex and age on seasonal changes in haematological variables of tropical goat kids

Buhari Habibu[1], Mohammed Kawu[1], Hussaina Makun[2], Tagang Aluwong[1], Lukman Yaqub[1], Tavershima Dzenda[1], and Hajarah Buhari[3]

[1]Department of Veterinary Physiology, Ahmadu Bello University, Zaria, Nigeria
[2]Small Ruminant Research Programme, National Animal Production Research Institute, Zaria, Nigeria
[3]Samaru College of Agriculture, Division of Agricultural Colleges, Ahmadu Bello University, Zaria, Nigeria

Correspondence to: Buhari Habibu (buharihabibu@rocketmail.com)

Abstract. The influences of breed, sex and age on seasonal changes in haematological variables of kids (1–4 months old) belonging to Red Sokoto ($n = 60$) and Sahel ($n = 60$) goats were studied at the peaks of the cold-dry, hot-dry and rainy seasons in a West Africa Guinea savanna climate. The results showed that, during the hot-dry season, Sahel goat kids had significantly higher ($P < 0.05$) packed cell volume (PCV) and red blood cell (RBC) count but lower ($P < 0.05$) mean corpuscular volume (MCV), mean corpuscular haemoglobin (MCH), mean corpuscular haemoglobin concentration (MCHC) and total leucocyte count than Red Sokoto kids. Similarly, younger kids of both breed had significantly higher ($P < 0.05$) PCV and RBC but lower MCV, MCH and MCHC compared with the older kids during the hot-dry season. Younger kids (1–2 months) of both breeds exhibited significantly ($P < 0.05$) high PCV and RBC but low MCH and MCHC during the hot-dry season as compared with the cold-dry and rainy seasons, with the magnitude of the change being greater in Sahel kids. More remarkable seasonal fluctuations in haematological parameters were observed in buck kids than doe kids of both breeds. Multivariate analysis revealed a clear distinction between the change in haematological parameters during the cold-dry as compared with the hot-dry and rainy seasons, with MCV having the strongest discriminating power (0.91*). In conclusion, breed, age and sex variations in haematological variables of goats were more pronounced in the hot-dry season, during which the seasonal changes were more dramatic in kids of Sahel goats, as well as in younger and male kids of both breeds. These findings may be useful in the management of tropical goat kids in different seasons.

1 Introduction

In many developing countries of the world, different breeds of goats play vital roles in the economy, religion, nutrition and tradition of poor livestock owners. In Nigeria, the three distinct indigenous breeds of goats are Red Sokoto (Maradi or Savanna Brown), Sahel (West African Long-legged or White Bornu) and West African Dwarf (Egwu et al., 1995). Distribution of the goat breeds indicates that the Red Sokoto goats are common in the Sudanian Savanna region of the Republic of Niger and the Guinea and Sudanian Savanna regions of Nigeria; the Sahel or desert goats are found in the Sahel belt of West Africa (Fig. 1), while the West African

Dwarf goats are largely restricted to the coast of West and Central Africa (Gall, 1996; Habibu et al., 2016a). In recent years, adaptation of these breeds of goats, particularly the Sahel goats, to new geographical regions of the country, including the northern Guinea savanna zone and rainforest zone has been observed (Habibu et al., 2017).

The Guinea savanna climate of West Africa is found north of the rainforest zone, while the Sahel climate is located south of the Sahara (Fig. 1). The northern Guinea savanna zone of Nigeria is a semi-arid region with three distinct seasons, namely the cold-dry (CDS), hot-dry (HDS) and rainy (RAS) seasons (Igono and Aliu, 1982; Dzenda et al., 2015). Like the young offspring of other livestock species,

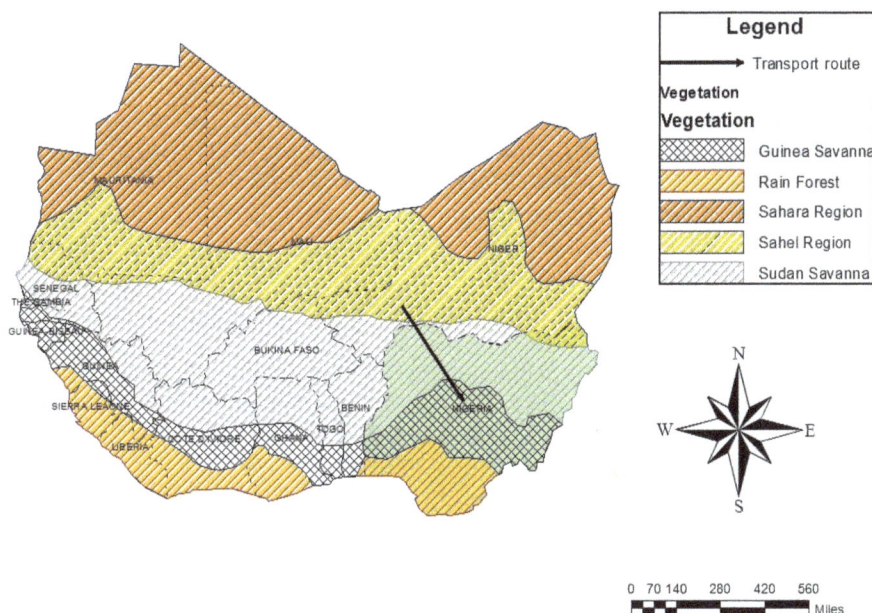

Figure 1. Map of West Africa indicating the geographical zones and transportation of Sahel goats. The map was carved and modified using a section of the map of Africa obtained from Bioclimatic Regions Map (https://eros.usgs.gov/westafrica/node/147) as a guide.

goat kids are confronted with the challenge of extra-uterine environment that is entirely different from that in the uterus, which makes them metabolically unstable and may increase their susceptibility to perinatal diseases, resulting in high neonatal mortality, most especially in thermally stressful seasons (Piccione et al., 2007). In West Africa, pre- and post-weaning kid mortality has been reported to be as high as 10 and 23.1 %, respectively (Turkson et al., 2004). The new-borns that survive usually adapt by developing functional changes in almost all organs and systems through a series of tissue-based metabolic and morphologic remodelling (Saddiqi et al., 2011), which greatly affects the survival of the newborn and, if effective, may reduce the incidence of newborn mortality (Nowak et al., 2000). The adaptive response of the newborn to the novel environments may differ due to differences in breed (Sawalha et al., 2007). Since most tropical breeds of goats are non-seasonal breeders, they cycle year-round and parturition may occur in any season of the year. Accordingly, kids may be exposed to all the distinct seasons of the year.

Like other tissues, the blood may undergo series of adaptive remodelling on exposure to the extra-uterine environment in different seasons. Haematological parameters are generally used to monitor and evaluate the health of small ruminants (Fazio et al., 2016). Due to the economic importance of goats, a lot of information is available on their haematology (Tibbo et al., 2004; Zumbo et al., 2011; Piccione et al., 2014). Breed, sex, age and season may greatly influence haematological profile of goats, with season classically increasing or decreasing haematological variables (Egbe-Nwiyi et al., 2000; Tibbo et al., 2004; Zumbo et al., 2011;

Habibu et al., 2017). However, there is a paucity of information on the haematological responses of different breeds, sexes and age groups of goat kids in different seasons of the tropics. Findings from the current study may provide valuable information on variations in physiological variables of tropical goats in relation to different endogenous and exogenous factors. Similarly, the study may provide information that could be useful in the management of goat kids in different seasons and in the design of a breeding programme to ensure parturition occurs in seasons that are less stressful. Therefore, the aim of the present study was to evaluate the influence of breed, sex and age on the seasonal changes in haematological variables of goat kids in the northern Guinea savanna zone of Nigeria.

2 Materials and methods

2.1 Experimental site and animal management

The study was carried out in the Small Ruminant Research Section of the National Animal Production Research Institute (NAPRI), Ahmadu Bello University, Shika, Zaria, Nigeria, located at 11°12' N, 7°33' E and an altitude of 610 m above sea level. The animals were housed in well-ventilated pens each with dimensions of 6.1, 6.1 and 2.2 m for length, width and height, respectively. The roofs of the pens were made of galvanized metal sheet, while the floor was covered with concrete. The pens had a stocking density of 35 kids. The health status of the goats was evaluated based on behaviour, rectal temperature (38.00–39.76 °C), respiratory rate (22–34 breaths per minute), heart rate (60.36–

90.34 beats per minute), appetite and faecal consistency. Animals on the farm were routinely screened by a trained staff member for helminths (using floatation and sedimentation tests) and haemoparasites. Only clinically healthy animals were used for the study. Both breeds of goats were housed within the same environment during all the seasons. Adult Red Sokoto and Sahel does were oestrous-synchronized using CIDR and mated by proven Red Sokoto and Sahel bucks, respectively. Kids born from Red Sokoto and Sahel does that were homogeneous for age (4 to 5 years), live weight (25 ± 1.6 kg) and body condition score (2.9 ± 0.1) were used for the study. Body condition was scored by the same person adopting a six-point scale method (Santucci et al., 1991). The kids were weaned at 2 months of age and provided with milk (by directly suckling their dams) and concentrate – ration of ground maize (30 %), cotton seed cake (36 %), maize offal (20 %), wheat offal (10 %), bone meal (2.5 %) and salt (1.5 %) at 3 % body weight per day – depending on their age. Good-quality drinking water was provided ad libitum.

2.2   Experimental design

The goat kids were sampled at the peak of the CDS, HDS and RAS during the months of December, April and July, respectively. A total of 120 goat kids, aged between 1 and 4 months and with body weight ranging from 2 to 8 kg, were used throughout the study period. In each season, 40 different sets of kids (20 kids from each of the two breeds) of both sexes were sampled as follows: Red Sokoto buck kids ($n = 10$), Red Sokoto doe kids ($n = 10$), Sahel buck kids ($n = 10$) and Sahel doe kids ($n = 10$). The kids from each breed were further divided into younger (1–2 months) and older (3–4 months) kids. The ratios of younger to older Red Sokoto kids were $11:9$, $8:12$ and $7:13$ during the CDS, HDS and RAS, respectively, while those for Sahel kids were $10:10$ (CDS), $9:11$ (HDS) and $13:7$ (RAS).

The study protocol and animal experimentation were reviewed and approved by the Animal Use and Welfare Committee of Ahmadu Bello University, Zaria, and the research followed the international guidelines for animal welfare.

2.3   Blood sample collection and analysis

Five millilitres (5 mL) of blood was obtained from each kid once in every season during the morning hours, between 08:00 and 10:00 (GMT+1), through the jugular vein into Vacutainer tubes containing potassium ethylenediaminetetra acetic acid (K_3EDTA). The blood sample was transported to the laboratory in an ice pack and analysed an hour after collection. Counting of erythrocytes and leukocytes was done manually using a haemocytometer. Packed cell volume (PCV) was measured using the microhaematocrit method (Jain, 1986). The centrifugation time was increased ($3000 \times g$ for 15 min) to ensure complete packing because of the small size of caprine erythrocytes. Haemoglobin (Hb)

concentration was determined using a haemoglobin meter (XF-1C haemoglobin meter, China). Only the counts of lymphocytes and neutrophils were determined, since the study was mainly aimed at evaluating the response of the kids to thermal stress. The mean corpuscular volume (MCV), mean corpuscular haemoglobin (MCH) and mean corpuscular haemoglobin concentration (MCHC) were then derived using standard formulae (Jain, 1986).

2.4   Meteorological parameters

The morning (09:00–10:00 h) and afternoon (13:00–14:00 h) values of air temperature and relative humidity, as well as duration of sunshine (h) and total rainfall (mm), recorded daily during the week the study was conducted, were obtained from the Meteorological Unit, Institute for Agricultural Research, Ahmadu Bello University, Zaria, located about 5 km away from the experimental site. The values of the morning and afternoon temperature–humidity index (THI) were used to evaluate the level of heat stress induced by the environment. The formula reported by Ravagnolo et al. (2000) was used to calculate THI:

$$THI = (1.8 \times T + 32) - \{(0.55 - 0.0055 \cdot RH) \\ \cdot (1.8 \times T - 26)\},$$

where T is ambient temperature ($^\circ$C) and RH is relative humidity (%).

2.5   Data analysis

The Statistical Package for Social Sciences (SPSS) version 20 was used to analyse the data. Values obtained were expressed as mean ($\pm$ SEM). Comparisons between breeds, sexes and age groups were done using Student's t test, while univariate one-way analysis of variance (ANOVA) followed by the Duncan multiple range test were used to compare values between the three seasons. Discriminant analysis was performed to identify variables with greater discriminatory power between the three seasons. This was followed by the discriminant analysis option of variable classification. Values of $P < 0.05$ were considered significant.

3   Results

3.1   Meteorological parameters

Mean values of meteorological parameters for each season are presented in Table 1. The peak minimum and maximum ambient temperatures were recorded during the HDS in the morning and afternoon, respectively, while the corresponding nadir values were obtained during the CDS. Values of relative humidity at both morning and afternoon hours were lowest during the CDS but highest in the RAS. Duration of sunshine was longest and shortest during the CDS and RAS, respectively. Rainfall was recorded only during the RAS.

Table 1. Mean values of meteorological parameters during the study period.

Seasons	Ambient temperature (°C)		Relative humidity (%)		Temperature–humidity index		Sunshine (h)	Rainfall (mm)
	Morning	Afternoon	Morning	Afternoon	Morning	Afternoon		
Cold-dry	16.53	32.32	20.26	15.65	59.44	73.66	8.58	0
Hot-dry	22.26	38.11	31.56	20.70	72.19	84.15	6.71	0
Rainy	20.03	30.87	85.19	70.87	70.85	82.01	6.16	334.9

Source: Meteorological Unit, Institute for Agricultural Research, Ahmadu Bello University, Zaria.

Table 2. Effect of season on haematological parameters of Red Sokoto and Sahel buck kids.

Parameter	Cold-dry season		Hot-dry season		Rainy season	
	Red Sokoto	Sahel	Red Sokoto	Sahel	Red Sokoto	Sahel
PCV (%)	33.00 ± 2.29	34.00 ± 4.77^{b}	40.58 ± 3.89^{y}	$56.28 \pm 5.87^{a,x}$	30.16 ± 2.4	31.42 ± 2.65^{b}
RBC (10^{6}/muL)	11.47 ± 0.63	10.01 ± 1.56^{b}	10.70 ± 1.2^{y}	$19.05 \pm 0.24^{a,x}$	7.43 ± 0.87	7.73 ± 0.66^{b}
Hb (g/dL)	10.00 ± 0.23	9.25 ± 1.35	10.00 ± 0.23	10.58 ± 0.38	9.48 ± 0.26	9.51 ± 0.48
MCV (fL)	20.96 ± 0.20	20.74 ± 0.16	20.92 ± 0.52	20.28 ± 0.33	22.88 ± 0.49	21.56 ± 0.67
MCH (pg)	6.63 ± 0.06	6.61 ± 0.05^{b}	$5.98 \pm 0.31^{a,y}$	$3.32 \pm 0.33^{a,x}$	7.38 ± 0.39^{b}	6.72 ± 0.30^{b}
MCHC (g/dL)	33.12 ± 0.22	33.29 ± 0.03^{b}	26.60 ± 2.31^{y}	$16.61 \pm 1.66^{a,x}$	33.62 ± 1.89	31.47 ± 1.72^{b}
TLC (10^{3}/μL)	14.81 ± 0.68	13.41 ± 1.83	15.28 ± 0.90^{y}	11.15 ± 0.70^{x}	16.01 ± 1.55	13.08 ± 0.90
LYMP (%)	59.54 ± 2.30	65.98 ± 3.21	67.10 ± 3.26^{a}	62.24 ± 2.27	54.36 ± 2.06^{b}	59.17 ± 2.98
NEUT (%)	40.46 ± 2.30	34.02 ± 3.21	32.90 ± 3.26	37.76 ± 2.27	45.64 ± 2.06	40.83 ± 2.98
N : L	0.74 ± 0.072	0.70 ± 0.18	0.53 ± 0.066	0.53 ± 0.11	0.86 ± 0.079	0.73 ± 0.09

Mean values of the same parameter with superscripts [a,b] (within breed, but between seasons) and [x,y] (between breeds and within season) are significantly different ($P < 0.05$). PCV, packed cell volume; RBC, red blood count; Hb, haemoglobin concentration; MCV, mean corpuscular volume; MCH, mean corpuscular haemoglobin; MCHC, mean corpuscular haemoglobin concentration; TLC, total leucocyte count; LYMP, lymphocyte count (%); NEUT, neutrophil count (%); N : L, neutrophil–lymphocyte ratio.

3.2 Erythrocytic parameters

Results on the effect of season on haematological variables of buck kids, doe kids and both sexes are presented in Table 2, 3 and 4 respectively, while the effect of season on haematological variables in different age groups of Red Sokoto and Sahel goat kids are presented in Tables 5 and 6, respectively. Irrespective of sex, significantly higher ($P < 0.05$) PCV and RBC but lower MCH and MCHC were observed in Sahel than Red Sokoto kids during the HDS. Similarly, during the HDS, the Sahel goats had significantly higher ($P < 0.05$) PCV and RBC but lower MCH and MCHC compared with the CDS and RAS, with the changes being consistent irrespective of the sex. In Red Sokoto goats, however, the MCHC was significantly lower ($P < 0.05$) during the HDS as compared with the CDS and RAS (Table 4), while the RBC was significantly higher during the CDS compared with the RAS in doe kids. Values of MCV in both breeds were significantly lower in the HDS compared to the other seasons, with the value in Sahel goats being remarkably lower ($P < 0.05$) than that in Red Sokoto kids (Table 4). There was no significant breed difference ($P > 0.05$) in erythrocytic variables during the CDS and RAS.

Although no sex difference in haematological parameters of both breeds was observed in any season, sex variation was observed in the pattern of haematological changes in different seasons. Except for the significantly lower RBC during the HDS in buck kids of Red Sokoto, haematological variables in different sexes of Red Sokoto kids were not affected by season. This was unlike the Sahel goats, in which the haematological parameters of buck kids were significantly ($P < 0.05$) influenced by seasonal changes than those of doe kids. In the buck kids, significantly ($P < 0.05$) higher PCV and RBC but lower MCH and MCHC were observed, while only MCH and MCHC were significantly ($P < 0.05$) lower during the HDS compared with the CDS and RAS.

In both breeds of goats, the increase in PCV and RBC and decrease in MCH and MCHC during the HDS were observed to be more marked ($P < 0.05$) in younger kids (Tables 5 and 6). During the HDS, younger Red Sokoto kids had significantly lower ($P < 0.05$) MCV values compared to their older counterparts. Throughout the study, Hb did not differ significantly between seasons, sexes and breeds.

Table 3. Effect of season on haematological parameters of Red Sokoto and Sahel doe kids.

Parameter	Cold-dry season		Hot-dry season		Rainy season	
	Red Sokoto	Sahel	Red Sokoto	Sahel	Red Sokoto	Sahel
PCV (%)	34.11 ± 0.51	38.33 ± 3.84	36.68 ± 2.61^y	$55.80 \pm 8.48^{a,x}$	30.99 ± 2.61	36.38 ± 2.22^b
RBC ($10^6/\mu L$)	13.37 ± 0.45^a	12.41 ± 0.78	9.56 ± 1.02^y	$15.59 \pm 2.52^{a,x}$	7.79 ± 0.69^b	9.53 ± 0.57^b
Hb (g/dL)	10.46 ± 0.19	9.94 ± 0.54	9.90 ± 0.34	9.80 ± 0.34	9.86 ± 0.34	10.69 ± 0.31
MCV (fL)	20.35 ± 0.39	20.20 ± 0.58	21.20 ± 0.50	21.24 ± 0.52	21.34 ± 0.80	20.74 ± 0.55
MCH (pg)	6.58 ± 0.08	6.50 ± 0.13^b	6.30 ± 0.26^y	$3.33 \pm 0.52^{a,x}$	6.94 ± 0.35	6.20 ± 0.24^b
MCHC (g/dL)	33.54 ± 0.26	33.77 ± 0.32^b	28.50 ± 1.59^y	$15.82 \pm 1.97^{a,x}$	32.66 ± 1.49	30.10 ± 1.47^b
TLC ($10^3/\mu L$)	14.85 ± 1.56	14.74 ± 1.47	17.23 ± 0.75^y	11.15 ± 1.46^x	17.40 ± 1.39	15.27 ± 1.85
LYMP (%)	60.35 ± 2.10	57.17 ± 3.57	61.98 ± 2.26^a	59.99 ± 3.61	54.54 ± 2.61^b	56.30 ± 2.76
NEUT (%)	39.65 ± 2.10	42.83 ± 3.57	38.02 ± 2.26	40.01 ± 3.61	45.45 ± 2.61	43.70 ± 2.76
N : L	0.85 ± 0.079	0.84 ± 0.17	0.68 ± 0.056	0.63 ± 0.14	0.86 ± 0.098	0.77 ± 0.08

Mean values of the same parameter with superscripts [a,b] (within breed, but between seasons) and [x,y] (between breeds and within season) are significantly different ($P < 0.05$). PCV, packed cell volume; RBC, red blood count; Hb, haemoglobin concentration; MCV, mean corpuscular volume; MCH, mean corpuscular haemoglobin; MCHC, mean corpuscular haemoglobin concentration; TLC, total leucocyte count; LYMP, lymphocyte count (%); NEUT, neutrophil count (%); N : L, neutrophil–lymphocyte ratio.

Table 4. Overall effect of season on haematological parameters of male and female Red Sokoto and Sahel goat kids.

Parameters	Cold-dry season		Hot-dry season		Rainy season	
	Red Sokoto	Sahel	Red Sokoto	Sahel	Red Sokoto	Sahel
PCV (%)	33.45 ± 1.35	36.17 ± 3.06^a	38.18 ± 2.18^y	$56.04 \pm 4.98^{x,b}$	30.57 ± 1.62	33.90 ± 1.62^a
RBC ($10^6/\mu L$)	12.42 ± 0.45^a	11.21 ± 0.99^a	10.13 ± 0.72^y	$17.31 \pm 1.44^{x,b}$	7.61 ± 0.48^b	8.63 ± 0.48^a
Hb (g/dL)	10.23 ± 0.29	9.59 ± 0.75	9.95 ± 0.20	10.19 ± 0.21	9.67 ± 0.21	10.10 ± 0.30
MCV (fL)	20.80 ± 0.20	20.50 ± 0.28^a	$21.10 \pm 0.28^{y,b}$	$20.76 \pm 0.32^{x,b}$	22.14 ± 0.48^a	21.14 ± 0.34^a
MCH (pg)	6.61 ± 0.05	6.56 ± 0.06^a	$6.19 \pm 0.19^{y,b}$	$3.33 \pm 0.28^{x,b}$	7.61 ± 0.31^a	6.46 ± 0.20^a
MCHC (g/dL)	33.33 ± 0.17^a	33.53 ± 0.16^a	$27.55 \pm 1.30^{x,b}$	$16.22 \pm 1.23^{x,b}$	33.14 ± 1.19^a	30.78 ± 1.57^a
TLC ($10^3/\mu L$)	14.83 ± 0.76	14.08 ± 1.15	16.26 ± 0.48^y	11.15 ± 0.75^x	16.71 ± 0.97	14.18 ± 1.02
LYMP (%)	59.95 ± 1.54	61.58 ± 3.70	64.54 ± 1.73^b	61.11 ± 3.01	54.45 ± 1.67^a	57.74 ± 2.05
NEUT (%)	40.05 ± 1.54	38.43 ± 3.70	35.46 ± 1.73	38.89 ± 3.01	45.55 ± 1.67	42.26 ± 2.04
N : L	0.80 ± 0.079	0.77 ± 0.097	0.61 ± 0.045	0.58 ± 0.093	0.86 ± 0.055	0.75 ± 0.059

Mean values of the same parameter with superscripts [a,b] (within breed, but between seasons) and [x,y] (between breeds and within season) are significantly different ($P < 0.05$). PCV, packed cell volume; RBC, red blood count; Hb, haemoglobin concentration; MCV, mean corpuscular volume; MCH, mean corpuscular haemoglobin; MCHC, mean corpuscular haemoglobin concentration; TLC, total leucocyte count; LYMP, lymphocyte count (%); NEUT, neutrophil count (%); N : L, neutrophil–lymphocyte ratio.

3.3 Leucocytic parameters

During the HDS, values of TLC in both sexes of Sahel goats were significantly lower ($P < 0.05$) than those in Red Sokoto kids. The percentage lymphocyte count (LYMP) was significantly ($P < 0.05$) lower during the RAS in comparison to the HDS in both sexes of Red Sokoto kids. The Red Sokoto buck kids had higher ($P < 0.05$) LYMP but lower ($P < 0.05$) percentage neutrophil counts (NEUT) than their female counterparts. Older Red Sokoto kids had significantly lower NEUT in the HDS compared with values recorded in the RAS. No significant difference ($P > 0.05$) was observed in neutrophil–lymphocyte ratio (N : L) among groups in the different seasons. The decrease in TLC during the HDS in Sahel goats was not significant ($P > 0.05$) when compared with the other seasons.

3.4 Discriminant analysis

Results on the discriminating powers of the variables are presented in Table 7. The variable with the greatest seasonal discriminating power was MCV, followed by MCHC, MCH, PCV and RBC. Both the clusters formed in the canonical plot representation (Fig. 2) and classification of goats into their predicted group membership (Table 8) showed distinct separation in haematological variables between the CDS and the other two seasons, and close similarity in haematological variables between the HDS and RAS. The classification error was relatively low, while 89.7 % of the kids were correctly classified into their respective groups. The only misclassification observed was between HDS and RAS.

Table 5. Effect of season on haematological parameters of younger (1–2 months) and older (3–4 months) Red Sokoto goat kids.

	Cold-dry season		Hot-dry season		Rainy season	
	Young	Old	Young	Old	Young	Old
N	10	10	9	11	13	7
PCV (%)	31.00 ± 1.67	33.00 ± 1.55	$54.54 \pm 3.24^{a,y}$	34.9 ± 2.66^{x}	31.23 ± 2.86^{b}	28.83 ± 2.48
RBC (10^6/µL)	11.45 ± 0.84	12.89 ± 1.08^{a}	$16.16 \pm 0.78^{a,x}$	9.08 ± 0.78^{y}	7.07 ± 0.63^{b}	8.19 ± 0.81^{b}
Hb (g/dL)	9.22 ± 0.61	10.17 ± 0.56	9.93 ± 0.33	9.27 ± 0.42	9.58 ± 0.50	9.87 ± 0.25
MCV (fL)	20.94 ± 0.16	21.20 ± 0.25	$19.08 \pm 0.74^{x,a}$	21.98 ± 0.26^{y}	23.12 ± 0.49^{b}	21.58 ± 0.73
MCH (pg)	6.58 ± 0.08^{b}	6.60 ± 0.17	$3.38 \pm 0.27^{a,x}$	6.24 ± 0.40^{y}	7.54 ± 0.22^{b}	7.52 ± 0.42
MCHC (g/dL)	32.36 ± 0.43^{b}	33.82 ± 0.43	$17.94 \pm 1.82^{a,x}$	29.09 ± 1.58^{y}	32.78 ± 2.12^{b}	33.61 ± 1.62
TLC (10^3/µL)	13.86 ± 1.87	12.73 ± 1.85	14.72 ± 1.23	15.58 ± 069	18.65 ± 1.68	16.80 ± 1.27
LYMP (%)	67.23 ± 1.83	60.10 ± 3.13	64.60 ± 3.52	59.39 ± 4.47	53.64 ± 3.81	54.88 ± 1.50
NEUT (%)	32.87 ± 1.83	39.90 ± 3.13	33.40 ± 3.52	37.86 ± 2.33^{a}	46.36 ± 3.81	46.22 ± 1.50^{b}
N : L	0.51 ± 0.53	0.70 ± 0.07	0.60 ± 0.06	0.67 ± 0.05	0.67 ± 0.05	0.88 ± 0.04

Mean values of the same parameter with superscripts [a,b] (between seasons) and [x,y] (within season) are significantly different ($P < 0.05$). N, sample size; PCV, packed cell volume; RBC, red blood count; Hb, haemoglobin concentration; MCV, mean corpuscular volume; MCH, mean corpuscular haemoglobin; MCHC, mean corpuscular haemoglobin concentration; TLC, total leucocyte count; LYMP, lymphocyte count (%); NEUT, neutrophil count (%); N : L, neutrophil–lymphocyte ratio.

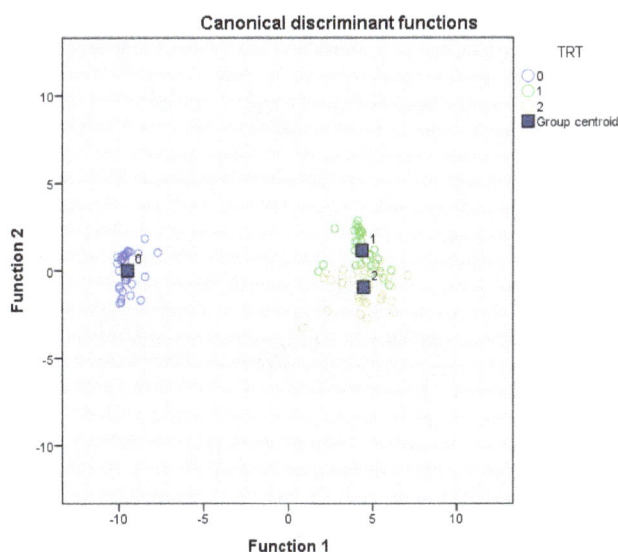

Figure 2. Canonical discriminant functions for the three seasons: 0, CDS; 1, HSD; and 2, RAS.

4   Discussion

In comparison to the thermoneutral zone for goats (THI: 65–75; Hamzaoiu et al., 2013), the kids were exposed to cold stress during the morning hours of the CDS (THI: 59.44) and heat stress during the afternoon hours of the HDS (THI: 84.15) and RAS (THI: 82.01). Generally, the haematological parameters recorded in the present study were within the normal range for goats reported by other authors (Jain, 1986; Addass et al., 2010; Zumbo et al., 2011), except for some deviations observed in PCV, RBC, MHC and MCHC during the HDS, particularly in the Sahel goat kids.

The high PCV and RBC recorded during the HDS of the current study, particularly in the Sahel and younger kids, agreed with the findings of previous studies (Sejian et al., 2013; Hashem, 2014). Severe dehydration has been reported in livestock exposed to thermal stress, which ultimately led to an increase in PCV, RBC and Hb (McManus et al., 2009; Sejian et al., 2014; Habibu et al., 2017). Thus, heat stress during the HDS was likely responsible for the high PCV and RBC observed in the current study, apparently due to increased thermoregulatory loss of body fluids. In contrast, other studies reported a decrease in PCV, RBC and Hb during heat stress, which was believed to be provoked by compensatory excess water intake (Mazzullo et al., 2014; Singh et al., 2016).

Since the increase in PCV during the HDS of the present study was not accompanied by a corresponding increase in Hb, decreases in haemoglobin content (MCH) and the cellular concentration (MCHC) were observed. The decrease in MCH and MCHC during the HDS affected both breeds, but with a more than 50 % decrease in Sahel kids during this season compared with the other seasons. Several studies have reported a decrease in blood Hb parameters during the hot season or amid experimental heat exposure. For instance, MCH in goats (Abdelatif et al., 2009) and MCHC in Saanen goat kids (Temizel et al., 2009) and Niamey sheep (Al-Haidary, 2004) were reported to decrease due to high environmental temperature. Oxidative stress due to high ambient temperature may denature and precipitate haemoglobin molecules in the erythrocytes followed by their degradation (Pacifici et al., 1993; Giulivi et al., 1994), contributing to the low MCH and MCHC, particularly in the younger kids and kids of Sahel goats. The significantly low MCH and MCHC may also be indicative of low nutritional status as feed intake has been reported to decrease during heat stress (Srikandakumar et al.,

Table 6. Effect of season on haematological parameters of younger (1–2 months) and older (3–4 months) Sahel goat kids.

	Cold-dry season		Hot-dry season		Rainy season	
	Young	Old	Young	Old	Young	Old
N	11	9	8	12	7	13
PCV (%)	42.33 ± 7.21^b	30.33 ± 6.47	$59.53 \pm 3.90^{a,y}$	30.97 ± 4.66^x	32.30 ± 3.02^b	36.66 ± 2.45
RBC ($10^6/\mu L$)	13.11 ± 1.05^b	12.10 ± 2.76	$19.19 \pm 0.96^{b,x}$	7.25 ± 1.38^y	8.19 ± 0.81^a	9.56 ± 0.60
Hb (g/dL)	11.00 ± 1.15	9.60 ± 2.11	10.14 ± 0.29	9.30 ± 0.78	9.76 ± 0.46	10.79 ± 0.32
MCV (fL)	20.86 ± 0.05	20.82 ± 0.01	20.22 ± 0.29	22.66 ± 0.85	21.62 ± 0.84	26.78 ± 0.55
MCH (pg)	6.55 ± 0.01^b	6.44 ± 0.20	$3.31 \pm 0.24^{a,x}$	6.99 ± 0.75^y	6.71 ± 0.35^b	6.24 ± 0.25
MCHC (g/dL)	33.24 ± 0.02^b	33.95 ± 0.62	$16.56 \pm 1.21^{a,x}$	30.83 ± 2.83^b	31.34 ± 2.08^b	30.23 ± 1.64
TLC ($10^3/\mu L$)	11.79 ± 2.13	12.38 ± 1.95	14.14 ± 0.29	15.30 ± 0.78	13.85 ± 0.93	10.98 ± 0.91
LYMP (%)	61.99 ± 6.66	59.23 ± 4.70	64.33 ± 3.12	54.90 ± 2.98	58.27 ± 3.08	55.03 ± 2.91
NEUT (%)	36.01 ± 6.66	41.38 ± 4.70	35.67 ± 3.12	45.10 ± 2.98	41.73 ± 3.08	44.97 ± 2.91
N : L	0.61 ± 0.19	0.67 ± 0.17	0.47 ± 0.08	0.84 ± 0.01	0.75 ± 0.10	0.80 ± 0.08

Mean values of the same parameter with superscripts [a,b] (between seasons) and [x,y] (within season) are significantly different ($P < 0.05$). N, sample size; PCV, packed cell volume; RBC, red blood count; Hb, haemoglobin concentration; MCV, mean corpuscular volume; MCH, mean corpuscular haemoglobin; MCHC, mean corpuscular haemoglobin concentration; TLC, total leucocyte count; LYMP, lymphocyte count (%); NEUT, neutrophil count (%); N : L, neutrophil–lymphocyte ratio.

Table 7. Structural matrix with discriminant function for the three seasons.

Variables	Functions	
	1	2
MCV	0.913*	−0.143
TLC	0.167*	−0.132
MCHC	−0.065	−0.671*
MCH	0.309	−0.640*
PCV	0.026	0.591*
RBC	−0.053	0.524*
LYMP	0.014	0.373*
N : L	−0.020	−0.359*
NEUT	−0.006	−0.266*
Hb	0.006	−0.117*

Asterisks (*) indicate functions with significant loading. PCV, packed cell volume; RBC, red blood count; Hb, haemoglobin concentration; MCV, mean corpuscular volume; MCH, mean corpuscular haemoglobin; MCHC, mean corpuscular haemoglobin concentration; TLC, total leucocyte count; LYMP, lymphocyte count (%); NEUT, neutrophil count (%); N : L, neutrophil–lymphocyte ratio.

Table 8. Percentage of kids classified into the three seasons.

Seasons	CDS	HDS	RAS
Cold-dry	40	0	0
Hot-dry	0	32	8
Rainy	0	4	36
Error (%)	0	10.3	20.6

2003; Sejian et al., 2014). Although there was a decrease in MCV, MCH and MCHC during the HDS in the current study, this may not be due to iron deficiency, since the Hb during the HDS was within the normal range for goats (Byers and Kramer, 2010) and comparable to other seasons.

In the current study, the values of MCV decreased during the HDS in comparison to the CDS and RAS, particularly in Sahel goats, and also showed the strongest discriminating power in distinguishing the kids based on the prevailing season. The finding agreed with those of Ribeiro et al. (2015)

in Brazilian Azul goats and Mazzullo et al. (2014) in cattle, who also reported a decrease in MCV in animals exposed to natural heat stress. However, the finding disagreed with that of Abdelatif et al. (2009), who reported higher MCV values in Nubian goats of Sudan during summer than in winter. The decrease in MCV (average size of the erythrocytes) during the HDS may be due to an increase in the number of erythrocyte membrane vesicles formed and shed from the erythrocytes as a result of elevation in body temperature, with a resultant decrease in the size of the parent erythrocytes (Foller et al., 2010; Moore et al., 2013; Habibu et al., 2017). Thus, it is logical to infer that a decrease in MCV may have the potential of being used as an indicator of heat stress in tropical goats naturally exposed to heat stress.

The dramatic change in some haematological parameters of the younger goat kids compared to the relatively stable parameters in older kids observed during the HDS in the current study indicates that the former were more susceptible to heat stress than the latter. The finding agrees with that of Haque et al. (2013), in which the effect of thermal stress was higher in young buffaloes, with an alteration in TLC of young buffaloes, while the value was stable in the adults.

Previous studies in tropical (Banerjee et al., 2015; Singh et al., 2016) and arid (Srikandakumar et al., 2003) climates demonstrated diverse physiological responses of different breeds of goats and sheep to changes in meteorological parameters with some breeds showing better thermoregulation then others. Similarly, results of the current study revealed remarkable influence of breed in the seasonal changes in erythrocytic variables in both sexes. The kids of Red Sokoto goats apparently employed more efficient adaptive and survival strategies than Sahel goat kids, particularly during the HDS, which prevented the dramatic change in haematological parameters that was observed in Sahel goats. The higher magnitude of the increase in PCV and RBC in Sahel compared with Red Sokoto kids during the HDS may imply higher thermoregulatory fluid loss in Sahel kids, probably due to higher respiratory rate (Habibu et al., 2016b). This difference in breed response may be due to variation in thermoregulatory plasma fluid dynamics, as higher plasma volume, indicated by decrease in RBC and Hb has been reported in adult bucks of Red Sokoto, while an increase in PCV and RBC was observed in Sahel goats during the HDS when compared with CDS and RAS (Habibu et al., 2017). The ability of Red Sokoto goats to maintain relatively higher plasma volume, despite the expected thermoregulatory fluid loss during heat stress, indicates a higher level of adaptation to the tropical savanna and may, at least in part, explain why they naturally survive in diverse geographic locations (Sudan and Guinea savannas) and are the predominant breed of goats in Nigeria. The relatively stable haematological response demonstrated by Red Sokoto goats in the current study was probably because the breed is indigenous to the northern Guinea savanna region of Nigeria. This is unlike the Sahel goats, which are indigenous to the Sahel region of West Africa. Both regions have high ambient temperatures, but the Sahel region has a lower relative humidity due to its closeness to the Sahara (Fig. 1; Nicholson, 1995). Although a previous study reported higher erythrocyte membrane integrity in Sahel than Red Sokoto kids during the HDS (Habibu et al., 2016a), the higher relative humidity in the Guinea savanna of Nigeria may be a negative factor influencing thermoregulation of Sahel kids in the geographical area and may be responsible for the changes observed in erythrocytic parameters observed during the HDS (Habibu et al., 2017).

The current study revealed a remarkable influence of sex on the seasonal changes in erythrocytic variables in both breeds. The seasonal fluctuations in erythrocytic variables were better profound in buck kids than doe kids of both breeds, suggesting that the latter adapted more to the prevailing ambient conditions. Similarly, a previous study reported greater hazard during all postnatal periods in male compared with female lambs (Sawalha et al., 2007). In addition to other factors, a relatively poor homeostasis in the buck kids may, at least in part, contribute to the low survival rate generally reported in males compared with females among various breeds of goat kids (Awemu et al., 1999; El-Abid and Abu Nikhaila, 2009) and lambs (Turkson and Sualisu, 2005). Although more studies are needed to substantiate this hypothesis, these corroborative findings may imply that

the female newborns of small ruminants have more efficient mechanisms of responding to postnatal challenges than their male counterparts.

The low TLC in Sahel as compared with Red Sokoto kids during the HDS in the current study supports the findings of Haldar (2012), who reported a decrease in TLC in goats exposed to heat stress. Similarly, this agreed with the report of Fadare et al. (2012), who indicated that heat stress suppressed the immune status of sheep that were more susceptible to thermal stress in the hot-humid tropics. The finding further demonstrates that the Red Sokoto kids were better adapted to the northern Guinea savanna zone of Nigeria than the Sahel kids.

In the current study, lower LYMP was observed during the RAS as compared with the HDS in the Red Sokoto goat kids. In a related study, higher LYMP was reported in Dhofari goats during the summer compared with the winter (Al-Busaidi et al., 2008). The comparable level of NEUT in all groups of kids in the current study was similar to that reported in Nubian goats by Abdelatif et al. (2009), but disagreed with that of Temizel et al. (2009), who reported a decrease in NEUT in heat-stressed goat kids.

In agreement with the findings of a previous study in Garfagnina goats (Ribeiro et al., 2015), the current study also reported strong discriminant power in MCV and PCV. This finding indicates that such haematological parameters (MCV MCHC, MCH, PCV and RBC) are influenced by the three seasons through thermoregulatory mechanisms that cause different physiological effects, and thus they vary markedly in different seasons. However, both the canonical plot representation and the group membership prediction showed that the CDS imposed distinct effects on haematological variables from those by the HDS and RAS which were similar. This was probably due to the low THI during the CDS and high THI during the HDS and RAS, which may induce cold stress and heat stress, respectively (Habibu et al., 2016b).

5 Conclusion

Season significantly influenced haematological variables, with the impact of heat stress during the HDS being more profound in kids of Sahel goats, and in younger and male kids of both breeds. The marked seasonal fluctuations in haematological variables of Sahel kids compared to Red Sokoto kids may indicate that the Red Sokoto goats, which are indigenous to the study area, were better adapted to the seasons. Moreover, the TLC was lower in Sahel goats compared with the Red Sokoto goats during the HDS, indicating a higher susceptibility of Sahel goats to heat stress in the study area. These factors necessitate the need for additional support such as provision of ad libitum drinking water and protective shading to the young kids, and especially the kids of the Sahel goats, during the peak of the HDS. It is suggested that the breeding programme be designed in such a way that the hot season coincides with the time when the kids. The influence of season on haematological parameters was more pronounced than those of breed, sex and age, hence the need to consider the environmental conditions in which goats are reared in defining reference intervals of haematological parameters. The higher risk of postnatal period of buck kids needs further investigation so as to know whether there is the need for a special managerial provision for the buck kids during thermal stress.

Author contributions. All authors made essential contributions to all steps of the experimental design and manuscript preparation and also approved the final manuscript. However, special contribution was made by Buhari Habibu, Mohammed Kawu, Hussaina Makun and Tagang Aluwong in designing the work. In particular, Buhari Habibu, Lukman Yaqub and Hussaina Makun carried out the field work. Buhari Habibu, Lukman Yaqub and Hajarah Buhari did the laboratory work and statistical analysis. Buhari Habibu, Lukman Yaqub, Mohammed Kawu and Hajarah Buhari wrote the manuscript.

Competing interests. The authors declare that they have no conflict of interest.

Acknowledgements. The Management of NAPRI, Ahmadu Bello University, Shika, Zaria, granted us the permission to use the institution's animals. The staff of the Small Ruminant Research Programme, NAPRI, and Department of Veterinary Physiology, Ahmadu Bello University, Zaria, gave technical assistance. J. O. Ayo and M. Tauheed assisted in drafting the manuscript.

Edited by: M. Mielenz

References

Abdelatif, A.M., Ibrahim, Y. M., and Hassan, M. Y.: Seasonal variation in erythrocytic and leukocytic indices and serum proteins of female Nubian goats, Middle-East J. Sci. Res., 4, 168–174, 2009.

Addass, P. A., Midau, A., and Babale, D. M.: Haemato-biochemical findings of indigenous goats in Mubi Adamawa State, Nigeria, J. Agric. Soc. Sci., 6, 14–16, 2010.

Al-Busaidi, R., Johnson, E. H., and Mahgoub, O.: Seasonal variations of phagocytic response, immunoglobulin G (IgG) and plasma cortisol levels in Dhofari goats, Small Rum. Res., 79, 118–123, 2008.

Al-Haidary, A. A.: Physiological responses of Niamey sheep to heat stress, challenge under semi-arid environments, Int. J. Agric. Biol., 6, 307–309, 2004.

Awemu, E. M., Nwakalor, L. N., and Abubakar, B. Y.: Environmental influences on pre-weaning mortality and reproductive performance of Red Sokoto does, Small Rum. Res., 34, 161–165, 1999.

Banerjee, D., Upadhyay, R. C., Chaudhary, U. B., Kumar, R., Singh, S., Ashutosh, Das, T. K., and De, S.: Seasonal variations in

physio-biochemical profiles of Indian goats in the paradigm of hot and cold climate, Biol. Rhythm. Res., 46, 221–236, 2015.

Byers, S. R. and Kramer, J. W.: Schalm's Veterinary Hematology, in: Normal Hematology of Sheep and Goats, edited by: Weiss, D. J. and Wardrop, K. J., 6th Edn., John Wiley & Sons, Ltd., Iowa, USA, 836–842, 2010.

Dzenda, T., Ayo, J. O., Lakpini, C. A.M., and Adelaiye, A. B.: Diurnal, seasonal and sex influences on respiratory rate of African Giant rats (Cricetomys gambianus, Waterhouse) in a tropical Savannah, Wulfenia J., 22, 475–485, 2015.

Egbe-Nwiyi, T. N., Nwaosu, S. C., and Salami, H. A.: Haematological values of apparently healthy sheep and goats as influenced by age and sex in arid zone of Nigeria, Afr. J. Biomed. Res., 3, 109–115, 2000.

Egwu, G. O., Onyeyili, P. A., Chibuzo, G. A., and Ameh, J. A.: Improved productivity of goats and utilisation of goat milk in Nigeria, Small Rum. Res., 16, 195–201, 1995.

El-Abid, K. E. and Abu Nikhaila, A. M. A.: A Study on Some Factors Affecting Mortality Rates in Sudanese Nubian Kids, Int. J. Dairy Sci., 4, 74–79, 2009.

Fadare, A. O., Peters, S. O., Yakubu, A., Sonibare, A. O., Adeleke, M. A., Ozoje, M. O., and Imumorin, I. G.: Physiological and haematological indices suggest superior heat tolerance of white-coloured West African Dwarf sheep in the hot humid tropics, Trop. Anim. Health Prod., 45, 157–165, 2012.

Fazio, F., Giangrosso, G., Marafioti, S., Zanghì, E., Arfuso, F., and Piccione, G.: Blood haemogram in ovis aries and capra hyrcus: Effect of storage time, Can. J. Anim. Sci., 96, 32–36, 2016.

Foller, M., Braun, M., Qadri, S. M., Lang, E., Mahmud, H., and Lang, F.: Temperature sensitivity of suicidal erythrocyte death, Eur. J. Clin. Invest., 40, 534–540, 2010.

Gall, C.: Goat breeds of the world, Margraf Publishing, Weikersheim, Germany, 1996.

Giulivi, C., Pacifici, R. E., and Davies, K. J. A.: Exposure of hydrophobic moieties promotes the selective degradation of hydrogen peroxide-modified hemoglobin by the multicatalytic proteinase complex, proteasome, Arch. Biochem. Biophys., 311, 329–341, 1994.

Habibu, B., Kawu, M. U., Makum, H. J., Buhari, H. U., and Hussaini, M.: Breed and seasonal variations in erythrocyte osmotic fragility of goat kids raised in semi-arid savannah, Comp. Clin. Pathol., 25, 1309–1312, 2016a.

Habibu, B., Kawu, M. U., Makun, H.J., Aluwong, T., and Yaqub, L. S.: Seasonal variation in body mass index cardinal physiological variables and serum thyroid hormones profiles in relation to susceptibility to thermal stress in goats kids, Small Rum. Res., 145, 20–27, 2016b.

Habibu, B., Kawu, M. U., Makum, H. J., and Aluwong, T.: Influence of seasonal changes on physiological variables, haematology and serum thyroid hormones profile in male Red Sokoto and Sahel goats, J. Appl. Anim. Res., 45, 508–516, 2017.

Haldar, K. C.: Correlation between peripheral melatonin and general immune status of domestic goat, Capra hircus: A seasonal and sex dependent variation, Small Rum. Res., 107, 147–156, 2012.

Hamzaoui, S., Salama, A. A. K., Albanell, E., Such, X., and Caja, G.: Physiological responses and lactational performances of late-lactation dairy goats under heat stress conditions, J. Dairy Sci., 96, 6355–6365, 2013.

Haque, N., Ludri, A., Hossain, S. A., and Ashutosh, M.: Impact on hematological parameters in young and adult murrah buffaloes exposed to acute heat stress, Buffalo Bulletin, 32, 321–326, 2013.

Hashem, A. L. S.: Effect of Summer Shearing on Thermoregulatory, Hematological and Cortisol Responses in Balady and Damascus Goats in Desert of Sinai, Egypt, World Appl. Sci. J., 30, 521–533, 2014.

Igono, M. O. and Aliu, Y. O.: Environmental profile and milk production of Friesian-Zubucrosses in Nigerian Guinea Savannah, Int. J. Biometeorol., 26, 115–120, 1982.

Jain, N. C.: Schalm's Veterinary Hematology, 4th Edn., Lea and Febiger, Philadelphia, USA, 1986.

Mazzullo, G., Rifici, C., Cammarata, F., Caccamo, G., Rizzo, M., and Piccione, G.: Effect of different environmental conditions on some haematological parameters in cows, Ann. Anim. Sci., 14, 947–954, 2014.

McManus, C., Paludo, G. R., Louvandini, H., Gugel, R., Sasaki, C. B. L., and Paiva, S. R.: Heat tolerance in Brazilian sheep: physiological and blood parameters, Trop. Anim. Health Prod., 41, 95–101, 2009.

Moore, T., Sorokulova, I., Pustovyy, O., Globa, L., and Vodyanoy, V.: Microscopic evaluation of vesicles shed by rat erythrocytes at elevated temperatures, J. Therm. Biol., 38, 487–492, 2013.

Nicholson, S. E.: Sahel, West Africa, in: Encyclopaedia of Experimental Biology, Vol. 3, Academic Press Inc., USA, 1995.

Nowak, R., Portej, R. H., Levy, F., Orgeur, P., and Schaal, B.: Role of mother-young interactions in the survival of offspring in domestic mammals, Rev. Reprod., 5, 153–163, 2000.

Pacifici, R. E., Kono, Y., and Davies, K. J. A.: Hydrophobicity as the signal for selective degradationon of hydroxyl radical-modified hemoglobin by the multicatalytic proteinase complex, proteasome, J. Biol. Chem., 268, 15405–15411, 1993.

Piccione, G., Monteverde, V., Rizzo, M., Vazzana, I., Assenza, A., Zumbo, A., and Niutta, P. P.: Haematological parameters in Italian goats: comparison between girgentana and aspromontana breeds, J. Appl. Anim. Res., 42, 434–439, 2014.

Piccione, G., Borruso, M., Fazio, F., Giannetto, C., and Caola, G.: Physiological parameters in lambs during the first 30 days post-partum, Small Rum. Res., 72, 57–60, 2007.

Ravagnolo, O., Misztal, I., and Hoogenboom, G.: Genetic component of heat stress in dairy cattle, development of heat index function, J. Dairy Sci., 83, 2120–2125, 2000.

Ribeiro, N. L., Pimenta Filho, E. C., Arandas, J. K. G., Ribeirob, M. N., Saraiva, E. P., Bozzi, R., and Costa, R. G.: Multivariate characterization of the adaptive profile in Brazilian and Italian goat population, Small Rum. Res., 123, 232–237, 2015.

Saddiqi, H. A., Nisa, M., Mukhtar, N., Shahzad, M. A., Jabbar, A., and Sarwar, M.: Documentation of physiological parameters and blood profile in newly born Kajli lambs, Asian-Aust. J. Anim. Sci., 24, 912–918, 2011.

Santucci, P. M., Branca, A., Napoleone, M., Bouche, R., Aumont, G., Poisot, F., and Alexandre, G.: Body condition scoring of goats in extensive conditions. Goat nutrition, edited by: Morand-Fehr, P., EAAP Publication, Pudoc III, Wageningen, 240–256, http://prodinra.inra.fr/record/92203 (last access: 1 March 2017), 1991.

Sawalha, R. M., Conington, J., Brotherstone, S., and Villanueva, B.: Analyses of lamb survival of Scottish Blackface sheep, Animal, 1, 151–157, 2007.

Sejian, V., Indu, S., and Naqvi, S. M. K.: Impact of short term exposure to different environmental temperature on the blood biochemical and endocrine responses of Malpura ewes under semiarid tropical environment, Indian J. Anim. Sci., 83, 1155–1160, 2013.

Sejian, V., Singh, A. K., Sahoo, A., and Naqvi, S. M. K.: Effect of mineral mixture and antioxidant supplementation on growth, reproductive performance and adaptive capability of Malpura ewes subjected to heat stress, J. Anim. Physiol. Anim. Nutr., 98, 72–83, 2014.

Singh, K. M., Singh, S., Ganguly, I., Ganguly, A., Nachiappan, R. K., Chopra, A., and Narula, H. K.: Evaluation of Indian sheep breeds of arid zone under heat stress condition, Small Rum. Res., 141, 113–117, 2016.

Srikandakumar, A., Johnson, E. H., and Mahgoub, O.: Effect of heat stress on respiratory rate, rectal temperature and blood chemistry in Omani and Australian Merino sheep, Small Rum. Res., 49, 193–198, 2003.

Temizel, E. M., Senturk, S., and Kasap, S.: Clinical, haematological and biochemical findings in Saanen goat kids with naturally occurring heat stroke, Tierarztl. Prax. Großtiere, 37, 236–241, 2009.

Tibbo, M., Jibril, Y., Woldemeskel, M., Dawo, F., Aragaw, K., and Rege, J. E. O.: Factors affecting hematological profiles in three Ethiopian indigenous goat breeds, Int. J. Appl. Res. Vet. Med., 2, 297–305, 2004.

Turkson, P. K. and Sualisu, M.: Risk factors for lamb mortality in Sahelian sheep on a breeding station in Ghana, Trop. Anim. Health Prod., 37, 49–64, 2005.

Turkson, P. K., Antiri, Y. K., and Baffuor-Awuah, O.: Risk factors for kid mortality in West African Dwarf goats under an intensive management system in Ghana, Trop. Anim. Health Prod., 36, 353–364, 2004.

Zumbo, A., Sciano, S., Messina, V., Casella, S., di Rosa, A. R., and Piccione, G.: Haematological profile of messinese goat kids and their dams during the first month post-partum, Anim. Sci. Pap. Rep., 29, 223–230, 2011.

4

Progesterone (P4), luteinizing hormone (LH) levels and ovarian activity in postpartum Santa Inês ewes subject to a male effect

José Carlos Ferreira-Silva[1], Tracy Anne Burnett[2], Paulo Francisco Maciel Póvoas Souto[1],
Paulo Castelo Branco Gouveia Filho[1], Lucas Carvalho Pereira[3], Mariana Vieira Araujo[3],
Marcelo Tigre Moura[1], and Marcos Antonio Lemos Oliveira[1]

[1]Laboratório de Biotécnicas Aplicadas à Reprodução, Universidade Federal Rural de Pernambuco,
Recife, PE, Brasil
[2]Faculty of Land and Food Systems. University of British Columbia, Vancouver, BC, V6T 1Z4, Canada
[3]Laboratório Nordeste In Vitro, Maceió, AL, Brasil

Correspondence to: José Carlos Ferreira-Silva (carlos.ztec@gmail.com)

Abstract. This study aimed to establish P4 and luteinizing hormone (LH) levels and ovarian activity as approaches to monitor the cyclicity of ewes under postpartum anestrus after the male effect approach. Santa Inês females ($n = 66$) were evenly distributed into experimental groups where they were brought into contact with an intact male during the postpartum period of 35 to 40 days (T1) and 55 and 60 days postpartum (T2). Ewes were isolated from males for 30 days before the onset of the experiment. Estrus events were detected in 93.30% (T1) and 100% (T2) of females. Mean P4 concentrations ($\eta\,g\,mL^{-1}$) before and after mating were 0.53 ± 0.17 and 4.55 ± 0.24 (T1) and 0.73 ± 0.06 and 4.90 ± 0.11 (T2), respectively, and concentrations were found to be lower ($P < 0.05$) before contact between genders. Preovulatory peaks of LH ($\eta\,g\,mL^{-1}$) were evaluated at 42 (T2) and at 80 h (T1) after exposure to males, with mean ovulatory follicles of 7.90 ± 0.31 (T1) and 8.50 ± 0.30 mm (T2) and a mean number of ovulations of 1.50 ± 0.54 (T1) and 1.60 ± 0.51 (T2). Pregnancy rates were 85.70% (T1) and 93.3% (T2), with no difference ($P > 0.05$) between groups. Results showed that the male effect was effective for inducing and concentrating the occurrence of estrus in postpartum ewes that had lambed within 35 to 60 days previously. Moreover, preovulatory LH peaks occurred within 80 h after physical contact between genders, which led to follicle luteinization and increased P4 concentration, without compromising pregnancy rates. The male effect can be used with postpartum ewes that had lambed within 35 to 40 days previously in order to decrease time between deliveries.

1 Introduction

Traditional sheep production in South America is predominantly sustained by production systems in extensive areas with low reproductive efficiency (Rubianes and Ungerfeld, 2002). Moreover, seasonal reproduction systems and long postpartum anestrus periods are factors that affect sheep production by reducing overall lambing rates. Shorter periods of sexual inactivity, the anticipation of the onset of the breeding season and increased prolificacy are economically relevant factors that allow profitability in such commercial settings

(Oliveira et al., 2015; Tenório Filho et al., 2016; Ferreira-Silva et al., 2016). These factors are particularly true when carried out under adequate management practices and when those technologies are of low cost and simply executed (Simplício, 2008).

Hair sheep breeds have the potential to produce high-quality meat and skin; they are also hardy animals and of great adaptive capacity to tropical regions, such as the northeast of Brazil (Machado et al., 1999). From a productive standpoint, the Santa Inês breed, which originated from crosses of exotic and Brazilian native breeds, is characterized by a large size, where ewes weigh from 40 to 60 kg and rams from 80 to 100 kg, although they may reach 120 kg (Figueiredo et al., 1983). From a reproductive standpoint, ewes are continuously polyestrous, with prolificacy varying from 1.3 to 1.4, and have maternal ability (Ferreira-Silva et al., 2016, 2017).

Sheep production can be improved by genetic selection for ewes of improved reproductive efficiency, with lower age at first lambing, shorter time between deliveries and increased prolificacy (Notter and Copenhaver, 1980; Azzarini, 2004). The male effect alone (Sasa et al., 2011; Caldas et al. 2015a, b) or combined with pharmaceuticals (Knights et al., 2001; Ungerfeld et al., 2004; Monreal et al., 2009; Ferreira-Silva et al., 2017) is a simple, effective and low-cost approach to reducing the duration of postpartum anestrus without affecting pregnancy rates (Caldas et al., 2015a, b).

The sudden introduction of males into a female flock of sheep where the animals are in anestrus results in an immediate increase in luteinizing hormone (LH) pulse frequency and amplitude at intervals of 54–72 h (Knight et al., 1983) followed by ovulation (Martin et al., 1983; Moraes, 1991; Thimonier, 2000). This variation in LH pulses may be due to silent estrus behavior in ewes under anestrus (Knight et al., 1983; Thimonier, 2000). Thus, depending upon the resulting type of corpus luteum that is formed, females may have displayed short or normal estrous cycles (Caldas et al., 2015a, b), accompanied by an increase in P4 concentration (Sasa et al., 2011).

In face of the ongoing demand to reduce the interval between deliveries to increase sheep reproductive efficiency, the male effect could be used to shorten anestrus postpartum in ewes. Furthermore, to our knowledge, there is no report that used the male effect for this purpose in hair sheep, and the hormone levels and ovarian activity within this period remain to be described. This study aimed to evaluate changes in the cyclicity of anestrus postpartum Santa Inês ewes after the introduction of an intact male by determining P4 and LH profiles and ovarian activity.

2 Material and methods

Experiments were conducted in Escada, Pernambuco state, Brazil. The geographic coordinates are 08°21′33″ S, 35°13′25″ W; the altitude is 109 m, the mean annual temperature 24.4 °C and the mean annual rainfall 1763 mm^3. Climactic conditions are tropical and semi-humid, with a rainy period from May to August.

Animals were raised on pastures during daylight hours and kept in a barn during the late afternoon. Nutrition was based upon cultivated (*Brachiaria humidicola, Pennisetum purpureum*) and native pastures (*Paspalum maritimum, Chloris orthonton, Cynodon dactylon, Brachiaria tunnergrass*). Mineral salt and water were offered ad libitum.

Postpartum ewes of 2 to 3 years of age were initially evaluated for body condition score and were preselected if they displayed a score of 2 or 3, as described by Caldas et al. (2015a, b). The anestrus condition was evaluated by reproductive tract ultrasonography following Santos et al. (2004). Selected ewes were identified with numbered plastic ear tags and colored neckband in order to ease management. Male effect preconditioning was established by isolating ewes from rams for 30 days at a distance of 10 m, which avoids physical contact between genders as previously described by Caldas et al. (2015b). Santa Inês rams ($n = 2$) were preselected according to fertility records and were subject to an andrology exam a week before the onset of the experiment.

Cyclicity status was determined based upon P4 concentrations in addition to ultrasonography. Blood samples were collected on days 10, 20 and 30 during the male effect preconditioning period. Blood plasma was stored before analysis at −20 °C. Progesterone (P4) concentration was determined in duplicates using radioimmunoassay. Females were considered in anestrus when they displayed serum concentrations lower than $1 \eta g\,mL^{-1}$ in two consecutive samples as described by Morales et al. (2003). After P4 concentration analysis, ewes ($n = 66$) were allocated to two groups: T1 (35–40 days after lambing) and T2 (55–60 days after lambing).

Rams were marked with a mixture of grease and ink (4 : 1) around the sternum and then introduced to the ewe flocks, with a male-to-female ratio of 1 : 33. Rams were marked with different ink colors and exchanged between ewe flocks on days 10, 20 and 30 of the breeding season (BS).

Blood plasma LH analysis was determined by duplicate double-antibody radioimmunoassay, following De St Jorre et al. (2012), and the limit of detection was $0.27 \eta g\,mL^{-1}$; the intra-assay and inter-assay coefficients of variation were 3.5 and 8.74 %, respectively. Variations in LH levels were considered significant when they increased by $20 \eta g\,mL^{-1}$ or more for two consecutive measurements, as suggested by Martin et al. (1983). Three ewes from each group were randomly chosen for LH analysis after the introduction of rams. Blood sampling for LH analysis was at 4 h intervals and performed during a period of 4 days after the introduction of rams, as described by Chanvallon et al. (2011) and Fabre-Nys et al. (2015, 2016). Moreover, ewes used for LH evaluation were excluded from further analysis in order to avoid any effect of excessive handling on estrus and pregnancy rates.

In order to estimate ovulation rates on ewes after mating, P4 concentration was also measured on days 10, 20 and 30 after the onset of the male effect as described above for LH analysis. Ewes were considered to have entered a cycle when they displayed a P4 concentration of $1 \eta g\,mL^{-1}$ or higher in two consecutive samples.

Ovarian activity was evaluated daily by ultrasonographic exams after estrous detection in six females of each group. The same technician performed all exams and the moment

Table 1. P4 concentrations in Santa Inês postpartum ewes in anestrus that had delivered within 35 to 40 (T1) and 55 to 60 days (T2) previously and were further subject to the male effect.

Group	Animals	P4 concentration (η g mL^{-1})	
	(n)	Before male effect ($\overline{x} \pm s$)	After male effect ($\overline{x} \pm s$)
T1	30	$0.53 \pm 0.17^{a,c}$	$4.55 \pm 0.24^{b,c}$
T2	30	$0.73 \pm 0.06^{a,d}$	$4.90 \pm 0.11^{b,d}$

Different superscripts [a,b] letters in the same row and superscripts letters [c,d] in the same column denote statistical difference ($P < 0.05$).

Table 2. Mean values for follicular diameter (mm) and ovulation in Santa Inês postpartum ewes in anestrus that had delivered within 35 to 40 (T1) and 55 to 60 (T2) days previously and were further subject to the male effect.

Group	Ovarian activity			
	Large follicle diameter ($\overline{x} \pm s$)	Second large follicle diameter ($\overline{x} \pm s$)	Follicular diameter ($\overline{x} \pm s$)	Ovulations ($\overline{x} \pm s$)
T1	7.9 ± 0.31^a	7.6 ± 0.41^a	7.8 ± 0.38^a	1.5 ± 0.54^a
T2	8.5 ± 0.30^b	8.0 ± 0.25^a	8.3 ± 0.37^b	1.6 ± 0.51^a

Different superscript letters on same column denote statistical difference ($P < 0.05$).

Table 3. Pregnancy rates of Santa Inês postpartum ewes in anestrus that had delivered within 35 to 40 (T1) and 55 to 60 (T2) days previously and were further subject to the male effect.

Group	Pregnancy per service		
	First n/n (%)	Second n/n (%)	Total n/n (%)
T1	18/20 (90.0)	6/8 (75.0)	24/28 (85.7)
T2	24/25 (96.0)	4/5 (80.0)	28/30 (93.3)

of ovulation was defined according to Tenório Filho et al. (2007).

Ewes were observed twice daily for estrus (06:00 and 16:00 LT) by trained technicians during a breeding season (BS) of 35 days. Estrus events were considered synchronized when detected within the initial 5 days of the BS. Pregnancy diagnosis was made by ultrasonography on days 35 and 60 after mating as described by Santos et al. (2004).

Parametric variables were submitted to analysis of variance (ANOVA) and compared by the Student–Newman–Keuls (SNK) test with the System for Statistical Analysis (SAEG) software, with results presented as means and standard deviation ($\overline{x} \pm s$). Nonparametric variables were evaluated using the chi-square test and presented as percentages (%). Differences were considered significant when displayed probabilities were lower than 5 %.

3 Results

Concentrations of P4, which were assessed before BS onset in order to determine the cyclicity status of postpartum ewes are summarized in Table 1. Mean P4 concentrations were lower than 1 η g mL^{-1}, indicating that all ewes were in anestrus before BS, irrespectively of the postpartum period. However, after the male effect, ewes cycled and ovulated, as demonstrated by increased P4 levels ($P < 0.05$).

Concentrations of LH were also measured after ewes were subject to the male effect. Despite varying postpartum periods, the male effect induced LH preovulatory peaks between 52 and 80 h after the onset of the BS (Fig. 1).

The incidence of estrus during the BS was also determined (Fig. 2). Estrus events were detected until day 33 of the BS. Estrus events were detected in 93.30 and 100 % of ewes in the T1 and T2 groups, respectively. An increased incidence of estrus was observed in T2 within the initial 10 days of the BS. First estrus occurred between days 1 and 33, with the mean for first estrus being 15.45 ± 10.36 (T1) and 9.25 ± 6.41 (T2) days post-lambing. The overall incidence of synchronized estrus was 23 % for all ewes, with 26 % in T1 and 20 % in T2, with no difference between groups ($P > 0.05$).

Ovarian activity was monitored in order to correlate it with estrus and pregnancy rates (Table 2). Large follicles ($P < 0.05$) were detected on T2, but the second large follicle did not differ between groups. The mean number of ovulations was similar between groups ($P > 0.05$). Pregnancy rates were determined after first and second mating throughout the BS (Table 3), but no difference was observed ($P > 0.05$) between groups.

4 Discussion

Monitoring reproductive activity by P4 concentration has been used to detect anestrus in ewes, in relation to seasonal factors or postpartum condition (Sasa et al., 2011). As described here, mean P4 concentrations before the male effect demonstrated that ewes were in anestrus, based upon P4 levels previously described by Morales et al. (2003). However, the exposure of ewes to the male effect, irrespective of the postpartum period, led to a significant increase in P4 levels, which is in accordance with previous reports (Sasa et al., 2002).

There is strong evidence correlating P4 concentrations to sexual behavior in anestrus ewes, as P4 stimulates receptivity to rams and increases estrous behavior (Fabre-Nys and Martin, 1991a, b; Caraty and Skinner, 1999). The total incidence of estrus was similar between groups. However, most estrus events within T2 were concentrated in the initial 10 days of the BS, in agreement with their higher P4 concentrations. Since all ewes were in anestrus prior to treatment, it was ex-

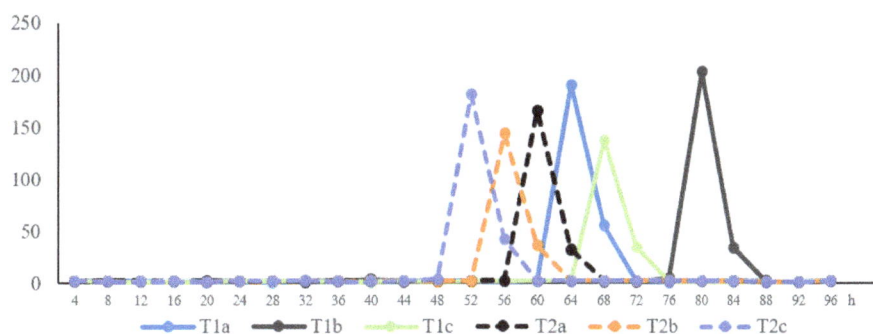

Figure 1. Individual LH concentration ($\eta\,\mathrm{g\,mL}^{-1}$) during the period of 4 to 96 h after the onset of the male effect in Santa Inês postpartum ewes in anestrus ($n = 6$) that had delivered within 35 to 40 days (T1) and 55 to 60 days (T2) previously.

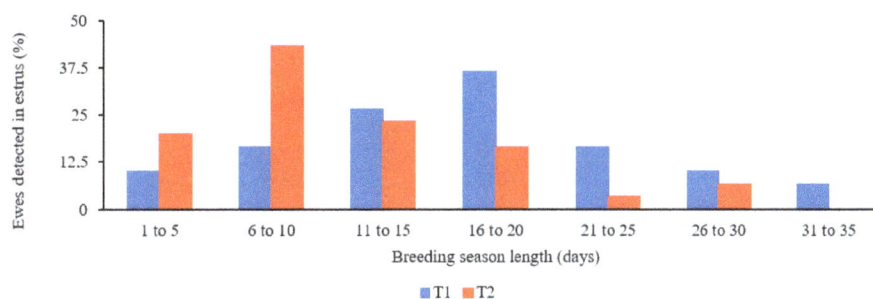

Figure 2. Distribution of estrus events in Santa Inês postpartum ewes in anestrus after being subject to the male effect and 35 to 40 days (T1) and 55 to 60 days (T2) after delivery.

pected that, irrespective to the number of days postpartum, ewes would display a similar responsiveness to the male effect. However, more encouraging results were described with cycling ewes (Caldas et al., 2015a, b) for estrus incidence within the initial 10 days of the BS and estrus synchronization within the initial 5 days of the BS.

The presence of P4 is required to induce GnRH production (Caraty and Skinner, 1999), which is responsible for the subsequent LH preovulatory peak (Martin et al., 1983). The data described here show that the LH preovulatory peak was induced approximately 30 h in advance in ewes that showed higher P4 concentrations before the male effect. Moreover, these ewes were under prolonged postpartum anestrus, a fact that may have contributed to this observation, since these ewes were expected to show higher basal LH levels than ewes under more recent postpartum periods. This hypothesis is in accordance with Martin et al. (1980), who found that higher basal LH levels before interaction between genders increased responsiveness to the male effect. Despite differences in LH preovulatory kinetics, mean LH concentrations were similar between groups and all ewes showed LH preovulatory peaks within 80 h after the male effect onset, as previously described by Oldham et al. (1979). This similarity between results is in agreement with similar incidences of short estrus cycles and the number of ovulations between groups. Due to these facts, the determination of LH levels at 4 h intervals

was efficient to capture the preovulatory peak as described in other reports (Chanvallon et al., 2011; Fabre-Nys et al., 2015).

Although LH preovulatory peaks occurred later in ewes with more recent delivery, it may not have contributed to oocyte competence during follicle growth or luteinization, even after the use of exogenous P4 in association with the male effect (Skinner et al., 2000). Accordingly, the large follicle diameter was large in ewes under prolonged postpartum period conditions and did not affect pregnancy viability, since P4 produced after follicle luteinization was sufficient to maintain full-term pregnancies at similar rates.

5   Conclusions

The male effect is efficient to induce and concentrate estrus within a small window of time in postpartum ewes within 35 to 60 days after lambing. Under these conditions, the LH preovulatory peak occurs within 80 h, which ultimately leads to follicle luteinization and further increases in P4 synthesis. Moreover, postpartum ewes can be subjected to the male effect in order to reduce the time between deliveries, with no effect on overall pregnancy rates.

Competing interests. The authors declare that there is no conflict of interest and are available to provide any clarification.

Acknowledgements. The authors are grateful to CNPq for financial support of this study.

This research was performed after evaluation and approval of the Ethics Committee of the Faculdade Pio Décimo, Aracaju-Se, Brazil, with protocol no. 08/12.

Edited by: M. Mielenz

References

Azzarini, M.: Potencial reproductivo de los ovinos, Produc. Ovin., 16, 5–17, 2004.

Caldas, E. L. C., Ferreira-Silva, J. C., Freitas Neto, L. M., Veloso Neto, H. F., Moura, M. T., Lima, P. F., Santos, M. H. B., and Oliveira, M. A. L.: Male effect associated with suckling interruption on the reproductive performance of santa inês ewes, B. Indústr. Anim., 72, 117–123, doi:10.17523/bia.v72n2p117, 2015a.

Caldas, E. L. C., Freitas Neto, L. M., Almeida-Irmão, J. M., Ferreira-Silva, J. C., Silva, P. G. C., Veloso Neto, H. F., Neves, J. P., Moura, M. T., Lima, P. F., and Oliveira, M. A. L.: The influence of separation distance during the preconditioning period of the male effect approach on reproductive performance in sheep, Vet. Sci. Dev., 4, 1–9, doi:10.4081/vsd.2015.5515, 2015b.

Caraty, A. and Skinner, D. C.: Progesterone priming is essential for the full expression of the positive feedback effect of estradiol in inducing the preovulatory gonadotropin-releasing hormone surge in the ewe, Endocrinology, 140, 165–170, doi:10.1210/endo.140.1.6444, 1999.

Chanvallon, A., Sagot, L., Pottier, E., Debus, N., François, D., Fassier, T., Scaramuzzi, R. J., and Fabre-Nys, C.: New insights into the influence of breed and time of the year on the response of ewes to the 'ram effect', Animal, 5, 1594–1604, doi:10.1017/S1751731111000668, 2011.

De St Jorre, T. J., Hawken, P., and Martin, G. B.: Role of male novelty and familiarity in male-induced LH secretion in female sheep, Reprod. Fertil. Dev., 24, 523–530, doi:10.1071/RD11085, 2012.

Fabre-Nys, C. and Martin, G. B.: Hormonal control of proceptive and receptive sexual behavior and the preovulatory LH surge in the ewe: reassessment of the respective roles of estradiol, testosterone, and progesterone, Horm. Behav., 25, 295–312, doi:10.1016/0018-506X(91)90003-Z, 1991a.

Fabre-Nys, C. and Martin, G. B.: Roles of progesterone and oestradiol in determining the temporal sequence and quantitative expression of sexual receptivity and the preovulatory LH surge in the ewe, J. Endocrinol., 130, 367–379, doi:10.1677/joe.0.1300367, 1991b.

Fabre-Nys, C., Chanvallon, A., Dupont, J., Lardic, L., Lomet, D., Martinet, S., and Scaramuzzi, R. J.: The "Ram Effect": A "Non-Classical" Mechanism for Inducing LH Surges in Sheep, PloS One, 11, e0158530, doi:10.1371/journal.pone.0158530, 2016.

Fabre-Nys, C., Kendrick, K. M., and Scaramuzzi, R. J.: The "ram effect": new insights into neural modulation of the gonadotropic axis by male odors and socio-sexual interactions, Front. Neurosci., 9, 111, doi:10.3389/fnins.2015.00111, 2015.

Ferreira-Silva, J. C., Chaves, M. S., Tenório Filho, F., Moura, M. T., Neto, L. F., Caldas, E. L. C., and Oliveira, M. A. L.: Reproductive efficiency of non-cycling postpartum ewes submitted to the male effect under tropical semi humid conditions, Livestock Res. Rural. Dev., 28, 171, 2016.

Ferreira-Silva, J. C., Basto, S. R. L., Tenório Filho, F., Moura, M. T., Silva Filho, M. L., and Oliveira, M. A. L.: Reproductive performance of postpartum ewes treated with insulin or progesterone hormones in association with ram effect have been received, Reprod. Domest. Anim., in press, 2017.

Figueiredo, E. D., Oliveira, E. D., Bellaver, C., and Simplicio, A. A.: Hair sheep performance in Brazil, in: Hair sheep of Western Africa and the Americas, edited by: Fitzhugh, H. A. and Bradford, G. E., Westview Press, Boulder, Colorado, 125–140, 1983.

Knight, T. W., Tervit, H. R., and Lynch, P. R.: Effect of boar pheromones, ram's wool and presence of bucks on ovarian activity in anovular ewes early in the breeding season, Anim. Reprod. Sci., 6, 129–134, doi:10.1016/0378-4320(83)90017-9, 1983.

Knights, M., Hoehn, T., Lewis, P. E., and Inskeep, E. K.: Effectiveness of intravaginal progesterone inserts and FSH for inducing synchronized estrus and increasing lambing rate in anestrous ewes, J. Anim. Sci., 79, 1120–1131, doi:10.2527/2001.7951120x, 2001.

Machado, J. B. B., Fernandes, A. A. O., Selaive-Villarroel, A. B., Costa, A. L., Lima, R. N., and Lopes, E. A.: Parâmetros reprodutivos de ovinos deslanados Morada Nova e Santa Inês mantidos em pastagem cultivada no estado do Ceará, Rev. Cient. Prod. Anim., 1, 81–87, 1999.

Martin, G. B., Oldham, C. M., and Lindsay, D. R.: Increased plasma LH levels in seasonally anovular Merino ewes following the introduction of rams, Anim. Reprod. Sci., 3, 125–132, doi:10.1016/0378-4320(80)90039-1, 1980.

Martin, G. B., Scaramuzzi, R. J., and Henstridge, J. D.: Effects of oestradiol, progesterone and androstenedione on the pulsatile secretion of luteinizing hormone in ovariectomized ewes during spring and autumn, J. Endocrinol., 96, 181–193, doi:10.1677/joe.0.0960181, 1983.

Monreal, A. C. D., Carneiro, L. O. H. B., and Redondo, M. V. S.: Efeito macho associado ao emprego de progesterona intravaginal em ovelhas, sob latitude 20°52′, South. Agrarian, 2, 143–152, 2009.

Moraes, J. C. F.: Emprego do "efeito carneiro" na indução e manipulação do ciclo estral em ovelhas durante o anestro, A Hora Veterinária, 11, 32–34, 1991.

Morales, J. U., Váquez, H. G. G., and Andrade, B. M. R.: Influencia del pastoreo restringido en el efecto macho em cabras em baja condición corporal durante la estación de anestro, Téc. Pecu. Méx., 41, 251–260, 2003.

Notter, D. R. and Copenhaver, J. S.: Performance of Finnish Landrace crossbred ewes under accelerated lambing. II. Lamb growth and survival, J. Anim. Sci., 51, 1043–1050, doi:10.2527/jas1980.5151043x,1980.

Oldham, C., Martin, G. B., and Knight, T. W.: Stimulation of seasonally anovular Merino ewes by rams. I. Time from introduction of the rams to the preovulatory LH surge and ovulation, Anim. Reprod. Sci., 1, 283–290, doi:10.1016/0378-4320(79)90013-7, 1979.

Oliveira, L. R. S., Ferreira-Silva, J. C., Chaves, M. S., Freitas Neto, L. M., Moura, M. T., Caldas, E. L. C., Lima, P. F., and Oliveira, M. A. L.: Male Effect and Breeding Season Shortening Under Contrasting Climatic Conditions Upon Reproduction of Nulliparous Anglo-Nubians, Ciênc. Vet. Tróp., 18, 43–48, 2015.

Rubianes, E. and Ungerfeld, R.: Perspectivas de la investigacíon sobre reproducción ovina em América Latina en El marco de las actuales tendências productivas, Arch. Latinoam. Prod. Anim., 10, 117–125, 2002.

Santos, M. H. B., Oliveira, M. A. L., Moraes, E. P. B. X., Moura, R. T. D., Lima, P. F., and Reichenbach, H.-D.: Diagnóstico de gestação por ultra-sonografia de tempo real, in: Diagnóstico de gestação na cabra e na ovelha, edited by: Santos, M. H. B., Oliveira, M. A. L., and Lima, P. F., Varela, São Paulo, 97–116, 2004.

Sasa, A., Teston, D. C., Rodrigues, P. D. A., Coelho, L. D. A., and Schalch, E.: Concentrações plasmáticas de progesterona em borregas lanadas e deslanadas no período de abril a novembro, no Estado de São Paulo, R. Bras. Zootec., 31, 1150–1156, doi:10.1590/S1516-35982002000500011, 2002.

Sasa, A., Nonaka, K. O., Balieiro, J. C. C., and Coelho, L. A.: Progesterona plasmática de ovelhas submetidas ao efeito-macho e mantidas sob diferentes condições nutricionais, Arq. Bras. Med. Vet. Zootec., 63, 1066–1072, doi:10.1590/S0102-09352011000500004, 2011.

Simplício, A. A.: Estratégias de manejo reprodutivo como ferramenta para prolongar o período de oferta de carnes caprina e ovina no Brasil, Rev. Tecnol. Ciên. Agropec., 2, 29–39, 2008.

Skinner, D. C., Harris, T. G., and Evans, N. P.: Duration and amplitude of the luteal phase progesterone increment times the estradiol-induced luteinizing hormone surge in ewes, Biol. Reprod., 63, 1135–1142, doi:10.1095/biolreprod63.4.1135, 2000.

Tenório Filho, F., Santos, M. H. B., Carrazzoni, P. G., Bartolomeu, C. C., Lima, P. F., and Oliveira, M. A. L.: Follicular dynamics in Anglo-Nubian goats using transrectal and transvaginal ultrasound, Small Ruminant Res., 72, 51–56, doi:10.1016/j.smallrumres.2006.08.007, 2007.

Tenório Filho, F., Ferreira-Silva, J. C., Nascimento, P. S., Freitas Neto, L. M., Moura, M. T., Almeida Irmão, J. M., and Oliveira, M. A. L.: Ação do efeito macho sobre a eficiência reprodutiva de ovelhas nulíparas das raças Santa Inês e Morada Nova criadas em diferentes regiões, Acta Scientiae Veterinariae, 44, 1353, 2016.

Thimonier, J.: Détermination de l'état physiologique des femelles par analyse des niveaux de progestérone, INRA Produção Animal, 13, 177–183, 2000.

Ungerfeld, R., Forsberg, M., and Rubianes, E.: Overview of the response of anoestrous ewes to the ram effect, Reprod. Fert. Dev., 16, 479–490, doi:10.10371/RD04039, 2004.

5

The relationships between transforming growth factors β and free thyroxine and progesterone in the ovarian cysts, preovulatory follicles, and the serum of sows

Tomasz Stankiewicz

West Pomeranian University of Technology, Szczecin, Faculty of Biotechnology and Animal Husbandry,
Department of Animal Reproduction Biotechnology and Environmental Hygiene,
29 Klemensa Janickiego Street, 71-270 Szczecin, Poland

Correspondence to: Tomasz Stankiewicz(tomasz.stankiewicz@zut.edu.pl)

Abstract. The aim of the study was to determine the relationships between bone morphogenetic protein 15 (BMP-15) and growth differentiation factor 9 (GDF-9) concentrations and free thyroxine (FT$_4$) and progesterone (P$_4$) concentrations in follicular cysts, preovulatory follicles, and the serum of sows (cyst-bearing ($n = 26$) and non-cyst-bearing ($n = 26$)). FT$_4$ and P$_4$ concentrations were higher in the cystic fluid than in the fluid of preovulatory follicles ($p < 0.01$ and $p < 0.05$ respectively). BMP-15 and GDF-9 concentrations were higher in the serum of cyst-bearing sows than non-cyst-bearing sows ($p < 0.05$) and higher in the cystic fluid than in the follicular fluid ($p < 0.05$). In the cysts and preovulatory follicles, GDF-9 concentration was higher than in serum ($p < 0.01$). FT$_4$ concentration in the serum of cystic sows was correlated with BMP-15 ($r = 0.50$, $p < 0.05$) and GDF-9 ($r = 0.62$, $p < 0.01$) concentrations in serum. In the serum of non-cyst-bearing sows, a positive correlation between P$_4$ concentration and BMP-15 concentration ($r = 0.60$, $p < 0.01$) was detected.

These data will help provide insight into the role of BMP-15, GDF-9, FT$_4$, and P$_4$ during cyst formation in sows.

1 Introduction

Ovarian cysts account for a major proportion of ovarian dysfunction (Cech and Dolezel, 2007; Szulańczyk-Mencel et al., 2010). In sows, polycystic ovaries cause reproductive disorders, reduce reproductive performance, and can result in their culling (Heinonen et al., 1998; Szulańczyk-Mencel et al., 2010). Despite much research, the etiopathogenesis of ovarian cysts is not yet fully understood. However, it is known that disturbances in hormonal regulation can cause ovarian cysts (Kozłowska et al., 2013). Thyroid hormones, as well as gonadotropins and steroid hormones, have important roles in the regulation of the porcine ovarian follicle function (Maruo et al., 1987; Gregoraszczuk et al., 1998) and are found in porcine ovarian follicular fluid (Stankiewicz et al., 2008). In in vitro experiments, thyroid hormones have been shown to affect steroidogenesis in porcine thecal and granulosa cells (Gregoraszczuk and Skalka, 1996). The participation of thy-

roid hormones in the synthesis of steroid hormones is noted by influencing the activity of aromatase (Gregoraszczuk et al., 1998). It has been reported that thyroid hormones increase the impact of the follicle-stimulating hormone on the functional differentiation of cultured porcine granulosa cells (Maruo et al., 1987). In addition, thyroid hormone receptors, their mRNA, or both have been identified in porcine granulosa cells from preovulatory antral follicles (Maruo et al., 1992). Thyroid status has also been implicated in ovarian cyst formation in gilts (Fitko et al., 1995, 1996). Fitko et al. (1995, 1996) have shown that hypothyroidism increases the exogenous gonadotropin formation of cysts and weakens the steroidogenesis activity of ovaries in gilts. However, exactly how thyroid hormones contribute to the pathogenesis of ovarian cysts is unknown and might be based on interactions in the ovaries and the central or peripheral interrelations. It is possible that such interactions with thyroid hormones involve a bone morphogenetic protein 15 (BMP-15) and growth differentiation factor 9 (GDF-9), which belong to the transforming growth factor β (TGF-β) superfamily. It is supposed that BMP-15 and GDF-9 are involved in hormonal regulation of the hypothalamic–pituitary–ovary axis (Paulini and Melo, 2011). These factors, depending on the stage of follicular development, may increase or weaken the influence of gonadotropins on the ovarian follicle (Knight and Glister, 2006; Crawford and McNatty, 2012). In addition, they influence the proliferation and differentiation of somatic cells of the follicle, steroidogenesis, deposition of the extracellular matrix, ovulation, and luteinization (Su et al., 2008; Orisaka et al., 2009; Peng et al., 2010). Also, recent work has suggested a role for BMP-15 and GDF-9 in the pathogenesis of follicular cysts in gilts and sows (Stankiewicz and Błaszczyk, 2014, 2016). However, the exact activity of BMP-15 and GDF-9 in the pathogenesis of follicular cysts is unknown.

Therefore, the objective of the research was to identify differences and dependencies between BMP-15 and GDF-9 concentrations and the concentrations of free thyroxine (FT$_4$) and progesterone (P$_4$) in follicular cysts, preovulatory follicles, and the serum of sows (cyst-bearing and non-cyst-bearing).

2 Material and methods

The study was carried out with Polish Large White × Polish Landrace crossbreds (from 2 to 3 years old) slaughtered at a local slaughterhouse. All sows were kept in a modern farm and then slaughtered in a modern, Polish slaughterhouse according to national legislation and in line with European Union legislation. At the time of slaughter, blood was collected from each sow's cervical vein into a serum separator tube. The blood was centrifuged at $1000 \times g$ for 15 min, and the resulting serum was stored at $-20\,^\circ$C until analysed. During an ongoing slaughtering process, 52 sows were chosen for further examinations, including 26 sows with follicular cysts (cyst-bearing sows) and 26 sows without ovarian cysts but having preovulatory follicles (non-cyst-bearing sows). Each sow was assigned one fluid sample (from only one cyst or only one preovulatory follicle) and a corresponding serum sample. In the experiment, the following groups were distinguished for further research: follicular cysts ($n = 26$), preovulatory follicles ($n = 26$), serum of cyst-bearing sows ($n = 26$), and serum of non-cyst-bearing sows ($n = 26$). In order to obtain a uniform material, only bilateral polycystic ovaries were examined in the experiment. In polycystic ovaries, there were no corpora lutea and the current follicular structures with thin walls were filled with fluid and had a diameter of more than 21 mm (Heinonen et al., 1998; Cech and Dolezel, 2007; Sun et al., 2011). Ovaries in the preovulatory phase were identified as those having at least several follicles of the appropriate colour and composition, and they were 7 to 9 mm in diameter (Hunter et al., 2004; Paradis et

al., 2009). In these ovaries a single corpus haemorrhagicum or corpus luteum was present. Ovaries with corpora albicantia were eliminated from the study (Babalola and Shapiro, 1988). The follicular/cystic fluid was aspirated with a needle and syringe from the preovulatory follicle/cyst. Fluid samples were centrifuged at $3000 \times g$ for 10 min to remove cellular material, and the supernatant was stored at $-20\,^\circ$C until being analysed.

2.1 BMP-15 and GDF-9 assay

Specimen-specific kits were used to determine the concentration of BMP-15 (Porcine BMP-15 ELISA kit, Novateinbio Biosciences, Cat. No. POR10362) and GDF-9 (Porcine GDF-9 ELISA kit, Novateinbio Biosciences, Cat. No. BG-POR11087). The test sensitivities were $0.1\,\mathrm{ng\,mL^{-1}}$ and $0.1\,\mathrm{pg\,mL^{-1}}$ for BMP-15 and GDF-9 respectively. The intra- and inter-assay coefficients of variation were $< 10\,\%$ for BMP-15 and GDF-9. The measurement was conducted using a Wallac fluorometer 1420 VICTOR^2 (Wallac Oy, Turku, Finland). All assays were carried out in duplicate.

2.2 FT$_4$ and P$_4$ assay

FT$_4$ concentrations were measured by fluoroimmunoassay using the Delfia® FT$_4$ kit (Perkin-Elmer, Wallac Oy, Finland) (Błaszczyk et al., 2006). The intra- and inter-assay coefficients of variation were 3.3 and 4.7 % respectively. The sensitivity of the assay was $1.56\,\mathrm{pg\,mL^{-1}}$. The P$_4$ concentration was also determined by fluoroimmunoassay using the Delfia® P$_4$ kit (Perkin-Elmer, Wallac Oy, Finland) (Stankiewicz et al., 2008, 2009; Błaszczyk et al., 2009). The intra- and inter-assay coefficients of variation were 4.9 and 6.9 % respectively. The sensitivity of the assay was $0.25\,\mathrm{ng\,mL^{-1}}$.

All assays were carried out in duplicate. The measurements were made using a Wallac fluorometer 1420 VICTOR^2 (Wallac Oy, Turku, Finland).

2.3 Statistical analysis

The data are presented as an average $\pm$ standard deviation of the mean and presented in the tables. Analysis of variance (ANOVA) and a post hoc test was done to identify statistically significant differences. Duncan's multiple range test was used to verify the significance of differences at $p < 0.01$ and $p < 0.05$. In addition, correlations between the analysed parameters were calculated with the Spearmen's rank correlation coefficient. Statistical analyses were conducted using the STATISTICA version 7.1, Stat Soft, Poland.

3 Results

Table 1 shows the mean concentrations of FT$_4$, P$_4$, BMP-15, and GDF-9 in cyst-bearing and non-cyst-bearing sows.

Table 1. FT$_4$, P$_4$, BMP-15, and GDF-9 concentrations in the ovarian cystic fluid, preovulatory follicular fluid, and the serum of sows.

Hormones		Sows cysts-bearing ($n = 26$)		Sows non-cysts-bearing ($n = 26$)		
		Cystic fluid	Serum	Follicle fluid	Serum	
		1	2	3	4	Statistical differences
FT$_4$ (pg mL^{-1})	mean $\pm$ SD	9.02 ± 2.11	8.46 ± 1.92	6.95 ± 1.31	7.05 ± 1.48	1, 3; $p < 0.01$
P$_4$ (ng mL^{-1})	mean $\pm$ SD	564.77 ± 127.03	3.49 ± 1.71	475.52 ± 170.45	2.54 ± 1.43	1, 2; $p < 0.01$
						1, 3; $p < 0.05$
						3, 4; $p < 0.01$
BMP-15 (ng mL^{-1})	mean $\pm$ SD	1.95 ± 0.98	1.47 ± 0.39	1.08 ± 0.70	0.89 ± 0.34	1, 3; $p < 0.05$
						2, 4; $p < 0.05$
GDF-9 (pg mL^{-1})	mean $\pm$ SD	23.99 ± 4.87	17.25 ± 2.99	19.08 ± 3.23	14.75 ± 1.61	1, 2; $p < 0.01$
						1, 3; $p < 0.05$
						2, 4; $p < 0.05$
						3, 4; $p < 0.01$

Table 2. Correlation coefficients (r) between FT$_4$ and P$_4$ concentrations and the concentration of the transforming growth factors (BMP-15 and GDF-9) in the cystic fluid and serum of cyst-bearing sows ($n = 26$).

Parameter	FT$_4$ in serum	FT$_4$ in cystic fluid	P$_4$ in serum	P$_4$ in cystic fluid
BMP-15 in serum	0.50*	0.23	−0.09	0.12
BMP-15 in cystic fluid	0.02	0.06	−0.18	0.14
GDF-9 in serum	0.62**	0.13	0.12	0.20
GDF-9 in cystic fluid	0.11	0.18	−0.17	0.11

Values marked * and ** are significant at $p < 0.05$ and $p < 0.01$ respectively.

The FT$_4$ concentration in the serum of cyst-bearing sows was not statistically different from its in-cyst concentration. Also, FT$_4$ concentration did not differ between the serum and preovulatory follicles of non-cyst-bearing sows nor were any differences found between the FT$_4$ concentration in the serum of cyst-bearing sows and non-cyst-bearing sows. However, the FT$_4$ concentration in the cystic fluid was significantly higher than in the fluid of preovulatory follicles.

In cyst-bearing and non-cyst-bearing sows, the P$_4$ concentration was significantly higher in the ovary structures (cysts and preovulatory follicles) than in the serum. Also, P$_4$ concentration was significantly higher in cysts than in preovulatory follicles.

BMP-15 and GDF-9 concentrations were significantly higher in the serum of cyst-bearing than in that of non-cyst-bearing sows. BMP-15 and GDF-9 concentrations were also higher in the cystic fluid than in follicular fluid. BMP-15 concentration in the preovulatory follicles did not differ significantly from its concentration in serum. However, these values did differ for GDF-9. In the preovulatory follicles, GDF-9 concentration was significantly higher than in the serum. Similar differences were found in cyst-bearing sows. BMP-15 concentration in cysts did not differ significantly from its concentration in serum. However, GDF-9 concentration in cysts was significantly higher than in serum.

Tables 2 and 3 show the correlation coefficients between FT$_4$, P$_4$, and transforming growth factors β in cyst-bearing and non-cyst-bearing sows. In the serum of cystic sows, the FT$_4$ concentration was positively and significantly correlated with BMP-15 and GDF-9 concentrations. In non-cyst-bearing sows, a positive and significant correlation was found between serum P$_4$ and BMP-15 concentrations.

4 Discussion

Studies on the pathogenesis of ovarian cysts must consider many potential contributing factors (Fitko et al., 1995, 1996; Kozłowska et al., 2013; Pierre et al., 2016). As such, it is not yet possible to exclude contributions by BMP-15 and GDF-9 during cyst formation (Stankiewicz and Błaszczyk, 2014, 2016). Increased BMP-15 and GDF-9 concentrations were found in the follicular cysts of gilts and sows (Stankiewicz and Błaszczyk, 2014, 2016). Here, I detected higher BMP-15 and GDF-9 concentrations in the cystic fluid than in follicular fluid. However, these differences were smaller than in earlier work (Stankiewicz and Błaszczyk, 2014, 2016). Nevertheless, the presence of BMP-15 and GDF-9 in the follicular and cystic fluid confirms that these factors create the proper follicular microenvironment and can participate in the development of follicular cysts. In addition, BMP-15 and GDF-9 concentrations are positively correlated in follicular fluid and

Table 3. Correlation coefficients (r) between FT_4 and P_4 concentrations and the concentration of the transforming growth factors (BMP-15 and GDF-9) in the preovulatory follicle fluid and serum of non-cysts-bearing sows ($n = 26$).

Parameter	FT_4 in serum	FT_4 in follicle fluid	P_4 in serum	P_4 in follicle fluid
BMP-15 in serum	0.28	0.20	0.60*	0.19
BMP-15 in follicle fluid	0.39	0.35	0.24	−0.18
GDF-9 in serum	0.02	−0.05	−0.01	−0.24
GDF-9 in follicle fluid	−0.14	−0.11	0.31	0.40

Values marked * are significant at $p < 0.01$.

serum (Stankiewicz and Błaszczyk, 2016). Thus, I cannot exclude their participation in the peripheral control of folliculogenesis.

As shown here, the concentration of BMP-15 and GDF-9 in serum is higher in cyst-bearing than non-cyst-bearing sows, which is in line with previous studies (Stankiewicz and Błaszczyk, 2016). Unlike in previous studies, in the current study it was also found that GDF-9 concentration was higher in the cystic and follicular fluid than in serum. The higher concentrations in the cystic and follicular fluids show, that the synthesis of transforming growth factors such as GDF-9 and BMP-15 occurs in ovarian follicles at various stages of development. According to Fitzpatrick et al. (1998) and Knight and Glister (2006), the expression of GDF-9 mRNA was found in the hypothalamus, pituitary, and uterus of different mammalian species, so that theses extraovarian organs can also influence the serum levels. The exact role of BMP-15 and GDF-9 in the pathogenesis of follicular cysts is unknown. Therefore, in the present study, I aimed to define the relationships between these and other factors that might participate in the formation of follicular cysts. One of such factor is the steroidogenic activity of the ovarian follicle. Disturbed follicular steroidogenesis is either an effect or a cause of ovarian disorders, such as ovarian cysts (Babalola and Shapiro, 1990; Szulańczyk-Mencel et al., 2010). Here, higher levels of P_4 were found in the cystic fluid than in preovulatory follicle fluid, which is in line with previous studies (Babalola and Shapiro, 1990; Kozłowska et al., 2013) and could be due to luteinization spontaneously beginning in non-ovulating, cystic follicles. However, the absence of any difference between the serum concentrations of P_4 in the serum of cyst-bearing and non-cyst-bearing sows confirms that the hormonal abnormalities in the follicular cysts are not reflected in the serum profiles of this steroid (Babalola and Shapiro, 1990). It is interesting that the concentration of P_4 was positively correlated with BMP-15 concentration in the serum of sows without follicular cysts. Based on this finding, I propose that the interactions between BMP-15 and progesterone are involved in the control of folliculogenesis in sows.

Thyroid hormones have also been suggested as participating in folliculogenesis and/or the formation of follicular cysts (Fitko et al., 1996; Błaszczyk et al., 2006; Stankiewicz et al., 2008). I tested the concentration of FT_4, which is a more

reliable indicator of thyroid status than total thyroxine concentration (Nowak, 1983). Here, I found no differences between the concentration of FT_4 in the serum of cyst-bearing and non-cyst-bearing sows, and the recorded concentration of this hormone in the serum of the examined sows was similar to concentrations observed in pigs (Spiegel et al., 1993). Thus, neither the non-cyst-bearing nor the cyst-bearing sows in this study displayed hypo- or hyperthyroidism. In contrast, Fitko et al. (1995) reported that the hypothyroid status of gilts intensifies the cyst-formative actions of extrapituitary gonadotropins, whereas hyperfunctioning of the thyroid significantly reduces this response. The authors of that study also suggested that the mechanism of these antagonist relations may be based on the interaction between receptors for thyroid hormones and gonadotropins in the ovary and/or on the central or peripheral relationship between thyroid hormone and estrogens (Fitko et al., 1995). The results of this study indicate a possible local activity of thyroid hormones during the formation of ovarian cysts because the concentration of FT_4 was significantly higher in the cysts than in preovulatory follicles. Moreover, despite a similar concentration of FT_4 in the serum of cyst-bearing and non-cyst-bearing sows, I cannot exclude peripheral effects of thyroid hormones in the pathogenesis of follicular cysts in pigs. In the present study shown positive correlations between FT_4 and BMP-15 and GDF-9 in the serum of cyst-bearing sows have been. The participation of BMP-15 and GDF-9 in the pathogenesis of follicular cysts in pigs has been suggested (Stankiewicz and Błaszczyk, 2014, 2016). In addition, these factors, depending on the stage of folliculogenesis, can either strengthen or weaken the influence of gonadotropins on the ovarian follicle (Knight and Glister, 2006; Crawford et al., 2012). Also, the interactive effect of thyroid hormones and gonadotropins on the formation of follicular cysts has been shown in gilts (Fitko et al., 1995, 1996). Therefore, the relationships between BMP-15, GDF-9, and FT_4 in cyst-bearing sows may be associated with a control of gonadotropins. This is a possible pathway for the action of BMP-15, GDF-9, and FT_4 in the mechanism of the formation of follicular cysts in sows.

5 Conclusion

The data presented here will be useful for investigations into the potential roles of the transforming growth factor β, thyroxine, and progesterone during the formation of follicular cysts in sows.

Competing interests. The author declares that he has no conflict of interest.

Edited by: M. Mielenz

References

Babalola, G. O. and Shapiro, B. H.: Correlation of follicular steroid hormone profiles with ovarian cyclicity in sows, J. Reprod. Fert., 84, 79–87, 1988.

Babalola, G. O. and Shapiro, B. H.: Sex steroid changes in porcine cystic ovarian disease, Steroids, 55, 319–324, 1990.

Błaszczyk, B., Stankiewicz, T., Udała, J., and Gączarzewicz, D.: Plasma progesterone analysis by a time-resolved fluorescent antibody test to monitor estrous cycles in goats, J. Vet. Diagn. Invest., 21, 80–87, 2009.

Błaszczyk, B., Stankiewicz, T., Udała, J., Gączarzewicz, D., Lasota, B., Błaszczyk, P., Szymańska, A., and Szymańska-Pasternak, J.: Free thyroid hormones and cholesterol in follicular fluid of bovine ovaries, Bull. Vet. Inst. Pulawy., 50, 189–193, 2006.

Cech, S. and Dolezel, R.: Treatment of ovarian cysts in sows – a field trial, Vet. Med.-Czech., 52, 413–418, 2007.

Crawford, J. L. and McNatty, K. P.: The ratio of growth differentiation factor 9: bone morphogenetic protein 15 mRNA expression is tightly coregulated and differs between species over a wide range of ovulation rates, Mol. Cell. Endocrinol., 348, 339–343, 2012.

Fitko, R., Kucharski, J., and Szlezyngier, B.: The importance of thyroid hormone in experimental ovarian cyst formation in gilts, Anim. Reprod. Sci., 39, 159–168, 1995.

Fitko, R., Kucharski, J., Szlezyngier, B., and Jana, B.: The concentration of GnRH in hypothalamus, LH and FSH in pituitary, LH, PRL and sex steroids in peripheral and ovarian venous plasma of hypo- and 15 hyperthyroid, cysts-bearing gilts, Anim. Reprod. Sci., 45, 123–138, 1996.

Fitzpatrick, S. L., Sindoni, D. M., Shughrue, P. J., Lane, M. V., Merchenthaler, I. J., and Frail D. E.: Expression of growth differentiation factor-9 messenger ribonucleic acid in ovarian and nonovarian rodent and human tissues, Endocrinology, 139, 2571–2578, 1998.

Gregoraszczuk, E. L. and Skalka, M.: Thyroid hormone as a regulator of basal and human chorionic gonadotrophin-stimulated steroidogenesis by cultured porcine theca and granulosa cells isolated at different stages of the follicular phase, Reprod. Fertil. Dev., 8, 961–971, 1996.

Gregoraszczuk, E. L., Słomczynska, M., and Wilk, R.: Thyroid hormone inhibits aromatase activity in porcine thecal cells cultured alone and in coculture with granulosa cells, Thyroid, 8, 1157–1163, 1998.

Heinonen, M., Leppävuori, A., and Pyörälä, S.: Evaluation of reproductive failure of female pigs based on slaughterhouse material and herd record survey, Anim. Reprod. Sci., 52, 235–244, 1998.

Hunter, M. G., Robinson, R. S., Mann, G. E., and Webb, R.: Endocrine and paracrine control of follicular development and ovulation rate in farm species, Anim. Reprod. Sci., 82–83, 461–477, 2004.

Knight, P. G. and Glister, C.: TGF-β superfamily members and ovarian follicle development, Reproduction (Cambridge, England), 132, 191–206, 2006.

Kozłowska, A., Wojtkiewicz, J., Majewski, M., and Jana, B.: The noradrenergic innervation and steroidogenic activity of porcine cystic ovaries, Physiol. Res., 62, 421–433, 2013.

Maruo, T., Hayashi, M., Matsuo, H., Yamamoto, T., Okada, H., and Mochizuki, M.: The role of thyroid hormone as a biological amplifier of the actions of follicle-stimulating hormone in the functional differentiation of cultured porcine granulosa cells, Endocrinology, 121, 1233–1241, 1987.

Maruo, T., Hiramatsu, S., Otani, T., Hayashi, M., and Mochizuki, M.: Increase in the expression of thyroid hormone receptors in porcine granulosa cells early in follicular maturation, Acta Endocrinol. (Copenh.), 127, 152–160, 1992.

Nowak, G.: Free thyroid hormone levels during the postnatal period in the pig, Biol. Neonate., 43, 164–171, 1983.

Orisaka, M., Jiang, J. Y., Orisaka, S., Kotsuji, F., and Tsang, B. K.: Growth differentiation factor 9 promotes rat preantral follicle growth by upregulating follicular androgen biosynthesis, Endocrinology, 150, 2740–2748, 2009.

Paradis, F., Novak, S., Murdoch, G. K., Dyck, M. K., Dixon, W. T., and Foxcroft, G. R.: Temporal regulation of BMP2, BMP6, BMP15, GDF9, BMPR1A, MPR1B, BMPR2 and TGFBR1 mRNA expression in the oocyte, granulosa and theca cells of developing preovulatory follicles in the pig, Reproduction, 138, 115–129, 2009.

Paulini, F. and Melo, E.: The role of oocyte-secreted factors GDF9 15 and BMP15 in follicular development and oogenesis, Reprod. Dom. Anim., 46, 354–361, 2011.

Peng, X., Yang, M., Wang, L., Tong, C., and Guo, Z.: In vitro culture of sheep lamb ovarian cortical tissue in a sequential culture medium, J. Assist. Reprod. Genet., 27, 247–257, 2010.

Pierre, A., Estienne, A., Racine, C., Picard, J.-Y., Fanchin, R., Lahoz, B., Alabart, J. L., Folch, J., Jarrier, P., Fabre, S., Monniaux, D., and di Clemente, N.: The bone morphogenetic protein 15 up regulates the anti-Müllerian hormone receptor expression in granulosa cells, J. Clin. Endocrinol. Metab., 101, 2602–2611, 2016.

Spiegel, C., Bestetti, G. E., Rossi, G. L., and Blum, J. W.: Normal circulating triiodothyronine concentrations are maintained despite severe hypothyroidism in growing pigs fed rapeseed presscake meal, J. Nutr., 123, 1554–1561, 1993.

Stankiewicz, T. and Błaszczyk, B.: Concentrations of bone morphogenetic protein-15 (BMP-15) and growth differentiation factor-9 (GDF-9) in follicular cysts, mono- and polyoocyte follicles in gilts, Acta Vet.-Beograd., 64, 24–32, 2014.

Stankiewicz, T. and Błaszczyk, B.: Relationship between the concentration of bone morphogenetic protein-15 (BMP-15) and growth differentiation factor-9 (GDF-9) in pre-ovulatory follicles, ovarian cysts, and serum in sows, Anim. Prod. Sci., 56, 141–146, 2016.

Stankiewicz, T., Błaszczyk, B., Lasota, B., Gączarzewicz, D., and Udała, J.: Saisonabhängige Veränderungen der Ovargröße sowie Konzentration von Steroidhormonen und Thyroxin in der Follikelflüssigkeit beim Schwein, Tierarztl. Prax., 36, 99–103, 2008.

Stankiewicz, T., Błaszczyk, B., and Udała, J.: A Study on the occurrence of polyovular follicles in porcine ovaries with particular reference to intrafollicular hormone concentrations, quality of oocytes and their in vitro fertilization, Anat. Histol. Embryol., 38, 233–239, 2009.

Su, Y. Q., Sugiura, K., Wigglesworth, K., O'Brien, M. J., Affourtit, J. P., Pangas, S. A., Matzuk, M. M., and Eppig, J. J.: Oocyte regulation of metabolic cooperativity between mouse cumulus cells and oocytes: BMP-15 and GDF-9 control cholesterol biosynthesis in cumulus cells, Development, 135, 111–121, 2008.

Sun, Y. L., Ping, Z. G., Li, C. J., Sun, Y. F., Yi, K. L., Chen, L., Li, X. Y., Wang, X. L., and Zhou, X.: Comparative proteomic analysis of follicular fluids from normal and cystic follicles in sows, Reprod. Dom. Anim., 46, 889–895, 2011.

Szulańczyk-Mencel, K., Rząsa, A., and Bielas, W.: Relationships between ovarian cysts and morphological and hormonal state of ovarian cortex in sows, Anim. Reprod. Sci., 121, 273–278, 2010.

6

Effects of nanoencapsulated aloe vera, dill and nettle root extract as feed antibiotic substitutes in broiler chickens

Amir Meimandipour[1], Ali Nouri Emamzadeh[2], and Ali Soleimani[1]

[1]Department of Animal Biotechnology, National Institute of Genetic Engineering and Biotechnology (NIGEB), Tehran, Iran
[2]Department of Animal Science, Garmsar Branch, Islamic Azad University, Garmsar, Iran

Correspondence to: Amir Meimandipour (meimandi@nigeb.ac.ir)

Abstract. Aloe vera, nettle and dill are herbs that have been used in the poultry diet as feed additives to utilise their benefits in improving performance, immune response and health of broiler chickens. However, reactive and volatile properties of bioactive compounds in herbal extracts cause limitations on direct usage of them in the diet. The use of chitosan (CS) nanoparticles for the entrapment of active components has gained interest in the last few years due to its mucous adhesiveness, non-toxicity, biocompatibility and biodegradability. This study was an effort to evaluate effects of nanoencapsulated extracts of aloe vera, dill and nettle root used in diet on performance, carcass traits and serum immunoglobulin (IgM and IgY) concentrations in broiler chickens. Chitosan nanoparticles were prepared by using ionotropic gelation principle. After nanogel preparation of herbal extracts, a total of 240 Ross (308) broiler chicks were divided into eight treatments, with three replicates of 10 birds. The eight dietary treatments consisted of control (no additives), antibiotic (bacitracin $500 \, \mathrm{g \, t^{-1}}$), non-encapsulated and nanoencapsulated extracts of aloe vera, dill and nettle root. In each experimental period, non-encapsulated (free extracts) and nanoencapsulated extracts of aloe vera, dill and nettle roots were added in amounts of 0.02, 0.025 and 0.05 % to starter, grower and finisher diets, respectively. Birds in different treatments received the same diets during the experimental periods. Results revealed that increasing both non-encapsulated and nanoencapsulated herbal extracts to 0.05 % in finisher diets improved body weight gain in the time period of 28–42 days and consequently the whole time from 1 to 42 days. However, in these periods, birds fed a diet containing nanoencapsulated dill extract had a significantly ($P < 0.05$) higher body weight gain compared with the antibiotic group, while non-encapsulated dill extract treatment was intermediate. The addition of nanoencapsulated nettle extract in diet significantly ($P < 0.05$) improved feed conversion efficiency in the 28–42-day period compared with the antibiotic group. In comparison with the antibiotic group, nanoencapsulation of dill extract could profoundly improve growth performance and can therefore be used as a substitute for antibiotics in the diet of broiler chickens.

1 Introduction

The modern poultry sector today is a specialised one and it concentrates more on the use of high-performance birds. Antibiotic growth promoters have been helpful in improving growth performance and feed conversion efficiency in poultry (Dibner and Richards, 2005). However, constant treatment of poultry with antibiotics may result in residue of these substances in poultry products and bacterial resistance against treatments in humans (Hughes and Heritage, 2007). Due to these threats to human health, use of antibiotics in poultry is banned. Therefore, there is a major demand for producing high-quality poultry meat at a low price without relying on antibiotics and other medicines in poultry feed and water.

Aloe vera, nettle and dill are herbs with various antibacterial, antiseptic, anti-inflammatory, nematocidal and immunomodulatory properties. Along with using these herbs for medicinal preparations, they can also be included in the poultry diet as feed additives to utilise their benefits in improving performance, immune response and health of broiler chickens (Mehala and Moorthy, 2008; Darabighanea et al., 2012; Safamehr et al., 2012; Bahadori and Irani, 2013).

However, there are limitations on direct usage of the extracts in poultry diet. These limitations are related to hydrophobic, highly active, reactive and volatile properties of bioactive compounds in the extracts. In presence of air oxygen, the compounds are also sensitive to peroxidation and oxidative damage. Therefore, these undesirable reactions decrease efficiency of the bioactive compounds and also change natural odour and taste of the extracts. Conversely, using high amounts of the extracts directly in the diet decreases palatability and feed intake due to change in feed taste (Lee et al., 2003). With the addition of a protective layer, these substances and their properties are retained.

Even though the concepts of micro-encapsulation or nanoencapsulation and controlled release are relatively old, attention to these concepts with respect to plant extracts has only been recent. Nanocapsules exhibit a membrane-wall structure and bioactive substances are entrapped in the core or adsorbed onto their exterior. Most nanocapsules prepared from water-insoluble polymers involve heat, organic solvent or high shear force that can be harmful to the effective material stability (Allemann et al., 1993). In contrast, water-soluble polymers offer mild and simple preparation methods without the use of organic solvent and high shear force. Among available water-soluble polymers, chitosan (a polyglucosamine derived from chitin), a cellulose-like polymer located in the exoskeletons of arthropods such as crabs, shrimps, lobsters and insects (Furda, 1983), is one of the most extensively studied. This is because chitosan possesses some ideal properties of polymeric carriers for nanoparticles, such as biocompatibility, biodegradability, no toxicity, and low cost (Hejazi and Amiji, 2003). Furthermore, it possesses a positive charge and exhibits absorption enhancing effects. These properties render chitosan a very attractive material as a drug or effective material delivery carrier (LeHoux and Grondin, 1993). The positive charge in chitosan molecules also decreases possession of negative charge due to bacteria growth (Razdan and Petterson, 1994; Tiyaboonchai, 2003).

Therefore, the present study was an effort to utilise nanoencapsulated extracts of aloe vera, dill and nettle root as natural growth promoters and effective substitutions for feed antibiotics in broiler chickens. Our preliminary experiment (unpublished) shows that increasing the herbal extract concentration over the experimental period improves broiler performance. This might be due to its adaptation and/or the birds' increasing requirement for herbal extract to improve their performance. Thus, in order to overcome this requirement, herbal extract concentration was raised over the course of the experiment.

2 Materials and methods

Herbal extracts were purchased from the traditional medicine department of the Barij Essence Pharmaceutical Company (Kashan, Iran). Ethanol and acetic acid were provided by Merck. The medium molecular weight chitosan, derived from crab shell and sodium triphosphate pentabasic (TPP), was purchased from Sigma-Aldrich. All the other reagents used in the experiment were of analytical grade.

2.1 Nanoencapsulation of herbal extracts

Nanoencapsulation of the herbal extract was performed by ionic gelation according to the procedure of Stoica et al. (2013). Chitosan was dissolved at a concentration of $1\,\mathrm{mg\,mL^{-1}}$ in $1\,\%$ (w/v) acetic acid and sonicated before the solution became transparent. The dropwise addition of $10\,\mathrm{mL}$ TPP solution $(1\,\mathrm{mg\,mL^{-1}})$ to a $25\,\mathrm{mL}$ CS solution $(\mathrm{pH}=5)$, under constant stirring at room temperature, produced the formation of CS-TPP nanoparticles by ionic gelation. For the preparation of CS-TPP nanoparticles loaded with herbal extracts, $20\,\%$ (w/v) extracts were added to the chitosan solution before adding the TPP solution.

2.2 Management and experimental design

A total of 240 1-day-old broiler chicks (Ross 308) were procured from a local hatchery and transferred to a research poultry house. At 1 day of age, the chicks were individually weighed and randomly divided into 24 pens $(100 \times 120 \times 75\,\mathrm{cm})$ with an equal initial weight of $132 \pm 2\,\mathrm{g}$. The pens were distributed into eight treatments of three replicates with 10 chicks in each under a completely randomised design. The chicks were reared in the litter system under standard management conditions in three experimental (starter, grower and finisher) periods until the sixth week. The basal diet was formulated according to the standards prescribed in Ross 308 Broiler Nutrition Specification, June 2007. The starter, grower and finisher diets presented in Table 1 were fed ad libitum to the birds from 1–14, 15–28 and 29–42 days of age respectively.

2.3 Experimental diets

The eight dietary treatments consisted of a negative control diet (basal diet without additives), a positive control diet (basal diet plus 500 g of antibiotic per tonne), three diets containing non-encapsulated extracts of aloe vera, dill or nettle root, and three diets containing nanoencapsulated extracts of aloe vera, dill or nettle root. In each experimental period, non-encapsulated and nanoencapsulated extracts of aloe vera, dill and nettle root were added in the amount of 0.02,

Table 1. Ingredients and chemical composition of basal diets in different periods.

Ingredients	Starter	Grower	Finisher
Corn	57.00	60.41	64.76
Soybean meal	37.14	34.22	29.7
Soybean oil	1.20	1.6	2
$CaCO_3$	1.26	1.05	1.02
Dicalcium phosphate	1.8	1.52	1.46
Common salt	0.21	0.2	0.2
$NaHCO_3$	0.16	0.17	0.17
DL-methionine	0.29	0.2	0.14
HCl-lysine	0.24	0.1	0.05
L-threonine	0.10	0.03	0
Vitamin & mineral premix[1]	0.6	0.5	0.5
Sum	100	100	100

Chemical composition

	Starter	Grower	Finisher
Metabolisable energy ($kcal\,kg^{-1}$)	2900	2970	3040
Crude protein (%)	21.09	19.8	18.05
Methionine (%)	0.45	0.4	0.35
Methionine + cysteine (%)	0.9	0.79	0.69
Lysine (%)	1.22	1.04	0.89
Threonine (%)	0.8	0.69	0.6
Tryptophan (%)	0.19	0.17	0.14
Arginine (%)	1.26	1.07	0.94
Calcium (%)	1.01	0.85	0.81
Available phosphorous (%)	0.48	0.42	0.4
Na (%)	0.15	0.15	0.15
Linoleic acid (%)	1.2	1.13	0.95

[1] The premix includes A: 10 000 IU, D_3: 5000, E: 50 IU, K: 3 mg, B_1: 2 mg, B2: 6 mg, niacin: 60 mg, pantothenic acid: 15 mg, B_6: 3 mg, biotin: 0.1 mg, folic acid: 1.75 mg, B_{12}: 0.016; Cu: 16 mg, I: 1.26, Fe: 40 mg, Mn: 120 mg, Se: 0.3 mg, and Zn: 100 mg.

0.025 and 0.05 % to starter, grower and finisher diets (Table 1) respectively.

2.4 Collection of data

Data on body weight and feed consumption were recorded at weekly intervals and mortality was recorded at occurrence. From the data above, body weight gain (BWG) and feed consumption per bird and feed conversion ratio were calculated.

2.5 Sample preparations and measurements

The blood samples collected from jugular veins (nine chicks per treatment) at 35 and 42 days of age were allowed to clot at a room temperature of $28 \pm 10\,°C$ for 2 h and were centrifuged at 3000 rpm for 5 min. The sera were then separated and frozen at $-40\,°C$ before assay. Serum levels of immunoglobulins (IgY and IgM) were determined by the ELISA (enzyme-linked immunosorbent assay) technique using a commercial kit of IgY and IgM (Bethyl Laboratory, Inc., USA) following manufacturer instructions with slight modification as described by Hamal et al. (2006). The samples were analysed in triplicate. At the end of the experimen-

tal period, after blood sampling, the birds were slaughtered following the standard procedures for stunning, exsanguination and de-feathering. Thereafter, carcass, breast, thigh, liver and intestines were immediately removed and weighed. The data were expressed as the percentage of total carcass weight.

2.6 Statistical analysis

The data collected on various parameters were subjected to statistical analysis using the GLM (general linear model) procedure of SAS software (SAS Institute, 1990) based on completely randomised design with three replications. The Duncan's test was used to elucidate differences between treatment means, with the 0.05 % level considered significant. Results are reported as the mean and pooled standard error of mean (SEM).

3 Results and discussion

Table 2 represents the effect of free herbal extracts and their nanoencapsulated counterparts in broiler diet on BWG. Results indicated that the addition of herbal extracts freely and nanoencapsulated in diet did not significantly ($P > 0.05$) affect BWG in the time periods of 1–14 and 14–28 days. Thereafter, however, the improved BWG in the time periods of 28–42 days and consequently 1–42 days was probably due to increase of both non-encapsulated and nanoencapsulated herbal extracts to 0.05 % in finisher diets. In these periods, birds fed a diet containing nanoencapsulated dill extract had a significantly ($P < 0.05$) higher BWG compared with the antibiotic group, while non-encapsulated dill extract treatment was intermediate. This study is the first report to evaluate the nanoencapsulated form of herbal extract in broiler diet. Previous studies using aloe vera extract (Swaim et al., 1992) and dill powder (Bahadori and Irani, 2013) in broiler chicken diet demonstrated higher BWG compared to the control. However, Mehala and Moorthy (2008) showed a total lack of impact on body weight gain when using different levels of aloe vera powder in the broiler chickens' diet during the whole experimental period. It is possible that the significant variability in the above-mentioned results is the combined effect of differences in dose, duration and processing of the medicinal plants, essential oils, and/or the different husbandry conditions. However, in the present study, effective delivery of dill extract via its nanoencapsulation and the antimicrobial properties of chitosan nanoparticles used to make capsules caused the higher body weight gain of broiler chickens as compared with the antibiotic group.

As indicated in Table 3, there was no significant difference in feed intake between treatment groups in different periods ($P > 0.05$). However, in the total experimental period (1–42 days), the highest and lowest amount of feed intake in chicks was obtained numerically ($P = 0.111$) by adding nanoencapsulated and non-encapsulated extract of nettle root to the diet, respectively. Bird feed intake responses to a diet containing

Table 2. Effect of dietary treatments on body weight gain (g) in broiler chickens.

Additive type	Experimental periods (days)			
	1–14	14–28	28–42	1–42
Negative control (no additive)	546	934	1060[c]	2540[c]
Positive control (antibiotic)	536	940	1136[bc]	2613[bc]
Non-encapsulated extract of nettle root	534	946	1100[c]	2580[bc]
Non-encapsulated extract of dill	562	938	1210[abc]	2710[abc]
Non-encapsulated extract of aloe vera	528	971	1177[abc]	2676[bc]
Nanoencaps. extract of nettle root	557	908	1293[ab]	2758[ab]
Nanoencaps. extract of dill	569	984	1317[a]	2870[a]
Nanoencaps. extract of aloe vera	554	919	1200[abc]	2673[bc]
P value	0.626	0.941	0.027	0.0127
SEM[1]	16.7	45.5	50.0	53.8

[a-c] In each column, means with different superscripts differ significantly ($P < 0.05$). [1] SEM: pooled standard error of mean.

Table 3. Effect of dietary treatments on feed intake (g) and feed conversion ratio in broiler chickens.

Additive type	Experimental periods (days)							
	1–42		14–28		28–42		1–14	
	FI[a]	FCR	FI	FCR	FI	FCR	FI	FCR
Negative control (no additive)	886	1.62	1647	1.76	2179	2.06[a]	4740	1.87
Positive control (antibiotic)	905	1.69	1552	1.65	2187	1.92[abc]	4643	1.78
Non-encapsulated extract of nettle root	883	1.65	1577	1.67	2153	1.96[ab]	4613	1.79
Non-encapsulated extract of dill	935	1.66	1673	1.78	2126	1.76[cd]	4734	1.75
Non-encapsulated extract of aloe vera	938	1.78	1672	1.72	2204	1.87[bcd]	4813	1.80
Nanoencaps. extract of nettle root	931	1.67	1623	1.79	2233	1.73[d]	4887	1.77
Nanoencaps. extract of dill	933	1.64	1633	1.66	2313	1.76[cd]	4880	1.70
Nanoencaps. extract of aloe vera	940	1.70	1629	1.77	2115	1.76[cd]	4684	1.75
P value	0.222	0.76	0.723	0.575	0.284	0.003	0.111	0.128
SEM[1]	19.2	0.059	68.7	0.092	54.6	0.058	71.9	0.034

[a-d] In each column, means with different superscripts differ significantly ($P < 0.05$). [1] SEM: pooled standard error of mean.

a herbal plant and its derivatives differ depending on extract dosage and the types of herbal plant. For example, no significant difference in mean feed intake was found among groups administered extract of aloe vera gel (Durrani et al., 2008), *Curcuma longa* (Ismail et al., 2004), *Berberis lycium* (Chand et al., 2005), and a mixture of cinnamon, oregano and pepper (Hernandez et al., 2004). Conversely, Durrani et al. (2007) on *Mentha longifolia* and Guo et al. (2004) on Chinese herb reported significant differences in feed consumption of broiler chickens fed various herbal extracts and antibiotics. In this study, numerically higher feed intake of nanoencapsulated treatments is probably due to the use of a very low concentration of herbal extracts in preparation of the formula.

With respect to Table 3, the effect of all tested extracts was not significant ($P > 0.05$) on FCR (feed conversion ratio) in broiler chickens during different experimental periods except for the 28–42-day period. In the 28–42-day period, feeding birds with all types of nanoencapsulated extracts and also non-encapsulated extracts of dill and aloe vera significantly improved FCR compared with the control group. However, in comparison with antibiotic treatment, only the administration of the nanoencapsulated form of nettle was significant ($P < 0.05$). Similar to growth, bird FCR responses to herbal extract supplementation are still controversial. However, the most important mechanisms of plant extracts for improving FCR are their digestion-stimulating properties and antimicrobial effects. As mentioned above, nanoencapsulation could develop these properties via effective delivery of the extracts and the antimicrobial properties of chitosan (Jain et al., 2008; Safamehr et al., 2013).

Table 4 shows the effect on broiler carcass traits of using non-encapsulated and nanoencapsulated extracts in the diet.

Table 4. Effect of dietary treatments on carcass traits in broiler chickens (expressed as a percentage)[1].

Additive type	Carcass	Breast	Thigh	Liver	Intestine	G. tract
Negative control (no additive)	58.5	42.5	30.6	2.67	6.79	15.27
Positive control (antibiotic)	62.7	44.1	30.4	2.87	6.14	13.69
Non-encapsulated extract of nettle root	59.0	44.0	30.5	2.70	6.83	14.43
Non-encapsulated extract of dill	60.5	44.9	30.3	4.07	6.54	14.30
Non-encapsulated extract of aloe vera	62.3	44.0	30.5	4.10	6.61	15.57
Nanoencaps. extract of nettle root	62.9	44.2	31.3	4.02	7.19	14.95
Nanoencaps. extract of dill	62.5	46.3	29.7	3.57	7.35	15.82
Nanoencaps. extract of aloe vera	59.3	46.2	32.6	3.83	6.26	14.04
P value	0.571	0.342	0.388	0.177	0.943	0.805
SEM[2]	7.62	1.13	0.81	0.829	0.761	1.038

[1] Means represent nine chicks per treatment. [2] SEM: pooled standard error of mean.

Table 5. Effect of dietary treatments on serum immunoglobulin (IgM and IgY) concentrations on the 35th and 42nd days[1].

Additive type	35 days		42 days	
	IgY ($\mu g\,mL^{-1}$)	IgM ($\mu g\,mL^{-1}$)	IgY ($\mu g\,mL^{-1}$)	IgM ($\mu g\,mL^{-1}$)
Negative control (no additive)	382[c]	117[d]	416[e]	153[e]
Positive control (antibiotic)	403[c]	134[c]	435[ed]	166[de]
Non-encapsulated extract of nettle root	435[b]	145[bc]	465[c]	175[cd]
Non-encapsulated extract of dill	425[b]	155[ab]	450[cd]	188[abc]
Non-encapsulated extract of aloe vera	426[b]	146[bc]	461[cd]	181[cd]
Nanoencaps. extract of nettle root	466[a]	162[a]	494[b]	186[bc]
Nanoencaps. extract of dill	479[a]	165[a]	518[ab]	203[a]
Nanoencaps. extract of aloe vera	483[a]	163[a]	527[a]	199[ab]
P value	0.001	0.0001	0.0001	0.0001
SEM[2]	7.2	4.1	8.9	5.1

[a-d] In each column, means with different superscripts differ significantly ($P < 0.05$). [1] Antibody levels are means based on nine chickens per treatment and triplicate analysis of each. [2] SEM: pooled standard error of mean.

The results revealed that there was no significant ($P > 0.05$) effect on different carcass traits and relative organ weight, including percentages of carcass, breast, thigh, liver, intestines and total gastrointestinal tract. Findings of the present study are in agreement with Ismail et al. (2004) and Durrani et al. (2008), who reported that abdominal fat deposition, breast, thigh, gizzard, liver, and heart percentages of broiler chickens were not affected by aloe vera extract. However, addition of all nanoencapsulated and non-encapsulated extracts to the diet numerically increased liver weight compared to both the control and antibiotic groups. A previous study by Debersac et al. (2001) showed that water-soluble extract enhanced hepatic metabolism and increased relative liver weight in rats. The highest breast percentage was also obtained numerically by administering nanoencapsulated extracts of dill and aloe vera.

Table 5 provides the results obtained from the analysis of serum immunoglobulin (IgM and IgY) concentrations on the 35th and 42nd days. Compared with other treatments, admin-istering nanoencapsulated form of the extracts significantly ($P < 0.05$) enhanced IgY and IgM at 35 and 42 days of age. These findings support previous research into this area of the brain and links aloe vera extract and humoral immunity (Akhtar et al., 2012). The authors reported that oral admin-istration of aloe vera extract resulted in higher numbers of anti-SRBC antibody (total Igs, IgM and IgG) titers. They at-tributed this activity to the stimulation of a humoral response to aloeride (a polysaccharide from aloe vera), which induces IL-6, a potent B-cell stimulant, to produce antibodies (Tan and Vanitha, 2004).

4 Conclusions

In conclusion, the results of the present study demon-strated that comparable to antibiotic treatment, the tested ex-tracts, except for non-encapsulated nettle root could improve growth performance in broiler chickens. However, nanoen-capsulation of herbal extracts with chitosan probably profoundly improves immune responses and increases broiler performance via a better delivery system and stimulation along with antimicrobial effects of chitosan nanoparticles. Further studies on nanoencapsulation of bioactive components and their influence on broiler chickens and other farm animals are underway in our lab.

Competing interests. The authors declare that they have no conflict of interest.

Acknowledgement. The authors gratefully acknowledge the financial support of the National Institute of Genetic Engineering and Biotechnology.

Edited by: M. Mielenz

References

Akhtar, M., Hai, A., Awais, M. M., Iqbal, Z., Muhammad, F., ul Haq, A., and Anwar, M. I.: Immunostimulatory and protective effects of aloe vera against coccidiosis in industrial broiler chickens, Vet. Parasitol., 186, 170–177, 2012.

Allemann, E., Gurny, R., and Doelker, E.: Drug-loaded nanoparticles-preparation methods and drug targeting issues, Eur. J. Pharm. Biopharm., 39, 173–191, 1993.

Bahadori, M. M. and Irani, M.: The Effects of Dill Powder in Diet on Some Blood Metabolites, Carcass Characteristics and Broiler Performance, Glob. Vet., 10, 500–504, 2013.

Chand, N., Durrani, F. R., Mian, M. A., and Durrani, Z.: Effect of different levels of feed added *Berberis lycium* on the performance of broiler chicks, Int. J. Biotechnol., 2, 971–974, 2005.

Darabighanea, B., Zareia, A., and Zare-Shahneh, A.: The effects of different levels of aloe vera gel on ileum microflora population and immune response in broilers: a comparison to antibiotic effects, J. Appl. Anim. Res., 40, 31–36, 2012.

Debersac, P., Vernevaut, M. F., Amiot, M. J., Suschetet, M., and Siess, M. H.: Effects of a water-soluble extract of rosemary and its purified component rosmarinic acid on xenobiotic metabolizing enzymes in rat liver, Food Chem. Toxicol., 29, 109–117, 2001.

Dibner, J. J. and Richards, J. D.: Antibiotic growth promoters in agriculture: History and mode of action, Poult. Sci., 84, 634–643, 2005.

Durrani, F. R., Sultan, A., Marri, M. L., Chand, N., and Durrani, Z.: Effect of Wild Mint (*Mentha longifolia*) Infusion on the Overall Performance of Broiler Chicks, Pak. J. Biol. Sci., 10, 1130–1133, 2007.

Durrani, F. R., Ullah, S., Chand, N., Durrani, Z., and Akhtar, S.: Using aqueous extract of aloe vera gel as anticoccidial and immunotimulant agent in broiler production, Sarhad J. Agri., 24, 665–669, 2008.

Furda, I.: Aminopolysaccharides – their potential as dietary fiber, in: unconventional sources of dietary fiber, edited by: Furda, I., ACS Symposium Series, 214, 105–122, 1983.

Guo, F. C., Kwakkel, R. P., Soede, J., Williams, B. A., and Verstegen, M. W.: Effect of a Chinese herb medicine formulation, as an alternative for antibiotics, on performance of broilers, Br. Poult. Sci., 45, 793–797, 2004.

Hamal, K. R., Burgess, S. C., Pevzner, I. Y., and Erf, G. F.: Maternal Antibody Transfer from Dams to Their Egg Yolks, Egg Whites, and Chicks in Meat Lines of Chickens, Poult. Sci., 85, 1364–1372, 2006.

Hejazi, R. and Amiji, M.: Chitosan-based gastro intestinal delivery systems, J. Control. Release, 89, 151–165, 2003.

Hernandez, F., Madrid, J., Garcia, V., Orengo, J., and Megias, M. D.: Influence of two plant extract on broiler performance, digestibility, and digestive organ size, Poult. Sci., 83, 169–174, 2004.

Hughes, P. and Heritage, J.: Antibiotic growth-promoters in food animals, Assessing quality and safety of animal feeds, available at: http://ftp.fao.org/docrep/fao/007/y5159e/y5159e05.pdf, 2007.

Ismail, M., Durrani, F.R., Amjad, M.,Suhail, S. M., and Chand, N.: Effect of different levels of feed added Curcuma longa on overall performance of broiler chicks, J. Agric. Biol. Sci., 1, 1–16, 2004.

Jain, M., Ganju, L., Katiyal, A., Padwad, Y., Mishra, K. P., Chanda, S., Karan, D., Yogendra, K. M., and Sawhney, R. C.: Effect of Hippophae rhamnoides leaf extract against Dengue virus infection in human blood-derived macrophages, Phytomedical., 15, 793–799, 2008.

Lee, K. W., Everts, H., Kappert, H. J., Frehner, M., Losa, R., and Beynen, A. C.: Effects of dietary essential oil components on growth performance, digestive enzymes and lipid metabolism in female broiler chickens, Br. Poult. Sci., 44, 450–457, 2003.

LeHoux, J. G. and Grondin, F.: Some effects of chitosan on liver function in the rat, Endocrinol., 132, 1078–1084, 1993.

Mehala, C. and Moorthy, M.: Production Performance of Broilers Fed with *Aloe vera* and *Curcuma longa* (Turmeric), Int. J. Poult. Sci., 7, 852–856, 2008.

Razdan, A. and Pettersson, D.: Effect of chitin and chitosan on nutrient digestibility and plasma lipid concentrations in broiler chickens, Br. J. Nut., 72, 211–288, 1994.

Safamehr, A., Mirahmadi, M., and Nobakht, A.: Effect of nettle (Urtica dioica) medicinal plant on growth performance, immune responses, and serum biochemical parameters of broiler chickens, Int. Res. J. Appl. Basic Sci., 3, 721–728, 2012.

Safamehr, A., Fallah, F., and Nobakht, A.: Growth performance and biochemical parameters of broiler chickens on diets consist of chicory (*Cichorium intybus*) and nettle (*Urtica dioica*) with or without muti-enzyme, Iranian J. Appl. Ani. Sci., 3, 131–137, 2013.

SAS Institute: SAS Stat User's Guide Release 6.08, SAS Institute Inc., Cary, NC, 1990.

Stoica, R., Şomoghi, R., and Ion, R. M.: Preparation of chitosan – tripolyphosphate nanoparticles for the encapsulation of polyphenols extracted form rose hips, Dig. J. Nano. Biostruct. P, 955–963, 2013.

Swaim, S. F., Riddell, K. P., and McGuire, J. A.: Effects of tropical medications on the healing of open pad wounds in dogs, J. Am. Anim. Hosp. Associate, 28, 499–502, 1992.

Tan, B. K. H. and Vanitha, J.: Immunomodulatory and antimicrobial effects of some traditional Chinese medicinal herbs: a review, Curr. Med. Chem., 11, 1423–1430, 2004.

Tiyaboonchai, W.: Chitosan Nanoparticles: A Promising System for Drug Delivery, Naresuan Uni. J., 11, 51–66, 2003.

7

Identification and expression analysis of *miR-144-5p* and *miR-130b-5p* in dairy cattle

Zhixiong Li[1,2], Hongliang Wang[3], Ling Chen[1], Mengxing Zhai[1], Si Chen[1], Na Li[1], and Xiaolin Liu[1]

[1]College of Animal Science and Technology, Northwest A&F University, Shaanxi Key Laboratory of Molecular Biology for Agriculture, Yangling, Shaanxi 712100, P.R. China
[2]College of Life Science and Technology, Southwest University for Nationalities, Chengdu 610041, P.R. China
[3]State Key Laboratory for Molecular Biology of Special Economic Animals, Institute of Special Animal and Plant Sciences, Chinese Academy of Agricultural Sciences, Changchun 130112, P.R. China

Correspondence to: Xiaolin Liu (liuxiaolin@nwsuaf.edu.cn)

Abstract. MicroRNAs (miRNAs) can coordinate the main pathways involved in innate and adaptive immune responses by regulating gene expression. To explore the resistance to mastitis in cows, *miR-144-5p* and *miR-130b-5p* were identified in bovine mammary gland tissue and 14 potential target genes belonging to the chemokine signaling pathway, the arginine and proline metabolism pathway and the mRNA surveillance pathway were predicted. Subsequently, we estimated the relative expression of *miR-144-5p* and *miR-130b-5p* in cow mammary tissues by using stem-loop quantitative real-time polymerase chain reaction. The results showed that the relative expression of *miR-144-5p* and *miR-130b-5p* in the mastitis-infected mammary tissues ($n = 5$) was significantly downregulated 0.14-fold ($p < 0.01$) and upregulated 3.34-fold ($p < 0.01$), respectively, compared to healthy tissues ($n = 5$). Our findings reveal that *miR-144-5p* and *miR-130b-5p* may have important roles in resistance to mastitis in dairy cattle.

1 Introduction

Bovine mastitis, defined as "an inflammation of the mammary gland", is a prevalent and complex infectious disease affected by genetics and pathogens that can result in significant dairy cattle losses (Nash et al., 2003). Mastitis can be caused by many bacteria, including *Staphylococcus aureus* and *Escherichia coli*. The primary defense against pathogens relies on the appropriate expression of antigen-presenting molecules triggering the release of effector molecules in the innate immune system. The immune system, as the central host determinant for dictating the outcome of intramammary infection, can defend against in-breaking pathogens as the first line once the pathogens penetrate the physical barrier (Bannerman et al., 2009).

Recent studies have shown that microRNAs (miRNAs) play important roles in regulating and modulating innate and adaptive immune responses (Zhou et al., 2012; Gu et al., 2012). Mature miRNAs are a class of small non-coding RNA molecules that are ~ 22 nucleotides (nt) long processed from ~ 70 nt long precursor miRNAs (pre-miRNAs) that form hairpin secondary structures and are evolutionarily conserved (Bartel et al., 2004; Cullen et al., 2004; Kim et al., 2005). miRNAs are post-transcriptional regulators that inhibit the translation or induce the degradation of protein-coding protein mRNAs that contain complementary sequences to miRNAs (Berezikov, 2011; Bartel et al., 2009).

miR-144 plays a crucial role in hemoglobin synthesis during primitive erythropoiesis and is associated with anemia severity in sickle-cell diseases (Fu et al., 2009; Sangokoya et al., 2010). *miR-130b* inhibits cell proliferation and invasion in pancreatic cancer through targeting *STAT3* and targets *DICER1* for aggression in endometrial cancer (Zhao et al., 2013; Li et al., 2013). *miR-130b* is associated with poor prognosis in colorectal cancer and is a prognostic marker (Colangelo et al., 2013). Our previous research showed two differentially expressed miRNAs matched to bta-mir-144 and bta-mir-130b that were detected in the peripheral blood of healthy and mastitis-infected dairy cattle (Li et al., 2014a).

Table 1. The primer sequences of the stem-loop qPCR experiments.

miRNA name	Primer	Primer sequence	Product size (bp)
miR-144-5p	Loop FW primer RW primer	GTCGTATCCAGTGCAGGGTCCGAGGTATTCGCACTGGATACGAC<u>CTTACAGT</u> CCG<u>GGATATCATCATATACTGTAAG</u> GTGCAGGGTCCGAGGT	59
miR-130b-5p	Loop FW primer RW primer	GTCGTATCCAGTGCAGGGTCCGAGGTATTCGCACTGGATACGAC<u>AGTAGTGC</u> CGG<u>ACTCTTTCCCTGTTGCACTACT</u> GTGCAGGGTCCGAGGT	59
18S-snRNA	FW primer RW primer	GTGGTGTTGAGGAAAGCAGACA TGATCACACGTTCCACCTCATC	79

Note: the underlined letters are the sequences from *miR-144-5p* and *miR-130b-5p*.

However, our knowledge of the differential expression of the two miRNAs in cattle mastitis resistance remains largely limited. The two miRNAs may have important roles in the development of the immune system against pathological changes in mammary tissue in dairy cattle.

The aims of this study were (1) to investigate whether *miR-144-5p* and *miR-130b-5p* are present in bovine tissues, (2) to predict the target genes *miR-144-5p* and *miR-130b-5p* and (3) to analyze whether *miR-144-5p* and *miR-130b-5p* are differentially expressed in healthy and mastitis-infected mammary tissues.

2   Materials and methods

2.1   Animal samples

Samples were collected from five healthy and five mastitis-infected Chinese Holstein cows of first lactation from a commercial bovine slaughter farm. The selection of mastitis-infected cows was carried out as previously described (Li et al., 2014b). A part of the 10 mammary tissue samples was collected and stored in liquid nitrogen for RNA isolation; others were used for the identification of the pathogen. All 10 mammary tissue samples were used for analysis of the expression profile of *miR-144-5p* and *miR-130b-5p*. The liver, heart, lung, kidney and spleen from five healthy samples were used to analyze the expression pattern. This study was approved by the Northwest A&F University Animal Care and Use Committee.

2.2   RNA extraction and cDNA synthesis

Total RNA was extracted using Trizol Reagent (Invitrogen, USA) following the manufacturer protocol. RNA purity was verified by measuring the absorbance at 260 and 280 nm with an ND-1000 spectrophotometer (NanoDrop Technologies, USA). First-strand cDNA synthesis was performed in a 20 μL volume using a PrimeScript RT reagent kit (Takara, Japan) following the manufacturer protocol with a specific

stem-loop primer. The primers for the reverse transcription polymerase chain reaction (RT-PCR) are shown in Table 1.

To identify *miR-144-5p* and *miR-130b-5p* expressed in the mammary tissue of dairy cattle, primers were designed according to the sequences previously detected (Li et al., 2014). PCR was performed in a total volume of 25 μL containing 50 ng of cDNA, 2.5 μL $10 \times$ PCR buffer, 2.1 mM $MgCl_2$, 0.1 mM dNTPs, 0.25 mM of each primer, 0.2 μL Easy Taq DNA polymerase and ddH$_2$O run for 32 cycles at 95° for 40 s, 60° for 30 s and 72° for 30 s, followed by incubation at 72° for 10 min. PCR products were ligated into the T-Vector pMD19 (Takara, Japan) after gel extraction and then transformed into competent *E. coli* DH5α. Finally, 10 randomly selected positive clones were sequenced. The experiment was repeated twice to confirm the result.

2.3   Sequence analysis

Sequence alignment was performed to verify *miR-144-5p* and *miR-130b-5p* using DNAman (version 6.0) software. To obtain the potential target genes of *miR-144-5p* and *miR-130b-5p*, the prediction of the target gene was performed with MIREAP software. The predicted target genes were classified by KEGG functional annotations; the identified pathways were actively regulated by *miR-144-5p* and *miR-130b-5p* in healthy and mastitis-infected dairy cattle.

2.4   Quantitative analysis of miRNAs

Stem-loop quantitative real-time polymerase chain reaction (stem-loop qPCR) was used to analyze miRNAs according to Chen et al. (2005). The stem-loop qPCR was performed in the Bio-Rad CFX96 Real-Time PCR Detection System using the SYBR Green PCR kit (Takara, Japan) according to the manufacturer instructions. 18S rRNA was used as the reference gene in the stem-loop qPCR detection of bovine miRNAs, and all reactions were run in triplicate. The relative expression level of *miR-144-5p* and *miR-130b-5p* was calculated according to the method of Livak and

(a)

G G A T A T C A T C A T A T A C T G T A A G

(b)

A C T C T T T C C C T G T T G C A C T A C T

(c)

```
bta-miR-144   ------------------------------------------------TACAGTATAGA
miR-144-5p    -----------GGATATCATCATATACTGTAAG----------------------
bta-mir-144   GGGCCCTGACTGGGATATCATCATATACTGTAAGTTTGCAACGAGACACTACAGTATAGA

bta-miR-144   TGATGTACTAG-------------
miR-144-5p    -----------------------
bta-mir-144   TGATGTACTAGTCCGGGTGCTCCC
```

(d)

```
miR-130b-5p   ------------ACTCTTTCCCTGTTGCACTACT----------------------------
bta-mir-130b  GGCCTGCCTGACACTCTTTCCCTGTTGCACTACTGTGCGCCCCTGGCAAGCAGTGCAATG
bta-miR-130b  ------------------------------------------------CAGTGCAATG

miR-130b-5p   --------------------
bta-mir-130b  ATGAAAGGGCATCGGTCAGGCC
bta-miR-130b  ATGAAAGGGCAT----------
```

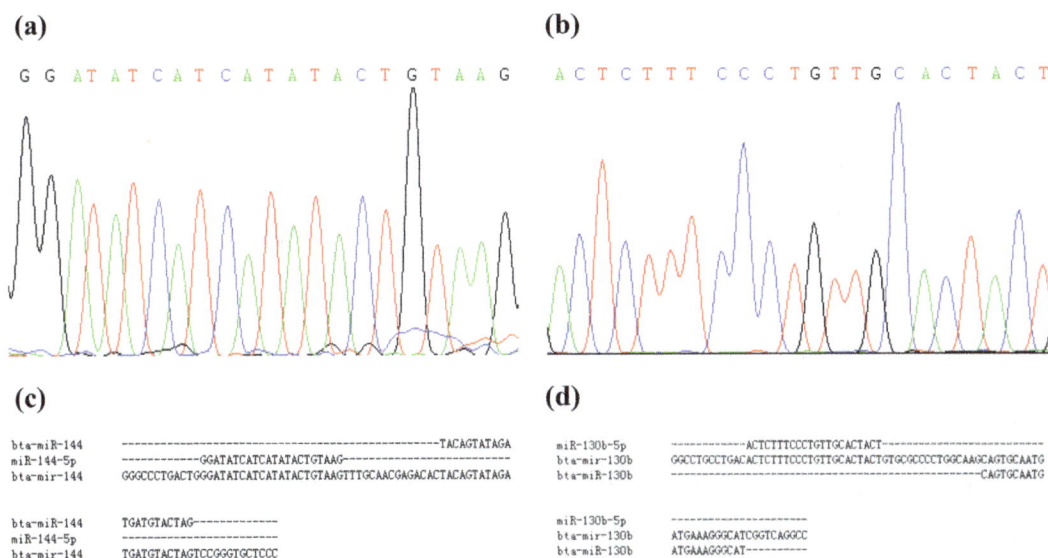

Figure 1. (a) Sequencing chromatograms for *miR-144-5p*. (b) Sequencing chromatograms for *miR-130b-5p*. (c) The bta-mir-144 sequence comparison with *bta-miR-144* and *miR-144-5p*. (d) The bta-mir-130b sequence comparison with *bta-miR-130b* and *miR-130b-5p*.

Schmittgen (2001). The primers for the qPCR are shown in Table 1.

2.5 Statistical analysis

The value of the relative quantity was presented as fold change. The means of two groups were compared by a Student's paired-samples t test. The analysis was performed with SPSS software (version 20.0); $p < 0.05$ was regarded as significant.

3 Results

3.1 Identification of *miR-144-5p* and *miR-130b-5p*

RT-PCR and sequencing were used to identify *miR-144-5p* and *miR-130b-5p* with specific primers (Fig. 1a and b). The miRNAs of *bta-miR-144* and *bta-miR-130b* from miR-Base (http://www.mirbase.org/) are matched to the 3′ of bta-mir-144 and bta-mir-130b. The cloned miRNAs are totally matched to the 5′ of bta-mir-144 and bta-mir-130b (Fig. 1c and d), so we called them *miR-144-5p* and *miR-130b-5p*.

3.2 Target gene prediction of *miR-144-5p* and *miR-130b-5p*

Twenty potential target genes of *miR-144-5p* and *miR-130b-5p* were predicted using MIREAP software (Table 2). Ten target genes of *miR-144-5p* and *miR-130b-5p* belong to the chemokine signaling pathway, which plays an important role in inflammatory responses and cancer (Coussens and Werb, 2002; Charo and Ransohoff, 2006; Baggiolini and Loetscher, 2000). Six target genes of *miR-144-5p* and *miR-130b-5p* belong to the arginine and proline metabolism pathway, which

Table 2. Target genes of *miR-144-5p* and *miR-130b-5p*.

miRNA name	Gene name	KEGG pathway name
miR-144-5p	CXCL2 CRK GNB5	Chemokine signaling pathway
	NOS2 ARG1	Arginine and proline metabolism
miR-130b-5p	AMD1 SAT2 ARG1 GLS2	Arginine and proline metabolism
	PPP2R2B PPP2CB SMG1 SAP18	mRNA surveillance pathway
	CHUK CXCL2 CXCL6 GNAT2 CCL11 NRAS BRAF	Chemokine signaling pathway

is closely involved in conceptus metabolism, growth and development. Four target genes of *miR-130b-5p* belong to the mRNA surveillance pathway, which ensures the viability and quality of mRNA.

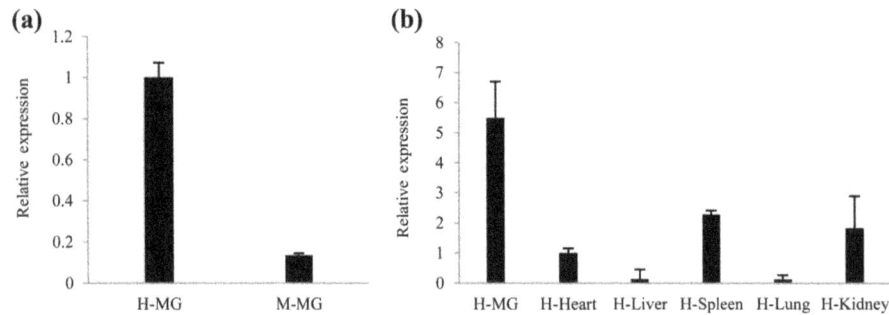

Figure 2. Relative expression of *miR-144-5p*. **(a)** Relative expression of *miR-144-5p* in healthy and mastitis-infected mammary gland tissues using qPCR. H-MG denotes healthy mammary gland and M-MG denotes mastitis-infected mammary gland. **(b)** Relative expression of *miR-144-5p* in a variety of healthy tissues using qPCR. H denotes the healthy cow group. The vertical bar represents the standard error.

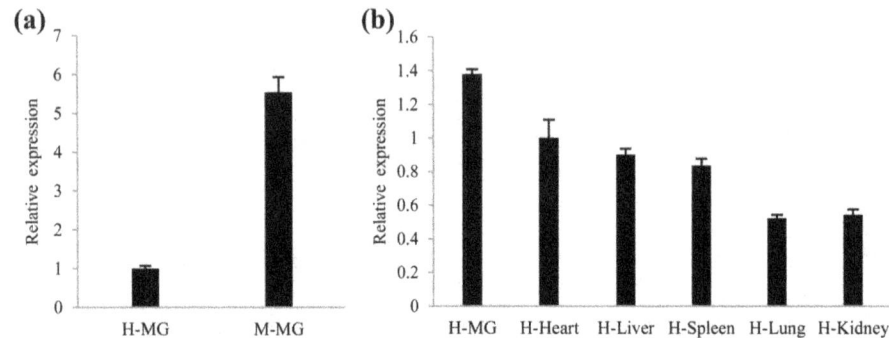

Figure 3. Relative expression of *miR-130b-5p*. **(a)** Relative expression of *miR-130b-5p* in healthy and mastitis-infected mammary gland tissues using qPCR. H-MG denotes healthy mammary gland and M-MG denotes mastitis-infected mammary gland. **(b)** Relative expression of *miR-130b-5p* in a variety of healthy tissues using qPCR. H denotes the healthy cow group. The vertical bar represents the standard error.

3.3 Expression of *miR-144-5p* and *miR-130b-5p* in healthy and mastitic cow tissues

The expression of *miR-144-5p* and *miR-130b-5p* in healthy and mastitis-infected mammary tissues was investigated using stem-loop qPCR. A lower expression of *miR-144-5p* was observed in the mastitis-infected mammary tissues compared to that in the healthy samples (Fig. 2a; $p < 0.01$). The expression of *miR-144-5p* in mammary glands was higher than that in other tissues, including the heart, liver, spleen, lung and kidney in the healthy cows (Fig. 2b). The expression of the *miR-130b-5p* was much higher in the mastitis-infected mammary gland tissues compared to that in the healthy samples (Fig. 3a; $p < 0.01$). The expression of *miR-130b-5p* in mammary glands was higher than that in other tissues, including the liver, heart, lung, kidney and spleen in the healthy cows (Fig. 3b). The findings suggest that *miR-144-5p* and *miR-130b-5p* are highly correlated with mastitis.

4 Discussion

Through sequencing alignment, *bta-miR-144* and *bta-miR-130b* from miRBase are actually *miR-144-3p* and *miR-130b-5p*; the cloned miRNAs in our study are *miR-144-5p* and

miR-130b-5p. In the present study, *miR-144-5p* and *miR-130b-5p* were identified in mammary, heart, liver, spleen, lung and kidney tissues at different expression levels, which may indicate functional differences.

Twenty potential target genes of *miR-144-5p* and *miR-130b-5p* were predicted, and 10 of them belonged to the chemokine signaling pathway, which plays an important role in orchestrating leukocyte migration under normal conditions and during inflammatory responses (Mellado et al., 2001), such as *CXCR4* (Lapteva et al., 2005; Balkwill, 2004) and *CXCR2* (Acosta et al., 2008). Another important pathway including six target genes is arginine and proline metabolism, which is known to be closely involved in conceptus metabolism (Wu et al., 2008), growth and development. It is also a potential treatment for intrauterine growth restriction (Wu et al., 2009), which is a significant problem in both human medicine and animal agriculture. Four of them belonged to the mRNA surveillance pathway, which assesses the quality of mRNAs to ensure that they are suitable for translation (Vasudevan et al., 2002). mRNA surveillance facilitates the detection and destruction of mRNAs that contain premature termination codons (Wagner and Lykke-Andersen, 2002). Whether these genes are regulated directly by *miR-144-5p* and *miR-130b-5p* needs to be confirmed in a further study.

5 Conclusions

In summary, the differential expression of *miR-144-5p* and *miR-130b-5p* in healthy and mastitis-infected mammary glands indicates that *miR-144-5p* and *miR-130b-5p* may play important roles in inflammation response. There could be a relationship between *miR-144-5p* and *miR-130b-5p* and mastitis in Chinese Holstein cattle. Our findings suggest that the differential expression of the post-transcriptional regulators *miR-144-5p* and *miR-130b-5p* may bind to complementary sequences of target mRNAs, resulting in different translational bovine repression. *miR-144-5p* and *miR-130b-5p* likely play critical roles in mastitis resistance in dairy cattle.

Competing interests. The authors declare that they have no conflict of interest.

Acknowledgements. This study was supported by the National 863 Program of China (2008AAl02144), the "13115" Sci-Tech Innovation Program of Shaanxi Province (2008ZDKG-11) and the Xi'an city science and technology project (NC09049-2). Comments from anonymous reviewers have greatly improved the paper.

Edited by: Steffen Maak

References

Acosta, J. C., O'Loghlen, A., Banito, A., Guijarro, M. V., Augert, A., Raguz, S., Fumagalli, M., Da Costa, M., Brown, C., Popov, N., Takatsu, Y., Melamed, J., d'Adda di Fagagna, F., Bernard, D., Hernando, E., and Gil, J.: Chemokine signaling via the CXCR2 receptor reinforces senescence, Cell, 133, 1006–1018, 2008.

Baggiolini, M. and Loetscher, P.: Chemokines in inflammation and immunity, Immunol. Today, 21, 418–420, 2000.

Balkwill, F.: The significance of cancer cell expression of the chemokine receptor CXCR4, Semin. Cancer Biol., 14, 171–179, 2004.

Bannerman, D. D.: Pathogen-dependent induction of cytokines and other soluble inflammatory mediators during intramammary infection of dairy cows, J. Anim. Sci., 87, 10–25, 2005.

Bartel, D. P.: MicroRNAs, genomics, biogenesis, mechanism, and function, Cell, 116, 281–297, 2004.

Bartel, D. P.: MicroRNAs, target recognition and regulatory functions, Cell, 136, 215–233, 2009.

Berezikov, E.: Evolution of microRNA diversity and regulation in animals, Nature Rev. Genet., 12, 846–860, 2011.

Charo, I. F. and Ransohoff, R. M.: The many roles of chemokines and chemokine receptors in inflammation, New Engl. J. Med., 354, 610–621, 2006.

Chen, C., Ridzon, D. A., Broomer, A. J., Zhou, Z., Lee, D. H., Nguyen, J. T., Barbisin, M., Xu, N. L., Mahuvakar, V. R., Andersen, M. R., Lao, K. Q., Livak, K. J., and Guegler, K. J.: Real-time quantification of microRNAs by stem–loop RT–PC R, Nucleic Acids Res., 33, e179, https://doi.org/10.1093/nar/gni178, 2005.

Colangelo, T., Fucci, A., Votino, C., Sabatino, L., Pancione, M., Laudanna, C., Binaschi, M., Bigioni, M., Maggi, C. A., Parente, D., Forte, N., and Colantuoni, V.: MicroRNA-130b promotes tumor development and is associated with poor prognosis in colorectal cancer, Neoplasia, 15, 1218–1231, 2013.

Coussens, L. M. and Werb, Z.: Inflammation and cancer, Nature, 420, 860–867, 2002.

Cullen, B. R.: Transcription and processing of human microRNA precursors, Mol. Cell, 16, 861–865, 2004.

Fu, Y. F., Du, T. T., Dong, M., Zhu, K. Y., Jing, C. B., Zhang, Y., Wang, L., Fan, H. B., Chen, Y., Jin, Y., Yue, G. P., Chen, S. J., Chen, Z., Huang, Q. H., Jing, Q., Deng, M., and Liu, T. X.: Mir-144 selectively regulates embryonic a-hemoglobin synthesis during primitive erythropoiesis, Blood, 113, 1340–1349, 2009.

Gu, Y., Li, M., Wang, T., Liang, Y., Zhong, Z., Wang, X., Zhou, Q., Chen, L., Lang, Q., He, Z., Chen, X., Gong, J., Gao, X., Li, X., and Lv, X.: Lactation-related MicroRNA expression profiles of porcine breast milk exosomes, PLoS One, 7, e43691, https://doi.org/10.1371/journal.pone.0043691, 2012.

Kim, V. N.: MicroRNA biogenesis, coordinated cropping and dicing, Nat. Rev. Mol. Cell Bio., 6, 376–385, 2005.

Lapteva, N., Yang, A. G., Sanders, D. E., Strube, R. W., and Chen, S. Y.: CXCR4 knockdown by small interfering RNA abrogates breast tumor growth in vivo, Cancer Gene Ther., 12, 84–89, 2005.

Li, B. L., Lu, C., Lu, W., Yang, T. T., Qu, J., Hong, X., and Wan, X. P.: miR-130b is an EMT-related microRNA that targets DICER1 for aggression in endometrial cancer, Med. Oncol., 30, 484–484, 2013.

Li, Z. X., Wang, H. L., Chen, L., Wang, L. J., Liu, X. L., Ru, C. X., and Song, A. L.: Identification and characterization of novel and differentially expressed microRNAs in peripheral blood from healthy and mastitis Holstein cattle by deep sequencing, Anim. Genet., 45, 20–27, 2014a.

Li, Z. X., Zhang, H. L., Song, N., Wang, H. L., Chen, L., Zhai, M. X., and Liu, X. L.: Molecular cloning, characterization and expression of miR-15a-3p and miR-15b-3p in dairy cattle, Mol. Cell. Probe., 28, 255–258, 2014b.

Livak, K. J. and Schmittgen, T. D.: Analysis of relative gene expression data using real-time quantitative PCR and the 2(-Delta Delta C(T)) method, Methods, 25, 402–408, 2001.

Mellado, M., Rodríguez-Frade, J. M., Mañes, S., Martínez-A, C.: Chemokine signaling and functional responses: The role of receptor dimerization and TK pathway activation, Annu. Rev. Immunol., 19, 397–421, 2001.

Nash, D. L., Rogers, G. W., Cooper, J. B., Hargrove, G. L., and Keown, J. F.: Heritability of intramammary infections at first parturition and relationships with sire transmitting abilities for somatic cell score, udder type traits, productive life, and protein yield, J. Dairy Sci., 86, 2684–2695, 2003.

Sangokoya, C., Telen, M. J., and Chi, J. T.: microRNA miR-144 modulates oxidative stress tolerance and associates with anemia severity in sickle cell disease, Blood, 116, 4338–4348, 2010.

Vasudevan, S., Peltz, S. W., and Wilusz, C. J.: Non-stop decay – a new mRNA surveillance pathway, Bioessays, 24, 785–788, 2002.

Wagner, E. and Lykke-Andersen, J.: mRNA surveillance: the perfect persist, J. Cell Sci., 115, 3033–3038, 2002.

Wu, G., Bazer, F. W., Datta, S., Johnson, G. A., Li, P., Satterfield, M. C., and Spencer, T. E.: Proline metabolism in the conceptus: implications for fetal growth and development, Amino Acids, 35, 691–702, 2008.

Wu, G., Bazer, F. W., Davis, T. A., Kim, S. W., Li, P., Marc Rhoads, J., Carey Satterfield, M., Smith, S. B., Spencer, T. E., and Yin, Y.: Arginine metabolism and nutrition in growth, health and disease, Amino Acids, 37, 153–168, 2009.

Zhao, G., Zhang, J. G., Shi, Y., Qin, Q., Liu, Y., Wang, B., Tian, K., Deng, S. C., Li, X., Zhu, S., Gong, Q., Niu, Y., and Wang, C. Y.: MiR-130b is a prognostic marker and inhibits cell proliferation and invasion in pancreatic cancer through targeting STAT3, PLoS One, 8, e73803, https://doi.org/10.1371/journal.pone.0073803, 2013.

Zhou, Q., Li, M., Wang, X., Li, Q., Wang, T., Zhu, Q., Zhou, X., Wang, X., Gao, X., and Li, X.: Immune-related MicroRNAs are abundant in breast milk exosomes, Int. J. Biol. Sci., 8, 118–123, 2012.

8

Behaviour and performance of suckling gilts and their piglets in single housing with different fixation times

Ralf Wassmuth[1], **Christoph Biestmann**[2], **and Heiko Janssen**[2]

[1]University of Applied Sciences, Am Kruempel 31, 49090 Osnabrück, Germany
[2]Chamber of Agriculture Lower Saxony, Mars-la-Tour-Str. 6, 26121 Oldenburg, Germany

Correspondence to: Ralf Wassmuth (r.wassmuth@hs-osnabrueck.de)

Abstract. The objective was to evaluate suckling performance and behaviour traits of gilts and piglets in two different single-housing farrowing systems under practical conditions. Performance data of 70 crossbred gilts and their 842 piglets were collected. The behavioural observation included 17 gilts and 211 piglets. Gilts of the control group (full-time crating, FTC) were fixed during farrowing and suckling (Pro Dromi® 1), and in the experimental group (short-time crating, STC) gilts were fixed for 6 days postpartum (p.p.) only (Pro Dromi® 1.5). Six farrowing crates were included in each group, and six replications were carried out. Performance data were collected and gilts' and piglets' behaviour was observed with 10 min scan samples and categorized by standing, walking (only in STC), sitting and lying (side and belly). The management and the housing systems were in accordance with the Tierschutz-Nutztierhaltungsverordnung (TierSchNutztV, 2017). No significant ($p > 0.05$) differences between FTC and STC were found in piglets born alive (13.2 and 13.9, respectively), loss of piglets (1.4 and 1.55, respectively) and weaned piglets including cross-fostering (12.0 and 12.4, respectively). Piglet loss due to crushing was 0.6 (FTC) and 0.64 (STC), with no significant difference ($p > 0.05$). Overall, 82.5 % of all piglets killed due to crushing were lost from farrowing to day 2 p.p. The daily gain of STC piglets was significantly higher than that of FTC piglets (205 g vs. 199 g, respectively; $p < 0.05$) during the suckling period (3 weeks). Concerning gilts' behaviour, significant differences ($p \leq 0.05$) were found in sitting duration only (FTC 5.8 % and STC 4.0 %, respectively). FTC piglets spent more time lying, sitting and standing (7.4 % vs. 4.4 %, 0.5 % vs. 0.4 %, 9.6 % vs. 8.4 %, respectively; $p \leq 0.05$). The reason could be the higher acceptance of the piglet nest in STC.

It was possible to conclude that gilts' welfare was improved by STC compared to FTC, and farrowing crates with loose single housing did not lead to higher piglet loss in the suckling period. An earlier end of the fixation period of the gilt at day 2 or 3 p.p. should be tested.

1 Introduction

During gestation sows must have the opportunity to move freely in a group of companions according to German legislation (TierSchNutztV, 2017). But in the suckling period sows are still kept on their own and full time in crates. This is not in accordance with a sow's behaviour under natural conditions. Sows live in a herd and leave the group approximately 24 h before farrowing (Jensen, 1986). On average 10 days postpartum (p.p.), sows return together with their piglets to the group and mix with it (Jensen and Redbo, 1987; Pitts et al., 2000). Hence, single housing is in accordance with the needs of a sow around farrowing (Von Borell et al., 2002), but fixation in crates during the whole suckling period is not related to sows' welfare (Baxter et al., 2011). Hence, short-time crating (STC) could be a possible alternative to the present commercial practice of full-time crating (FTC) (Bünger, 2002; Weber et al., 2009).

Studies from Bohnenkamp et al. (2013) showed that suckling duration was similar in single and group housing. But other studies concluded that loose single housing or even group housing led to higher piglet loss due to crushing

(Damm et al., 2005; Danholt et al., 2011; D'Eath, 2005; Maletinska and Spinka, 2001).

The aim of the present study was to evaluate suckling performance and behaviour of gilts and piglets in two different single-housing farrowing systems under practical conditions.

2 Material and methods

2.1 Animals and housing

The present investigation was performed in a newly built barn for 550 sows in the northern part of Germany, and it was not organized according to the rules of organic farming. The trial period lasted from February to May 2015, and it was possible to include BHZP (Bundes Hybrid Zucht Programm) Viktoria gilts mated to db77 boars. The gilts of the control group experienced FTC, which means conventional single housing. It was possible to include six farrowing crates, and gilts were fixated from 48 h prior to farrowing till weaning (Pro Dromi® 1). In the trial group, STC was practised and the gilts were fixated from 48 h prior to farrowing till the sixth day of suckling (Pro Dromi® 1.5). Six farrowing crates were a part of the investigation.

The single pens were identical in construction with an area of $6.5\,m^2$ in total. The moving area of a gilt was $3.3\,m^2$ (STC) and $1.3\,m^2$ (FTC), respectively. In FTC and STC, six experiments each were conducted, and gilts entered the pens 5 days before farrowing; for 3 days they were not fixated and could explore the pen. In each pen, a jute bag ($1.2\,m \times 0.6\,m$) was provided to allow nest-building behaviour. During the first few sow feeding times, piglets were fixated in their nest to reduce loss due to crushing. Cross-fostering was done within FTC and STC as far as possible.

The housing systems and the management used in the present study were in accordance with the Tierschutz-Nutztierhaltungsverordnung (TierSchNutztV, 2017).

2.2 Recorded traits

It was possible to analyse performance data of 70 gilts (first parity) and 842 piglets. Behavioural observations of 17 gilts and 211 piglets were performed.

Birth weight was measured 1 day after the main farrowing day and at the end of the first and second week of suckling. The final weight was recorded at day 18 p.p.

Behavioural parameters were recorded on video (Mobotix, 2017) with the scan sampling method (Hoy, 2009) on day -1, 0 (farrowing), $+1$, $+5$, $+7$ and $+16$. Videotapes were analysed with the instantaneous sampling (Hoy, 2009) method and an interval of 10 min.

2.3 Statistical analysis

Statistical analysis was performed with the statistical software package IBM SPSS, Version 22.0 (IBM Corp., 2013).

Table 1. Mean (SD as index; minimum and maximum in parenthesis) for fertility traits of gilts across the two housing systems.

Trait	FTC	STC
Number of gilts/farrows (n)	37	33
Piglets born alive (n)	$13.2_{4.0}$ (3–20)	$13.9_{3.6}$ (4–21)
Piglets born dead (n)	$0.9_{1.4}$ (0–6)	$1.1_{1.3}$ (0–5)
Loss of piglets (n)	$1.4_{1.4}$ (0–4)	$1.6_{1.4}$ (0–5)
Loss of piglets due to crushing (n)	$0.6_{0.9}$ (0–3)	$0.6_{0.9}$ (0–3)
Loss of piglets (%)	$9.8_{9.3}$ (0–28.6)	$10.2_{9.0}$ (0–29.4)

The statistical model of the piglets' performance traits and of behavioural observations of piglets and gilts included fixed effects as shown below. The model of the traits of gilts performance obtained the fixed effect of the housing system only.

$$Y_{ijkl} = \mu + H_i + B_j + S_k + b\left(W_{ijkl} - \overline{W}\right) + e_{ijkl},$$

where Y_{ijkl} is observation, μ is the sample mean, H_i is the fixed effect of housing system (FTC, STC), B_j is the fixed effect of the batch (1 to 6), S_k is the fixed effect of the piglets sex (male, female), $b(W_{ijkl} - \overline{W})$ is the linear covariate of piglets birth weight and e_{ijkl} is the residual random error.

3 Results

As shown in Table 1 the number of piglets born alive did not differ between FTC and STC, with 13.2 and 13.9, respectively. About one piglet was born dead per farrow, and the loss of piglets was low, with 1.4 in FTC and 1.6 in STC. The percentage of piglet loss due to crushing was the same in both systems and the proportion of lost piglets did not vary significantly (between 9.8 (FTC) and 10.2 (STC)). In STC only 1 out of a total of 21 piglets killed due to crushing were lost after the sixth day of suckling, the last day of gilts' fixation.

In STC, birth weight was lower than in FTC, but there was no difference in live weight at the 18th day p.p. between both housing systems (Table 2). Because of a significantly higher daily gain in week 2 and 3, STC piglets had a significantly higher daily gain during the suckling period with 205 g compared with FTC piglets reaching 199 g.

Focusing on the lying behaviour, no significant differences were found between FTC and STC (Table 3). Gilts were lying on their side 57 % (FTC) and 55.9 % (STC) and on their belly 22.5 % (FTC) and 23.4 % (STC) of the total observation time. In FTC gilts spent a higher time period sitting (5.8 %) than gilts in STC (4.0 %). The difference was statistically significant. Neither in standing nor in suckling behaviour were significant differences found between the two systems (Table 3).

Piglets' behaviour differed between the two systems as shown in Table 4. FTC piglets showed significantly higher lying, sitting and standing frequencies. STC piglets spent significantly more time in their nest (61.6 %) than piglets in FTC (55.9 %). The observed differences in suckling behaviour between FTC and STC were not significant (Table 4).

Table 2. LSQ (least-square) mean (SE as index; minimum and maximum in parenthesis) for growth traits of piglets across the two housing systems.

Trait	FTC	STC
Number of piglets (n)	432	410
Birth weight (kg)	$1.35^a_{0.015}$ (0.56–2.64)	$1.31^b_{0.016}$ (0.62–2.54)
Live weight (kg), 18th day p.p.	$4.71_{0.030}$ (1.76–7.48)	$4.77_{0.031}$ (1.55–7.93)
Daily gain (g day^{-1}), first week p.p.	$158_{2.3}$ (−53–358)	$153_{2.4}$ (−100–333)
Daily gain (g day^{-1}), second week p.p.	$212^a_{2.1}$ (36–360)	$223^b_{2.1}$ (34–423)
Daily gain (g day^{-1}), third week p.p.	$214^a_{2.6}$ (−2–408)	$225^b_{2.6}$ (4–436)
Total daily gain (g day^{-1})	$199^a_{1.7}$ (46–329)	$205^b_{1.8}$ (32–344)

[a,b] Significant differences ($p \leq 0.05$).

Table 3. LSQ mean (SE as index) for behaviour of gilts across the two housing systems.

Trait	FTC	STC
Number of gilts (n)	8	9
Lying (side, %)	$57.0_{1.85}$	$55.9_{1.72}$
Lying (belly, %)	$22.5_{1.45}$	$23.4_{1.35}$
Sitting (%)	$5.8^a_{0.56}$	$4.0^b_{0.52}$
Standing (%)	$14.7_{1.00}$	$14.8_{0.92}$
Walking (%)	–	1.9
Suckling (%)	$29.8_{2.05}$	$28.3_{1.90}$

[a,b] Significant differences ($p \leq 0.05$).

Table 4. LSQ mean (SE as index) for behaviour of piglets across the two housing systems.

Trait	FTC	STC
Number of piglets (n)	95	116
Lying (%)	$7.4^a_{1.04}$	$4.4^b_{0.96}$
Sitting (%)	$0.5^a_{0.05}$	$0.4^b_{0.04}$
Standing (%)	$9.6^a_{0.37}$	$8.4^b_{0.34}$
Suckling (%)	$26.6_{1.55}$	$25.2_{1.43}$
Time spending in the nest (%)	$55.9^a_{2.08}$	$61.6^b_{1.93}$

[a,b] Significant differences ($p \leq 0.05$).

4 Discussion

The investigation took place in a newly built barn with a capacity of 550 sows, and it was not organized according to the rules of organic farming. So the circumstances differed from the usual ones when short-time crating in single housing or even group housing during farrowing and suckling was practised.

The number of piglets born alive in the present study was 13.2 and 13.9 and was a bit higher than the number of 13.0 observed in German commercial herds (ZDS, 2013). The total loss was 9.8 and 10.2 % and was lower than reported in German commercial herds (14.5 %; ZDS, 2013). Baumgartner (2011) showed a live-born piglet mortality between 12.6 and 17.2 % in eight European countries (CH, DK, GE, NL, SE, UK, NO, AT). Hence, it was possible to conclude that the circumstances of the present investigation represented well-managed commercial herds.

The loss of piglets due to crushing was less than half of all loss, and it was the same in both systems (FTC, STC). Jungbluth et al. (2005) concluded that 60 % of all loss was caused by crushing. With 4.2 % (FTC) and 4.3 % (STC) crushed piglets in similar systems, Soede et al. (2014) had nearly the same results as the present study. By contrast with these results, Weber et al. (2007) found higher loss due to crushing in loose single housing (5.4 %) than in FTC (4.5 %). Even

Damm et al. (2005), Danholt et al. (2011), D'Eath (2005), and Maletinska and Spinka (2001) found higher loss due to crushing in loose single and even group housing.

One reason for the low crushing loss in STC in the present study could be that piglets spent about 60 % of the observed time in the nest. This could be due to the construction of the Pro Dromi® system were the piglets' nest was located in front of the head of the gilt and piglets could keep contact with their mother. Further, the jute bag used by the gilt to show nesting behaviour was put into the piglets' nest which made it more attractive for the piglets. Possibly this management led to a strong mother offspring bond.

Only 1 of 21 piglets was lost after ending the fixation of the gilt at day 6 p.p. This led to the conclusion that the ending of fixation at an earlier stage of suckling must not necessarily lead to higher loss. Research should be intensified to optimize STC systems.

In the present study suckling frequencies were a bit lower in STC, but the difference was not significant. The study of Bohnenkamp et al. (2013) confirmed this result because they found no differences in suckling frequencies between FTC and STC. Although the suckling frequencies were similar from a statistical point of view and the birth weight was higher in FTC, in the second and third week the daily gain of piglets was significantly higher in STC compared to FTC. This is in accordance with Soede et al. (2014), who found a significantly higher daily gain in STC ($250\,\text{g}\,\text{day}^{-1}$) compared to FTC ($237\,\text{g}\,\text{day}^{-1}$). It was possible to conclude that the housing system had an effect on daily gain.

Hence, piglets' welfare was not negatively influenced by STC, where the gilt has more opportunities to move than in FTC. Gilts showed sitting behaviour less often, which might lead to the conclusion that gilts' welfare was improved by STC compared to FTC.

5 Conclusions

We conclude the following:

1. The loss of piglets due to crushing was not significantly higher in STC compared to FTC.

2. The daily gain of piglets was slightly higher when gilts were fixated for a short time compared to full-time crating.

3. Sitting behaviour occurred more often in FTC, and it was possible to conclude that gilts' welfare was adversely influenced.

4. No differences between FTC and STC were found in suckling frequency, and a higher moving activity of piglets could be observed in STC.

Competing interests. The authors declare that they have no conflict of interest.

Edited by: M. Mielenz

References

Baumgartner, J.: Pig industry in CH, CZ, DE, DK, NL, NO, SE, UK, AT and EU, Report of the Free Farrowing Workshop Vienna 2011, 8–9 December, 2011, Vienna, Austria, University of Veterinary Medicine Vienna Austria, 3–7, http://www.vetmeduni.ac.at/fileadmin/_migrated/content_uploads/FFWV_2011-Report.pdf (last access: May 2017), 2011.

Baxter, E. M., Lawrence, A. B., and Edwards, S. A.: Alternative farrowing systems: design criteria for farrowing systems based on the biological needs of sows and piglets, Animal, 5, 580–600, 2011.

Bohnenkamp, A.-L., Meyer, C., Müller, K., and Krieter, J.: Group housing with electronically controlled crates for lactating sows. Effect on farrowing, suckling and activity behavior and piglets, Appl. Anim. Behav. Sci., 145, 37–43, 2013.

Bünger, B.: Einflüsse der Haltungsbedingungen von ferkelnden und ferkelführenden Sauen auf die Entwicklung der Ferkel: Eigene Studien und eine Bewertung der Literatur, Dtsch. Tierärztl. Wschr., 109, 277–289, 2002.

Damm, B. I., Forkman, B., and Pedersen, L. J.: Lying down and rolling behaviour in sows in relation to piglet crushing, Appl. Anim. Behav. Sci., 90, 3–20, 2005.

Danholt, L., Mousten, V. A., Nielsen, M. B. F., and Kristensen, A. R.: Rolling behaviour of sows in relation to piglet crushing on sloped versus level floor pens, Livestock Sci., 141, 59–68, 2011.

D'Eath, R. B.: Socialising piglets before weaning improves social hierarchy formation when pigs are mixed post-weaning, Appl. Anim. Behav. Sci., 93, 199–211, 2005.

Hoy, S. (Ed.): Methoden der Nutztierethologie, in: Nutztierethologie, Verlag Eugen Ulmer, Stuttgart, 2009.

IBM Corp: Released 2013, IBM SPSS Statistics for Windows, Version 22.0, IBM Corp., Armonk, NY, 2013.

Jensen, P.: Observations on the maternal behaviour of free-ranging domestic pigs, Appl. Anim. Behav. Sci., 16, 131–142, 1986.

Jensen, P. and Redbo, I.: Behaviour during nest leaving in free-ranging domestic pigs, Appl. Anim. Behav. Sci., 18, 355–362, 1987.

Jungbluth, T., Büscher, W., and Krause, T.: Technik Tierhaltung (Kapitel 6 – Verfahren der Schweinehaltung), Verlag Eugen Ulmer, Stuttgart, 2005.

Maletinska, J. and Spinka, M.: Cross-suckling and nursing synchronisation in group housed lactating sows, Appl. Anim. Behav. Sci., 75, 17–32, 2001.

Mobotix: Handbuch zur Kamera 'M25M Allround', http://mobotix.com/ger_DE/Produkte/Outdoor-Kameras, last access: May 2017.

Pitts, A. D., Weary, D. M., Pajor, E. A., and Fraser, D.: Mixing at young ages reduces fighting in unacquainted domestic pigs, Appl. Anim. Behav. Sci., 68, 191–197, 2000.

Soede, N., Schuttert, M., and Hoofs, A.: Resultate Forschung VIC Sterksel und Implementation in die Praxis, in: Vortrag auf dem Pro Dromi® Symposium, 14 February 2014, Arnheim, the Netherlands, http://www.prodromi.nl/afbeeldingen/Symposium-Pro-Dromi-Day.pdf (last access: May 2017), 2014.

TierSchNutztV: Tierschutznutztierhaltungsverordnung, veröffentlicht im BGBl, Teil I/2006, S. 2043, 22 August 2006, zuletzt geändert durch Artikel 1 der Verordnung veröffentlicht im BGBl, I/2014, S. 94, 5 February 2014, http://www.gesetze-im-internet.de/bundesrecht/tierschnutztv/gesamt.pdf, last access: May 2017.

Von Borell, E., Von Lengerken, G., and Rudovsky, A.: Tiergerechte Haltung von Schweinen – Grundlegende Anforderungen, in: Umwelt- und tiergerechte Haltung von Nutz,- Heim-, und Begleittieren, edited by: Methling, W. and Unshelm, J., Parey Buchverlag, Berlin, 2002.

Weber, R., Keil, N. M., Fehr, M., and Horat, R.: Praxisvergleich der Reproduktionsleistungen zwischen Abferkelbuchten mit und ohne Kastenständen, in: Tagungsband zur 8 Tagung Bau, Technik und Umwelt 2007 in der landwirtschaftlichen Nutztierhaltung, 8–10 October 2007, Kuratorium für Technik und Bauwesen in der Landwirtschaft, Bonn, Darmstadt, 2007.

Weber, R., Keil, N. M., Fehr, M., and Horat, R.: Factors affecting piglet mortality in loose farrowing systems on commercial farms, Livestock Sci., 124, 216–222, 2009.

ZDS: Zahlen aus der Deutschen Schweineproduktion 2012, 2013 Edn., Zentralverband der Deutschen Schweineproduktion e.V. (ZDS), Bonn, 2013.

9

Milk fatty acid composition as an indicator of energy status in Holstein dairy cows

Lana Vranković[1], **Jasna Aladrović**[1], **Daria Octenjak**[2], **Dušanka Bijelić**[2], **Luka Cvetnić**[2], **and Zvonko Stojević**[1]

[1]Department of Physiology and Radiobiology, University of Zagreb Faculty of Veterinary Medicine, Heinzelova 55, 10000 Zagreb, Croatia
[2]Students of Faculty of Veterinary Medicine, University of Zagreb, Heinzelova 55, 10000 Zagreb, Croatia

Correspondence to: Lana Vranković (lana.vrankovic@vef.hr)

Abstract. Transition dairy cows often enter a stage of negative energy balance during which the utilization of energy reserves is reflected in the milk fatty acid (FA) composition. In this study, metabolic status was evaluated by measuring milk FA, ruminal short-chain FA (RSCFA), and serum biochemical parameters in Holstein cows. Samples (milk, rumen contents, and blood) were collected around days 30 (early) and 150 (middle) of lactation, and rumen contents and blood samples were collected 30 days before calving (dry). Fatty acids were extracted and FA composition was determined. Glucose, triacylglycerols, total cholesterol, low-density lipoprotein cholesterol (LDL-C) and high-density lipoprotein cholesterol (HDL-C), beta-hydroxybutyric acid (BHB), and non-esterified fatty acid (NEFA) concentrations were determined in serum samples. Lower percentages of saturated FA in milk and higher percentages of monounsaturated FA, polyunsaturated FA, and C18:1n-9 were observed in early lactation compared to mid-lactation. In rumen higher concentrations of propionic acid were determined at mid-lactation compared to early lactation. Acetic and butyric acid concentrations showed no significant differences between sampling intervals. In serum higher glucose concentrations were observed during the dry period and mid-lactation than during early lactation. Lower BHB and higher NEFA concentrations were noted during early lactation compared to mid-lactation and the dry period. Total cholesterol, LDL-C, HDL-C and triacylglycerols showed no significant differences between sampling intervals. The results of the present study suggest that determination of milk FA is a potential indicator of energy status in dairy cows.

1 Introduction

Ruminal short-chain fatty acids (RSCFAs) are produced in the rumen of the cow through the microbial fermentation of carbohydrates and protein. They are subsequently absorbed through all compartments of the stomach (Leek, 2004). Acetic, propionic, and butyric acids are the predominant RSCFAs in rumen fluid. Concentrations and relative proportions of RSCFAs are associated with the level and composition of feed intake (Murphy et al., 1982). Ruminal short-chain fatty acids constitute the major source of energy in ruminants, providing 50–80 % of the total amount of digested energy (Thomas, 2012) and, thus, are of great importance in the production of milk in dairy cows. Both the total yield of RSCFAs and the types of RSCFAs formed can sig-

nificantly affect the utilization of absorbed nutrients in dairy cows, and can thus considerably affect milk volume and composition (Butler et al., 2008).

The peripartal period is the most demanding in terms of herd health management (Walsh et al., 2011; Fiore et al., 2014, 2015). The increased energy demands of foetal development and milk production are evident in transition dairy cows (Arfuso et al., 2016). Therefore, cows often enter a stage of negative energy balance (Walsh et al., 2011). The most important aspect in the assessment of energy status is centred on the determination of blood non-esterified fatty acids (NEFAs) and beta-hydroxybutyrate (BHB) levels (LeBlanc, 2010). NEFA levels can be increased through increased mobilization of stored lipids, which can thereby indicate metabolic disorders, such as ketosis and fatty liver syndrome (Duffield et al., 1998; Fiore et al., 2017). The first measurable indicator of ketogenesis is an increase in BHB concentration in the blood (Payne and Payne, 1987; Schulz et al., 2014) The utilization of energy reserves is also reflected in the milk fat content (Bauman et al., 2006), namely in the fatty acid (FA) composition and mutual ratios between individual FA groups (Ducháček et al., 2014). Milk FA and protein contents are correlated with metabolic status (Mulligan et al., 2006); therefore, measurement of these two parameters is usually recommended for the determination of negative energy balance (Gross et al., 2011). Several researchers (Berry et al., 2006; Soyeurt et al., 2006; Bastin et al., 2011) have determined the FA composition of milk to predict energy status.

It is well known that the utilization of fat reserves during the transition period affects the FA composition of milk. Currently, data on the RSCFA content and FA composition of milk as indicators of energy metabolism in Holstein dairy cows are scarce. The collection of milk samples is more easily obtained and is more suitable in terms of animal welfare considerations. Therefore, the aim of this study was to evaluate the energy status of Holstein dairy cows during different stages of lactation by investigating the milk FA composition, the RSCFA concentration, and serum biochemical parameters.

2 Material and methods

This study was approved by the Ethics Committee of the Faculty of Veterinary Medicine (251-61-01/139-13-2; 251-61-01/139-16-3).

2.1 Animals and diets

Prior to the selection of cows for this study, we contacted government authorities to get the list of farms. Afterwards, we contacted farm owners to ask whether they were willing to accept our research on their farm. After that we carried out a survey where we asked the owners about the number of animals on the farm, feeding regime, management system, and production of animals. Based on the survey results, we selected a farm which was acceptable for our research goals. The selected farm properly kept records about animals and comprised around 100 cows of Holstein breed. The selected farm was located near the town of Đakovo, Croatia ($45°18'35.989''$ N, $18°24'35.215''$ E). Twelve Holstein cows aged 2.5–7.8 years during lactations ranging from the first to the seventh were selected for the study. Selected cows were in a production phase acceptable for this research and were uniform depending on production phase as well as clinically healthy. Average body mass of cows was 680 ± 42 kg and body score condition (BSC) was as follows: in the dry period (3.42 ± 0.47), around days 30 and 150 of lactation (2.83 ± 0.51, 3.10 ± 0.45, respectively). Cows were kept freely in stalls without pasture. Data on milk compo-

Table 1. Composition of daily meal of Holstein cows based on production cycle.

Composition of food before calving	Amount (kg)
Straw	1.5
Hay	2.3
Beet pulp	5.0
Corn silage	12.0
Corn maize	0.5
Mineral–vitamin and energy supplements	5.4
Composition of food 30 days after calving	
Hay	4.5
Beet pulp	10.0
Corn silage	24.0
Corn maize	1.0
Mineral–vitamin and energy supplements	11.7
Composition of food 150 days after calving	
Straw	0.5
Hay	4.0
Beet pulp	10.0
Corn silage	25.0
Mineral–vitamin and energy supplements	11.8

sition and average milk production were obtained through the Croatian Agricultural Agency. Feed composition of the diet fed to the subjects of the study is shown in Table 1. The following mineral–vitamin and energy supplements administered to the subjects all had protected formulas: Rindamin MF Sauer, Schaumann energy (Schaumann Agri Ltd., Koprivnica, Croatia, respectively), Mycostop (INBERG Ltd., Belgrade, Serbia), Bovi top Elevate (Alltech, Dunboyne, Ireland), and glycerol. All animals were clinically healthy during the study.

2.2 Sampling and preparation of samples for analysis

The composite milk samples (pool of four quarters of the udder) were first collected in tubes, twice during the lactation period, around days 30 and 150, and stored at -20 °C until analysis. Samples of rumen contents and blood were collected 30 days before delivery and around days 30 and 150 of lactation. The rumen contents were sampled using a rumen fluid collector. First, 100 mL was discarded and the next 50 mL was stored in Falcon® tubes and immediately frozen in liquid nitrogen. Blood was sampled via jugular venepuncture into BD Vacutainer® tubes (BD Diagnostics, Plymouth, UK). After centrifugation at $1600 \times g$ for 20 min at 20 °C, blood serum samples were separated and stored at -20 °C, until analysis. All samples were analysed within 2–3 months after collection.

Table 2. Average milk production around days 30 and 150 of standard lactation and milk fat and protein percentage in Holstein cows.

| | Average milk production and composition (kg) | | | |
	Day 30 (approx.)	Day 150 (approx.)	SEM	P value
Milk production (kg)	40.14	32.65	2.36	0.002
Milk fat (%)	3.16	3.44	0.23	0.446
Protein (%)	2.97	3.35	0.09	0.011

Results presented as mean $\pm$ SEM.

2.3 Milk fatty acid analysis

Milk fat was extracted by the standard procedure of Hara and Radin (1978) using the solvent mixture of isopropanol and hexane in a ratio of 3 : 2. The FAs were converted to methyl esters via trans-esterification with a 20 % solution of boron trifluoride in methanol (Rule, 1997). Fatty acid composition was determined using a Shimadzu GC2010 Plus gas chromatograph (Shimadzu, Kyoto, Japan) equipped with a flame ionization detector and a ZB-WAX column (30 m length, Phenomenex, Torrance, CA, USA). The chromatographic conditions were as follows: carrier gas, helium; injection volume, 1 mL; injector temperature, 250 °C; detector temperature, 300 °C; oven temperature programme, initially 60 °C, then increased at 13 °C min^{-1} to 150 °C, at 2 °C min^{-1} to 220 °C, and at 2 °C min^{-1} to 240 °C. All experimental measurements were repeated three times and the average values were reported. Quantification was determined through area normalization, with an external standard mixture of fatty acid methyl esters (Sigma-Aldrich, Steinheim, Germany). Fatty acid composition was calculated as the percentage of each individual fatty acid relative to the total fatty acids.

2.4 Rumen fatty acid analysis

To determine RSCFAs, approximately 1 g of thawed digesta was diluted with 1 mL of ultrapure water and centrifuged. After centrifugation (10 min at $5000 \times g$), the supernatant was transferred to another tube, and 0.2 mL of 24 % metaphosphoric acid was added. The sample was then placed in an ice bath to allow the protein to settle completely. Finally, samples were centrifuged (10 min at $5000 \times g$, 20 °C) and the clear supernatant was analysed. The supernatants were analysed to determine the concentrations of acetic, propionic, and butyric acids, using a Shimadzu GC2010Plus gas chromatograph (Shimadzu, Kyoto, Japan), equipped with a model AOC 20i auto injector, a flame ionization detector (FID) and a Nukol™ column (30 m × 0.25 mm × 0.25 µm) (Supelco, Bellefonte, PA, USA). The chromatographic conditions employed were as follows: carrier gas, helium; split ratio, 1 : 100; injection volume, 1 µL; injector temperature, 220 °C; detector temperature, 230 °C; oven temperature programme, initially 100 °C and increased at 20 °C min^{-1} to

140 °C, and at 8 °C min^{-1} to 200 °C. Crotonic acid was used as the internal standard.

2.5 Serum biochemical analysis

Concentrations of the following serum biochemical parameters were determined using commercial reagents and a SABA 18 biochemistry analyser (Analyzer Medical System, Rome, Italy): triacylglycerols, total cholesterol, low-density lipoprotein cholesterol (LDL-C), high-density lipoprotein cholesterol (HDL-C), glucose, NEFAs, BHB, and activities of gamma-glutamyltransferase (GGT), aspartate aminotransferase (AST), alanine aminotransferase (ALT), and alkaline phosphatase (ALP).

2.6 Statistical data analysis

Cows were grouped according to the lactation period (days 30 and 150 of lactation and the dry period). Results are presented as arithmetic means $\pm$ SEM. The normality of distribution was checked using the Shapiro–Wilk W test. For cases in which the data were not normally distributed, the Box–Cox transformation was used. The significance of differences between groups was checked using repeated measured analysis of variance and the Tukey's HSD test. Differences were considered statistically significant at $P \leq 0.05$. Statistical analysis was performed using the statistical software package Statistica, version 12 (StatSoft, Inc., Tulsa, OK, USA).

3 Results

The average daily production, milk fat, and protein in 1 kg of milk around days 30 and 150 of lactation are shown in Table 2. A significantly higher percentage of protein was detected in the milk of cows around day 150 compared to day 30 of lactation ($P = 0.01$). However, no significant differences were observed in the milk fat percentage ($P > 0.05$). In Table 3 it can be seen that the most prevalent FAs around days 30 and 150 of lactation were SFAs, with palmitic acid being the most common (C16:0). Fatty acids with one double bond, or monounsaturated fatty acids (MUFAs), were the second most represented, with oleic acid (C18:1n-9) showing

Table 3. Distribution of milk fatty acids (%) around days 30 and 150 of lactation in Holstein cows.

	Day 30	Day 150	SEM	P value
SFA	63.05	65.28	0.48	0.001
Palmitic acid (C16:0)	35.92	37.71	0.27	0.001
MUFA	26.93	24.35	0.42	0.001
Oleic acid (C18:1n-9)	22.67	19.61	0.39	0.001
PUFA	2.78	1.99	0.07	0.001
Linoleic acid (C18:2n-6c)	2.25	1.63	0.07	0.001

Results presented as mean ± SEM; SFA: saturated fatty acids; MUFA: monounsaturated fatty acids; PUFA: polyunsaturated fatty acids.

Table 4. Fatty acid composition (%) of milk around days 30 and 150 of lactation in Holstein cows.

	Day 30	Day 150	SEM	P value
C10:0	2.67	3.00	0.06	0.007
C12:0	3.31	3.70	0.05	0.001
C14:0	11.84	12.03	0.08	0.226
C14:1	1.21	1.44	0.03	0.001
C15:0	1.72	1.63	0.02	0.005
C15:1	0.02	0.16	0.03	0.116
C16:0	35.92	37.71	0.27	0.001
C16:1	2.94	3.03	0.06	0.457
C17:0	0.65	0.55	0.03	0.133
C17:1	0.09	0.11	0.03	0.821
C18:0	6.86	6.67	0.11	0.415
C18:1n-9	22.67	19.61	0.39	0.001
C18:2n-6c	2.25	1.63	0.07	0.001
C18:3n-3	0.54	0.36	0.03	0.001
C 20:1	0.09	0.03	0.02	0.116

Results presented as mean ± SEM.

the highest percentage. The least represented were polyunsaturated fatty acids (PUFAs), with linoleic acid (C18:2n-6c) being the most common. Significantly lower percentages of C10:0, C12:0, C14:1, and C16:0 on day 30 compared to day 150 of lactation ($P = 0.007$, 0.001, 0.001, and 0.001, respectively) are shown in Table 4. Significantly higher percentages of C15:0, C18:1n-9, C18:2n-6c, and C18:3n-3 were observed around day 30 compared to day 150 of lactation ($P = 0.005$, $P = 0.001$, $P = 0.001$, and $P = 0.001$, respectively). The FA composition of milk showed no significant differences in the levels of C14:0, C15:1, C16:1, C17:0, C17:1, C18:0, and C20:1 among the various stages of lactation ($P > 0.05$).

A significantly higher proportion of C18:1 / C18:0 and higher unsaturated/saturated fatty acid (UFA / SFA) ratio around day 30 compared to day 150 of lactation ($P = 0.01$ and 0.001, respectively) are shown in Fig. 1.

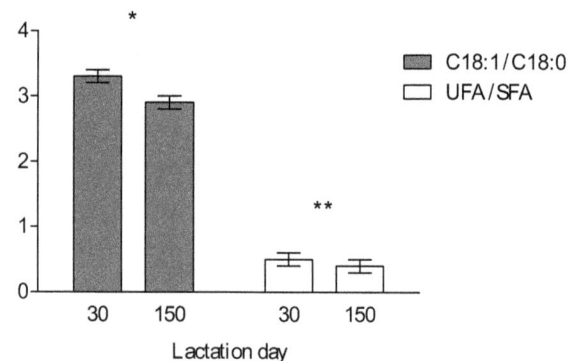

Figure 1. Ratio of C18:1 / C18:0 and UFA / SFA (unsaturated fatty acids / saturated fatty acids) in the milk of Holstein cows around days 30 and 150 of lactation; * significant difference between sampling periods at $P < 0.05$; ** significant difference between sampling periods at $P < 0.01$.

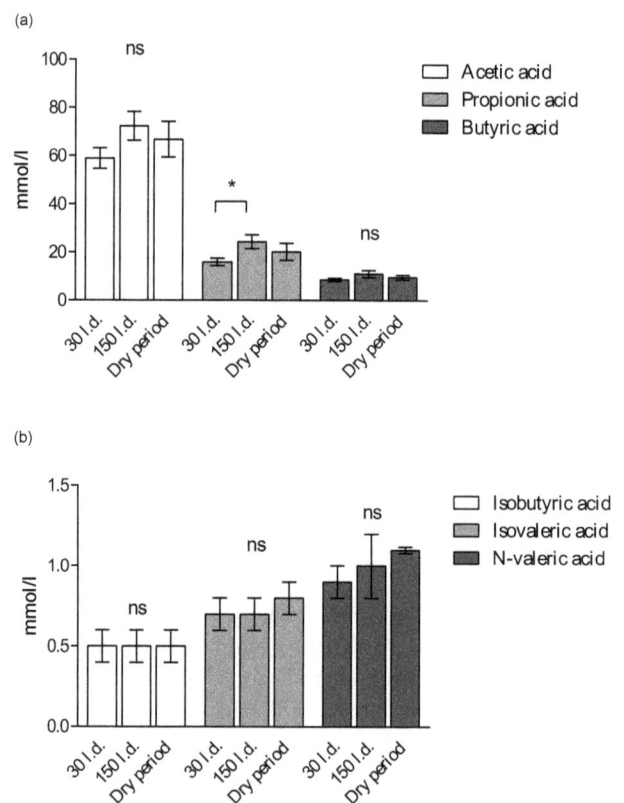

Figure 2. RSCFA concentrations in the rumen contents of Holstein cows during the production period: (a) concentrations of acetic, propionic, and butyric acids; (b) concentrations of isobutyric, isovaleric, and n-valeric acid. l.d.: lactation day; ns: non-significant; * significant difference between sampling periods at $P < 0.05$.

A significantly higher concentration of propionic acid around day 150 ($P = 0.04$) compared to day 30 of lactation is shown in Fig. 2a. Concentrations of acetic and butyric acids showed no significant differences among the periods

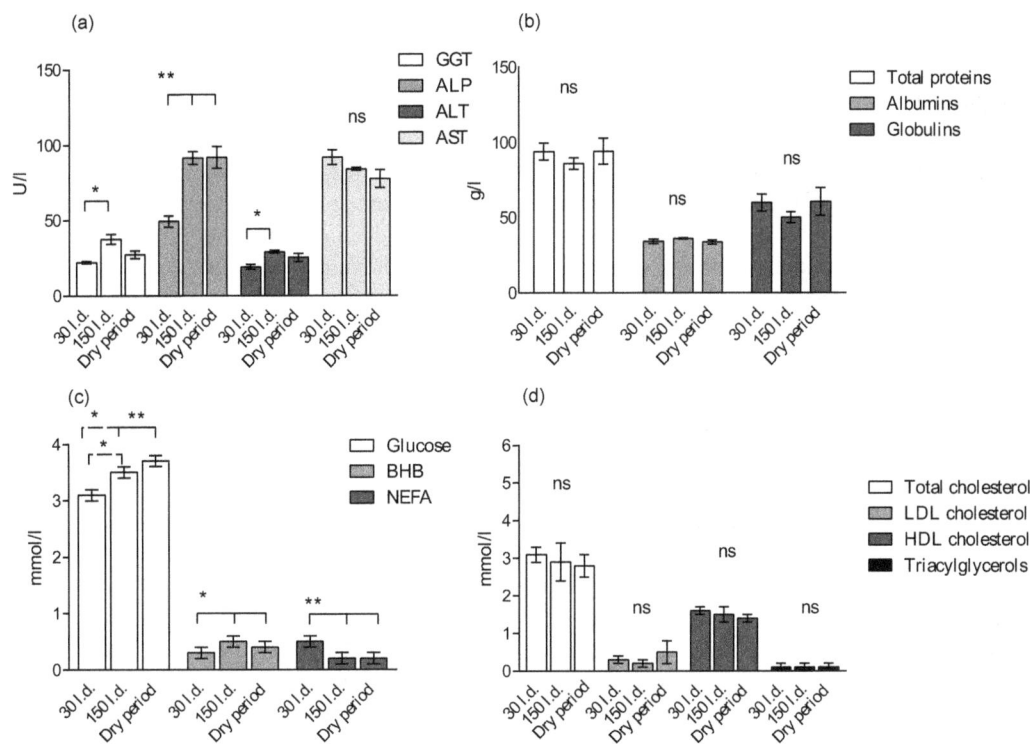

Figure 3. Serum biochemical parameters in Holstein cows during the production period: **(a)** activities of gamma-glutamyltransferase (GGT), aspartate aminotransferase (AST), alanine aminotransferase (ALT), and alkaline phosphatase (ALP); **(b)** concentrations of total proteins, albumins and globulins; **(c)** concentrations of glucose, beta-hydroxybutyric acid (BHB) and non-esterified fatty acids (NEFAs); **(d)** concentrations of total, low-density lipoprotein (LDL) and high-density lipoprotein (HDL) cholesterol, and triacylglycerols. l.d.: lactation day; ns: non-significant; * significant difference between sampling periods at $P < 0.05$; ** significant difference between sampling periods at $P < 0.01$.

under investigation. Figure 2b illustrates the concentrations of isobutyric, isovaleric, and *n*-valeric acids, with no significant differences among the periods under investigation.

Significantly higher levels of GGT ($P = 0.02$), ALP ($P = 0.001$), and ALT ($P = 0.02$) activity around day 150 compared to day 30 of lactation are shown in Fig. 3a. In addition, significantly higher ALP activity was observed around day 30 of lactation compared to the dry period ($P = 0.001$). Activity of AST showed no significant difference among the periods under investigation. The concentrations of total proteins, albumins and globulins in the serum of cows, with no significant differences among the periods under investigation, are shown in Fig. 3b. A significantly higher concentration of glucose during the dry period compared to that measured around days 30 ($P = 0.001$) and 150 ($P = 0.03$) of lactation is shown in Fig. 3c. The concentration of glucose was also significantly higher at day 150 compared to that measured at day 30 of lactation ($P = 0.03$). The concentration of BHB was significantly lower around day 30 compared to day 150 of lactation ($P = 0.01$) and during the dry period ($P = 0.02$). The serum NEFA concentration was significantly higher at day 30 of lactation compared to that measured at day 150 ($P = 0.002$) and during the dry pe-

riod ($P = 0.002$). The concentrations of total, LDL and HDL cholesterol, and triacylglycerols, with no significant differences among the periods under investigation, are shown in Fig. 3d.

4 Discussion

Fatty acid composition of milk in the present study is in accordance with results obtained by Hanuš et al. (2016) for Holstein cows reared in a silage-based feeding system. In the present study, the most represented group of FAs in milk was SFAs, followed by MUFAs and PUFAs, which is consistent with the results of Stádník et al. (2015). Grummer (1991) found that cow's milk normally contains 70 % SFAs, 25 % MUFAs, and 5 % PUFAs, whereas in the present study, the SFAs and PUFAs accounted for 64.3 ± 1.7 and 2.8 ± 0.2 % of the FAs, respectively. At the beginning of lactation, the majority of high-producing dairy cows are in negative energy balance, resulting in increased mobilization of adipose FAs and the incorporation of these FAs in the milk (Palmquist et al., 1993). In ruminant adipose tissue, FAs stored as triglycerides comprise mainly C16:0, C18:0, and C18:1 cis-9 (Chilliard et al., 2000). When lipolysis is high, the FA composition of milk has a much higher proportion of C18:0 (Barber et al., 1997). The high uptake of long-chain FAs by mammary gland tissue inhibits de novo synthesis of FAs through the inhibition of acetyl-coenzyme A carboxylase (Palmquist et al., 1993), as almost all C4:0 to C14:0 FAs and approximately half of C16:0 FAs in milk are derived from de novo synthesis (Grummer, 1991). The increase in C18:1n-9, C18:2n-6c, and C18:3n-3 in the present study indicates an increase in fat mobilization from reserves during early lactation. Lower percentages of short-chain FAs (C10:0–C16:0), as well as higher percentages of long-chain FAs, indicate negative energy balance, which is consistent with the findings of Bastin et al. (2011).

In the present study, a significantly higher percentage of SFAs around day 150 was associated with the later stages of lactation (Komprda et al., 2005), when the animals were no longer in negative energy balance (Ducháček et al., 2012; Stádník et al., 2015). Unlike the SFAs, significantly higher levels of MUFAs and PUFAs were observed around day 30 compared to day 150 of lactation. Furthermore, Stádník et al. (2015) reported a higher percentage of MUFAs during early lactation, with a gradual decline that was accompanied by a less severe negative energy balance. Lower levels of MUFAs indicate a well-balanced energy intake in cows. Reports of milk PUFA content in the literature are inconsistent. Specifically, Stádník et al. (2015) reported no significant differences in the PUFA content of milk during five lactation weeks in Czech Fleckvieh cows, whereas Bastin et al. (2011) reported that Holstein cows had a lower percentage of PUFAs in milk in early lactation. In the present study, reduced SFAs and increased MUFA + PUFA (UFA) around day 30 of lactation indicated a negative energy balance, which can lead to disturbances in the development and maturation of follicles, reduced frequency of ovulation, weaker development of the corpus luteum and a delayed onset of the sexual cycle after parturition (Tamadon et al., 2011). Furthermore, in the present study, the ratio of C18:1 / C18:0 around day 150 was 3.3, in comparison to 2.9 around day 30 of lactation, which may indicate lower levels of desaturase activity in later lactation (DePeters et al., 1995). In the present study, a higher ratio of UFA / SFA was detected in early lactation, which may indicate a disruption in reproductive activity; Stádník et al., 2015) These results suggest the possibility of the use of FA contents in milk to monitor fertility in dairy cows.

The amounts and ratio of RSCFAs can vary, depending on the composition of the feed, extent of intestinal absorption, and the time of sampling. In the present study, animals were fed hay, haylage, silage, and a source of cellulose, which are metabolized in the rumen to acetic acid (Seymour et al., 2005). The highest concentration of acetic acid in the present study was observed around day 150 of lactation, although this was not significant. The second highest concentration was that of propionic acid and the lowest was that of butyric acid, which was consistent with RSCFA levels reported by Thomas (2012). Although not statistically significant, the

highest concentration of butyric acid measured in the present study was around day 150 of lactation. A major part of the butyrate that is produced in the rumen is oxidized to BHB during absorption across the ruminal epithelium (Weigand et al., 1972; Kristensen et al., 2012), which is consistent with the results of the present study that reflected the highest BHB concentration around day 150 of lactation. When rumen fermentation declines, the absorption of butyrate across the ruminal epithelium is reduced; thus, as the epithelial formation of BHB is also reduced, this leads to lower levels in the plasma (Agenäs et al., 2003).

The most important parameters in the assessment of energy status are the levels of BHB and NEFAs in the blood (LeBlanc, 2010). Beta-hydroxybutyrate originates in the ruminal epithelium during absorption and oxidation of butyrate (Weigand et al., 1972; Kristensen et al., 2012). In the present study, the highest BHB concentration was observed around day 150 of lactation, which is consistent with the period around which the highest concentration of butyrate was observed. Blood NEFA has been shown to reflect lipolysis in cattle (Laarveld et al., 1981). When the concentration of NEFA increases, this is indicative of negative energy balance (Vernon, 2005). In this study, significantly higher concentrations of NEFA were detected around day 30 compared to day 150 of lactation and the dry period. Mobilization of fat stores in adipose tissue to provide the mammary gland with FAs for milk fat synthesis is mirrored by an increase in plasma NEFAs during feed deprivation (Agenäs et al., 2003). The higher concentration of NEFAs observed in the present study around day 30 of lactation is indicative of mild lipolysis, which is not typically followed by an increase in the levels of BHB.

5 Conclusions

In conclusion, the results of the present study suggest the possible use of milk FAs, particularly long-chain FAs, as indicative of energy status in dairy cows. Although milk FAs are not always fully applicable to the prediction of specific metabolic states, milk samples are more easily obtained and are more suitable with respect to animal welfare considerations.

Author contributions. LV carried out the research, performed the statistical analysis, analysed and interpreted the data, participated in the design of the manuscript, and gave final approval of the version to be published; JA carried out the research and contributed to conception and design of the manuscript; DO, DB, and LC contributed to design of the manuscript; and ZS revised the manuscript for critically important intellectual content and designed the research.

Competing interests. The authors declare that they have no conflict of interest.

Acknowledgements. The results of this study were achieved through projects financed by the Ministry of Science, Education and Sports of the Republic of Croatia titled "Mineral metabolism of domestic animals in high production and stress" (053-1080229-2104), for which the authors express their gratitude. The authors also wish to thank Robert Martinec and Hrvoje Ciganović for assisting with sample collection.

Edited by: Manfred Mielenz

References

Agenäs, S., Dahlborn, K., and Holtenius, K.: Changes in metabolism and milk production during and after feed deprivation in primiparous cows selected for different milk fat content, Livest. Prod. Sci., 83, 153–164, 2003.

Arfuso, F., Fazio, F., Levanti, M., Rizzo, M., Di Pietro, S., Giudice, E., and Piccione, G.: Lipid and lipoprotein profile changes in dairy cows in response to late pregnancy and the early postpartum period, Arch. Anim. Breed., 59, 429–434, https://doi.org/10.5194/aab-59-429-2016, 2016.

Barber, M. C., Clegg, R. A., Travers, M. T., and Vernon, R. G.: Lipid metabolism in the lactating mammary gland, Biochim. Biophys. Acta – Lipids Lipid Metab., 1347, 101–126, 1997.

Bastin, C., Gengler, N., and Soyeurt, H.: Phenotypic and genetic variability of production traits and milk fatty acid contents across days in milk for Walloon Holstein first-parity cows, J. Dairy Sci., 94, 4152–4163, 2011.

Bauman, D. E., Mather, I. H., Wall, R. J., and Lock, A. L.: Major advances associated with the biosynthesis of milk, J. Dairy Sci., 89, 1235–1243, 2006.

Berry, D. P., Veerkamp, R. F., and Dillon, P.: Phenotypic profiles for body weight, body condition score, energy intake, and energy balance across different parities and concentrate feeding levels, Livest. Sci., 104, 1–12, 2006.

Butler, S. T., Pelton, S. H., Knight, P. G., and Butler, W. R.: Follicle-stimulating hormone isoforms and plasma concentrations of estradiol and inhibin A in dairy cows with ovulatory and non-ovulatory follicles during the first postpartum follicle wave, Domest. Anim. Endocrinol., 35, 112–119, 2008.

Chilliard, Y., Ferlay, A., Mansbridge, R., and Doreau, M.: Ruminant milk fat plasticity: nutritional control of saturated, polyunsaturated, trans and conjugated fatty acids, Ann. Zootech., 49, 181–205, 2000.

DePeters, E. J., Medrano, J. F., and Reed, B. A.: Fatty acid composition of milk fat from three breeds of dairy cattle, Can. J. Anim. Sci., 75, 267–269, 1995.

Ducháček, J., Stádník, L., Ptáček, M., Beran, J., Okrouhlá, M., Čítek, J., and Stupka, R.: Effect of cow energy status on the hypercholesterolaemic fatty acid proportion in raw milk, Czech. J. Food. Sci., 32, 273–279, 2014.

Ducháček, J., Vacek, M., Stádník, L., Beran, J., and Okrouhlá, M.: Changes in Milk Fatty Acid Composition in Relation To Indicators of Energy Balance in Holstein Cows, Acta Univ. Agric. et Silvic. Mendel. Brun., 1, 29–38, 2012.

Duffield, T. F., Sandals, D., Leslie, K. E., Lissemore, K., McBride, B. W., Lumsden, J. H., Dick, P., and Bagg, R.: Effect of prepartum administration of monensin in a controlled-release capsule on postpartum energy indicators in lactating dairy cows, J. Dairy Sci., 81, 2354–2361, 1998.

Fiore, E., Gianesella, M., Arfuso, F., Giudice, E., Piccione, G., Lora, M., Stefani, A., and Morgante, M.: Glucose infusion response on some metabolic parameters in dairy cows during transition period, Arch. Anim. Breed., 57, 1–9, 2014.

Fiore, E., Piccione, G., Gianesella, M., Praticò, V., Vazzana, I., Dara, S., and Morgante, M.: Serum thyroid hormone evaluation during transition periods in dairy cows, Arch. Anim. Breed., 58, 403–406, https://doi.org/10.5194/aab-58-403-2015, 2015.

Fiore, E., Piccione, G., Perillo, L., Barberio, A., Manuali, E., Morgante, M., and Gianesella, M.: Hepatic lipidosis in high-yelding dairy cows during the transition period: haematochemical and hostopathological findings, Anim. Prod. Sci., 57, 74–80, 2017.

Gross, J., van Dorland, H. A., Bruckmaier, R. M., and Schwarz, F. J.: Milk fatty acid profile related to energy balance in dairy cows, J. Dairy Res., 78, 479–488, 2011.

Grummer, R. R.: Effect of feed on the composition of milk fat, J. Dairy Sci., 74, 3244–3257, 1991.

Hara, A. and Radin, N. S.: Lipid extraction of tissues with a low-toxicity solvent, Anal. Biochem., 90, 420–426, 1978.

Hanuš, O., Křížová, L., Samková, E., Špicka, J., Kucera, J., Klimešová, M., Roubal, P., and Jedelská, R.: The effect of cattle breed, season and type of diet on the fatty acid profile of raw milk, Arch. Anim. Breed., 59, 373–380, https://doi.org/10.5194/aab-59-373-2016, 2016.

Komprda, T., Dvořák, R., Fialová, M., Šustová, K., and Pechová, A.: Fatty acid content in milk of dairy cows on a diet with high fat content derived from rapeseed, Czech J. Anim. Sci., 7, 311–319, 2005.

Kristensen, N. B., Gäbel, G., Pierzynowski, S. G., and Danfær, A.: Portal recovery of short-chain fatty acids infused into the temporarily–isolated and washed reticulo-rumen of sheep, Br. J. Nutr., 84, 477–482, 2012.

Laarveld, B., Christensen, D. A., and Brockman, R. P.: The effect of insulin on net metabolism of glucose and amino acids by the bovine mammary gland, Endocrinology, 108, 2217–2221, 1981.

LeBlanc, S.: Monitoring metabolic health of dairy cattle in the transition period, J. Reprod. Dev., 56, 29–35, 2010.

Leek, B. F.: Digestion in the Ruminant Stomach, in: Duke's Physiology of Domestic Animals, edited by: Reece, W. O., Cornell University Press, Ithaca, NY, USA, 438–474, 2004.

Mulligan, F. J., O'grady, L., Rice, D. A., and Doherty, M. L.: A herd health approach to dairy cow nutrition and production diseases of the transition cow, Anim. Reprod. Sci., 96, 331–353, 2006.

Murphy, M. R., Baldwin, R. L., and Koong, L. J.: Estimation of stoichiometric parameters for rumen fermentation of roughage and concentrate diets, J. Anim. Sci., 55, 411–421, 1982.

Palmquist, D. L., Beaulieu, A. D., and Barbano, D. M.: Feed and Animal Factors Influencing Milk Fat Composition, J. Dairy Sci., 76, 1753–1771, 1993.

Payne, J. M. and Payne, S.: The metabolic profile test, Oxford University Press, NY, USA, 1987.

Rule, D. C.: Direct transesterification of total fatty acids of adipose tissue, and of freeze-dried muscle and liver with boron-trifluoride in methanol, Meat. Sci., 46, 23–32, 1997.

Schulz, K., Frahm, J., Meyer, U., Kersten, S., Reiche, D., Rehage, J., and Dänicke, S.: Effects of prepartal body condition score and peripartal energy supply of dairy cows on postpartal lipolysis, energy balance and ketogenesis: an animal model to investigate subclinical ketosis, J. Dairy Res., 81, 257–266, 2014.

Seymour, W. M., Campbell, D. R., and Johnson, Z. B.: Relationships between rumen volatile fatty acid concentrations and milk production in dairy cows: A literature study, Anim. Feed. Sci. Technol., 119, 155–169, 2005.

Soyeurt, H., Dardenne, P., Gillon, A., Croquet, C., Vanderick, S., Mayeres, P., Bertozzi, C., and Gengler, N.: Variation in fatty acid contents of milk and milk fat within and across breeds, J. Dairy Sci., 89, 4858–4865, 2006.

Stádník, L., Ducháček, J., Beran, J., Toušová, R., and Ptáček, M.: Relationships between milk fatty acids composition in early lactation and subsequent reproductive performance in Czech Fleckvieh cows, Anim. Reprod. Sci., 155, 75–79, 2015.

Tamadon, A., Kafi, M., Saeb, M., and Ghavami, M.: Association of milk yield and body condition score indices with the commencement of luteal activity after parturition in high producing dairy cows, Iran J. Vet. Res., 12, 184–191, 2011.

Thomas, H. H.: Gastrointestinal physiology and metabolism, in: Cunningham's Textbook of Veterinary Physiology, edited by: Klein, B. G., Elsevier, Health Sciences Division, PA, USA, 305–382, 2012.

Vernon, R. G.: Lipid metabolism during lactation: a review of adipose tissue-liver interactions and the development of fatty liver, J. Dairy Res., 72, 460–469, 2005.

Walsh, S. W., Williams, E. J., and Evans, A. C. O.: A review of the causes of poor fertility in high milk producing dairy cows, Anim. Reprod. Sci., 123, 127–138, 2011.

Weigand, E., Young, J. W., and McGilliard, A. D.: Extent of butyrate metabolism by bovine ruminoreticulum epithelium and the relationship to absorption rate, J. Dairy Sci., 55, 589–597, 1972.

10

Genome-wide QTL mapping results for regional DXA body composition and bone mineral density traits in pigs

Sophie Rothammer[1], **Maren Bernau**[2], **Prisca V. Kremer-Rücker**[2,3], **Ivica Medugorac**[1], **and Armin M. Scholz**[2]

[1]Chair of Animal Genetics and Husbandry, LMU Munich, 80539 Munich, Germany
[2]Livestock Center of the Faculty of Veterinary Medicine, LMU Munich, 85764 Oberschleissheim, Germany
[3]University of Applied Sciences Weihenstephan-Triesdorf, 91746 Weidenbach, Germany

Correspondence to: Sophie Rothammer (s.rothammer@gen.vetmed.uni-muenchen.de)

Abstract. In a previous study, genome-wide mapping of quantitative trait loci (QTL) for five body composition traits, three bone mineral traits and live weight was performed using whole-body dual-energy X-ray absorptiometry (DXA) data. Since QTL for bone mineral traits were rare, the current study aimed to clarify whether the mapping results were influenced by the analysed body regions. Thus, the same material (551 pigs) and methods as in the whole-body QTL mapping study were used. However, for evaluation of the DXA scans, we manually defined two body regions: (i) from the last ribs to the pelvis (A) and (ii) including the pelvis and the hind limbs (P). Since live weight was not affected by the regional analysis, it was omitted from the QTL mapping design.

Our results show an overall high consistency of mapping results especially for body composition traits. Two thirds of the initial whole-body QTL are significant for both A and P. Possible causes for the still low number of bone mineral QTL and the lower consistency found for these traits are discussed. For body composition traits, the data presented here show high genome-wide Pearson correlations between mapping results that are based on DXA scans with the time-saving "whole-body standard setting" and mapping results for DXA data that were obtained by time-consuming manual definition of the regions of interest. However, our results also suggest that whole-body or regional DXA scans might generally be less suitable for mapping of bone mineral traits in pigs. An analysis of single reference bones could be more useful.

1 Introduction

In a previous study, whole-body dual-energy X-ray absorptiometry (DXA) data were used to perform a genome-wide mapping of quantitative trait loci (QTL) for the following nine traits: fat mass, fat percentage (Fat, FatPC), soft lean tissue mass, soft lean tissue percentage (Lean, LeanPC), live weight (Weight), soft tissue X-ray attenuation coefficient (R), bone mineral content, bone mineral percentage (BMC, BMCPC), and bone mineral density (BMD) (Rothammer et al., 2014). In total, 72 QTL were mapped and promising candidate genes (e.g. *ZNF608*) could be identified. From these 72 QTL, only seven QTL were associated with bone min-

eral traits (two for BMC and BMD, another five for only BMC and none at all for BMCPC). Moreover, only two of these seven QTL were not simultaneously associated with body composition traits. Since it has been extensively shown that both weight and obesity correlate with bone metabolism in humans (reviewed by Cao, 2011; Zhao et al., 2008), associations of QTL with bone mineral traits could actually be indirectly caused by associations of these loci with body composition traits. The number of real bone mineral QTL would thus be reduced even more. However, this would be an unexpected result for two reasons: in humans, bone mineral traits were found to be strongly influenced by genetic aspects according to observations in families and twins (e.g.

Pocock et al., 1987; Soroko et al., 1994; Hernandez-de Sosa et al., 2014). Furthermore, several QTL and candidate genes for bone mineral traits have already been identified in different species (e.g. Ames et al., 1999; Beamer et al., 2001; Xiao et al., 2012; Yang et al., 2012; Willing et al., 2003; Kaufman et al., 2008), including pigs (Laenoi et al., 2011, 2012; Rangkasenee et al., 2013). In pigs, there has been evidence for a genetic component in the multifactorial leg weakness syndrome, which is associated with osteochondrosis and bone fracture risk and thus with bone mineral traits (Laenoi et al., 2011, 2012; Rangkasenee et al., 2013). This syndrome affects animal welfare and profitableness in swine production (Fukawa and Kusuhara, 2001). The identification of QTL for bone mineral traits might therefore contribute to solving this serious problem (Laenoi et al., 2011, 2012; Rangkasenee et al., 2013). In contrast to our previous study (Rothammer et al., 2014), bone mineral measurements in humans by DXA scan are typically not conducted using a whole-body mode but restricted to distinct sites, e.g. the femoral neck or the lumbar spine (Wright et al., 2014). In this context, it has been shown that measured values can vary "substantially from site to site" (Arlot et al., 1997). This, in turn, raised the question of whether the whole-body mode that was used in the previous study had caused an underestimation of actual QTL numbers for bone mineral and possibly even body composition traits. In order to clarify this question, we repeated the mapping procedure for two manually defined body regions and compared the results among themselves and with the whole-body results of the previous study.

2   Materials and methods

Since the same material and methods as in the previous study (Rothammer et al., 2014) were used, the animal set comprised 551 pigs that were between 160 and 200 days old and that were a mixture of the following breeds in various proportions: Large Black, Pietrain, Duroc, Schwäbisch-Hällisch, Cerdo Iberico (Lampiño variety), European wild boar, and Hampshire. For all 551 animals, genotype data (Illumina's PorcineSNP60 Genotyping BeadChip) and DXA data were available. Based on these data, a genome-wide combined linkage disequilibrium and linkage analysis (Meuwissen et al., 2002) was conducted using 44 611 20-SNP sliding windows. Variance component analyses were carried out at the midpoint of each sliding window using ASReml (Gilmour et al., 2009). The underlying mixed linear model included random QTL and polygenic and fixed effects of sex, housing, season and age. Finally, the likelihood ratio test statistic (LRT) was calculated using the logarithm of the likelihood estimated by ASReml for the model with (logLP) and without (logL0) QTL effects ($\text{LRT} = -2(\text{logL0} - \text{logLP})$). For more details on material and methods, see Rothammer et al. (2014).

In contrast to the previous study, DXA scans were not evaluated using the standard "whole-body" (WB) setting. Instead, two distinct regions were manually defined as shown in Fig. 1: (i) the region from the last ribs to the pelvis, called "A" for "abdominal" in the following, and (ii) the region including the pelvis and the hind limbs, named "P" for "pelvic". As can be seen in Fig. 1, region A includes only the lumbar spine as the bony fraction and is thus comparable to bone mineral measurements in humans (according to the International Society for Clinical Densitometry (ISCD), the lumbar spine is an adequate and often used site for the diagnosis of osteoporosis or low bone mass in humans; Schousboe et al., 2013; Wright et al., 2014; El Maghraoui and Roux, 2008).

Moreover, since live weight is not affected by the region-of-interest analysis, this study concentrated on only eight traits for each body region (A and P): Fat, FatPC, Lean, LeanPC, R, BMC, BMCPC and BMD.

Mapping results for regions A, P and WB were evaluated in two ways: (i) according to the (missing) overlap of significant QTL and (ii) using genome-wide Pearson correlations that were calculated based on the LRT values of all 44 611 20-SNP sliding windows for each combination of DXA regions (WB × A, WB × P and A × P). Calculation of genome-wide Pearson correlations was done for all eight traits separately as well as for three distinct combinations (bone mineral traits combined, body composition traits combined and all traits combined).

In addition, a linear regression analysis was performed to calculate the phenotypic relationship between whole-body DXA BMD and whole-body DXA fat mass using "Proc Reg" of SAS® 9.3 software (©SAS Institute Inc., SAS Campus Drive, Cary, North Carolina 27513, USA).

3   Results and discussion

As can be seen in Fig. 2 and in the complete overview that is given in Fig. S1–8 in the Supplement, the mapping results (peak positions) were overall very similar for WB, A and P even though peak heights varied slightly. In the previous study (Rothammer et al. 2014) 72 QTL were detected in total. Eight of these QTL, however, were significant for only live weight. Thus, 64 QTL were detected for traits that were also investigated in the current study. From these, 59 QTL could be confirmed either by A (5), P (8) or both (46). Besides the QTL that had already been mapped for WB, additional QTL were identified for A (32), P (5) and the combination of both (8) (see Table S1 in the Supplement for unconfirmed WB QTL and newly identified A or P QTL). In general, these discrepancies are in most cases not due to the complete absence of former or the existence of new QTL but rather to the fact that the peaks of these QTL are close to the significance threshold (Bonferroni corrected P value < 0.001). Thus, differences in peak size cause formerly suggestive QTL ($P < 0.05$) to become significant and

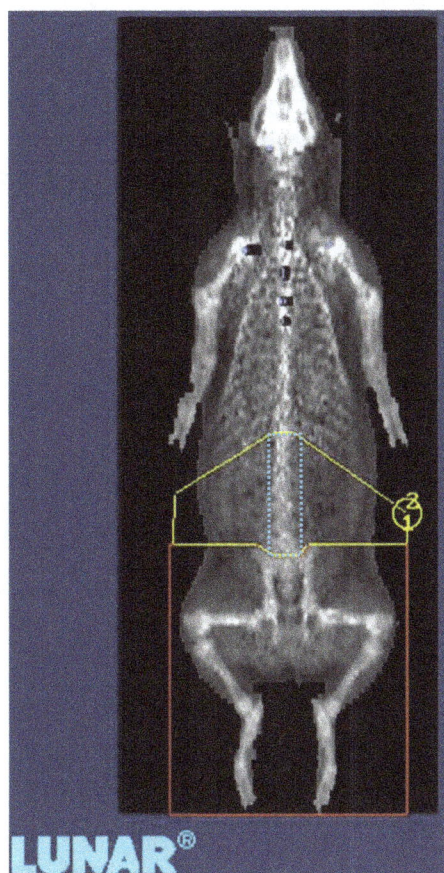

Figure 1. DXA scan showing the manually defined body regions. This screenshot of a DXA scan evaluation shows the manually defined body regions A (yellow) and P (red). While region A extends from the costal arch to the pelvis region, P includes the pelvis and the hind limbs. Moreover, region A$_B$, which was used for an additional scan and represents A restricted to the bony fraction only, is marked with a blue dotted box.

Table 1. Genome-wide Pearson correlations of QTL mapping results between DXA regions. This table shows the correlations between the genome-wide LRT values of the DXA regions "whole-body" (WB), "pelvic" (P) and "abdominal" (A) for all traits combined, bone mineral traits combined (bone), body composition traits combined (body) and single traits.

Pearson	WB × A	WB × P	A × P
All traits	0.894	0.946	0.846
Body	0.922	0.971	0.896
Bone	0.664	0.838	0.571
Fat	0.906	0.969	0.885
FatPC	0.979	0.978	0.959
Lean	0.651	0.908	0.543
LeanPC	0.982	0.986	0.970
R	0.979	0.987	0.967
BMC	0.720	0.920	0.670
BMCPC	0.570	0.582	0.421
BMD	0.639	0.824	0.511

vice versa (Table S1). The overall high degree of accordance between the mapping results was also confirmed by genome-wide Pearson correlations (Table 1, Fig. 3). Measured over all traits, the genome-wide Pearson correlation was 0.95 for WB and P, and 0.89 for WB and A. The genome-wide Pearson correlation between A and P was, although still high, somewhat lower (0.85). For all traits but FatPC, the same relations were found when traits were considered separately (Table 1). For FatPC, the genome-wide Pearson correlations between WB and A and between WB and P were almost identical (0.979; 0.978). Generally, it can be assumed that the use of the (rather) time-saving WB DXA data provides as reliable and substantial results as the use of the time-consuming manually defined body regions.

However, of the 32 "new" QTL that were mapped for A, 12 (detected for the phenotypes Lean, Fat or R) did not show a suggestive peak in either the WB data or in P (Table S1). For these QTL, two explanations are possible: (i) they rep-

resent QTL with only a local effect or (ii) the DXA scans and consequently the mapping results were biased by organs and/or intestinal contents that naturally occupy an enormous volume in region A, thus increasing the number of false positives. Although (ii) cannot be ruled out without detailed studies of these QTL, the results of a literature study for all genes annotated to these 12 QTL supports assumption (i). For 2 of the 12 regions, only uncharacterised genes were annotated. The decision about a possible candidate status was thus hindered. However, in 7 of the 10 remaining regions, possible candidate genes could be identified: *WBSCR17* (International Mouse Phenotyping Consortium), *SRSF7* (International Mouse Phenotyping Consortium), *USP28* (Valero et al., 2001), *NNMT* (Trammell and Brenner, 2015), *ZBED6* (Clark et al., 2015; Markljung et al., 2009), *NMNAT2* (Hicks et al., 2012), *ABHD6* (Thomas et al., 2013), *PDHB* (Sasaki et al., 2006; Serao et al., 2011) and *PSMA2* (Sakamoto et al., 2009). These are listed in Table 2. These QTL that were found exclusively for region A might indeed represent QTL with exclusively or at least predominantly local effects. Thus, in cases where there is a special interest in a particular region of the body, manually analysed regions of interest might contribute additional knowledge about locally acting QTL.

In accordance with the results for WB bone mineral traits, no QTL was found for BMCPC for the manually defined regions A and P. The number of QTL that were detected for BMC and BMD, however, differed markedly for regions A and P. While 11 QTL were found for region P, only one QTL was found for region A.

The QTL for region A was positioned on chromosome 13 and was verified three times as it was associated with bone mineral traits for not only region A (BMD; at 20 010 946–20 621 104 bp) but also region P (BMC, BMD; at 19 832 456–21 955 312 bp) and WB (BMC; at 19 850 150–

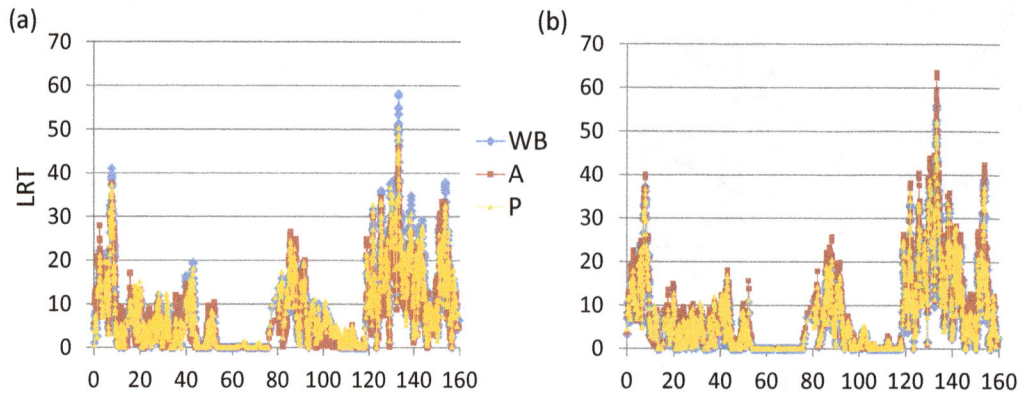

Figure 2. Comparison of local mapping results of whole-body data with manually defined regions. For chromosome 2 (*x* axis in Mb), this figure exemplarily shows the high consistency of the QTL mapping results for the three analysed regions (the whole body (WB), the abdominal (A) and the pelvic (P) region). Chromosome 2 was chosen for illustration as it harbours the genome-wide most significant LRT values for all three regions at position 133 237 232 bp. At this position, significant peaks (LRT > 31.275) were detected for the traits Fat, FatPC, LeanPC, and R (and also Weight for WB) in all analysed regions. However, while Fat was the most significant trait in WB data, R was most significant for A and P. Thus, this figure shows the mapping results for both **(a)** Fat and **(b)** R (soft tissue X-ray attenuation coefficient).

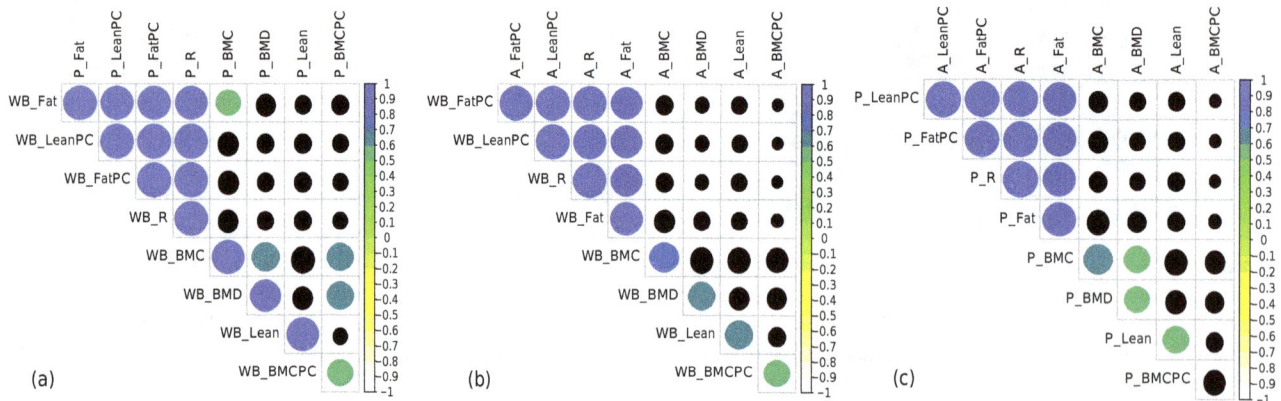

Figure 3. Graphical representations of the genome-wide Pearson correlation matrices for WB × P, WB × A and P × A. This visualisation reflects the correlations between different traits and DXA regions. Based on the diagonal, it is obvious that the correlation between WB and P is even higher than the correlation between WB and A, and between P and A (see also Table 1). Moreover, it is clearly visible that the body composition traits Fat, FatPC, LeanPC and R are highly correlated independent of the investigated region. Confidence matrices and plots were constructed using R (R Development Core Team, 2008) and the package corrplot (Wei, 2013). The order of the traits within the matrix differs according to the first principal components (parameter *order* set to "FPC"). Besides coloration, dot size also indicates the extent of correlation (dot size proportional to absolute value of correlation).

20 339 924 bp). In close proximity to the common region defined by these QTL (20 010 946–20 339 924 bp) lies the gene *CMTM8* (20 414 847–20 549 120 bp). This gene acts as a candidate tumour suppressor gene (Both et al., 2014) and is associated with the EGFR (epidermal growth factor receptor) signalling pathway (Jin et al., 2005) which, in turn, is involved in the suppression of osteoblast differentiation and inhibition of osteoblastic transcription factor expression (Zhu et al., 2011). *CMTM8* can thus be considered as a candidate gene for this QTL.

Concerning region P, all 11 QTL overlapped with QTL that had previously been identified for WB (Rothammer et

al., 2014). Four of these WB QTL, however, had only been identified for body composition traits, not for bone mineral traits (Rothammer et al., 2014). To check the plausibility of these four QTL, a second literature study was conducted. By thorough screening of all positional candidate genes, we were able to identify possible bone mineral candidate gene(s)/hypotheses for each QTL:

The region defined by the QTL for BMC and Fat at chr3: 11 090 781–11 719 832 bp is part of the critical region associated with Williams–Beuren syndrome (WBS) (Francke, 1999). This syndrome in humans is caused by heterozygous deletions of about 1.6 Mb, and one of the numerous symp-

Table 2. Candidate genes for QTL detected for region A only.

Chr	QTL position	Phenotype	GeneID	Description of association
3	14 538 518–15 532 954	Fat	WBSCR17	decreased lean body mass in mutant mice (International Mouse Phenotyping Consortium)
3	107 672 432–108 880 944	Lean	SRSF7	decreased lean body mass in heterozygous mutant mice (International Mouse Phenotyping Consortium)
9	45 9277 52–46 989 664	Lean	USP28	preferential expression in heart and muscle (Valero et al., 2001)
			NNMT	correlates with adiposity (Trammell and Brenner, 2015)
9	70 779 512–71 704 720	Lean	ZBED6	regulates IGF2 expression and muscle growth in pigs (Clark et al., 2015; Markljung et al., 2009)
9	136 173 120–137 019 760	Lean	NMNAT2	axonal growth → missing innervation → reduction in total skeletal muscle mass (Hicks et al., 2012)
13	43 748 912–44 426 072	Fat	ABHD6	selective knockdown protects mice from high-fat-diet-induced obesity (Thomas et al., 2013)
			PDHB	Candidate gene for intramuscular fat deposition in pigs and cattle (Sasaki et al., 2006; Serao et al., 2011)
18	56 485 824–57 128 560	R	PSMA2	negative correlation with male BMI (Sakamoto et al., 2009)

toms in about 15 % of the cases is hypercalcemia (Cagle et al., 2004). Thus, at least one of the genes in this region must have an influence on blood calcium levels. In two severe cases of hypercalcemia, patients could successfully be treated by calcitonin, which increases the renal clearance of calcium and decreases the activity of osteoclasts, thereby increasing the amount of calcium that is bound in the bones (Cagle et al., 2004). Consequently, if certain a gene(s) in this region is (are) important for correct calcitonin release or function, an influence on bone mineral traits can be hypothesised.

For the QTL affecting BMC and Lean at chr6: 24 770 684–26 159 346 bp, the genes *EDC4* and *NFATC3* may serve as candidate genes for bone mineral traits for the following reasons: according to Seto et al. (2015), the assembly of EDC4 and Dcp1a into processing bodies is critical for the translational regulation of IL-6, which plays a dual role in bone remodelling and bone tumours (Franchimont et al., 2005; Blanchard et al., 2009). Additionally, heterozygous *EDC4* knockout mice showed a change in bone density, though it was not significant (Origins of bone and cartilage disease project, 2015). *NFATC3* regulates the expression of the NF-κB ligand receptor activator in osteoblasts and is therefore involved in osteoclastogenesis (Lee et al., 2011).

The gene *SNCA* is a plausible candidate gene for the QTL associated with BMC at chr8: 138 766 256–139 846 368 bp as it was one of the top three genes correlated with bone mass in rat lines (Alam et al., 2010).

The QTL affecting BMD and Lean at chr9: 26 965 738–28 366 584 bp includes the gene *FAT3*, which was among 75 genes that were differentially expressed in human osteosarcoma cell lines (Luk et al., 2011). Thus, an involvement of *FAT3* in bone homeostasis can at least be assumed.

None of these four P QTL were detected for bone mineral traits in WB data. However, there was also one bone mineral QTL that was mapped only in WB data and not in either A or P. This QTL affected BMC and Lean and was positioned at chr12: 41 715 472–44 560 156 bp. Within this region, there are a number of cytokine genes (*CCL1*, *CCL2*, *CCL8*, *CCL11*). It is generally known that immune cells and cytokines are involved in regulating bone turnover (summarised in Nicolaidou et al., 2012). Of particular importance for bone remodelling is *CCL2* (also MCP-1), which is associated, for example, with the recruitment of monocytes to areas of bone formation and bone resorption (reviewed in Yadav et al., 2010). Moreover, this QTL reached suggestive LRT values in both A (30.18) and P (26.46). Of the four QTL that were significant for only region P, three showed suggestive LRT values for WB data. These facts and the promising results of the literature search thus affirm the basic ability of our methodology to map bone mineral traits.

Nevertheless, the mapping results and Pearson correlations for bone mineral traits, especially BMCPC and BMD, obviously show less accordance than those for body composition traits. Explanations for these findings can be (i) bone mineral traits are less influenced by genetics; (ii) for reliable bone mineral data, body regions must be even more restricted, e.g. to single bones; and (iii) the DXA method might not be perfectly suitable for measuring bone mineral traits in pigs in general.

Since bone mineral traits, osteoporosis and/or osteochondrosis have already been associated with distinct genomic regions, genes and/or variants in diverse species including humans, mice and pigs (Ames et al., 1999; Beamer et al., 2001; Willing et al., 2003; Xiao et al., 2012; Yang et al., 2012; Rangkasenee et al., 2013), explanation (i) seems to be rather unlikely. However, a plausible explanation might be the existence of numerous QTL with only small effects that could hinder their effective mapping (Bian and Holland, 2015; Heffner et al., 2009). To assess the probability of explanation (ii), we restricted region A to the bony fraction only (A_B; see Fig. 1) and determined DXA values for 20 arbitrarily chosen animals. We then estimated phenotypic Pearson correlations of region A and A_B and found very high correlations of 0.91 for BMD and of 0.97 for BMC. Consequently, explanation (ii) is invalid for the lumbar region. Post-mortem QTL studies on bone mineral traits for ulna and radius, however, resulted in associated candidate genes like *MMP3* (Laenoi et al., 2012) and *KRT8* (Rangkasenee et al., 2013). Although the total number of QTL for bone mineral traits was also low here, scanning single bones might nonetheless be advantageous. At a first glance, explanation (iii) also seems implausible, especially when taking into account that DXA scans are still the gold standard for bone mineral trait measurements in humans. However, there are important differences between humans and pigs. To begin with, humans can be precisely positioned in a standardised way, while the same is much more complicated with anesthetised pigs. Since DXA reduces three-dimensional bodies to two-dimensional images (BMD is measured in $\mathrm{g\,cm^{-2}}$, whereas volume is defined as $\mathrm{g\,cm^{-3}}$), it is sensitive to positioning errors and also bone size (Carter et al., 1992; Vogl et al., 2011; El Maghraoui and Roux, 2008). Furthermore, osteoporosis in humans is not directly diagnosed by BMD but by the T score, which considers the mean BMD values and standard deviations of young adults of the same gender and ethnicity (El Maghraoui and Roux, 2008). It can thus be hypothesised that T scores might serve as more reliable phenotypes for QTL mapping than BMD values. However, since similar values for estimation of T scores based on BMD are missing for pigs, this hypothesis has not been verified yet. Based on the fact that the mechanical properties of bones are determined by BMD and bone microarchitecture, another shortcoming of the DXA method is its inability to differentiate between cortical and trabecular bone (Ulrich et al., 1999; Sornay-Rendu et al., 2006). Moreover, it has already been shown that increased body fat causes deviations in BMD measurements by DXA scans and, to a much lesser extent, also in quantitative computed tomography (QCT) (Yu et al., 2012). This finding is further supported by the observation of a medium phenotypic relationship between BMD and fat mass ($R^2 = 0.32$; RMSE $= 0.057\,\mathrm{g\,cm^{-2}}$) in our own material. Whole-body BMD increases with growing body fat mass (DXA BMD $= 0.9978 + 5.397 \times 10^{-6} \times$ DXA Fat). Since the breeding objectives of some breeds that were used in this study differ strongly with regard to fatness and leanness, measurements of bone mineral content and bone mineral density could indeed be biased and thus hinder effective QTL mapping. Consequently, it might be useful for mapping of bone mineral QTL if only crossbreeds between two breeds (e.g. Duroc × Pietrain as shown in Laenoi et al., 2011, 2012, and Rangkasenee et al., 2013) or at least similar breeds were used.

4   Conclusion

All in all, it can be stated that, especially for body composition traits, the results from a mapping with whole-body DXA data that were gathered using the time-saving standard setting were in high agreement with the results obtained for manually defined body regions of interest. Thus, time-consuming manual DXA analyses can be restricted to studies in which local QTL are of particular interest. For certain bone mineral traits in pigs, however, it might be more useful to study reference bones by DXA or, for example, alternatively by QCT instead of using whole-body or regional DXA analysis.

Author contributions. S. Rothammer designed and coordinated the mapping study, performed data analysis, interpreted data and drafted the manuscript. M. Bernau performed DXA analyses, blood sampling, veterinary control, and revised the manuscript. P. V. Kremer-Rücker performed DXA analyses, blood sampling, veterinary control, and revised the manuscript. I. Medugorac designed the mapping study, contributed analysis tools and revised the manuscript. A. M. Scholz designed and coordinated the study and critically revised the manuscript. All authors read and approved the final manuscript.

Competing interests. The authors declare that they have no conflict of interest.

Acknowledgements. The authors thank their colleagues for stimulating discussion, the stockmen for assistance with the DXA analysis, the laboratory staff for technical assistance and E. Kunz for proofreading.

Edited by: S. Maak

References

Alam, I., Carr, L. G., Liang, T., Liu, Y., Edenberg, H. J., Econs, M. J., and Turner, C. H.: Identification of genes influencing skeletal phenotypes in congenic P/NP rats, J. Bone Miner. Res., 25, 1314–1325, doi:10.1002/jbmr.8, 2010.

Ames, S. K., Ellis, K. J., Gunn, S. K., Copeland, K. C., and Abrams, S. A.: Vitamin D receptor gene Fok1 polymorphism predicts calcium absorption and bone mineral density in children, J. Bone Miner. Res., 14, 740–746, doi:10.1359/jbmr.1999.14.5.740, 1999.

Arlot, M. E., Sornay-Rendu, E., Garnero, P., Vey-Marty, B., and Delmas, P. D.: Apparent pre- and postmenopausal bone loss evaluated by DXA at different skeletal sites in women: the OFELY cohort, J. Bone Miner. Res., 12, 683–690, doi:10.1359/jbmr.1997.12.4.683, 1997.

Beamer, W. G., Shultz, K. L., Donahue, L. R., Churchill, G. A., Sen, S., Wergedal, J. R., Baylink, D. J., and Rosen, C. J.: Quantitative trait loci for femoral and lumbar vertebral bone mineral density in C57BL/6J and C3H/HeJ inbred strains of mice, J. Bone Miner. Res., 16, 1195–1206, doi:10.1359/jbmr.2001.16.7.1195, 2001.

Bian, Y. and Holland, J. B.: Ensemble Learning of QTL Models Improves Prediction of Complex Traits, G3 (Bethesda), 5, 2073–2084, doi:10.1534/g3.115.021121, 2015.

Blanchard, F., Duplomb, L., Baud'huin, M., and Brounais, B.: The dual role of IL-6-type cytokines on bone remodeling and bone tumors, Cytokine Growth F. R., 20, 19–28, doi:10.1016/j.cytogfr.2008.11.004, 2009.

Both, J., Krijgsman, O., Bras, J., Schaap, G. R., Baas, F., Ylstra, B., and Hulsebos, T. J.: Focal chromosomal copy number aberrations identify CMTM8 and GPR177 as new candidate driver genes in osteosarcoma, PloS one, 9, e115835, doi:10.1371/journal.pone.0115835, 2014.

Cagle, A. P., Waguespack, S. G., Buckingham, B. A., Shankar, R. R., and Dimeglio, L. A.: Severe infantile hypercalcemia associated with Williams syndrome successfully treated with intravenously administered pamidronate, Pediatrics, 114, 1091–1095, doi:10.1542/peds.2003-1146-L, 2004.

Cao, J. J.: Effects of obesity on bone metabolism, J. Orthop. Surg. Res., 6, doi:10.1186/1749-799x-6-30, 2011.

Carter, D. R., Bouxsein, M. L., and Marcus, R.: New approaches for interpreting projected bone densitometry data, J. Bone Miner. Res., 7, 137–145, doi:10.1002/jbmr.5650070204, 1992.

Clark, D. L., Clark, D. I., Beever, J. E., and Dilger, A. C.: Increased prenatal IGF2 expression due to the porcine intron3-G3072A mutation may be responsible for increased muscle mass, J. Anim. Sci., 93, 2546–2558, doi:10.2527/jas.2014-8389, 2015.

El Maghraoui, A. and Roux, C.: DXA scanning in clinical practice, QJM, monthly journal of the Association of Physicians, 101, 605–617, doi:10.1093/qjmed/hcn022, 2008.

Franchimont, N., Wertz, S., and Malaise, M.: Interleukin-6: An osteotropic factor influencing bone formation?, Bone, 37, 601–606, doi:10.1016/j.bone.2005.06.002, 2005.

Francke, U.: Williams-Beuren syndrome: genes and mechanisms, Hum. Mol. Genet., 8, 1947–1954, 1999.

Fukawa, K. and Kusuhara, S.: The genetic and non-genetic aspects of leg weakness and osteochondrosis in pigs – Review, Asian Austral. J. Anim., 14, 114–122, 2001.

Gilmour, A. R., Gogel, B. J., Cullis, B. R., and Thompson, R.: ASReml User Guide (Release 3.0), available at: http://vsni.de/downloads/asreml/release3/UserGuide.pdf (last access: 27 February 2014), 2009.

Heffner, E. L., Sorrells, M. E., and Jannink, J. L.: Genomic Selection for Crop Improvement, Crop. Sci., 49, 1–12, doi:10.2135/cropsci2008.08.0512, 2009.

Hernandez-de Sosa, N., Athanasiadis, G., Malouf, J., Laiz, A., Marin, A., Herrera, S., Farrerons, J., Soria, J. M., and Casademont, J.: Heritability of bone mineral density in a multivariate family-based study, Calcified Tissue Int., 94, 590–596, doi:10.1007/s00223-014-9852-9, 2014.

Hicks, A. N., Lorenzetti, D., Gilley, J., Lu, B., Andersson, K. E., Miligan, C., Overbeek, P. A., Oppenheim, R., and Bishop, C. E.: Nicotinamide mononucleotide adenylyltransferase 2 (Nmnat2) regulates axon integrity in the mouse embryo, PloS one, 7, e47869, doi:10.1371/journal.pone.0047869, 2012.

International Mouse Phenotyping Consortium: available at: http://www.mousephenotype.org, last access: 25 November 2016.

Jin, C., Ding, P., Wang, Y., and Ma, D.: Regulation of EGF receptor signaling by the MARVEL domain-containing protein CKLFSF8, FEBS letters, 579, 6375–6382, doi:10.1016/j.febslet.2005.10.021, 2005.

Kaufman, J. M., Ostertag, A., Saint-Pierre, A., Cohen-Solal, M., Boland, A., Van Pottelbergh, I., Toye, K., de Vernejoul, M. C., and Martinez, M.: Genome-wide linkage screen of bone mineral density (BMD) in European pedigrees ascertained through a male relative with low BMD values: evidence for quantitative trait loci on 17q21-23, 11q12-13, 13q12-14, and 22q11, J. Clin. Endocr. Metab., 93, 3755–3762, doi:10.1210/jc.2008-0678, 2008.

Laenoi, W., Uddin, M. J., Cinar, M. U., Grosse-Brinkhaus, C., Tesfaye, D., Jonas, E., Scholz, A. M., Tholen, E., Looft, C., Wimmers, K., Phatsara, C., Juengst, H., Sauerwein, H., Mielenz, M., and Schellander, K.: Quantitative trait loci analysis for leg weakness-related traits in a Duroc x Pietrain crossbred population, Genet. Sel. Evol., 43, 1–7, doi:10.1186/1297-9686-43-13, 2011.

Laenoi, W., Rangkasenee, N., Uddin, M. J., Cinar, M. U., Phatsara, C., Tesfaye, D., Scholz, A. M., Tholen, E., Looft, C., Mielenz, M., Sauerwein, H., Wimmers, K., and Schellander, K.: Association and expression study of MMP3, TGFbeta1 and COL10A1 as candidate genes for leg weakness-related traits in pigs, Mol. Biol. Rep., 39, 3893–3901, doi:10.1007/s11033-011-1168-5, 2012.

Lee, H. L., Bae, O. Y., Baek, K. H., Kwon, A., Hwang, H. R., Qadir, A. S., Park, H. J., Woo, K. M., Ryoo, H. M., and Baek, J. H.: High extracellular calcium-induced NFATc3 regulates the expression of receptor activator of NF-kappaB ligand in osteoblasts, Bone, 49, 242–249, doi:10.1016/j.bone.2011.04.006, 2011.

Luk, F., Yu, Y., Dong, H. T., Walsh, W. R., and Yang, J. L.: New gene groups associated with dissimilar osteoblastic differentiation are linked to osteosarcomagenesis, Cancer genomics & proteomics, 8, 65–75, 2011.

Markljung, E., Jiang, L., Jaffe, J. D., Mikkelsen, T. S., Wallerman, O., Larhammar, M., Zhang, X., Wang, L., Saenz-Vash, V., Gnirke, A., Lindroth, A. M., Barres, R., Yan, J., Stromberg, S.,

De, S., Ponten, F., Lander, E. S., Carr, S. A., Zierath, J. R., Kullander, K., Wadelius, C., Lindblad-Toh, K., Andersson, G., Hjalm, G., and Andersson, L.: ZBED6, a novel transcription factor derived from a domesticated DNA transposon regulates IGF2 expression and muscle growth, PLoS Biol., 7, e1000256, doi:10.1371/journal.pbio.1000256, 2009.

Meuwissen, T. H. E., Karlsen, A., Lien, S., Olsaker, I., and Goddard, M. E.: Fine mapping of a quantitative trait locus for twinning rate using combined linkage and linkage disequilibrium mapping, Genetics, 161, 373–379, 2002.

Nicolaidou, V., Wong, M. M., Redpath, A. N., Ersek, A., Baban, D. F., Williams, L. M., Cope, A. P., and Horwood, N. J.: Monocytes induce STAT3 activation in human mesenchymal stem cells to promote osteoblast formation, PloS one, 7, e39871, doi:10.1371/journal.pone.0039871, 2012.

Origins of bone and cartilage disease project: Gene of the month June 2015: Edc4: available at: http://www.boneandcartilage.com/GOTMarchive.html (last access: 25 November 2016), 2015.

Pocock, N. A., Eisman, J. A., Hopper, J. L., Yeates, M. G., Sambrook, P. N., and Eberl, S.: Genetic determinants of bone mass in adults. A twin study, J. Clin. Invest., 80, 706–710, doi:10.1172/JCI113125, 1987.

R: A Language and Environment for Statistical Computing: available at: http://www.R-project.org (last access: 13 February 2017), 2008.

Rangkasenee, N., Murani, E., Brunner, R., Schellander, K., Cinar, M. U., Scholz, A. M., Luther, H., Hofer, A., Ponsuksili, S., and Wimmers, K.: KRT8, FAF1 and PTH1R gene polymorphisms are associated with leg weakness traits in pigs, Mol. Biol. Rep., 40, 2859–2866, doi:10.1007/s11033-012-2301-9, 2013.

Rothammer, S., Kremer, P. V., Bernau, M., Fernandez-Figares, I., Pfister-Schar, J., Medugorac, I., and Scholz, A. M.: Genome-wide QTL mapping of nine body composition and bone mineral density traits in pigs, Genet. Sel. Evol., 46, 1–11, doi:10.1186/s12711-014-0068-2, 2014.

Sakamoto, K., Sato, Y., Shinka, T., Sei, M., Nomura, I., Umeno, M., Ewis, A. A., and Nakahori, Y.: Proteasome subunits mRNA expressions correlate with male BMI: implications for a role in obesity, Obesity (Silver Spring), 17, 1044–1049, doi:10.1038/oby.2008.612, 2009.

Sasaki, Y., Nagai, K., Nagata, Y., Doronbekov, K., Nishimura, S., Yoshioka, S., Fujita, T., Shiga, K., Miyake, T., Taniguchi, Y., and Yamada, T.: Exploration of genes showing intramuscular fat deposition-associated expression changes in musculus longissimus muscle, Anim. Genet., 37, 40–46, doi:10.1111/j.1365-2052.2005.01380.x, 2006.

Schousboe, J. T., Shepherd, J. A., Bilezikian, J. P., and Baim, S.: Executive Summary of the 2013 International Society for Clinical Densitometry Position Development Conference on Bone Densitometry, J. Clin. Densitom., 16, 455–466, doi:10.1016/j.jocd.2013.08.004, 2013.

Serao, N. V., Veroneze, R., Ribeiro, A. M., Verardo, L. L., Braccini Neto, J., Gasparino, E., Campos, C. F., Lopes, P. S., and Guimaraes, S. E.: Candidate gene expression and intramuscular fat content in pigs, J. Anim. Breed. Genet., 128, 28–34, doi:10.1111/j.1439-0388.2010.00887.x, 2011.

Seto, E., Yoshida-Sugitani, R., Kobayashi, T., and Toyama-Sorimachi, N.: The Assembly of EDC4 and Dcp1a into Processing Bodies Is Critical for the Translational Regulation of IL-6, PLoS one, 10, e0123223, doi:10.1371/journal.pone.0123223, 2015.

Sornay-Rendu, E., Boutroy, S., Munoz, F., and Delmas, P. D.: Alterations of cortical and trabecular architecture are associated with fractures in postmenopausal women, independently of decreased bmd, The OFELY study, Osteoporosis Int., 17, S97–S97, 2006.

Soroko, S. B., Barrett-Connor, E., Edelstein, S. L., and Kritz-Silverstein, D.: Family history of osteoporosis and bone mineral density at the axial skeleton: the Rancho Bernardo Study, J. Bone Miner. Res., 9, 761–769, doi:10.1002/jbmr.5650090602, 1994.

Thomas, G., Betters, J. L., Lord, C. C., Brown, A. L., Marshall, S., Ferguson, D., Sawyer, J., Davis, M. A., Melchior, J. T., Blume, L. C., Howlett, A. C., Ivanova, P. T., Milne, S. B., Myers, D. S., Mrak, I., Leber, V., Heier, C., Taschler, U., Blankman, J. L., Cravatt, B. F., Lee, R. G., Crooke, R. M., Graham, M. J., Zimmermann, R., Brown, H. A., and Brown, J. M.: The serine hydrolase ABHD6 Is a critical regulator of the metabolic syndrome, Cell reports, 5, 508–520, doi:10.1016/j.celrep.2013.08.047, 2013.

Trammell, S. A. and Brenner, C.: NNMT: A Bad Actor in Fat Makes Good in Liver, Cell Metabolism, 22, 200–201, doi:10.1016/j.cmet.2015.07.017, 2015.

Ulrich, D., van Rietbergen, B., Laib, A., and Ruegsegger, P.: The ability of three-dimensional structural indices to reflect mechanical aspects of trabecular bone, Bone, 25, 55–60, 1999.

Valero, R., Bayes, M., Francisca Sanchez-Font, M., Gonzalez-Angulo, O., Gonzalez-Duarte, R., and Marfany, G.: Characterization of alternatively spliced products and tissue-specific isoforms of USP28 and USP25, Genome Biology 2001, 2, research0043.1–0043.10, 2001.

Vogl, T. J., Reith, W., and Rummeny, E. J.: Diagnostische und interventionelle Radiologie, Springer-Verlag Berlin Heidelberg, 2011.

Wei, T.: corrplot: Visualization of a correlation matrix, available at: http://CRAN.R-project.org/package=corrplot (last access: 13 February 2017), 2013.

Willing, M. C., Torner, J. C., Burns, T. L., Janz, K. F., Marshall, T., Gilmore, J., Deschenes, S. P., Warren, J. J., and Levy, S. M.: Gene polymorphisms, bone mineral density and bone mineral content in young children: the Iowa Bone Development Study, Osteoporosis international: a journal established as result of cooperation between the European Foundation for Osteoporosis and the National Osteoporosis Foundation of the USA, 14, 650–658, doi:10.1007/s00198-003-1416-1, 2003.

Wright, N. C., Looker, A. C., Saag, K. G., Curtis, J. R., Delzell, E. S., Randall, S., and Dawson-Hughes, B.: The recent prevalence of osteoporosis and low bone mass in the United States based on bone mineral density at the femoral neck or lumbar spine, J. Bone Miner. Res., 29, 2520–2526, doi:10.1002/jbmr.2269, 2014.

Xiao, S. M., Gao, Y., Cheung, C. L., Bow, C. H., Lau, K. S., Sham, P. C., Tan, K. C., and Kung, A. W.: Association of CDX1 binding site of periostin gene with bone mineral density and vertebral fracture risk, Osteoporosis international: a journal established as result of cooperation between the European Foundation for Osteoporosis and the National Osteoporosis Foundation of the USA, 23, 1877–1887, doi:10.1007/s00198-011-1861-1, 2012.

Yadav, A., Saini, V., and Arora, S.: MCP-1: chemoattractant with a role beyond immunity: a review, Clin. Chim. Acta, 411, 1570–1579, doi:10.1016/j.cca.2010.07.006, 2010.

Yang, T. L., Guo, Y., Liu, Y. J., Shen, H., Liu, Y. Z., Lei, S. F., Li, J., Tian, Q., and Deng, H. W.: Genetic variants in the SOX6 gene are associated with bone mineral density in both Caucasian and Chinese populations, Osteoporosis international: a journal established as result of cooperation between the European Foundation for Osteoporosis and the National Osteoporosis Foundation of the USA, 23, 781–787, doi:10.1007/s00198-011-1626-x, 2012.

Yu, E. W., Thomas, B. J., Brown, J. K., and Finkelstein, J. S.: Simulated increases in body fat and errors in bone mineral density measurements by DXA and QCT, J. Bone Miner. Res., 27, 119–124, doi:10.1002/jbmr.506, 2012.

Zhao, L. J., Jiang, H., Papasian, C. J., Maulik, D., Drees, B., Hamilton, J., and Deng, H. W.: Correlation of obesity and osteoporosis: Effect of fat mass on the determination of osteoporosis, J. Bone Miner. Res., 23, 17–29, doi:10.1359/Jbmr.070813, 2008.

Zhu, J., Shimizu, E., Zhang, X., Partridge, N. C., and Qin, L.: EGFR signaling suppresses osteoblast differentiation and inhibits expression of master osteoblastic transcription factors Runx2 and Osterix, J. Cell. Biochem., 112, 1749–1760, doi:10.1002/jcb.23094, 2011.

11

The effects of in ovo feeding of glutamine in broiler breeder eggs on hatchability, development of the gastrointestinal tract, growth performance and carcass characteristics of broiler chickens

Mehdi Salmanzadeh, Yahya Ebrahimnezhad, Habib Aghdam Shahryar, and Jamshid Ghiasi Ghaleh-Kandi

Departments of Animal Science, Shabestar branch, Islamic Azad University, Shabestar, Iran

Correspondence to: Yahya Ebrahimnezhad (ebrahimnezhad@gmail.com)

Abstract. The aim of the present study was to investigate the effect of in ovo feeding (IOF) of glutamine on hatchability, development of the gastrointestinal tract, growth performance and carcass characteristics of broiler chickens. Fertilized eggs were subjected to injections with glutamine (Gln) (10, 20, 30, 40 or 50 mg dissolved in 0.5 mL of dionized water) on day 7 of incubation. Hatchability, growth performance, carcass characteristics (carcass weight and relative weights of breast, thigh, heart, liver, gizzard, abdominal fat, intestine, pancreas and spleen) and jejunal morphometry (measurement of villus height and width and crypt depth) were determined during the experiment. The weight of newly hatched chickens was significantly greater in groups with Gln injection than in control and sham groups. But IOF caused lower hatchability than in the control group (non-injected eggs) ($p < 0.05$). Chickens from IOF of Gln showed better weight gain and feed conversion ratio (0–42 days of age), when compared to chickens hatched from control and sham groups. The IOF of Gln significantly increased villus height, villus width and crypt depth at hatch period and villus height at 42 days of age. In addition, carcass weights and relative weights of breast, thigh and gizzard were also markedly increased in chickens treated in ovo with Gln; whereas heart, liver, abdominal fat, intestine, pancreas and spleen were not significantly altered at the end of the experimental period. These data suggest that the IOF of Gln may improve jejunum development, leading to an increased nutrient assimilation and consequently to greater performance in broiler chickens.

1 Introduction

In ovo feeding (IOF) is a method of supplementing exogenous nutrients into amnion of the avian embryo (Uni and Ferket, 2003), which can improve the performance of chicks from injected eggs (Salmanzadeh, 2011; Salmanzadeh et al., 2012; Dong et al., 2013). Previous studies have been conducted to assess nutrients for in ovo injection in broiler embryos, such as carbohydrates (Tako et al., 2004a; Uni et al., 2005; Smirnov et al., 2006), amino acids (Al-Murrani, 1982; Ohta et al., 1999, 2001, 2004), vitamins (Nowaczewski et al., 2012) and minerals (Tako et al., 2004b). However, there are few studies on IOF of Gln in broiler breeder hen eggs.

Glutamine (Gln) is the principle metabolic fuel for development of the gastrointestinal tract (Andrew and Griffiths, 2002), and is considered as a non-essential amino acid since most animal cells can synthesize it, which can play an important role in the synthesis of many biologically important molecules (Souba, 1993). These results are in agreement with the findings of Samli et al. (2007), who reported that Gln is an important amino acid for utilization as an energy source for the development of the gastrointestinal tract and stimulates intestinal cell proliferation, which leads to increasing the absorptive source of the gastrointestinal mucosa and consequently the access to nutrients. In parallel, previous experiments, showed that supplementing the diet with

Table 1. Composition and calculated contents of nutrients of broiler diet.

Item	Diet		
	Starter 0–10 days	Grower 11–26 days	Finisher 27–42 days
Ingredient (%)			
Corn	60.36	65.44	66.8
Soybean meal (44 % CP)	34.12	28.62	26.33
Vegetable fat	1.23	1.74	2.84
Dicalcium phosphate	1.83	1.8	1.67
Oyster sell-ground	1.22	1.19	1.13
Salt	0.35	0.3	0.3
Sodium bicarbonate	0.11	0.07	0.07
Vitamin premix[1]	0.25	0.25	0.25
Trace mineral premix[2]	0.25	0.25	0.25
DL-Met	0.17	0.18	0.18
L-Lys	0.11	0.16	0.18
Calculated analysis			
ME (kcal kg^{-1})	2894	2987	3176
CP (%)	20.3	18.3	18
Ca (%)	1	0.96	0.9
Available P (%)	0.50	0.48	0.45
Met (%)	0.46	0.44	0.43
Met + cystine (%)	0.89	0.84	0.82
Lys (%)	1.20	1.10	1.05

[1] Vitamin premix provided the following per kilogram of diet: vitamin A, 11 013 IU; vitamin D3, 3525 IU; vitamin E, 33 IU; vitamin K, 2.75 mg; riboflavin, 7.7 mg; pantothenic acid, 17.6 mg; niacin, 55.1 mg; choline, 478 mg; vitamin B12, 0.028 mg; pyridoxine, 5.0 mg; thiamine, 2.2 mg; folic acid, 1.1 mg; biotin, 0.22 mg. [2] Trace mineral premix provided the following per kilogram of diet: manganese, 64 mg; zinc, 75 mg; iron, 40 mg; copper, 10 mg; iodine, 1.85 mg; and selenium, 0.3 mg.

Gln increased intestinal villus height and consequently, improved growth performance in broilers (Bartell and Batal, 2007; Yi et al., 2005; Jazideh et al., 2014), turkey poults (Yi et al., 2001; Salmanzadeh and Shahryar, 2013a) quails (Salmanzadeh and Shahryar, 2013b) and weanling pigs (Kitt et al., 2002), compared to control groups. Chen et al. (2009) demonstrated that, body weights measured at the 7-days old post-hatching were significantly modified in ducks treated in ovo injection with Gln.

All these beneficial actions of Gln in particular, make it an amino acid deserving of scientific and technical attention.

It was hypothesized that IOF of Gln can improve the development of the gastrointestinal tract, by increasing the absorptive surface area that consequently promotes the nutrient assimilation and improves growth performance in broiler chickens. Thus, the aims of this research were to investigate the effects of IOF of Gln in broiler breeder eggs in hatchability, development of the gastrointestinal tract, growth performance and carcass characteristics of broiler chickens.

2 Material and method

2.1 Incubation and in ovo feeding

Hatching eggs of approximately similar weights (65 ± 1 g) were obtained from broiler breeder strain (Cobb 500) at 42 weeks of age. A total of 1400 fertile eggs were numbered, fumigated with formaldehyde gas, and incubated at 37.7 °C and 64 % RH. On the 6th day of incubation, the eggs were candled, and the infertile ones or those containing only dead embryos were removed. On the 7th day, fertile eggs were randomly allotted to seven treatments with four replicate per treatment and 50 eggs per replicate. Eggs were injected with Gln (10, 20, 30, 40, and 50 mg, respectively) dissolved in 0.5 mL of deionized water whereas in the sham group, eggs were only treated with deionized water (0.5 mL) and those of the control group received no injection.

Before injection, each egg was candled to identify the location of the future injection. The injection hold area was cleaned with ethyl alcohol (70 %) and the blunt end was punched using a 22 gauge needle. Then, the solutions were injected into albumen (0.5 mL egg^{-1}) using another disposable syringe equipped with a 22 gauge needle to a depth of about 13 mm (Salmanzadeh et al., 2012). After injection, the eggs were sealed with cellophane tape, and transferred to the incubator. Control eggs were removed from the incubator together with the treated groups, and kept in the same environment. The group of eggs designated as sham-injected controls were injected with 0.5 mL of deionized water. Deionized water injections were included as sham controls primarily to rule out a possible negative response caused by the stress of injection and handling. L-Glutamine was supplied from Sigma® Co (anhydrous $\geq 99\%$, CAS Number: 56-85-9). All of the treatment solutions were prepared in autoclaved water.

2.2 Birds and data collection

After hatching, chickens were transferred to an experimental house and reared for 42 days with the same ration according to the requirements of broiler as recommended by the catalog of Cobb 500 broilers (Table 1). Each chicken according to the treatment group was identified by the neck tag and recorded. All chickens and treatments were randomly assigned to 1 of 28 pens. Each open was provided with water and an individual feeder. Room temperature was maintained at 32 °C from 0 to 4 and then gradually reduced from 32 to 21 °C. All experimental protocols and procedures were approved by the institutional Animal Care of Iran.

Upon hatching, the number of hatchlings was determined to calculate the hatchability of fertile eggs. The weight of newly hatched chickens was determined by weighing all chickens hatched one by one. In each pen, bird body weight and food intake were recorded on days 1, 10, 26, and 42 posthatching and thereafter mean body weight gain, food intake, and food conversion ratio were calculated for each pen (replicate) between 0 and 10, 11 and 26, 27 and 42, and 1 and 42 days. In each time period, body weight gain was calculated and expressed as grams per bird. Food intake (g of food intake/bird) over the entire grow-out period was calculated by totalling food consumption in each time interval between each bird sampling. Food conversion ratio (g of food intake/g of body weight gain) was calculated by dividing total food intake by total weight gain in each pen.

2.3 Morphometric indices of the jejunum

At hatching and 10 days of age, eight birds from pretreatment were euthanized by cervical dislocation. Then, gastrointestinal tract was carefully excised. One cross-section for jejunum was fixed with formalin solution and was prepared using standard paraffin embedding procedures by sectioning at 5 mm thickness, and staining with hematoxylin and eosin. Villus height (VH), villus width (VW) and crypt depth (CD) were determined using an image processing and analyzing system (Image Pro plus) and were expressed as micrometers (Touchette et al., 2002).

Table 2. Effects of IOF of Gln on weight and hatchability in newly hatched chickens.

Groups	Hatchability (%)	Weight (g)
Control	89.58[a]	43.41[e]
Sham group	72.22[c]	43.31[e]
Gln[1] 10 mg	75.69[bc]	43.82[cd]
Gln 20 mg	75.00[bc]	43.71[d]
Gln 30 mg	76.38[bc]	44.09[ab]
Gln 40 mg	79.16[b]	43.93[bc]
Gln 50 mg	77.77[bc]	44.21[a]
SEM	1.75	0.05
P Value	0.0001	0.0001

[1] Gln = Glutamine; [a–e] averages in a column with different superscript letters are significantly different.

Table 3. Effects of IOF of Gln on body weight gain (BWG), food intake (FI) and feed conversion ratio (FCR) of broilers to 1–10 days of age.

Groups	1–10 days of age (g)		
	BWG	FI	FCR
Control	199.09[bc]	251.25	1.26[ab]
Sham group	196.03[c]	255.78	1.30[a]
Gln 10 mg	206.21[ab]	251.98	1.22[abc]
Gln 20 mg	207.87[a]	245.00	1.17[bc]
Gln 30 mg	209.98[a]	246.97	1.17[bc]
Gln 40 mg	206.49[ab]	250.14	1.21[abc]
Gln 50 mg	213.70[a]	243.24	1.14[c]
SEM	2.53	6.49	0.03
P Value	0.0010	0.8358	0.0237

[a–c] Averages in a column with different superscript letters are significantly different.

2.4 Carcass measurements

On day 42, two broilers per pen were selected (close to the mean weight of each cage), weighed and killed by cervical dislocation, and then the abdominal cavity was opened. Weights of the eviscerated hot carcass, breast muscle, thigh, liver, heart, gizzard, abdominal fat, intestine, pancreas and spleen were recorded and the corresponding percentages (% of live body weight) were calculated.

2.5 Statistical analysis

Analyses of variance were performed using the GLM procedure of SAS Institute Inc. (2005) as a completely randomized design. Results are presented as mean ± SEM. The significantly different treatment means were investigated using Duncan's new multiple rang test. Differences were considered significant when $p < 0/05$.

Table 4. Effects of IOF of Gln on body weight gain (BWG), food intake (FI) and feed conversion ratio (FCR) of broilers to 11–26 days of age.

Groups	11–26 days of age (g)		
	BWG	FI	FCR
Control	647.74[bc]	915.24	1.41[a]
Sham group	639.75[c]	917.80	1.43[a]
Gln 10 mg	673.12[ab]	909.36	1.35[bc]
Gln 20 mg	669.22[abc]	907.66	1.35[b]
Gln 30 mg	686.50[a]	904.98	1.31[bc]
Gln 40 mg	679.96[a]	892.87	1.31[bc]
Gln 50 mg	692.19[a]	895.13	1.29[c]
SEM	9.68	8.23	0.01
P Value	0.0074	0.2981	0.0001

[a–c] Averages in a column with different superscript letters are significantly different.

Table 5. Effects of IOF of Gln on body weight gain (BWG), food intake (FI) and feed conversion ratio (FCR) of broilers to 27–42 days of age.

Groups	27–42 days of age (g)		
	BWG	FI	FCR
Control	1403.42[b]	3152.50	2.24[a]
Sham group	1397.97[b]	3143.50	2.24[a]
Gln 10 mg	1452.57[a]	3139.52	2.16[b]
Gln 20 mg	1448.25[a]	3135.03	2.16[b]
Gln 30 mg	1455.59[a]	3124.25	2.14[bc]
Gln 40 mg	1472.97[a]	3115.05	2.11[bc]
Gln 50 mg	1483.72[a]	3124.99	2.10[c]
SEM	11.41	11.54	0.01
P Value	0.0001	0.3229	0.0001

[a–c] Averages in a column with different superscript letters are significantly different.

Table 6. Effects of IOF of Gln on body weight gain (BWG), food intake (FI) and feed conversion ratio (FCR) of broilers to 1–42 days of age.

Groups	1–42 days of age (g)		
	BWG	FI	FCR
Control	2250.26[c]	4318.99	1.91[a]
Sham group	2233.74[c]	4317.08	1.93[a]
Gln 10 mg	2331.91[ab]	4300.86	1.84[b]
Gln 20 mg	2325.34[b]	4287.69	1.84[b]
Gln 30 mg	2352.08[ab]	4276.20	1.81[bc]
Gln 40 mg	2359.43[ab]	4258.06	1.80[bc]
Gln 50 mg	2389.61[a]	4263.36	1.78[c]
SEM	19.69	22.60	0.01
P Value	0.0001	0.3558	0.0001

[a–c] Averages in a column with different superscript letters are significantly different.

Table 7. Effects of IOF of Gln on villus height, villus width and crypt depth in jejunum of newly hatched chickens.

Groups	Villus height (μm)	Villus width (μm)	Crypt depth (μm)
Control	342.24[ef]	49[bc]	50.50[d]
Sham group	338.19[f]	48[c]	49.25[d]
Gln 10 mg	357.89[bcd]	55[ab]	60.00[bc]
Gln 20 mg	355.48[bcde]	58[a]	59.00[c]
Gln 30 mg	352.25[cde]	56[a]	65.25[abc]
Gln 40 mg	360.49[ab]	59[a]	68.00[ab]
Gln 50 mg	372.07[a]	60[a]	70.00[a]
SEM	4.37	2.13	2.71
P Value	0.0001	0.0012	0.0001

[a–e] Averages in a column with different superscript letters are significantly different.

3 Results

The effects of IOF with different levels of Gln solutions at 7 days of incubation on hatchability and weights of newly hatched chickens are presented in Table 2. IOF with Gln and deionized water reduced the hatchability compared with the control group (not injection eggs). But, the mean body weights of newly hatched chickens injected by Gln were increased more than the control groups. IOF of Gln improved body weight gain and feed conversion ratio through the whole experimental period but feed intake was not significantly altered (Tables 3–6).

As seen in Tables 7 and 8, villus height, villus width and crypt depth in jejunum were significantly increased in chickens treated in ovo by Gln compared to the non-injected and sham controls in newly hatched and 10-day old chickens.

On the 42nd day, weights of the carcass, breast, thigh and gizzard were also significantly increased in broilers from injected eggs with Gln compared to the control groups ($p <$ 0.05); the highest value was observed in the group receiving 50 mg Gln. By contrast, the liver, heart, abdominal fat, intestine, pancreas and spleen weights were not significantly altered in broilers from Gln treated eggs (Tables 9, 10).

4 Discussion

Concerning hatchability, Uni et al. (2005) showed that the positive effects of IOF at late-term chicken embryos may include increased hatchability. On the other hand, Pedroso et al. (2006a) reported that IOF of Gln in the amniotic fluid of embryos on day 18 of incubation had no significant effect on the hatchability. In parallel, Dos Santos et al. (2010) demonstrate that, IOF of 0.5 mL of a 10 % Gln solution did not affect hatching of newly hatched chickens. In contrast, Ohta

Table 8. Effects of IOF of Gln on villus height, villus width and crypt depth in jejunum of broilers when they were 10-days old.

Groups	Villus height (μm)	Villus width (μm)	Crypt depth (μm)
Control	656.25[c]	139.00[d]	86[b]
Sham group	647.50[c]	141.75[cd]	85[b]
Gln 10 mg	677.00[b]	152.75[abc]	94[a]
Gln 20 mg	687.50[ab]	162.50[a]	98[a]
Gln 30 mg	684.25[ab]	151.75[abc]	97[a]
Gln 40 mg	691.00[ab]	155.75[ab]	96[a]
Gln 50 mg	702.75[a]	158.50[a]	95[a]
SEM	6.99	3.70	1.87
P Value	0.0001	0.0009	0.0001

[a–c] Averages in a column with different superscript letters are significantly different.

Table 9. Effects of IOF of Gln on weight of carcass, breast and thigh of broiler chickens when they were 42-days old (based on percentage of live body weight).

Groups	Carcass (%)	Breast (%)	Thigh (%)
Control	67.56[d]	23.92[c]	19.96[c]
Sham group	67.71[d]	24.21[c]	19.82[c]
Gln 10 mg	68.58[c]	24.94[b]	20.68[b]
Gln 20 mg	68.48[c]	24.89[b]	20.64[b]
Gln 30 mg	68.71[bc]	25.55[a]	21.35[a]
Gln 40 mg	68.97[ab]	25.78[a]	21.48[a]
Gln 50 mg	69.18[a]	25.86[a]	21.61[a]
SEM	0.08	0.11	0.09
P Value	0.0001	0.0001	0.0001

[a–d] Averages in a column with different superscript letters are significantly different.

et al. (1999) showed that the hatchability was significantly reduced when injecting amino acids at day 0 of incubation. Chen et al. (2010) stated that hatchability of control, sources and maltose (DS), L-alanyl–L-glutamine (Ala–Gln), Sucrose, maltose and L-alanyl–L-glutamine (DS + Ala–Gln), groups were 85, 65, 70 and 82 % respectively. Moreover, DS and Ala–Gln decreased the hatchability by 24 and 18 % ($p < 0.01$), respectively. In an experiment by Chen et al. (2010), the pH of all the injection solution was adjusted to about 6.8 and the osmolarity was 154, 2264 and 30 mOsm for DS, Ala–Gln and DS + Ala–Gln solution respectively. These researchers showed that decreasing in hatchability was due to varied osmolarity of the injection. Previous work demonstrated that the osmolarity of injection solution was a critical factor in affecting the hatchability of the chick (Ferket et al., 2005). In parallel, Pedroso et al. (2006b) also found that decreased hatchability was observed when chick embryos received glucose in ovo injection at 16 days of incubation. The significant difference in hatchability among the in ovo injected groups indicated that nutrient specificity might result in the differing response of embryos. According to the US Patent (6592878) of Uni and Ferket (2003) the importance of the osmolarity of the in ovo feeding solution on hatchability of chicks was illustrated. Acceptable hatchability of chicks was observed when eggs were injected with solutions having an osmolarity ranging below 800 mOsm with an optimum hatchability observed at about 400–600 mOsm. Unacceptable hatching rates were observed when the in ovo feeding solution exceeded 800 mOsm. In the present study osmolality of all the injection of solutions was adjusted to about 450 mOsm. Thus, osmolarity could not be the reason for the decreasing of the hatchability. This is consistent with the results obtained in the present study, in which the injection of different levels of Gln and deionized water caused significant decrease in the hatchability. Probably the reducing hatchability was due to the IOF into the albumin. One of the possible reasons for the decreasing rate of hatching is the allergic cav-

ity under the air sac causing the respiration of developing embryo to stop and die. Heiblum et al. (2001) showed that IOF of glucocorticoid at day 7 of incubation resulted in 35 % decline of rate of hatching. Also, Salmanzadeh et al. (2012) observed lower hatchability when 7-day old broiler embryos were injected with glucose and magnesium. Thus, it seems that any IOF at early embryonic life can be harmful for the internal environment susceptibility and have negative effects on rate of hatching. This effect may largely independent depend on injected Gln effect.

Data of this study showed that the IOF of Gln into the albumin can be seen as an effective tool to improve the mean body weights of newly hatched chickens. Previous studies showed that to stimulate the development of the chick embryo is an important factor in increasing the weight of newly hatched chickens (Tako et al., 2004a; Uni and Ferket, 2004; Uni et al., 2005; Smirnov et al., 2006). Al-Murrani (1982) showed that IOF of amino acid into fertile goose eggs at day 7 of incubation increased the weight of newly hatched geese. In turkeys, Foye et al. (2006) observed that in ovo injection of egg white protein at day 23 of incubation increased hatching weight. Ohta et al. (2001) reported that IOF of amino acids into the air cell of 7-day old embryos improved utilization of amino acid by the embryo and consequently increased hatching weight. On the other hand, Dos Santos et al. (2010) showed that IOF with 10 % Gln at 18 days of incubation did not influence hatch weight. In another study, Chen et al. (2009) showed that in ovo administration of Gln and carbohydrates in combination in the duck did not affect the body weight at 25 days of incubation, weight of hatch and weight of ducks on day 3 post-hatching. According to the past studies and our present observations, the late-term embryo and neonatal chicken depends on gluconeogenesis from amino acids, resulting in the depletion of muscle protein reserves and the reduction of hatching weight. To reduce the depletion of muscle protein, we carried out IOF Gln into albumen prior to hatching, which would support the en-

Table 10. Effects of IOF of Gln on weight of heart, liver, gizzard, abdominal fat, intestine, pancreas and spleen of broiler chickens when they were 42-days old (based on percentage of live body weight).

Groups	Heart (%)	Liver (%)	Gizzard (%)	Abdominal fat (%)	Intestine(%)	Pancreas (%)	Spleen (%)
Control	0.58	2.09	1.70[b]	2.13	4.49	0.143	0.126
Sham group	0.60	2.11	1.69[b]	2.16	4.45	0.148	0.121
Gln 10 mg	0.64	2.13	2.01[a]	2.23	4.68	0.158	0.129
Gln 20 mg	0.64	2.18	1.99[a]	2.34	4.62	0.156	0.133
Gln 30 mg	0.68	2.24	1.96[a]	2.32	4.85	0.160	0.135
Gln 40 mg	0.69	2.18	2.09[a]	2.36	4.92	0.164	0.141
Gln 50 mg	0.73	2.21	2.11[a]	2.42	5.04	0.170	0.142
SEM	0.06	0.09	0.05	0.08	0.16	0.009	0.004
P Value	0.7562	0.9168	0.0001	0.1251	0.2058	0.6085	0.4787

[a–b] Averages in a column with different superscript letters are significantly different.

ergy status of the hatching by moderating the use of muscle that consequently increases body weight at hatching. In addition, Foye et al. (2006) showed that the supply of nutrient by IOF improved avian energy status, which spared energy used for metabolism and, consequently, increased postnatal performance. Today, whole embryonic life is almost 35 % of the productive life of broilers. Thus, to stimulate the development of the chick embryo is an important factor to increase the weight of newly hatched chicks. Also, previous studies showed that weight of newly hatched chickens is a major predictor for marketing weight in modern broilers. Wilson (1991) stated that each 1 g of improvement in weight of newly hatched chickens leads to 8–13 g of improvement in body weight at marketing. In this study we demonstrate that a 1 g difference in weight of newly hatched chickens due to IOF resulted in 109–130 g increase in body weight at day 42.

Based on the results of this study, the growth performance of broiler chickens linearly increased with the improved development of the gastrointestinal tract. Thus, improved growth performance was dependent upon the development of the gastrointestinal tract and IOF of Gln used. Previous studies showed that the dietary supplementation with Gln stimulates the intestine development and improved the growth performance of turkey poults first week after hatch (Yi et al., 2001), weanling pigs (Kitt et al., 2002), broilers (Bartell and Batal, 2007) and Japanese quails (Salmanzadeh and Shahryar, 2013b).

It is well demonstrated that Gln appears to be a conditionally essential amino acid nutrient as exogenous Gln can be a potential candidate in improving intestinal morphology and digestive function (Alverdy et al., 1992; Helton et al., 1990; Shizuka et al., 1990). On the other hand, from the conclusions above, it is concluded that the increase in intestinal villus height reported in animals fed with Gln supplemented diets may explain the improvement in growth performance. Therefore, the increased villus height may amplify nutrient absorption and utilization of nutrients leading to improved growth performance of broiler chickens. On the

contrary, Chen et al. (2009) reported that body weights measured at the 25th day of incubation, at hatch (Day 0), and at the 3rd day post-hatching were not significantly modified in ducks treated in ovo with Gln, whereas IOF of Gln improved small intestine development and body weight than the control ducks when they were 7-days old.

In the present study, the relative weights of carcass, breast, thigh and gizzard were also significantly improved in broiler chickens injected with Gln. Chen et al. (2009) demonstrated that the in ovo injection of Gln at day 21 of incubation improved breast weight of ducks at 25 d of incubation until day 7 of post-hatch, whereas gizzard, proventriculus and liver weight were not significantly altered. Salmanzadeh and Shahryar (2013b) showed that dietary Gln supplementation (0, 20, 30, 40 and 50 mg kg^{-1} of diet) significantly increased the relative weights of carcass and breast compared to the control quails whereas liver and gizzard weights were not significantly modified.

Dai et al. (2009) demonstrated that dietary Gln addition (5 and 10 g kg^{-1}) improved growth performance and carcass traits of broilers. Furthermore, in another study, increases in carcass weight and relative weight of breast and thigh were also significantly improved in broilers supplemented with Gln at 5 g kg^{-1} of diet (Dai et al., 2011).

5 Conclusions

As a conclusion, IOF of Gln into broiler breeder eggs stimulated development of gastrointestinal tract and consequently, improved the growth performance and carcass characteristics whereas hatchability significantly depressed in all injected eggs compared to the not injected ones.

Author contributions. M. Salmanzadeh carried out the experiment and prepared the manuscript. Y. Ebrahimnezhad, H. Aghdam Shahryar, J. Ghiasi Ghaleh-Kandi designed and coordinated the experiment and helped to modify the manuscript.

Acknowledgements. This article is a part of PhD thesis in animal science, Islamic Azad University, Shabestar Branch (thesis supervisors: Y. Ebrahimnezhad and H. A. Shahryar, consulting advisor: J. Ghiasi Ghaleh-Kandi). Also, the authors have highly appreciated the cooperation of Saeid Ashrafi and Ayob heshmati sis for taking care of birds during the experiment period and helping in the process of experimental work.

Edited by: S. Maak

References

Al-Murrani, W. K.: Effect of injecting amino acids into the egg on embryonic and subsequent growth in the domestic fowl, Br. Poult. Sci., 23, 171–174, 1982.

Alverdy, J. A., Aoys, E., Weiss-Carrington, P., and Burke, D. A: The effect of glutamine-enriched TPN on gut immune cellularity, J. Surg. Res., 52, 34–38, 1992.

Andrews, F. J. and Griffiths, R. D.: Glutamine: essential for immune nutrition in the critically ill, Br. J. Nutr., 1, 3–8, 2002.

Bartell, S. M. and Batal, A. B.: The effect of supplemental glutamine on growth performance, development of the gastrointestinal tract, and humoral immune response of broilers, Poult. Sci., 86, 1940–1947, 2007.

Chen, W., Wang, R., Wan, H. F., Xiong, X. L., Peng, P., and Peng, J.: Influence of in ovo injection of glutamine and carbohydrates on digestive organs and pectoralis muscle mass in the duck, Br. Poult. Sci., 50, 436–442, 2009.

Dai, S. F., Wang, L. K., Wen, A. Y., and Wang, L. X.: Protection of glutamine supplementation to performance, intestinal enzyme activity and morphosis in broiler under heat stress, J. Chin. Cereal. Oils Assoc., 24, 103–107, 2009.

Dai, S. F., Gao, F., Zhang, W. H., Song, S. X., Xu, X. L., and Zhou, G. H.: Effects of dietary glutamine and gamma-aminobutyric acid on performance, carcass characteristics and serum parameters in broilers under circular heat stress, Anim. Feed. Sci. Technol., 168, 51–60, 2011.

Dong, X. Y., Jiang, Y. J., Wang, M. Q., Wang, Y. M., and Zou, X.T.: Effects of in ovo feeding of carbohydrates on hatchability, body weight, and energy status in domestic pigeons, Poult. Sci., 92, 2118–2123, 2013.

Dos Santos, T. T., Corzo, A., Kidd, M. T., McDaniel, C. D., Torres Filho, R. A., and Araújo, L. F.: Influence of in ovo inoculation with various nutrients and egg size on broiler performance, J. Appl. Poult. Res., 19, 1–12, 2010.

Ferket, P., De, J., Ghane, A., and Uni, Z.: Effects of in ovo feeding solution osmolality on hatching turkey, Poult. Sci., 84, 117–121, 2005.

Foye, O. T., Uni, Z., and Ferket, P. R.: Effect of in ovo feeding egg white protein, β-hydroxy-β-methylbutyrate, and carbohydrateson glycogen status and neonatal growth of turkeys, Poult. Sci., 85, 1185–1192, 2006.

Heiblum, R., Arnon, E., Chazan, G., Robinzon, B., Gvaryahu, G., and Snapir, N.: Glucocorticoid administration during incubation: Embryo mortality and post-hatch growth in chickens, Poult. Sci., 80, 1357–1363, 2001.

Helton, W. S., Jacobs, D. O., Bonner-Weir, S., Bueno, R., Smith, R. J., and Wilmore, D. W.: Effects of glutamine-enriched parenteral nutrition on the exocrine pancreas, J. Parenter. Enteral. Nutr., 14, 344–352, 1990.

Jazideh, F., Farhoomand, P., Daneshyar, M., and Najafi, G.: The effects of dietary glutamine supplementation on growth performance and intestinal morphology of broiler chickens reared under hot conditions, Turk. J. Vet. Anim. Sci., 38, 264–270, 2014.

Kitt, S. J., Miller, P. S., Lewis, A. J., and Fischer, R. L.: Effects of glutamine on growth performance and small intestine villus height in weanling pigs, Nebraska Swine Report, 82, 29–32, 2002.

Nowaczewski, S., Kontecka, H., and Krystianiak, S.: Effect of in ovo injection of vitamin C during incubation on hatchability of chickens and ducks, Folia Biologica (Kraków), 60, 93–97, 2012.

Ohta, Y., Tsushima, N., Koide, K., Kidd, M. T., and Ishibashi, T.: Effect of amino acid injection in broiler breeder eggs on embryonic growth and hatchability of chicks, Poult. Sci., 78, 1493–1498, 1999.

Ohta, Y., Kidd, M. T., and Ishibashi, T.: Embryo growth and amino acid concentration profiles of broiler breeder eggs, embryos, and chicks after in ovo administration of amino acids, Poult. Sci., 80, 1430–1436, 2001.

Ohta, Y., Yoshida, T., and Tsushima, T.: Comparison between broilers and layers for growth and protein use by embryos, Poult. Sci., 83, 783–787, 2004.

Pedroso, A. A., Chaves, L. S., Café, M. B., Leandro, N. S. M, Stringhini, J. H., and Menten, J. F. M.: Glutamine as broilers embryos nutrient, Braz. J. Avian. Sci. Camp., 8, 43–49, 2006a.

Pedroso, A. A., Chaves, L. S., Lopes, K. L. D. M., Leandro, N. S. M., Café, M. B., and Stringhini, J. H.: Nutrient inoculation in eggs from heavy breeders, Rev. Bras. Zootecn., 35, 2018–2026, 2006b.

Salmanzadeh, M.: The effects of in-ovo injection of glucose on hatchability, hatching weight and subsequent performance of newly-hatched chicks, Braz. J. Poult. Sci., 14, 137–140, 2011.

Salmanzadeh, M. and Shahryar, H. A.: Effects of dietary supplementation with glutamine on growth performance, small intestinal morphology and carcass traits in turkey poults under heat stress, Revue. Méd. Vét., 164, 476–480, 2013a.

Salmanzadeh, M. and Shahryar, H. A.: Effects of dietary glutamine addition on growth performance, carcass characteristics and development of the gastrointestinal tract in Japanese quails, Revue. Méd. Vét., 164, 471–475, 2013b.

Salmanzadeh, M., Ebrahimnezhad, Y., Shahryar, H. A., and Beheshti, R.: The effects of in ovo injection of glucose and magnesium in broiler breeder eggs on hatching traits, performance, carcass characteristics and blood parameters of broiler chickens, Arch. Geflügelk., 76, 277–284, 2012.

Samli, H. E., Senkoylu, N., Koc, F., Kanter, M., and Agma, A.: Effects of Enterococcus faecium and dried whey on broiler performance, guthistomorphology and microbiota, Arch. Anim. Nutr., 61, 42–49, 2007.

SAS Institute, SAS Users guide: Statistics, Version 9.12, SAS Institute Inc., Cary, NC, USA, 2005.

Shizuka, F., Vasupongsotorn, S., Kido, Y., and Kish, K.: Comparative effect of intravenously or intragastrically administered glutamine on small intestinal function of the rat, Tokushima, J. Exp. Med., 37, 49–57, 1990.

Smirnov, A., Tako, E., Ferket, P. R., and Uni, Z.: Mucin gene expression and mucin content in the chicken intestinal goblet cells are affected by *in ovo* feeding of carbohydrates, J. Poult. Sci., 85, 669–673, 2006.

Souba, W. W.: Glutamine and cancer, Ann. Surg., 218, 715–728, 1993.

Tako, E., Ferket, P. R., and Uni, Z.: Effects of *in ovo* feeding of carbohydrates and β- Hydroxy-β-Methylbutyrate on the development of chicken intestine, Poult. Sci., 83, 2023–2028, 2004a.

Tako, E., Ferket, P. R., and Uni, Z.: Changes in chicken intestinal zinc exporter mRNA expression and small intestinal functionality following intra-amniotic zinc-methionine administration, J. Nutr. Biochem., 16, 339–346, 2004b.

Touchette, K. J., Carroll, J. A., Allee, G. L., Matteri, R. L., Dyer, C. J., Beausang, L. A., and Zannelli, M. E.: Effect of spray-dried plasma and lipopolysaccharide exposure on weaned pigs: I. Effects on the immune axis of weaned pigs, J. Anim. Sci., 80, 494–501, 2002.

Uni, Z. and Ferket, P. R.: Enhancement of development of oviparous species by *in ovo* feeding, US Regular Patent US 6, 592, 878, 2003.

Uni, Z. and Ferket, R. P.: Methods for early nutrition and their potential, World's Poult. Sci. J., 60, 101–111, 2004.

Uni, Z., Ferket, P. R., Tako, E., and Kedar, O.: *In ovo* feeding improves energy status of lateterm chicken embryos, Poult. Sci., 84, 764–770, 2005,

Wilson, J. H.: Bone strength of caged layers as affected by dietary calcium and phosphorus concentrations, reconditioning, and ash content, Br. Poult. Sci., 32, 501–508, 1991.

Yi, G. F., Allee, G. L., Spencer, J. D., Frank, J. W., and Gaines, A. M.: Impact of glutamine, menhaden fish meal and spray-dried plasma on the growth performance and intestinal morphology of turkey poults, Poult. Sci., 80, 201–204, 2001.

Yi, G. F., Allee, G. L., Knight, C. D., and Dibner, J. J.: Impact of glutamine and oasis hatchling supplement on growth performance, small intestinal morphology, and immune response of broilers vaccinated and challenged with Eimeria maxima, Poult. Sci., 84, 283–293, 2005.

12

Effects of gender and diet on back fat and loin area ultrasound measurements during the growth and final stage of fattening in Iberian pigs

Ana González[1], Dolores Ayuso[2], Francisco Peña[1], Andrés L. Martínez[1], and Mercedes Izquierdo[2]

[1]Department of Animal Production, University of Córdoba, Córdoba, Spain
[2]Departament of Animal Production, CICYTEX, Badajoz, Spain

Correspondence to: Ana González (v32gomaa@uco.es)

Abstract. Reared in extensive parcels, 50 castrated or spayed Iberian pigs were fattened with conventional or high-oleic-concentrate diets to evaluate the effect of diet and sex on the measurements of the loin area depth, back fat thickness and its layers during the last 6 months before slaughter at eight time points in two anatomical locations by serial scans. The scan was the factor that had the greatest influence, followed by sex and diet. Back fat thickness at 10th rib level was higher than at 14th rib level. The thickness of the outer, middle and inner layers progressively increased over the study period. Throughout the experimental period, the differences between two successive scans of the M. longissimus area did not differ significantly, obtaining the lowest value at the third scan and the highest at the fifth scan. The ultrasound back fat depth was affected by sex, being greater in females and in animals with a high-oleic diet. Positive significant correlations were observed for measurements assessed. The R^2 values for the regression equations to estimate M. longissimus area were lower than the values found for the prediction of fat measurements, and they differed between sex and diet. The relative back fat growth was higher than M. longissimus area, not permitting the establishment of a similar growth pattern for fat and muscle. The sex and diet was taken into account in the predictive models. The subcutaneous adipose layers in Iberian pigs grow at different rates during the last 6 months before slaughter; with the ultrasound serial scan, it is possible to show these changes. The change in diet and the sex affect the adipose tissue development, being more noticeable in the middle layer of back fat at 10th rib level and the inner layer at 14th rib level. However, the sex and the use of an enriched oleic acid diet do not affect to loin development. As the middle layer of back fat shows more growth, this layer could be the best to be included in predictive models. The middle layer of back fat could also be good to be included in predictive models. Back fat thickness at the eighth scan can be predicted with moderate accuracy from corresponding measurements taken 30 days earlier and with less accuracy as the interval between measurements increases.

1 Introduction

The autochthonous Iberian pig breed is traditionally bred in southwestern Spain and has a specific conformation and high adipogenetic potential, influenced by a diet based on grass and acorns. This pig breed is reared under different conditions, all of which are regulated by the Spanish Ministry of Agriculture (BOE, 2014) according to the features of the last fattening phase: intensive system, *Recebo*, *Cebo a campo* and *Montanera*.

In the last decades the pork industry has found features to improve meat quality. In this sense, the main trait of Iberian pig meat is its high content of intramuscular fat, which is desirable in obtaining dry-cured products (mainly ham, foreleg and loin) and important for flavour and to provide slow dehydration during the curing process (Fernandez et al., 2003). According to Lopez-Bote (1998), in the last phase of fattening (around 150 kg of body weight) the Iberian pig can reach 60 % carcass fat, 15 cm of back fat depth and 10–13 % intramuscular fat content. Factors such as diet or sex are the most investigated in Iberian pig production, with the purpose of improving intramuscular fat content (Ayuso et al., 2004).

Back fat layer thickness is an important parameter at all stages of pig production. It is used as a tool for the evaluation of dietary requirements in order to optimize growth and determine the price (McEvoy et al., 2007). Measurements of thickness of subcutaneous adipose tissue are useful for monitoring the production process in order to optimize growth and carcass composition. These measurements can be obtained using ultrasound techniques, the use of which has become widespread for the Iberian pig in recent decades either for descriptive studies or to predict carcass composition (Ayuso et al., 2013). Likewise, these measurements can be very useful in improving the response of genetic selection in economic traits (Moeller et al., 1998). It has been proved that back fat thickness, loin area and intramuscular fat content are good features to predict pig carcass characteristics (Ayuso et al., 2014).

The goal of this study was to evaluate, under outdoor intensive conditions, the effect of diet and sex on the measurements of the loin area depth, back fat thickness and its layers at eight time points in two anatomical locations by serial scans before and during the final fattening phase in Iberian pigs.

2　Material and methods

2.1　Animal management and experimental design

The experimental procedures used in this study were in compliance with the Spanish guidelines for the care and use of animals in research (BOE, 2007). The data were collected from 50 castrated or spayed Iberian pigs (25 males and 25 females). Pigs in the study were progeny from Iberian purebred sires and dams. The trial was conducted at Valdesequera Farm, a central Iberian swine test station (Badajoz, Spain). From 15 days of age to weaning (42–49 days old), piglets had free access to a commercial pre-starter feed. At weaning, all piglets were moved to an open-air fenced-in area where they had full access to feed and water. The pigs were reared in extensive parcels according to the Spanish legislation (BOE, 2007), without the presence of oaks to avoid the intake of acorns, and they received the same standard diet based on commercial concentrate feed. A gradual change from a starter diet (~ 23 kg live weight) to subsequently receiving a growth diet (~ 70 kg live weight) occurred. At the beginning of the fattening period (~ 90 kg live weight), the males were castrated and the females were ovariectomized under anesthesia following the Spanish regulations. Then pigs of the same sex were sorted by weight into eight pens of 3–4 pigs each and were then randomly assigned to one of two dietary treatments: conventional concentrate feed (C) or concentrate diet containing a high level of oleic acid (HO). The pens had similar characteristics. The chemical composi-

Table 1. The chemical composition of dietary ingredients.

Analysed composition[a]	Concentrate[b]	Concentrate high oleic[b]
Crude protein %	16.10	15.00
Crude fat %	2.44	5.66
Crude fibre %	3.99	5.07
Ash %	5.80	5.10
Oleic %	0.54	3.30

[a] Expressed as percentage of dry matter. [b] By formulation.

tion of each feed is presented in Table 1. The diet was offered ad libitum.

2.2　Ultrasound measurements

The pigs were weighed and ultrasonically scanned throughout the fattening period. To obtain the ultrasound images, pigs were immobilized and restrained by the head in a squeeze chute and the image sites were determined by physical palpation to accurately ascertain the scanning sites. The animals were held manually, avoiding any abnormal situation that could stress the animal, and they were only scanned in a relaxed posture, permitting accurate measurements. A mix of Eco Gel and isopropyl alcohol was used as a sound conducting material to allow a better acoustic contact surface between the probe and the skin. An Aloka 500V real-time ultrasound machine (Aloka Holding Europe, Switzerland) equipped with a 12.5 cm, 3.5 MHz linear array transducer (Aloka Holding Europe, Switzerland) was used. The adjust gain settings were 90 for overall gain, -25 for near gain and 2.1 for far gain. The focus zones were 1 and 2 on the Aloka 500V for all images collected. Captured ultrasonic images were recorded and interpreted later with Biosoft® software (Biotronics Inc., Ames, IA, USA). Ultrasound images for measurement were taken along the dorsal midline at 10th and 14th rib projection with the transducer centred perpendicularly at the anatomical site. A cross-sectional image of the loin area and back fat thickness and its layers (outer, middle and inner) on the right side of the pig at the 10th intercostal space and just behind the last rib (14th rib) were obtained using an ultrasound stand-off guide (Superflab®) mounted on the linear probe and conforming to the curvature of the pig's back and ham. Images were digitalized and stored in a computer for later analysis with Biosoft® software (Biotronics Inc., Ames, IA, USA). Determination of the anatomical location of the 10th and 14th ribs was made based on presence of muscle systems viewed on the ultrasound image. Back fat thickness was measured at a midline of the distance along the loin muscle area from skin with a perpendicular line to skin. The loin area was measured to draw the outline around the loin muscle.

2.3 Statistical analyses

All the statistical analyses were performed using SAS 9.3 (SAS Institute Inc., Cary, NC). Since the study included repeated measurements (serial ultrasound scanning), the mixed procedure assuming non-equal covariances among serial samples was used to compare the eight serial ultrasound scanning measurements of the different traits (back fat thickness and loin area at the 10th and 14th rib levels). The model included sex, feed and scan as fixed effects and body weight as a covariate. Ultrasound scan was included as a repeated measurement in the repeated statement of the mixed procedures. Several interactions were also included in the model. The least significant difference test was used to compare least squares means. Correlations between the ultrasound measurements at 10th rib level and the same ultrasound measurements at 14th rib level were investigated in the fourth to seventh scans. Similarly, total back fat thickness and loin area regression coefficients relating the ultrasound measurements at the eighth scan and corresponding measurements at the previous scan were obtained too.

The allometric equation was used for the depiction of differential growth of muscle and fatty tissue. This model describes a part-to-whole relationship and has the following form:

$$\log Y = \log a + b \times \log X,$$

where is the intercept on the Y axis, b is the allometric growth coefficient (slope), X is the body weight, and Y is the total back fat thickness and its layers or loin area.

The allometric coefficients were t tested to determine if the allometric coefficient defined a fast, slow or equivalent pattern of growth. The t tests were computed as $t = (b - Bo)/SEb$, which distributes as Student's t.

3 Results

Least square means (±standard errors) of the ultrasound measurements are listed in Table 2. As expected, the scan was the factor that had the greatest influence, followed by sex and diet.

Table 3 shows the least squares means (±standard errors) for the body weight, back fat thickness and M. longissimus area at the 10th and 14th rib levels. Table 3 also shows the thickness of individual back fat layers: outer, middle and inner. The live weight ranged from 65.6 ± 1.5 kg at the start to 161.6 ± 1.4 kg at the end of the experimental period. This represents an increase of 96.0 kg and an average daily gain of 458 g day^{-1} (360 and 720 g day^{-1} at the start and end of the trial, respectively). Increases in the ultrasound measurements with increased body weight were expected. BF10 and BF14 (see underneath Table 2 for an explanation of abbreviations) increased significantly ($p < 0.05$) during the studied period. In all scans, back fat thickness at 10th rib level was higher than that recorded at 14th rib level, although this relationship

changed significantly ($p < 0.05$) as body weight increased (1.97 and 1.28 mm in the first and last scans, respectively) as a result of higher back fat growth at 14th rib level (152.8 vs. 1225.5 %). The thickness of the outer, middle and inner layers progressively increased over the period of study, representing 152.80, 115.8, 119.4, 135.3, 184.0 and 135.9 % of the initial value for BFO10, BFM10, BFI10, BFO14, BFM14 and BFI14, respectively. The greatest increase occurred at the 14th rib level. Regarding the back fat layers, the middle and outer layers showed the highest and lowest growth, respectively. In the data set, the outer, middle and inner layers of the fat represented 20.33, 55.28 and 24.73 % at 10th rib level and 22.97, 46.10 and 32.22 % at 14th rib level. As a proportion of the total back fat depth, a decrease appeared in BFO10 (from 57.05 to 53.94 %), BFO14 (from 23.43 to 20.76 %) and BFI14 (from 34.42 to 30.62 %), and an increase in BFM10 (from 52.88 to 56.68 %) and BFM14 (from 41.49 to 49.97 %), while BFI10 was relatively static over the time period (from 25.20 to 24.60 %).

The change in back fat thickness per unit of body weight was 0.456 and 0.308 mm kg^{-1} for BF10 and BF14, respectively, recording the highest values in the first scan (0.635 and 0.390 mm kg^{-1}, respectively). Similarly, to what was observed for back fat thickness, the M. longissimus area at 10th rib level was higher than at 14th rib level. The differences in the M. longissimus area between two successive scans are not significant ($p > 0.05$) throughout the experimental period, recording the lowest value (0.04 and 0.92 cm^2 for 10th and 14th levels, respectively) for the fourth scan and the highest for the first scan (1.41 cm^2). The M. longissimus area at both levels increased throughout the study period, representing 138.6 % for LA10 at the last scan and 123.6 % for LA14 at the first scan. The M. longissimus area per kilogram of body weight ratio decreased significantly as body weight increased: from 0.223 to 0.126 mm kg^{-1} for LA10 and from 0.209 to 0.105 mm kg^{-1} for LA14 for the first and last scans, respectively.

The least squares means for the ultrasound measurements by sex are showed in Figs. 1 and 2. Sex did not significantly affect ($p > 0.05$) the body weight gain (95.7 and 96.2 kg in males and females, respectively) for the entire study period. All ultrasound measurements increased significantly as body weight increased, although at different rates. The back fat depth and rate of back fat growth were affected by sex ($p < 0.05$). The ultrasound back fat depth was greater in females, although the rates of back fat depth growth in the males were lower than those recorded in females (118.46 and 124.18 % for BF10 and 149.21 and 152.0 % for BF14 in males and females, respectively). The back fat thickness had higher growth in females (138.09 %) than the M. longissimus area (135.19 %); in males the opposite occurred (133.83 vs. 127.45 %, back fat thickness (BF) and M. longissimus area (LA), respectively). Also, it has been noted that the increase in back fat thickness was higher at 14th rib level, which caused the BF10 / BF14 ratio to de-

Table 2. Least squares means (±SE) for the data set of ultrasound traits studied by sex and feed and analysis of variance test.

	Sex		Feed		Effects						
	Male	Female	C	HO	Weight	Sex	Feed	Scan	Sex × feed	Feed × scan	Sex × scan
BF10 (mm)	45.19 ± 0.79	47.67 ± 0.86	44.72 ± 0.90	48.14 ± 0.84	< 0.0001	0.0177	0.0101	< 0.0001	0.2421	< 0.0001	0.2701
BFO10 (mm)	9.21 ± 0.20	9.67 ± 0.22	9.25 ± 0.25	9.64 ± 0.23	< 0.0001	0.0325	0.3369	< 0.0001	0.2346	0.0060	0.3713
BFM10 (mm)	24.08 ± 0.56	27.25 ± 0.61	24.75 ± 0.64	26.58 ± 0.60	< 0.0001	0.0045	0.0263	< 0.0001	0.0885	< 0.0001	0.0077
BFI10 (mm)	11.08 ± 0.25	11.88 ± 0.27	11.12 ± 0.30	11.84 ± 0.29	< 0.0001	0.5469	0.0542	0.0841	0.0533	0.0010	0.0148
LA10 (cm²)	17.80 ± 0.21	18.06 ± 0.23	18.11 ± 0.24	17.76 ± 0.22	0.0067	0.2602	0.2209	< 0.0001	0.4333	0.0495	0.0001
BF14 (mm)	31.60 ± 0.72	32.36 ± 0.75	30.51 ± 0.92	33.44 ± 0.88	< 0.0001	0.0362	0.0608	< 0.0001	0.6659	< 0.0001	0.1668
BFO14 (mm)	7.25 ± 0.20	7.43 ± 0.21	7.22 ± 0.24	7.47 ± 0.22	< 0.0001	0.5736	0.4512	< 0.0001	0.0259	0.0670	0.2191
BFM14 (mm)	14.25 ± 0.43	15.23 ± 0.45	14.17 ± 0.55	15.31 ± 0.52	< 0.0001	0.0258	0.2150	< 0.0001	0.3102	0.0001	0.6585
BFI14 (mm)	10.23 ± 0.30	10.37 ± 0.32	9.63 ± 0.37	10.97 ± 0.35	< 0.0001	0.1164	0.0250	< 0.0001	0.1732	0.0263	0.4353
LA14 (cm²)	15.55 ± 0.20	16.03 ± 0.22	15.75 ± 0.23	15.83 ± 0.22	< 0.0001	0.0251	0.9739	< 0.0001	0.1482	0.5185	0.6610

C: concentrated feed system; HO: high oleic system; BF10: ultrasonic back fat depth at 10th rib; BFO10: ultrasonic outer layer depth at 10th rib; BFM10: ultrasonic middle layer depth at 10th rib; BFI10: ultrasonic inner layer depth at 10th rib; LA10: M. longissimus area at 10th rib; BF14: ultrasonic back fat depth at 14th rib; BFO14: ultrasonic outer layer depth at 14th rib; BFM14: ultrasonic middle layer depth at 14th rib; BFI14: ultrasonic inner layer depth at 14th rib; LA14: M. longissimus area at 14th rib.

Table 3. Least square means (±SE) for body weight, ultrasonic back fat depth and loin area in pigs of the Iberian breed.

Scan	First	Second	Third	Fourth	Fifth	Sixth	Seventh	Eighth
N	50	50	50	50	50	50	50	50
BW (kg)	65.59 ± 1.54	79.05 ± 1.58	88.27 ± 1.52	95.24 ± 1.52	110.78 ± 1.48	124.00 ± 1.43	139.91 ± 1.43	161.57 ± 1.43
BF10 (mm)	41.63 ± 1.73	41.98 ± 1.30	43.90 ± 1.00	46.24 ± 0.83	47.46 ± 0.79	48.34 ± 0.93	50.89 ± 1.36	50.98 ± 2.02
BFO10 (mm)	7.08 ± 0.43	7.95 ± 0.33	8.80 ± 0.27	9.34 ± 0.22	10.01 ± 0.21	10.73 ± 0.23	10.86 ± 0.36	10.77 ± 0.52
BFM10 (mm)	23.75 ± 1.09	23.66 ± 0.83	24.61 ± 0.63	26.17 ± 0.52	25.91 ± 0.56	26.23 ± 0.58	27.49 ± 0.86	27.50 ± 1.24
BFI10 (mm)	10.50 ± 0.54	10.79 ± 0.42	11.02 ± 0.32	11.55 ± 0.264	11.59 ± 0.22	11.77 ± 0.32	12.10 ± 0.44	12.54 ± 0.72
LA10 (cm²)	14.66 ± 0.54	16.07 ± 0.42	16.89 ± 0.33	18.01 ± 0.28	18.55 ± 0.21	19.17 ± 0.28	19.77 ± 0.40	20.32 ± 0.63
BF14 (mm)	25.59 ± 1.38	26.85 ± 1.06	28.39 ± 0.89	31.60 ± 0.80	32.77 ± 0.86	34.81 ± 0.99	36.71 ± 1.28	39.10 ± 1.73
BFO14 (mm)	6.00 ± 0.39	6.33 ± 0.29	6.78 ± 0.23	7.51 ± 0.22	7.57 ± 0.20	8.14 ± 0.24	8.31 ± 0.32	8.12 ± 0.47
BFM14 (mm)	10.62 ± 0.79	11.18 ± 0.60	12.24 ± 0.48	14.13 ± 0.48	15.22 ± 0.53	16.67 ± 0.59	18.33 ± 0.77	19.54 ± 1.04
BFI14 (mm)	8.81 ± 0.69	9.43 ± 0.52	9.96 ± 0.41	10.63 ± 0.37	10.53 ± 0.32	10.31 ± 0.37	10.80 ± 0.57	11.97 ± 0.89
LA14 (cm²)	13.69 ± 0.51	14.56 ± 0.37	15.48 ± 0.29	16.28 ± 0.24	16.32 ± 0.23	16.45 ± 0.24	16.63 ± 0.36	16.92 ± 0.55

BW: body weight; BF10: ultrasonic back fat depth at 10th rib; BFO10: ultrasonic outer layer depth at 10th rib; BFM10: ultrasonic middle layer depth at 10th rib; BFI10: ultrasonic inner layer depth at 10th rib; LA10: M. longissimus area at 10th rib; BF14: ultrasonic back fat depth at 14th rib; BFO14: ultrasonic outer layer depth at 14th rib; BFM14: ultrasonic middle layer depth at 14th rib; BFI14: ultrasonic inner layer depth at 14th rib; LA14: M. longissimus area at 14th rib.

crease from 1.63 and 1.61 for males and females, respectively, in the first control to 1.29 and 1.31 in the last scan. For M. longissimus area no significant difference ($P > 0.05$) between locations was found, with a LA10 / LA14 ratio of 1.07 in the first scan for both sexes and 1.23 and 1.16 in the last scan for males and females, respectively. No significant differences were observed in the percentages of each back fat layer and total thickness between the sexes (54.15 and 56.16 % for BFO10, 20.20 and 20.48 % for BFM10, 25.23 and 24.17 % for BFI10, 23.68 and 22.28 % for BFO14, 45.27 and 46.84 % for BFM14, and 32.35 and 32.13 % for BFI14 in males and females, respectively). The back fat layer growth was different in both sex and anatomy levels (143.06 and 152.66 for BFO10, 137.63 and 130.36 for BFO14, 108.35 and 122.44 for BFM10, 178.88 and 181.76 for BFM14, 126.43 and 116.81 for BFI10, and 130.88 and 135.89 for BFI14 in males and females, respectively.)The back fat thickness / body weight ratio increased as body weight increased, with females showing a higher increase (0.378 vs. 0.405 mm kg^{-1} for BF10 in males and 0.404 vs. 0.448 mm kg^{-1} in females, 0.195 vs. 0.316 mm kg^{-1} for BF14 in males and 0.206 vs. 0.350 mm kg^{-1} in fe-

males). By contrast, the M. longissimus area / live weight ratio showed a significant ($p < 0.05$) decrease for LA10 in males and females (0.194 vs. 0.137 cm^2 kg^{-1} and 0.213 vs. 0.134 cm^2 kg^{-1}, respectively) and LA14 in males (0.167 vs. 0.121 cm^2 kg^{-1}) instead of an increase for LA14 in females (0.182 vs. 0.124 cm^2 kg^{-1}). These results are in line with allometric coefficients that were obtained (Table 4).

The least squares means for the ultrasound measurements by diet are shown in Figs. 3 and 4. The growth was affected ($p < 0.05$) by feeding system. Total weight gain of pigs on a C ration was slightly lower (91.6 vs. 99.4 kg) than that recorded in pigs on a HO ration. The back fat thickness increased more (110.25 and 133.44 % for BF10, and 134.71 and 144.82 % for BF14 in C and HO groups, respectively) than the M. longissimus area (139.74 and 124.83 % for LA10, and 134.71 and 122.52 % for LA14 in C and HO groups, respectively). The back fat layers showed a higher increase at 14th rib level than 10th rib level except for the outer layer in the C group. The layer with the lowest growth was the medium layer at 10th rib level (106.22 and 125.14 % for C and HO, respectively) and the inner layer for C (116.09 %) and the outer layer for HO, both at 14th rib level.

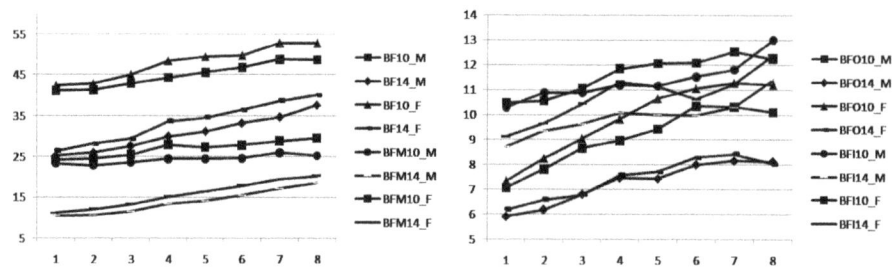

Figure 1. Changes in ultrasonic back fat depth and its layers as functions of scans in pigs of the Iberian breed.

Table 4. Allometric coefficients (R^2) for ultrasonic back fat depth and loin area in pigs of the Iberian breed.

	Scans 1 to 8			Scans 1 to 4	Scans 5 to 8	
		Males	Females		C	HO
BF10	1.09 (0.88)	1.08 (0.86)	1.12 (0.89)	1.09 (0.67)	1.10 (0.65)	1.03 (0.79)
OBF10	0.99 (0.75)	1.03 (0.79)	0.98 (0.76)	1.13 (0.58)	0.70 (0.46)	0.61 (0.28)
MBF10	1.2 (0.81)	1.20 (0.80)	1.26 (0.87)	1.20 (0.56)	1.29 (0.59)	1.23 (0.79)
IBF10	1.10 (0.79)	1.08 (0.78)	1.14 (0.82)	1.17 (0.58)	1.00 (0.51)	1.08 (0.61)
LA10	0.52 (0.77)	0.56 (0.79)	0.48 (0.78)	0.53 (0.51)	0.43 (0.51)	0.40 (0.42)
BF14	1.56 (0.88)	1.54 (0.90)	1.61 (0.88)	1.74 (0.75)	1.37 (0.59)	1.45 (0.82)
OBF14	1.35 (0.79)	1.37 (0.84)	1.35 (0.75)	1.63 (0.63)	1.16 (0.58)	0.96 (0.45)
BF14	1.9 (0.85)	1.87 (0.87)	1.91 (0.87)	2.16 (0.72)	1.52 (0.50)	1.63 (0.76)
IBF14	1.46 (0.76)	1.42 (0.77)	1.52 (0.76)	1.84 (0.61)	1.39 (0.49)	1.34 (0.61)
LA14	0.58 (0.77)	0.61 (0.81)	0.56 (0.76)	0.68 (0.57)	0.55 (0.60)	0.43 (0.40)

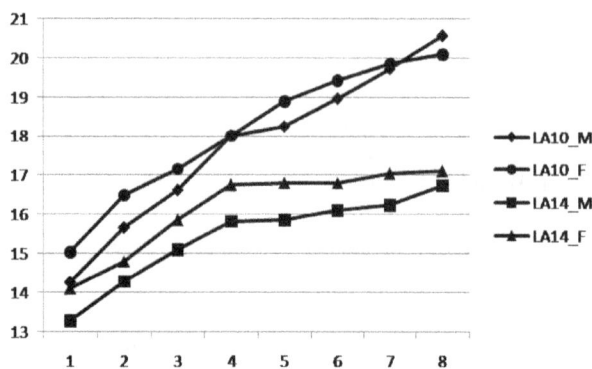

Figure 2. Changes in ultrasonic M. longissimus area as functions of scans in pigs of the Iberian breed.

These results are in line with the allometric coefficients obtained both in adipose tissue and muscle being higher in the back fat measurements. In relation to adipose tissue was the medium layer of back fat at 14th rib level with the highest growth rates and the outer back fat layer at 10th rib level with the lowest development speed. The M. longissimus area showed an allometric coefficient lower than 1 in all cases. Also, the allometric coefficients of ultrasound back fat depth measurements were slightly higher in females, while the allometric coefficients of M. longissimus area were higher in males. Although the differences were not statistically signifi-

cant ($p > 0.05$). However, in relation to the diet, the C group presented a higher growth level.

Correlation coefficients between ultrasound measurements at the eighth scan with corresponding measurements at the previous scan (only those that showed statistically significant values) are shown in Table 5. Positive and in most cases significant correlations were observed for the measurements assessed, except for M. longissimus area at the 10th rib, which was not significantly ($p > 0.05$) correlated with the fourth, fifth and sixth scans and was positively significantly ($p > 0.05$) correlated with the seventh scan. These coefficients ranged from 0.18 to 0.89 ($p < 0.05$ when $r \geq 0.43$) and showed a tendency to increase this relationship as we approach the eighth scan. The correlations between M. longissimus area measurements were lower than for the other ultrasound measurements.

The R^2 values for the regression equations (Table 6) to estimate M. longissimus area were lower (0.15 to −0.01) than the values found for the prediction of fat measurements (0.62 to −0.02). As the time between scans increased, the regression coefficient decreased.

4 Discussion

There are no previous studies on serial scans of adipose and muscular tissue during growth and the final fattening phase in adipogenetic breeds such as the Iberian pig. However,

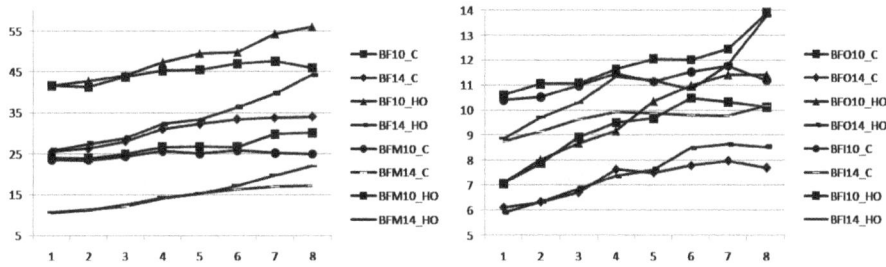

Figure 3. Changes in ultrasonic back fat depth and its layers as functions of feed in pigs of the Iberian breed.

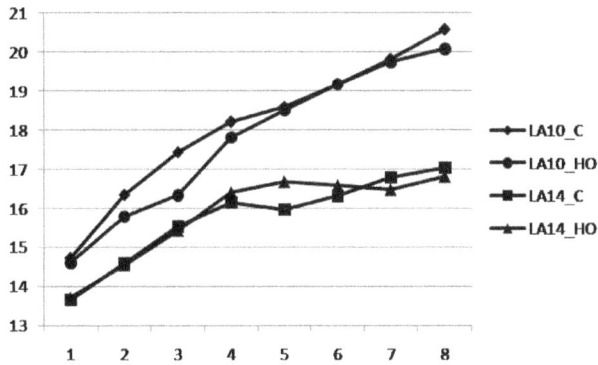

Figure 4. Changes in ultrasonic M. longissimus area as functions of feed in pigs of the Iberian breed.

Table 5. Correlation coefficients between ultrasound measurements at the eighth scan and previous scans.

Scan	Seventh	Sixth	Fifth	Fourth
BF10-8	0.81***	0.58*	0.54*	0.59*
BFO10-8	0.84***	0.68**	0.68**	0.55*
BFM10-8	0.80***	0.62**	0.62**	0.63**
BFI10-8	0.77***	0.55*	0.56*	0.49*
LA10-8	0.46*	0.24	0.22	0.18
BF14-8	0.82***	0.67**	0.57*	0.53*
BFO14-8	0.82***	0.66**	0.54*	0.45*
BFM14-8	0.89***	0.74***	0.69**	0.61**
BFI14-8	0.75***	0.60**	0.55*	0.46*
LA14-8	0.63**	0.55*	0.43*	0.45*

BF10-8: ultrasonic back fat depth at 10th rib in the eighth scan; BFO10-8: ultrasonic outer layer depth at 10th rib in the eighth scan; BFM10-8: ultrasonic middle layer depth at 10th rib in the eighth scan; BFI10-8: ultrasonic inner layer depth at 10th rib in the eighth scan; LA10-8: M. longissimus area at 10th rib in the eighth scan; BF14-8: ultrasonic back fat depth at 14th rib in the eighth scan; BFO14-8: ultrasonic outer layer depth at 14th rib in the eighth scan; BFM14-8: ultrasonic middle layer depth at 14th rib in the eighth scan; BFI14-8: ultrasonic inner layer depth at 14th rib in the eighth scan; LA14-8: M. longissimus area at 14th rib in the eighth scan. * $p < 0.05$. ** $p < 0.01$. *** $p < 0.001$.

there is consensus in the literature on the influence of breed on back fat thickness (Rybarczyk et al., 2011). Warriss et al. (1990) recorded average values of 11.0 and 23.0 mm in Pietrain and Large Black pigs, respectively. Also, Serrano

Table 6. Coefficients (R^2) for the regression function relating the ultrasound measurements at the eighth scan to corresponding measurements at the previous scan.

Scan	First	Second	Third	Fourth	Fifth	Sixth	Seventh
BF10_8	−0.03	−0.02	0.03	0.18**	0.08*	0.24***	0.53***
BF14_8	−0.02	−0.03	0.04	0.02	−0.02	0.47***	0.62***
LA10_8	0.09*	0.06	0.08*	−0.01	0.07*	0.00	−0.01
LA14_8	0.15**	0.08	−0.01	0.06	0.10*	0.04	0.12**

BF10-8: ultrasonic back fat depth at 10th rib in the eighth scan; LA10-8: M. longissimus area at 10th rib in the eighth scan; BF14-8: ultrasonic back fat depth at 14th rib in the eighth scan; LA14-8: M. longissimus area at 14th rib in the eighth scan.
* $p < 0.05$. ** $P 0.01$. *** $p < 0.001$.

et al. (2008) recorded values of 50, 51.1 and 80.1 mm for back fat thickness in Danish Duroc, Spanish Duroc and Retinto Iberian pig breeds, respectively. The means of back fat thickness at slaughter were similar to results in previous studies carried out on Iberian pigs (De Pedro, 1987; Dobao et al., 1987; Daza et al., 2005, 2006; Rey et al., 2006; Serrano et al., 2008; Ayuso et al., 2014). As expected, these values were higher than those reported for local or autochthonous breeds and improved breeds in previous studies (Minelli et al., 2013; Franco et al., 2014). The values were also higher than those reported for Celta pigs (38.0 ± 7.9 mm; a Spanish native breed adapted to the extensive production system) (Temperan et al., 2014). These results are not surprising since the Iberian is a local breed from the southwestern region of Spain with a distinct adipogenetic nature and shows a high subcutaneous adiposity, contributed to by their diet because as indicated by Cunningham et al. (1973), pigs fed with a low-protein diet were fatter than pigs fed with a high-protein diet. These differences could be considered as a consequence of selection for growth efficiency of lean meat or fat. The differences between breeds for back fat thickness remained when comparing the thickness / body weight ratio (mm kg^{-1}): 0.379 mm kg^{-1} in our study vs. 0.177 to 0.298 (Cunningham et al., 1973) or 0.271 mm kg^{-1} (Moeller et al., 1998). However, some local breeds have values close to those found in our study: 0.327 mm kg^{-1} in Nero Siciliano (Pugliese et al., 2003), 0.363 mm kg^{-1} in Cinta Senese (Franci et al., 2005), 0.308 mm kg^{-1} in Casertana, and 0.359 mm kg^{-1} in Mora (Fortina et al., 2005). This could be primarily due to the differences in slaughter weight, breed and feeding of the animals, factors that significantly influence the back fat thickness (Schinckel et al., 2002; Serrano et al., 2008).

The differences found between the back fat thickness at the 10th and 14th rib levels in our work demonstrate the lack of uniformity in the distribution of subcutaneous fat throughout the body of pigs, consistent with previous studies (Correa et al., 2006; Mas et al., 2010; Aro and Akinjokun, 2012). However, the back fat thickness at 10th rib level was higher in the first scans, but was higher at 14th rib level for the whole study period. In this regard, Fortin et al. (1980) indicated that fat over the back of the pork carcass was not evenly distributed when evaluating back fat thickness at various anatomical locations and positions. Comparison of the results of our work (Iberian pigs) with those obtained by Fortin (1986) in Yorkshire pigs clearly verifies the effect of breed (fattest vs. leanest) on back fat depth growth. In our work, we find a relative increase over live weight in back fat thickness during the control period studied (0.512 and 0.279 mm kg^{-1} at the first and last scans, respectively), whereas Yorkshire pigs descended from 20 to 120 kg live weight (0.51 and 0.282 mm kg^{-1}, respectively). McKay et al. (1984) also recorded declines in dorsal fat thickness / body weight ratio between 22.5 and 90.0 kg of live weight in Minnesotan (0.551 to 0.413 mm kg^{-1}), Pietrain (0.443 to 0.328 mm kg^{-1}) and Yorkshire pigs (0.412 to 0.308 mm kg^{-1}).

The increase in back fat thickness recorded by Swantek et al. (2014) in Berkshire pigs (286.5 %) and Greer et al. (1987) in crossbreed pigs (250 %) was similar to that obtained in our study for BF10 (266.3 %) but much lower than the increase recorded for BF14 (409.3 %). The lowest increase recorded by Richmond and Berg (1971) in Duroc × Yorkshire, Hampshire × Yorkshire and Yorkshire × Yorkshire (143.4 %) pigs may in part be because the animals used belonged to a lean breed and also because a different weight range was considered (25.3 kg at baseline and 80.5 kg at the end), as was found in the work of Fortin (1980).

Subcutaneous fat tissue in pigs consists of two, and sometimes three, fat layers (Fortin, 1986). In the present study, the outer, middle and inner layers represented 21.65, 50.69 and 28.48 %, respectively, of the total back fat thickness. However, they differed from the results of Newcom et al. (2004), who recorded values of 50.2, 29.67 and 20.11 % for the outer, middle and inner layers, respectively, in Duroc pigs and the results of Alfonso et al. (2005) in Basque pigs (33.7, 40 and 26.27 %, respectively). Likewise, the importance of each layer differs between studies: middle > inner > outer vs. outer > middle > inner. Meanwhile, Pugliese et al. (2003), recorded only the outer and inner layers, providing percentages of 44.94 and 55.32 %, respectively, in Nero Siciliano pigs, confirming the differences between breeds in this work. The differences found by Dobao et al. (1987) could be mainly due to slaughter weight (130–140 kg vs. 162 kg in our study). The total back fat depth also influences the percentage of each layer (Moody and Zobrisky, 1966). These authors found that increasing the total back fat thickness decreased the middle layer and increased the outer layer, as occurred in our case at 10th rib level, while at 14th rib level the middle layer is the one that grows. The unequal rate of development of adipose layers is in agreement with Fortin (1986) and McEvoy et al. (2007).

Similarly to what was observed for back fat thickness, we check the lack of uniformity in the distribution of the different layers along the body of pigs, in line with Pugliese et al. (2003), who recorded percentages of 35.48, 44.82, and 54.51 % for the outer layer and 65.52, 55.18, and 45.26 % for the inner layer of back fat on the dorsal midline opposite from the first and last ribs, and over the M. Gluteus medius.

In contrast to the values reported for the back fat thickness, values of M. longissimus area recorded in our study (21.99 and 19.79 cm^2, or 0.136 and 0.122 cm^2 kg^{-1} for LA10 and LA14, respectively) were lower than those recorded in lean breeds (Newcom et al., 2004; Stahl and Berg, 2003; Sullivan et al., 2007; Hsu et al., 2010; Mas et al., 2010; Rybarczyk et al., 2011), both in absolute values (30.3–56.7 cm^2) and relative to body weight (0.329–0.498 cm^2 kg^{-1}) or rate of deposition (0.088 cm^2 kg^{-1} in our study vs. 0.304 cm^2 kg^{-1} by Moeller et al. (1998). This difference may be explained in part by constant genetic improvement of this parameter.

Sex had a significant effect on back fat thickness, both overall and in each of the layers, except the inner layer of back fat, since the sexes differed significantly in distribution of back fat thickness. Females had thicker 10th and 14th rib level fat depth and bigger M. longissimus area than males. Kemster and Evans (1979) concluded that subcutaneous fat accumulated more ventrally in females and more dorsally in males. Sex differences in the present study are supported by previous reports for the effect of the sex (Serrano et al., 2008; Mas et al., 2010; Aro and Akinjokun, 2012; Minelli et al., 2013; Swantek et al., 2014; Suárez-Belloch et al., 2015). However, the differences in sex for back fat thickness were contrary to the observations of McKay et al. (1984). The higher adiposity of male carcasses resulted from the development of subcutaneous fat rather than internal fat tissue, and females having proportionally more fat depth over the shoulder and relatively less over the mid-back than males. Also, the higher fat depth in males is indicative of higher potential for fattening than females, in agreement with previous reports (Cisneros et al., 1996), which is likely related to the higher voluntary feed intake of males. Generally, females were leaner than males based on greater values for M. longissimus area, along with less back fat. However, in our study males were leaner than females and we did not find significant differences in M. longissimus area possibly because the animals were castrated. The effects of sex were probably reduced by castration of both males and females, as reported by Mayoral et al. (1999) for castrated male and female Iberian pigs.

Diet had significant effects on the back fat thickness, both overall and for each layer, while M. longissimus area was not affected by this factor, in agreement with Rey et al. (2006). Pigs fed a HO diet had significantly higher ($p < 0.05$) values for the total back fat thickness when compared with the group fed with the C diet. These results could be due in part to the growth rate, higher in the HO group, which suggests that the faster growth was a fatter growth, in agreement with Pugliese et al. (2003) and Millet et al. (2004). In contrast, Ayuso et al. (2014) indicated that *Montanera* animals showed similar back fat thickness compared to the high-oleic group, although the values were higher in the group fed a diet high in oleic; the M. longissimus area between those two groups was statistically different. Also, Martin et al. (2008), Guillevic et al. (2009) and Mas et al. (2010) working with crossbred pigs found no differences between swine fed elevated levels of monounsaturated fat and control animals for first-rib fat thickness and M. longissimus area.

The outer, middle and inner back fat layers measured at two locations (10th and 14th rib level) followed the same trend as the back fat thickness and showed significant ($p < 0.05$) differences between feeding systems. Conversely, in the inner back fat layer more marked effects than those observed in the total back fat, outer and middle layers were detected. These results could indicate that the outer and middle layers tended to have a more constant fat deposition when compared with the inner layer. Wood et al. (1975) in previous research found that the rate at which cell diameter increased in size with increasing live weight was greater in the inner layer than in the other layers.

The allometric coefficients of back fat depth in relation to live weight allow one to better understand its evolution with growth. The isauxesis is condition of a linear measurement with respect to another of third degree, such as live weight, and is expressed by the allometric coefficient of 0.333 (Walstra, 1980) instead of unity. Lower or higher coefficients indicate slower or faster relative growth, respectively. The back fat development and its layers both in castrated Iberian male pigs and spayed Iberian female pigs showed differences through allometric coefficients. Each differential value could be interpreted as a different rate of maturation in tissue. Also, Wood et al. (1978) found different grades of back fat layer development and Mersmann (1982) evaluated the back fat development by ultrasound technique in different points of spin, concluding that the growth was not homogeneous.

The three back fat layers have distinct patterns of growth and should be considered as three separate tissues (Fortin, 1986). At birth, the outer layer is predominant. Then, at a given time during growth, the middle layer, due to a faster rate of development, becomes thicker. Finally, at a later stage, the inner layer begins to develop (Fortin, 1986). Fortin (1986) and Schinckel et al. (2002) indicated that, at light weights, the outer layer was usually more predominant, whereas, at heavier weights, the middle layer became predominant. In our study, the middle layer was predominant throughout the

study period because the study began when the animals had a live weight of 64 kg.

Consistent with Fortin (1986) and Schinckel et al. (2002), in our study the allometric coefficients for the outer layer thickness of back fat ($b = 1.13$) were of lower magnitude than the middle layer thickness of back fat ($b = 1.20$). Contrary to the conclusions of Mersmann (1982) and according with Fortin (1986), the growth of the inner layer was intermediate between the outer and the middle. These results agree with those of McMeekan (1940), Sink and Miller (1962), and Sink et al. (1964), who reported that the inner (second) layer increased faster and had a greater thickness than the outer (first) layer as total back fat increased. Also, McEvoy et al. (2007) indicated that the change in thickness per unit change in body weight is greatest in the middle layer followed by the outer and inner layers.

The inner layer development is especially interesting because there are previous studies (Newcom et al., 2005; Ayuso et al., 2012) that relate thickness with a higher fat infiltration. We found remarkable differences in the allometric coefficient of the inner layer between the 10th and 14th rib levels. Thus, the level at which the measurement is recorded is important.

The M. longissimus area presented an allometric coefficient lower than 1, presenting a negative growth level of lean tissue as the fattening phase increased. This is in accordance with Virgili et al. (2003), who reported a decrease in lean yield of the loin from 143 to 183 kg live weight pigs and with Courchaine et al. (1996), who found that as age and body weight increased lean meat percentage decreased.

The Pearson correlation coefficients between ultrasound measurements at the fourth, fifth, sixth and seventh scans with the corresponding ultrasound measurements at the eighth scan ranged from 0.53 to 0.82 for back fat thickness and from 0.18 to 0.63 for M. longissimus area. These results were similar to those obtained by Tyra et al. (2001) in Polish Large White, Polish Landrace, Pietrain and Duroc breeds ($r = 0.567$–0.635 and $r = 0.552$ for back fat and loin area measurements, respectively) and higher than those recorded by Courchaine et al. (1999) in Yorkshire and crossbred pigs ($r = 0.639$ and $r = 0.453$, respectively), with initial and final average weights of 68.3 and 109.8 kg, respectively. As in the works cited, in our study the correlation coefficients between the traits of back fat thickness were greater than those obtained for the M. longissimus area. Robinson et al. (1987) recorded a correlation of 0.65 between ultrasound back fat at 17 and 20 weeks of age, and McLaren et al. (1989) reported moderate to high correlations of live animal ultrasound back fat measured throughout the nursery, grower and finisher stages to carcass back fat at slaughter. In line with the results of previous work on Iberian pigs (Ayuso et al., 2013), the highest correlation coefficients were recorded for the middle layer, followed by the outer and inner layers. This could be attributed, as indicated by these authors, to the lower accuracy in the measurement of the inner layer depth due to a wide variation in loin shape. As in the study of Kolstad et al. (1996), correlation coefficients decreased with increasing intervals between scans. Regression resulted as expected, thus corroborating the correlation results. Consistent with the results obtained by Courchaine et al. (1999) in Yorkshire barrows and Ayuso et al. (2012) in Iberian pigs, the regression coefficients for back fat measurements were higher than those obtained for M. longissimus area. However, the regression coefficients obtained from measurements taken between the seventh and eighth scans were higher than the regression coefficients between carcass and ultrasonic measurements for back fat layer variables and M. longissimus area at two anatomical locations obtained by the aforementioned authors in Iberian pigs.

5 Conclusions

The relative back fat growth was higher than M. longissimus area, not permitting the establishment of a similar growth pattern for fat and muscle. Also, the sex and diet must be taken into account in predictive models. The subcutaneous adipose layers in Iberian pigs grow at different rates during the last 6 months before slaughter, with the ultrasound serial scan it is possible to show these changes.

The changes in diet, in the fat development and the sex affect the adipose tissue development, being more noticeable in the middle layer of back fat at 10th rib level and the inner layer at 14th rib level. However, the use of an enriched oleic acid diet and the sex do not affect loin development.

As the middle layer of back fat shows more growth, this layer could be the best to be included in predictive models.

Back fat thickness at the eighth scan could be predicted with moderate accuracy from corresponding measurements taken 30 days earlier and with less accuracy as the interval between measurements increases. The M. longissimus area at the eighth scan can also be predicted, but with less accuracy.

Competing interests. The authors declare that they have no conflict of interest.

Acknowledgements. This research was supported by INIA RTA 2007-000-93-00-00 and FEDER.

Edited by: Steffen Maak

References

Alfonso, L., Mourot, J., Insausti, K., Mendizabal, J. A., and Arana, A.: Comparative description of growth, fat deposition, carcass and meat quality characteristics of Basque and Large White pigs, Anim. Res., 54, 33–42, 2005.

Aro, S. O. and Akinjokun, O. M.: Meat and carcass characteristics of growing pigs fed microbially enhanced cassava peel diets, Arch. Zootec., 61, 407–414, 2012.

Ayuso, D., Izquierdo, M., Hernández, F. I., Bazán, J., and Corral, J. M.: Ultrasonographic in vivo estimation of back fat depth and Longissimus dorsi area in Iberian pigs, in: 7th International Symposium on the Mediterranean Pig Zaragoza: CIHEA M, edited by: De Pedro, E. J., CABe, 309–313, 2012.

Ayuso, D., González, A., Hernández, F., Corral, J. M., and Izquierdo, M.: Prediction of carcass composition, ham and foreleg weights, and lean meat yields of Iberian pigs using ultrasound measurements in live animals, J. Anim. Sci., 91, 1884–1892, https://doi.org/10.2527/jas.2012-5357, 2013.

Ayuso, D., González, A., Hernández, F., Peña, F., and Izquierdo, M.: Effect of sex and final fattening on ultrasound and carcass traits in Iberian pigs, Meat. Sci., 96, 562–567, 2014.

BOE: REAL DECRETO 1469/2007, de 2 de noviembre, por el que se aprueba la norma de calidad para la carne, el jamón, la paleta y la caña de lomo ibéricos, Madrid, Spain, 45087–45104, 2007.

BOE: Real Decreto 4/2014, de 10 de enero, por el que se aprueba la norma de calidad para la carne, el jamón, la paleta y la caña de lomo ibérico, in: Ministerio de Agricultura ayMA (ed). Ministerio de Agricultura, alimentacion y Medio Ambiente edn, 2014.

Cisneros, F., Ellis, M., McKeith, F. K., McCaw, J., and Fernando, R. L.: Influence of slaughter weight on growth and carcass characteristics, commercial cutting and curing yields, and meat quality of barrows and gilts from two genotypes, J. Anim. Sci., 925–933, 1996.

Correa, J. A., Faucitano, L., Laforest, J. P., Rivest, J., Marcoux, M., and Gariépy, C.: Effects of slaughter weight on carcass composition and meat quality in pigs of two different growth rates, Meat. Sci., 72, 91–99, https://doi.org/10.1016/j.meatsci.2005.06.006, 2006.

Courchaine, J. K., Jones, R. D., Gasa, J., and Azain, M. J.: Use of real-time ultrasound in pigs during the early finishing phase to predict carcass composition at slaughter, The Professional Animal Scientist, 15, 100–105, https://doi.org/10.15232/S1080-7446(15)31736-8, 1999.

Cunningham, P. J., Socha, T. E., Peo, E. R., and Mandigo, R. W.: Gain, feed conversion and carcass traits of swine fed under two nutritional regimes, J. Anim. Sci., 37, 75–80, 1973.

Daza, A., Mateos, A., Rey, A. I., and López-Bote, C. J.: Feeding level in the period previous to the late fattening phase influences fat composition at slaughter in free-ranged Iberian pigs, Arch. Anim. Nutr., 59, 227–236, 2005.

Daza, A., Mateos, A., Carrasco, C. L., Rey, A., Ovejero, I., and López-Bote, C. J.: Effect of feeding system on the growth and carcass characteristics of Iberian pigs, and the use of ultrasound to estimate yields of joints, Meat. Sci., 72, 1–8, 2006.

De Pedro, E.: Estudio de los factores sexo y peso de sacrificio sobre las características de la canal del cerdo Ibérico, Cordoba, PhD Thesis, University of Cordoba, Spain, 181 pp., 1987.

Dobao, M. T., Rodrigañez, J., Silio, L., Toro, M. A., de Pedro, E., and Garcia de Siles, J. L.: Crecimiento y caracterisitcas de canal en cerdos Ibéricos, Duros-Jersey x Ibérico y Jianxing x Ibérico, Investigación Agraria Producción y Sanidad Animales, 2, 9–23, 1987.

Dourmad, J., Etienne, M., and Noblet, J.: Measuring backfat depth in sows to optimize feeding strategy, INRA Prod. Anim., 14, 41–50, 2001.

Fernández, A., De Pedro, E., Núñez, N., Silió, L., García-Casco, J., and Rodríguez, C.: Genetic parameters for meat and fat quality and carcass composition traits in Iberian pigs, Meat. Sci., 64, 405–410, 2003.

Fortin, A.: The effect of slaughter weight on the carcass characteristics of Yorkshire barrows and gilts, Can. J. Anim. Sci., 60, 265–274, 1980.

Fortin, A.: Development of backfat and individual fat layers in the pig and its relationship with carcass lean, Meat. Sci., 18, 255–270, https://doi.org/10.1016/0309-1740(86)90016-1, 1986.

Fortin, A., Sim, D. W., and Talbot, S.: Ultrasonic measurements of backfat thickness at different locations and positions on the warm pork carcass and comparisons of ruler and ultrasonic procedures, Can. J. Anim. Sci., 60, 635–641, 1980.

Fortina, R., Barbera, S., Lussiana, C., Mimosi, A., Tassone, S., Rossi, A., and Zanardi, E.: Performances and meat quality of two Italian pig breeds fed diets for commercial hybrids, Meat. Sci., 71, 713–718, https://doi.org/10.1016/j.meatsci.2005.05.016, 2005.

Franci, O., Bozzi, R., Pugliese, C., Acciaioli, A., Campodoni, G., and Gandini, G.: Performance of Cinta Senese pigs and their crosses with Large White, 1 Muscle and subcutaneous fat characteristics, Meat. Sci., 69, 545–550, https://doi.org/10.1016/j.meatsci.2004.10.005, 2005.

Franco, D., Vazquez, J. A., and Lorenzo, J. M.: Growth performance, carcass and meat quality of the Celta pig crossbred with Duroc, and Landrance genotypes, Meat. Sci., 96, 195–202, https://doi.org/10.1016/j.meatsci.2013.06.024, 2014.

Greer, E. B., Lowe, T. W., and Giles, L. R.: Comparison of ultrasonic measurement of backfat depth on live pigs and carcases with a digital recording instrument, Meat. Sci., 19, 111–120, https://doi.org/10.1016/0309-1740(87)90016-7, 1987.

Guillevic, M., Kouba, M., and Mourot, J.: Effect of a linseed diet or a sunflower diet on performances, fatty acid composition, lipogenic enzyme activities and stearoyl-CoA-desaturase activity in the pig, Livest. Sci., 124, 288–294, https://doi.org/10.1016/j.livsci.2009.02.009, 2009.

Hsu, W. L., Johnson, R. K., and van Vleck, L. D.: Effect of pen mates on growth, backfat depth, and longissimus muscle area of swine, J. Anim. Sci., 88, 895–902, https://doi.org/10.2527/jas.2009-1879, 2010.

Kempster, A. J.: Fat partition and distribution in the carcass of cattle, sheep and pigs: a review, Meat. Sci., 5, 83–98, 1980.

Kempster, A. J. and Evans, D. G.: A comparison of different predictors of the lean content of pig carcasses, 1. Predictors for use in commerical classification and grading, Anim. Prod., 28, 87–96, 1979.

Kolstad, K., Jopson, N. B., and Vangen, O.: Breed and sex differences in fat distribution and mobilization in growing pigs fed at maintenance, Livest. Prod. Sci., 47, 33–41, https://doi.org/10.1016/s0301-6226(96)01001-9, 1996.

Lopez-Bote, C. J.: Sustained utilization of the Iberian pig breed, Meat. Sci., 49, S17–S27, 1998.

Martin, D., Muriel, E., Gonzalez, E., Viguera, J., and Ruiz Carrascal, J.: Effect of dietary conjugated linoleic acid and monounsaturated fatty acids on productive, carcass and meat quality traits of pigs, Livest. Sci., 117, 155–164, 2008.

Mas, G., Llavall, M., Coll, D., Roca, R., Diaz, I., Gispert, M., Oliver, M. A., and Realini, C. E.: Carcass and meat quality characteristics and fatty acid composition of tissues from Pietrain-crossed barrows and gilts fed an elevated monounsaturated fat diet, Meat. Sci., 85, 707–714, https://doi.org/10.1016/j.meatsci.2010.03.028, 2010.

Mayoral, A. I., Dorado, M., Guillén, M. T., Robina, A., Vivo, J. M., Vazquez, C., and Ruiz, J.: Development of meat and carcass quality characteristics in Iberian pigs reared outdoors, Meat. Sci., 52, 315–324, 1999.

McEvoy, F. J., Strathe, A. B., Madsen, M. T., and Svalastoga, E.: Changes in the relative thickness of individual subcutaneous adipose tissue layers in growing pigs, ACTA Vet. Scand., 49, https://doi.org/10.1186/1751-0147-49-32, 2007.

McKay, R. M., Rempel, W. E., Cornelius, S. G., and Allen, C. E.: Differences in carcass traits of three breeds of swine and crosses at five stages of development, Can. J. Anim. Sci., 64, 293–304, 1984.

McLaren, D. G., McKeith, F. M., and Novakofski, J.: Prediction of carcass characteristics at market weight from serial real-time ultrasound measures of backfat and loin eye area in the growing pig, J. Anim. Sci., 67, 1657–1667, 1989.

McMeekan, C. P.: Growth and development in the pig with special reference to carcass quality characters, J. Agr. Sci., 30, 511–569, https://doi.org/10.1017/S002185960004822X, 1940.

Mersmann, H. J.: Ultrasonic determination of backfat depth and loin area in swine, J. Anim. Sci., 54, 268–275, 1982.

Millet, S., Hesta, M., Seynaeve, M., Ongenae, E., De Smet, S., Debraekeleer, J., and Janssens, G. P. J.: Performance, meat and carcass traits of fattening pigs with organic versus conventional housing and nutrition, Livest. Prod. Sci., 87, 109–119, https://doi.org/10.1016/j.livprodsci.2003.10.001, 2004.

Minelli, G., Macchioni, P., Ielo, M. C., Santoro, P., and lo Fiego, D. P.: Effects of dietary level of pantothenic acid and sex on carcass, meat quality traits and fatty acid composition of thigh subcutaneous adipose tissue in Italian heavy pigs, Ital. J. Anim. Sci., 12, 329–336, https://doi.org/10.4081/ijas.2013.e52, 2013.

Moeller, S. J., Christian, L. L., and Goodwin, R. N.: Development of adjustment factors for backfat and loin muscle area from serial real-time ultrasonic measurements on purebred lines of swine, J. Anim. Sci., 76, 2008–2016, 1998.

Moody, W. G. and Zobrisky, S. E.: Study of backfat layers of swine, J. Anim. Sci., 25, 809–813, 1966.

Newcom, D. W., Baas, T. J., Schwab, C. R., and Stalder, K. J.: Relationship between backfat depth and its individual layers and intramuscular fat percentage in swine, Animal Industry Report, ASL R1944, AS 650 p., 2004.

Newcom, D. W., Baas, T. J., Schwab, C. R, and Stalder, K. J.: Genetic and phenotypic relationships between individual subcutaneous backfat layers and percentage of longissimus intramuscular fat in Duroc swine, J. Anim. Sci., 83, 316–323, 2005.

Pugliese, C., Madonia, G., Chiofalo, V., Margiotta, S., Acciaioli, A., and Gandini, G.: Comparison of the performances of Nero Siciliano pigs reared indoors and outdoors, 1. Growth and carcass composition, Meat. Sci., 65, 825–831, 2003.

Rey, A. I., Daza, A., López-Carrasco, C., and López-Bote, C. J.: Feeding Iberian pigs with acorns and grass in either free-range or confinement affects the carcass characteristics and fatty acids and tocopherols accumulation in Longissimus dorsi muscle and backfat, Meat. Sci., 73, 66–74, 2006.

Richmond, R. J. and Berg, R. T.: Tissue development in swine as influenced by liveweight, breed, sex and ration, Can. J. Anim. Sci., 51, 31–40, 1971.

Robinson, T. F., Orme, L. E., and Park, R. L.: Growth characteristics of immature swine as determined by real-time linear array ultrasound, ASAS WS. P., 151, 1987.

Rybarczyk, A., Pietruszka, A., Jacyno, E., and Dvořák, J.: Carcass and meat quality traits of pig reciprocal crosses with a share of Pietrain breed, Czech. J. Anim. Sci., 56, 47–52, 2011.

Sather, A., Bailey, D., and Jones, S.: Real-time ultrasound image analysis for the estimation of carcass yield and pork quality, Can. J. Anim. Sci., 76, 55–62, 1996.

Schinckel, A. P., Mills, S. E., Weber, T. E., and Eggert, J. M.: A review of genetic and nutritional factors affecting fat quality and belly firmness, Proc. Natl. Swine Imp. Fed. Conf. Nashville, TN, available at: http://www.nsif.com/Conferences/2002/reviewgeneticnutritionalfactors.htm, 2002.

Serrano, M. P., Valencia, D. G., Nieto, M., Lázaro, R., and Mateos, G. G.: Influence of sex and terminal sire line on performance and carcass and meat quality of Iberian pigs reared under intensive production systems, Meat. Sci., 78, 420–428, 2008.

Sink, J. D. and Miller, R. C.: Fat deposition in swine, J. Anim. Sci., 21, 985, 1962.

Sink, J. D., Watkins, J. L., Ziegler, J. H., and Miller, R. C.: Analysis of fat deposition in swine by gas-liquid chromatography, J. Anim. Sci., 23, 121, 1964.

Stahl, C. A. and Berg, E. P.: Growth parameters and meat quality of finishing hogs supplemented with creatine monohydrate and a high glycemic carbohydrate for the last, 30 days of production, Meat. Sci., 64, 169–174, https://doi.org/10.1016/s0309-1740(02)00176-6, 2003.

Suárez-Belloch, J., Guada, J. A., and Latorre, M. A.: The effect of lysine restriction during grower period on productive performance, serum metabolites and fatness of heavy barrows and gilts, Livest. Sci., 171, 36–43, https://doi.org/10.1016/j.livsci.2014.11.006, 2015.

Sullivan, Z. M., Honeyman, M. S., Gibson, L. R., and Prusa, K. J.: Effects of triticale-based diets on finishing pig performance and pork quality in deep-bedded hoop barns, Meat. Sci., 76, 428–437, https://doi.org/10.1016/j.meatsci.2006.12.002, 2007.

Swantek, P. M., Roush, W. B., Stender, D. R., Mabry, J. W., and Honeyman, M. S.: Backfat Depth and Loin Eye Area Measurements of Purebred Berkshire Pigs Housed in Hoop Buildings in Iowa, Animal Industry Report, available at: http://lib.dr.iastate.edu/ans_air/vol660/iss1/101, ASL R2936, AS 660 p., 2014.

Temperan, S., Lorenzo, J. M., Castiñeiras, B. D., Franco, I., and Carballo, J.: Carcass and meat quality traits of celta heavy pigs, Effect of the inclusion of chestnuts in the finishing diet, Span. J. Agric. Res., 12, 694–707, https://doi.org/10.5424/sjar/2014123-5057, 2014.

Tyra, M., Szyndler-Nedza, M., and Eckert, R.: Possibilities of using ultrasonography in breeding work with pigs. Part II – Relationships between measurements obtained by different techniques and detailed dissection results, Ann. Anim. Sci., 11, 193–205, 2011.

Virgili, R., Degni, M., Schivazappa, C., Faeti, V., Poletti, E., Marchetto, G., Pacchioli, M. T., and Mordenti, A.: Effect of age at slaughter on carcass traits and meat quality of Italian heavy pigs, J. Anim. Sci., 81, 2448–2456, 2003.

Walstra, P.: Growth and carcass composition from birth to maturity in relation to feeding level and sex in Dutch Landrace pigs, PhD Thesis, Agricultural University of Wageningen, the Netherlands, 207 pp., 1980.

Warriss, P. D., Brown, S. N., Franklin, J. G., and Kestin, S. C.: The thickness and quality of backfat in various pig breeds and their relationship to intramuscular fat and the setting of joints from the carcasses, Meat. Sci., 28, 21–29, https://doi.org/10.1016/0309-1740(90)90017-z, 1990.

Wood, J. D., Enser, M. B., and Restall, D. J.: Fat cell size in Pietrain and Large White pigs, J. Agr. Sci., 84, 221–225, 1975.

Wood, J. D., Enser, M. B., Macfie, H. J., Smith, W. C., Chadwick, J. P., Ellis, M., and Laird, R.: Fatty acid composition of backfat in Large White pigs selected for low backfat thickness, Meat. Sci., 2, 289–300, 1978.

Selected quality traits of eggs and the productivity of newly created laying hen hybrids dedicated to an extensive rearing system

Justyna Batkowska and Antoni Brodacki

Department of Biological Basis of Animal Production, University of Life Sciences in Lublin,
13 Akademicka St., 20-950 Lublin, Poland

Correspondence to: Justyna Batkowska (justyna.batkowska@up.lublin.pl)

Abstract. The aim of the study was to evaluate the usefulness of hybrids derived from Greenleg Partridge cocks and Rhode Island Red (GPR) hens for an extensive rearing system (RS), with special emphasis on high-quality table eggs. Newly created hybrids were compared to Hy-Line Brown (HLB) hens. The experiment was carried out with a total of 2400 hens. Both hybrid types (H's) were divided into two equal groups according to the rearing system: intensive (I) and extensive (E). The traits analysed in hens were body weight at the 8th, 16th, and 33rd weeks of age; laying production; and feed intake. At the 33rd week of a bird's age, egg quality was evaluated. The results of the study showed different reaction of the hybrids to the rearing system. This may confirm better usefulness of GPR for extensive farming and HLB hens for intensive methods of rearing. Furthermore, it can be concluded that the extensive system had a positive impact on the productivity of GPR birds.

1 Introduction

The quality of table eggs may depend on many factors, such as genotype, age (Zita et al., 2009; Sarica et al., 2012), environmental conditions, feed supplements (Safaa et al. 2008), or rearing system (RS) of hens (Đukić-Stojčić et al., 2009; Batkowska et al., 2014). Due to a higher level of welfare and consumer preferences, the so-called organic, free-range, backyard, and extensive methods of poultry housing have become increasingly popular (Ferrante et al., 2009; Krawczyk, 2009; Tůmová et al., 2009). However, in the market there are still only a few gene sets of laying bird hybrids that would provide satisfactory production results in such kinds of housing systems. Additionally, because of the opinion that typical commercial laying hybrids under extensive housing conditions are not able to fully exploit their genetic potential, the use of local breeds and their hybrids for organic farming is suggested because of their better adaptation to changeable environmental conditions (Rizzi and Chiericato, 2005; Zita et al., 2009).

The aim of this study was to evaluate the usefulness of hybrids derived from Greenleg Partridge cocks and Rhode Island Red hens (GPR) for an extensive RS with special emphasis on obtaining high-quality table eggs.

Greenleg Partridge is the oldest native Polish breed of hen, developed at the end of the 19th century, and due to their occurrence mainly in south-eastern Poland, they were called "Galician hens". The breed was first exhibited with the name of Greenleg at a show in Lviv (Ukraine) in 1894. The breed standard was established in 1923 in Poland. Greenleg Partridge chickens have been maintained in a conservation flock since 1960 (Wójcik et al., 2012). Recently they have not been subjected to selection for productive trait improvement. They are characterized by low body weight (around 1700–1800 g), green legs, and grey, partridge-like plumage. They produce good-tasting eggs (≈ 56 g) with cream-coloured shells and a high percentage of yolk content (Krawczyk et al., 2011). Greenleg Partridge hens today are progressively used in free-range and organic production systems, and their eggs are marketed at high prices (Krawczyk and Sokołowicz, 2015). The choice of Greenleg Partridge was also dictated by the

Table 1. Nutritional value of feed in both systems of rearing used up to the 16th week of the laying hens' age.

Rearing system	Intensive		Extensive
Age of birds (weeks) Ingredients	0–8	9–16	9–16
Dry matter (%)	89.0	88.0	88.1
Metabolic energy (MJ kg^{-1})	12.0	12.3	12.0
Total protein (g)	180.0	155.0	156.0
Crude fat (g)	30.0	28.0	23.3
Crude fibre (g)	40.0	48.0	57.0
Crude ash (g)	40.0	57.0	41.5
Calcium (g)	8.0	9.0	6.7
Phosphorus (g)	5.8	6.0	6.0
Magnesium (g)	1.5	1.5	2.37
Sodium (g)	0.8	1.64	1.16
Lysine (g)	1.1	0.75	0.76
Methionine (g)	0.42	0.34	0.36

Table 2. Nutritional value of feed in both systems of rearing used after the 16th week of the laying hens' age.

Ingredients	Rearing system	
	Intensive	Extensive
Dry matter (%)	87.7	87.4
Metabolic energy (MJ)	11.51	11.63
Total protein (%)	17.99	15.45
Crude fat (%)	3.80	2.98
Crude fibre (%)	3.90	3.88
Crude ash (%)	4.92	3.78

fact that in popular opinion hens of this breed are characterized by beneficial traits of eggs, better disease resistance, and high resourcefulness in searching for food. All of these features are important in the extensive rearing of laying hens.

The breed Rhode Island Red was also created in the second half of the 19th century in the state of Rhode Island (USA) by crossing different hen breeds with Asian birds such as Cochins and Malay bantams and through selection for increased egg production. The Rhode Island Red (R-11) line was imported into Poland from Great Britain before 1939. As a dual-purpose type, by the mid-1970s it accounted for 50 % of the population of commercial farms and backyard flocks. The birds are characterized by yellow skin and good dressing percentage. Due to its genetically determined resistance to Marek's disease, the breed is especially suitable for backyard free-range farming (Połtowicz et al., 2004; Mohammed et al., 2013).

Newly created hybrids were compared to Hy-Line Brown (HLB) hens, which are currently the leading commercial hybrids in Poland, regardless of housing system.

2 Materials and methods

The research material consisted of 2400 laying hens: Greenleg Partridge and Rode Island Red hybrids (GPR) and Hy-Line Brown (HLB). Birds of both hybrid types (H's) were divided into two equal groups (four replications each containing 150 layer hens) according to the RS – intensive (I) and extensive (E). Up to the eighth week of age, the birds were maintained on deep litter. After this time the birds from E groups were transferred to special sheds with green runs and reared at stock density of 6 hens m^{-2} of hen house (on the run: 1 hen/4 m^2). The natural length of daylight was used (ca. 16 h). Microclimate parameters were not controlled in

this system. The so-called intensive groups were maintained on deep litter throughout the whole rearing period, at stock density of 9 hens m^{-2}, under regulated environmental parameters: temperature (21 °C), humidity (40–60 %), and lighting program (16 h of light, 5 lx).

The nutritional value of the feed used for birds in both housing systems up to the 16th week of age is shown in Table 1. From the 9th to the 16th weeks of rearing the hens kept extensively received additionally green fodder in the amount of 20 g/bird/day. From the 17th week of age the hens from E groups were given fodder in the following proportions: 60 % of balanced mixture (composed of wheat – 40 %, corn – 30 %, wheat bran – 10 %, soya meal – 21 %, limestone – 8 %, and premix – 1 %), 30 % of crushed wheat, 10 % of wheat bran, and 5 % of green forage of alfalfa, calculated on the basis of dry matter. Nutritional composition of green fodder was as follows (per 1000 g): dry matter 186.4 g, metabolic energy 259.6 kcal/1.09 MJ, crude protein 42.8 g, crude fat 7.2 g, crude fibre 40.5 g, and crude ash 23.4 g. The chemical composition of other feed is presented in Table 2. Feed was analysed according to procedures carrying out by National Laboratory of Feedingstuffs (National Research Institute of Animal Production, Balice near Kraków, Poland).

The following productivity traits were analysed in hens: body weight at the 8th, 16th, and 33rd weeks of age (50 birds per group were weighted individually), as well as the number of eggs, laying production percentage, and feed intake per 1 egg and per 1 hen day^{-1}. At the 33rd week of the birds' age, the evaluation of egg quality was performed. For this purpose, 60 eggs were collected from each replication group for 3 consecutive days. All eggs were collected at the same time (in the morning) from all groups. They were chosen randomly and evaluated on the same day. The following traits were evaluated: shape index, egg weight, and specific gravity (calculated on the basis of Archimedes principle), features describing the shell (colour, strength, weight, thickness, and density), albumen (weight, height, Haugh units), and yolk (colour, weight). Proportions of particular elements of eggs were calculated as well. The evaluation of the egg traits was performed using electronic-set egg quality measurements (TSS$^{®}$) and an Instron Mini 55 device.

Table 3. Effect of the hybrid type (H), rearing system (RS), and their interaction (H × RS) on the productivity traits in laying hens (values presented as means).

Rearing system (RS)	E		I		SEM	Probability		
Hybrid (H) Trait	GPR	HLB	GPR	HLB		H	RS	H × RS
Body weight at 8th week of life	–	–	646.2^a	720.0^b	6.643	0.006	–	–
Body weight at 16th week of life	1710^b	1396^a	1733^b	1375^a	17.70	0.008	0.695	0.241
Body weight at 33rd week of life	1872	1778	1899	1920	17.26	0.458	0.255	0.785
Average laying production (%)	79.8^b	80.4^b	72.9^a	84.7^c	0.842	0.000	0.340	0.000
Average feed conversion per 1 egg (g)	141.8^a	148.4^a	162.1^b	147.8^a	1.673	0.182	0.001	0.000
Average feed conversion per 1 hen day^{-1} (g)	113.0^a	116.3^b	116.5^b	109.6^a	4.912	0.160	0.202	0.001

RS is rearing system. E is extensive. I is intensive. H × RS is the interaction effect of hybrid type and rearing system. GPR is the newly created hybrid Greenleg Partridge × Rhode Island Red. HLB is the commercial hybrid Hy-Line Brown. $^{a, b, c, d}$ show values that are significantly different.

During the whole experiment principles of ethics and the welfare of the birds were maintained. Experimental procedures were approved by the III Local Ethics Committee for Experiments on Animals in Warsaw (approval no. 27/2009 of 16 April 2009). The data collected were analysed using the statistical package SPSS V21.0 (IBM, 2011). The normality of data was verified using the Kolmogorov–Smirnov test. Further two-way ANOVA was applied to assess the effect of the RS, H, and their interaction on the productivity and egg traits in the hens. Survivability of birds was statistically evaluated by a nonparametric χ^2 test.

3 Results

During the trial, 33 cases of death in birds were registered and the survivability in the groups was calculated as follows: 98.7 and 99.0 % in GPR E and GPR I and 98.3 and 98.5 % in HLB E and HLB I respectively. The dependence of this parameter on experimental groups was not statistically confirmed (χ^2 $P = 0.790$). This showed that uncontrolled and non-optimal environmental conditions of the extensive RS did not result in deterioration of the birds' health status regardless of hybrid type.

Table 3 shows the effect of the RS, H, and their interaction on the growth performance of the laying hens. Considerably higher body weight (approximately 10 %) at 8 weeks of age was displayed in HLB in comparison to GPR pullets. At 16 weeks of life these relations were reversed and higher body weight was achieved by GPR birds, regardless of the housing system ($P \leq 0.008$). The body weight of birds at the 33rd week of age was almost equal in all groups. There was no interaction between the experimental factors in regard to the body weight of the hens.

The average laying percentage ranged from 72 to 84 % and depended on the genotype and housing system of the birds. Birds from the different genotypes responded differently to extensive maintenance conditions ($P \leq 0.000$). GPR hybrids from the E group demonstrated a significantly higher value of

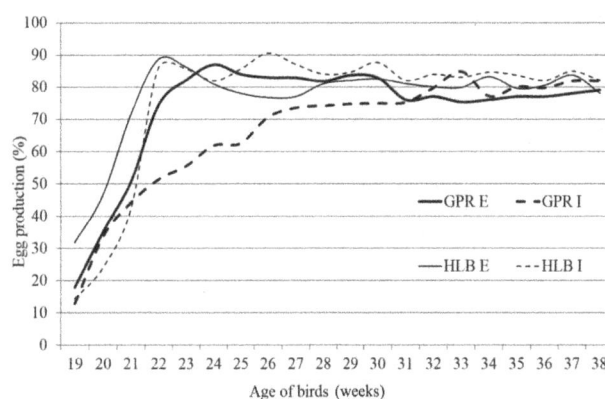

Figure 1. Laying curves of hens according to the hybrid type and rearing system (E is the extensive system of rearing, I is the intensive system of rearing, GPR is the newly created hybrid Greenleg Partridge × Rhode Island Red, and HLB is the commercial hybrid Hy-Line Brown).

egg production than birds in the I group, while in HLB hens, opposite relations were recorded. This could result from difficulties in adaptation to a closed housing system in GPR hybrids.

The average feed conversion per 1 egg (Table 3) was the lowest in GPR hens kept extensively and HLB birds maintained intensively. The value of this parameter indicates that the GPR hens are most suitable to keep under extensive conditions. HLB layers in both intensive and extensive RSs consumed a similar amount of feed per egg (148 g). Similar patterns were recorded for feed conversion per 1 bird day^{-1}. The lowest values of this indicator were noticed in GPR E and HLB I hens. Both parameters connected with feed conversion remained under the interactional influence of both experimental factors.

Figure 1 illustrates laying curves of hens depending on H and RS. The achieved body weight did not have a big impact on sexual maturity of birds; regardless of the group, first

Table 4. Effect of the hybrid type (H), rearing system (RS), and their interaction (H × RS) on the external traits and proportions of egg elements.

Rearing system (RS)		E		I		SEM	Probability		
Hybrid (H) Trait		GPR	HLB	GPR	HLB		H	RS	H × RS
Egg weight (g)		53.7^b	59.0^d	50.8^c	60.6^a	0.196	0.000	0.090	0.000
Egg specific gravity (g cm^{-3})		1.088^a	1.083^b	1.082^b	1.089^a	0.000	0.000	0.874	0.000
Shape index (width to length, %)		75.4^a	77.6^b	77.1^b	75.1^a	0.278	0.030	0.168	0.000
Proportions	yolk	29.2^c	24.0^b	23.0^a	28.8^c	0.135	0.000	0.000	0.000
(%)	albumen	58.6^a	62.6^c	63.8^d	59.5^b	0.140	0.000	0.000	0.000
	shell	13.1^b	13.4^b	13.2^a	11.6^a	0.085	0.000	0.003	0.000

RS is rearing system. E is extensive. I is intensive. H × RS is the interaction effect of hybrid type and rearing system. GPR is the newly created hybrid Greenleg Partridge × Rhode Island Red. HLB is the commercial hybrid Hy-Line Brown. [a, b, c, d] show values that are significantly different.

eggs appeared in approximately the 19th week of the layers' life. HLB E layers achieved 50 % egg production earliest, between the 20th and 21st weeks, followed by GPR E, GPR I, and HLB I. This could result from a longer lighting day, which was used for the groups kept extensively. HLB E layers also reached the peak production earliest; their productivity amounted to 90 % in the 22nd and 23rd weeks of rearing. In GPR hybrids the peak production was noticed slightly later, between the 24th and 25th weeks, but their further egg production did not differ from that recorded for commercial laying hybrids. The worst production results were recorded in GPR hens kept intensively, which can confirm poor adaptation of these birds for closed farming.

Table 4 shows the parameters characterizing the external features and proportions of particular morphological elements of eggs obtained from each group of birds as affected by the RS, genotype, and their interaction. The heaviest eggs were obtained from commercial HLB I laying hens (60.6 g) and the lightest ones from GPR hens kept under intensive housing conditions (50.8 g). Similarly, the highest egg specific gravity was recorded in eggs from HLB I hens and the lowest from GPR I; the difference amounted to more than 5 %. Both egg weight and egg density were influenced by both experimental factors. On the basis of the proportion of long and short egg axis, the egg shape index was calculated. In intensive breeding, eggs from GPR hens were significantly more elongated, while in extensive breeding, eggs from HLB were more elongated. Both types of hybrids showed a diversity of shape, depending on the RS.

The percentage of particular morphological egg elements in the extensive RS varied, depending on the H. The biggest proportion of egg content (88.3 %) and the smallest of shell (11.6 %) were observed in the eggs from HLB I birds. It should be emphasized that the GPR E hybrids obtained a substantially higher percentage of yolk (29.2 %) than other groups of layers. GPR E hybrids also had the biggest yolk content, 0.4–6.2 % in absolute values. The difference was higher by 21.2 %, in relative values, than in eggs from the

same hybrids kept intensively. In HLB layers these differences were slightly lower and amounted to 4.8 (in absolute values) and 16.6 % in relative values, but in favour of group I birds. For a proportion of all morphological elements of eggs, interaction of experimental factors was significant ($P \leq 0.000$).

Table 5 presents the characteristics of particular egg elements: yolk, albumen, and shell, depending on H and their RS. The farming system did not induce significant differences in the yolk weight within commercial HLB hybrids. GPR layers dominated in the extensive system in terms of this parameter. The yolks from GPR E hens were considerably darker in the intensive system. They were also darker than yolks in eggs from HLB birds, regardless of the RS. This could result from using Greenleg Partridge cocks in creating these hybrids. Hens of this breed are very resourceful in finding food, and darker colour of yolk may be the result of intensive foraging and thus consuming more carotenoids from green fodder and plants available on the fowl run. Hybrids from GPR E were characterized by a lower albumen weight than commercial HLB layers. The greater average height of albumen was observed in HLB eggs, probably due to higher albumen content. Eggs from extensively reared birds also showed the highest number of Haugh units (86.7), which may indirectly be indicative of albumen quality. Generally, both the housing system and H considerably modified the parameters of egg albumen. The significance of the experiment factor interaction was also demonstrated in all the analysed characteristics of egg shell. The smallest shell weight was found in eggs from HLB I hybrids (6.03 g). It was significantly lower (34 %) than in eggs from hybrids held extensively and in eggs from GPR hens ($P \leq 0.000$).

The eggs from GPR E hens were characterized by the highest percentage of reflected light, i.e. they had substantially lighter shell colour (51.82 %) than in other groups. The shell strength of eggs was similar (43 N) and it was not influenced by the RS. In the case of HLB layers, group I eggs required significantly greater force to crush the shell (49.4 N),

Table 5. Effect of the hybrid type (H), rearing system (RS), and their interaction (H × RS) on the internal traits of egg elements.

Rearing system (RS)		E		I		SEM	Probability		
Hybrid (H) Trait		GPR	HLB	GPR	HLB		H	RS	H × RS
yolk	weight (g)	15.32[c]	14.52[b]	13.78[a]	14.90[b]	0.058	0.000	0.000	0.000
	colour (pts)	9.31[c]	8.56[b]	8.00[a]	8.55[b]	0.037	0.000	0.000	0.000
albumen	weight (g)	30.79[a]	38.05[b]	38.32[b]	30.86[a]	0.164	0.000	0.587	0.000
	height (mm)	5.52[a]	7.64[c]	6.89[b]	5.47[a]	0.058	0.000	0.001	0.000
	Haugh units	76.14[a]	86.71[c]	80.59[b]	74.50[a]	0.379	0.000	0.000	0.000
shell	strength (N)	43.45[a]	46.32[b]	43.75[a]	44.38[a]	0.326	0.003	0.279	0.138
	colour (%)	51.82[d]	29.26[a]	31.99[b]	49.40[c]	0.409	0.000	0.799	0.132
	weight (g)	6.37[b]	8.09[c]	7.90[b, c]	6.03[a]	0.061	0.000	0.000	0.000
	thickness (mm)	0.284[a]	0.319[c]	0.331[d]	0.316[b]	0.002	0.000	0.001	0.000
	density (g cm^{-3})	97.8[c]	112.4[b, c]	110.6[b, c]	93.5[a]	0.751	0.000	0.000	0.000

RS is rearing system. E is extensive. I is intensive. H × RS is the interaction effect of hybrid type and rearing system. GPR is the newly created hybrid Greenleg Partridge × Rhode Island Red. HLB is the commercial hybrids Hy-Line Brown. [a, b, c, d] show values that are significantly different.

which was not confirmed in the thickness or density of shell; therefore, it may result from a more spherical shape of the eggs in comparison to the eggs from the other groups. The thickest shell was noticed in eggs laid by GPR I hens. The eggs from the laying hybrids reared under extensive housing conditions were characterized by a greater density than those from hens kept intensively.

4 Discussion

The observations concerning body weight of layers (especially in HLB hens) made in this study were opposed to those reported by Küçükyılmaz et al. (2012), who showed that birds from an extensive RS were slightly heavier than birds reared conventionally, while this parameter depended more on the genotype than the RS. However, in the studies of Singh et al. (2009), the birds kept less intensively were characterized by higher body weight. Also, the main cause of variation was the origin of hens. Miao et al. (2005) noticed that layers from open housing were lighter than those from intensive housing because of higher energy output associated with motor activity of birds kept with green run access.

When comparing the productivity of birds in two different housing systems, contradictory opinions can be found in the available references. Lukanov and Alexieva (2013) and Arbona et al. (2009) concluded that egg production is greater in an intensive RS in comparison to an extensive one. Şekeroğlu et al. (2010) stated that hens kept intensively reached sexual maturity 2 days earlier than birds in environmentally friendly farming. The observations made in this study are completely different and more compatible with those presented by Gerzilov et al. (2012). Despite the fact that conventionally reared hens finished the maturation process about

2 weeks earlier than the birds maintained extensively, they reached the period of maximum production more or less at the same time (23 weeks), regardless of the housing system. The egg production in an intensive system between 30 and 33 weeks of age was significantly lower than in extensive rearing and evened out in the following weeks. Birds kept extensively, regardless of the hybrid type, started to lay eggs earlier than I groups; therefore, the length of lighting day, different in these two RSs, seems to be fundamentally important in this case.

Similar to Tauson et al. (1999), in our study the different reaction of various hybrids to the maintenance of system intensity was observed in the case of production parameters. These authors also noticed an increase in egg numbers and laying production (%), as well as better feed intake and conversion, with increased rearing intensity. Englmaierová et al. (2014) found that the daily feed consumption and feed conversion ratio increased in aviary or litter systems in comparison with cages. Also, hen egg production per day was better in an intensive system than in alternative systems. According to Luiting (1990), the RS does not influence feed intake and the differences probably result from genetics, physical activity and condition, metabolic rate, and temperature. However, the above-mentioned factors might be due to the housing system.

Our results differ from the study of Golden et al. (2012), who found no effect of housing system on egg weight. Krawczyk (2009) indicated a 1.3 % difference in favour of extensive farming, but it was not statistically confirmed. In the research of Singh et al. (2009), the difference in favour of the floor system, in relation to hens kept in cages, amounted to more than 7 %. Similar results were reported by Pištěková et al. (2006), analysing hens in litter and cage RSs. Egg weight increased with the age of birds during the 9 months of the experiment in almost all terms of analysis (except the first and last month of rearing); considerably bigger eggs were laid by hens kept on the litter.

A statistically significant effect of RS on egg weight, egg shell, and albumen weight was shown by Farrante et al. (2009). The eggs from Hy-Line Brown hens kept on litter were characterized by a higher weight, heavier shell, and higher albumen content than eggs from hens raised in an organic system. Đukić-Stojčić et al. (2012) also evaluated the impact of various cage types on laying hens and the H on egg quality. They did not show any statistically significant interactions between the analysed experimental factors. However, genotype of birds considerably modified the number of eggs, shell strength, albumen height, and Haugh units. The types of cage also affected the number of eggs, shell thickness, and colour of yolk.

Comparing the impact of various RSs on egg quality parameters, Ahammed et al. (2014) noted the best technological characteristics of albumen in eggs from cages, which is the most intensive maintenance system for laying hens. However, the strongest shell and the best-stained yolk characterized eggs from hens kept in an aviary system. Roll et al. (2009) demonstrated changes in egg quality traits under the influence of RS in the case of yolk colour and shell characteristics, such as thickness, density, and mass, but they were statistically significant only in the period from the 60th to 78th weeks of the hens' lives. Eggs from hens kept in a floor system had a darker colour of yolk and characteristics of shell were better than in the eggs from hens kept in cages. Also, in a paper of Şekeroğlu et al. (2008), the effect of the free-range system was manifested only in darker stained yolks compared to eggs from the litter system.

According to Lewko and Gornowicz (2011), eggs from free-range rearing are characterized by the heaviest and thickest shell, while the shell was lightest and thinnest from the litter. Özbey and Esen (2007) did not confirm any differences in the shell thickness in eggs from the litter and cage systems; however, they showed that the mass of shell was significantly bigger in eggs from cages. Comparing the parameters of egg shell from layers kept under free-range and cage conditions, van den Brandt et al. (2004) and Đukić-Stojčić et al. (2009) did not find any differences between the systems in weight and thickness of shells. In the present study, the impact of housing system on the shell characteristics was clearly visible and statistically significant.

5   Conclusions

The results clearly show a completely different reaction of both types of laying hybrids to the applied extensive and intensive RSs. Most of the parameters that were decreased in the birds created by crossing Greenleg Partridge cocks and Rhode Island Red hens showed an increase in Hy-Line Brown hens and vice versa. This may confirm more usefulness of GPR birds for extensive farming and HLB hens for intensive methods of rearing.

The results of the study showed that the newly created hybrid Greenleg Partridge × Rhode Island Red is suitable for rearing under an extensive system. This was evidenced by such characteristics as maturity age, the average percentage of egg production, feed conversion per 1 egg, and the quality of egg traits. Based on the assessed traits, it can be concluded that the extensive RS positively influenced the birds and their production characteristics. We recommend using Greenleg Partridge × Rhode Island Red crosses for this system as a source of good-quality table eggs.

Competing interests. The authors declare that they have no conflict of interest.

Acknowledgements. Research was realized within the project "BIOŻYWNOŚĆ – innowacyjne, funkcjonalne produkty pochodzenia zwierzęcego" (BIOFOOD – innovative, functional products of animal origin) no. POIG.01.01.02-014-090/09 co-financed by the European Union from the European Regional Development Fund within the Innovative Economy Operational Programme 2007–2013.

Edited by: M. Mielenz

References

Ahammed, M., Chae, B. J., Lohakare, J., Keohavong, B., Lee, M. H., Lee, S. J., Kim, D. M., Lee, J. Y., and Ohh, S. J: Comparison of aviary, barn and conventional cage raising of chickens on laying performance and egg quality, Asian Australas, J. Anim. Sci., 27, 1196–1203, 2014.

Arbona, D. V., Hoffman, J. B., and Anderson, K. E.: A comparison of production performance between caged and free-range Hy-Line Brown layers, Poult. Sci., 88, 21–25, 2009.

Batkowska, J., Brodacki, A., and Knaga, S.: Quality of laying hen eggs during storage depending on egg weight and type of cage system (conventional vs. furnished cages), Ann. Anim. Sci., 14, 707–719, 2014.

Đukić-Stojčić, M., Perić, L., Bjedov, S., and Milošević, N.: The quality of table eggs produced in different housing systems, Biotechnol. Anim. Husb., 25, 1103–1108, 2009.

Đukić-Stojčić, M., Perić, L., Milošević, N., Rodić, V., Glamočić, D., Škrbić, Z., and Lukić, M.: Effect of genotype and housing system on egg production, egg quality and welfare of laying hens, J. Food Agric. Environ., 10, 556–559, 2012.

Englmaierová, M., Tůmová, E., Charvátová, V., and Skřivan, M.: Effects of laying hens housing system on laying performance, egg quality characteristics, and egg microbial contamination, Czech J. Anim. Sci., 59, 345–352, 2014.

Ferrante, V., Lolli, S., Vezzoli, G., and Cavalchini, L. G.: Effects of two different rearing systems (organic and barn) on production performance, animal welfare traits and egg quality characteristics in laying hens, Ital. J. Anim. Sci., 8, 165–174, 2009.

Gerzilov, V., Datkova, V., Mihaylova, S., and Bozakova, N.: Effect of poultry housing systems on egg production, Bulg. J. Agric. Sci., 18, 955–956, 2012.

Golden, J. B., Arbona, D. V., and Anderson, K. E.: A comparative examination of rearing parameters and layer production performance for brown egg-type pullets grown for either free-range or cage production, J. Appl. Poult. Res., 21, 95–102, 2012.

IBM: Corp. Released, IBM SPSS Statistics for Windows, Version 20.0, IBM Corp., Armonk, NY, 2011.

Krawczyk, J.: Quality of eggs from Polish native Greenleg Partridge chicken-hens maintained in organic vs. backyard production systems, Anim. Sci. Pap. Rep., 27, 227–235, 2009.

Krawczyk, J. and Sokołowicz, Z.: Effect of chicken breed and storage conditions of eggs on their quality, Acta Sci. Pol.-Zootech., 14, 109–118, 2015.

Krawczyk, J., Sokołowicz, Z., and Szymczyk, B.: Effect of housing system on cholesterol, vitamin and fatty acid content of yolk and physical characteristics of eggs from Polish native hens, Arch. Geflugelkd., 75, 151–157, 2011.

Küçükyılmaz, K., Bozkurt, M., Herken, E.N., Çınar, M., Çatlı, A.U., Bintaş, E. and Çöven, F.: Effects of rearing systems on performance, egg characteristics and immune response in two layer hen genotype, Asian Australas, J. Anim. Sci., 25, 559–568, 2012.

Lewko, L. and Gornowicz, E.: Effect of housing system on egg quality in laying hens, Ann. Anim. Sci., 11, 607–616, 2011.

Luiting, P.: Genetic variation of energy partitioning in laying hens: Causes of variation in residual feed consumption, World's Poult. Sci. J., 46, 132–152, 1990.

Lukanov, H. and Alexieva, D.: Trends in battery cage husbandry systems for laying hens. Enriched cages for housing laying hens, J. Agr. Sci. Tech., 5, 143–152, 2013.

Miao, Z. H., Glatz, P. C., and Ru, Y. J.: Free-range poultry production – A review, Asian Australas, J. Anim. Sci., 18, 113–132, 2005.

Mohammed, K. A. F., Sarmiento-Franco, L., Santos-Ricalde, R., and Solorio-Sanchez, J. F.: Egg production, egg quality and crop content of Rhode Island Red hens grazing on natural tropical vegetation, Trop. Anim. Health Prod., 45, 367–372, 2013.

Özbey, O. and Esen, F.: The effects of different breeding systems on egg productivity and egg quality characteristics of rock partridges, Poult. Sci., 86, 782–785, 2007.

Pištěková, V., Hovorka, M., Večerek, V., Straková, E., and Suchý, P.: The quality comparison of eggs laid by laying hens kept in battery cages and in a deep litter system, Czech J. Anim. Sci., 51, 318–325, 2006.

Połtowicz, K., Wężyk, S., Calik, J., and Paściak, P.: The use of native chickens breed in poultry meat production, in: Proceedings of the British Society of Animal Science, 14–15 October 2004, Cracow, 30–32, 2004.

Rizzi, C. and Chiericato, G. M: Organic farming production. Effect of age on the productive yield and egg quality of hens of two commercial hybrid lines and two local breeds, Ital. J. Anim. Sci., 4, 160–162, 2005.

Roll, V. F. B., Briz, R. C., and Levrino, G. A. M.: Floor versus cage rearing: effects on production, egg quality and physical condition of laying hens housed in furnished cages, Cienc. Rural, 39, 1527–1532, 2009.

Safaa, H. M., Serrano, M. P., Valencia, D. G., Arbe, X., Jiménez-Moreno, E., Lázaro, R., and Mateos, G. G.: Effects of the levels of methionine, linoleic acid, and added fat in the diet on productive performance and egg quality of brown laying hens in the late phase of production, Poul. Sci., 87, 1595–1602, 2008.

Sarica, M., Onder, H., and Yamak, U. S.: Determining the most effective variables for egg quality traits of five different hen genotypes, Int. J. Agric. Biol. Eng., 14, 235–240, 2012.

Şekeroğlu, A., Sarica, M., Demir, E., Ulutaş, Z., Tilki, M., and Saatcı, M.: The effects of housing system and storage length on the quality of eggs produced by two lines of laying hens, Arch. Geflugelkd., 72, 106–109, 2008.

Şekeroğlu, A., Sarica, M., Demir, E., Ulutaş, Z., Tilki, M., Saatcı, M. and Omed, H.: Effects of different housing systems on some performance traits and egg qualities of laying hens, J. Anim. Vet. Adv., 9, 1739–1744, 2010.

Singh, R., Cheng, K. M., and Silversides, F. G.: Production performance and egg quality of four strains of laying hens kept in conventional cages and floor pens, Poult. Sci., 88, 256–264, 2009.

Tauson, R., Wahlström, A., and Abrahamsson, P.: Effect of two floor housing systems and cages on health, production, and fear response in layers, J. Appl. Poult. Res., 8, 152–159, 1999.

Tůmová, E., Skřivan, M., Englmaierová, M. and Zita, L.: The effect of genotype, housing system and egg collection time on egg quality in egg type hens, Czech J. Anim. Sci., 54, 17–23, 2009.

van den Brand, H., Parmentier, H. K., and Kemp, B.: Effects of housing system (outdoor vs cages) and age of laying hens on egg characteristics, Brit. Poult. Sci., 45, 745–752, 2004.

Wójcik, E., Andraszek, K., Gryzińska, M., Witkowski, A., Pałyszka, M., and Smalec, E.: Sister chromatid exchange in Greenleg Partridge and Polbar hens covered by the gene-pool protection program for farm animals in Poland, Poult. Sci., 91, 2424–2430, 2012.

Zita, L., Tůmová, E., and Štolc, L.: Effects of genotype, age and their interaction on egg quality in brown-egg laying hens, Acta Vet. Brno, 78, 85–91, 2009.

14

Cytological quality of milk of primiparous cows kept in cubicles bedded with separated manure

Stanisław Winnicki[1], Zbigniew Sobek[2], Ryszard Kujawiak[3], Jerzy Jugowar[1],
Anna Nienartowicz-Zdrojewska[2], and Jolanta Różańska-Zawieja[2]

[1]Institute of Technology and Life Sciences, Poznań Branch, ul. Biskupińska 67, 60-463 Poznań, Poland
[2]Department of Genetics and Animal Breeding, Poznan University of Life Sciences,
ul. Wołyńska 33, 60-637 Poznań, Poland
[3]Sano Agrar Institut, ul. Lipowa 10, 64-541 Sękowo, Poland

Correspondence to: Jolanta Różańska-Zawieja (jolek@up.poznan.pl)

Abstract. A study was conducted on the effect of separated manure as bedding material on milk quality as manifested in the somatic cell count. Cows were maintained in a loose barn in cubicles bedded with fresh separated cattle manure (SCM) with 40 % solids content.

Analyses were conducted on 242 primiparous Polish Black and White Holstein-Friesian cows in the course of a 305-day lactation. Mean milk yield in that period amounted to over 9000 kg per cow. Somatic cell counts, daily milk yields and chemical composition of milk were analysed. Data were obtained from analyses of 2324 milk samples.

It was found that 93.3 % of samples contained less than 400 000 somatic cells per 1 mL milk. Approximately 4.3 % of milk samples contained the number of somatic cells indicating subclinical mastitis (200 000 cells mL^{-1}), while in 2.4 % it was clinical mastitis ($>$800 000 cells mL^{-1}). The incidence rate for both forms of mastitis was similar in the beginning and at the end of lactation. Mean daily milk yield of cows producing milk classified according to quality (SCC) grades 1 ($<$25 000 cells mL^{-1}) and 2 ($<$25 000; 50 000 $>$ cells mL^{-1}) was statistically significantly greater than the yields of other cows. For analysed milk constituents a relationship was found between SCC classes and contents of milk fat and solids.

Conducted analyses showed that SCM as bedding in cow cubicles had no effect on somatic cell counts in milk of primiparous cows. The study was conducted in a single holding, in one lactation, on cows calving in 2014.

1 Introduction

A significant factor determining efficient milk production is connected with the health status of udders in dairy cows. Mastitis is one of the most frequent diseases in dairy cows and a major cause for their culling (Malinowski and Kłossowska, 2002; Engels, 2015; De Vliegher et al., 2005a). Mastitis is a condition of multiple etiology and its incidence is determined by both genetic and environmental factors. Among environmental factors, a considerable effect on udder hygiene is observed for bedding due to the direct contact with teats and the long lying time of cows in the 24 h period.

Two types of cubicle surface management are distinguished – bedding and floor management. A conventional solution is provided by cubicles with straw bedding. Farms with a large area of grassland and maize crops experience a shortage of straw for bedding. Sand, peat and sawdust are used as straw substitutes (Schrade et al., 2008). In recent years studies have been conducted on the suitability of conventional compost and compost produced from biogas plant digestate (Schrade et al., 2008) as well as *Miscanthus* chaff (Hohenbrink et al., 2013). Separated cattle manure, also referred to as dry manure solids (DMS), is currently being used as an alternative bedding material. Studies on its applicability have been conducted in the United States (Harrison et al., 2008; Schwarz et al., 2010), Germany and Switzerland (Leifker, 2013; Schlueter, 2012; Schrade et al., 2008; Sontheimer, 2011; Zaehner et al., 2009a, b). These analyses assessed the effect of DMS on the lying comfort of cows, cleanliness of their body, the incidence of limb skin damage, the presence of microorganisms in the bedding and in bulk milk, and labour outlays connected with bedding replacement. Results of those studies indicated no microbiological hazard for cows or milk connected with the use of DMS as bedding and showed no negative effect of such bedding on animal comfort. However, no studies were found in the available literature presenting an analysis of the effect of DMS bedding throughout lactation on the variation in SCC levels in milk collected from individual cows in the herd. For this reason it was decided to investigate the variation in somatic cell counts in milk from individual cows in the course of a standard 305-day lactation. The proposed research hypothesis postulates that the application of DMS as bedding in individual cubicles may, after an extended period of time, result in an increased somatic cell count in milk and cause mastitis.

The aim of the study was to determine changes in SCC in milk of primiparous cows maintained in individual cubicles bedded with DMS during standard lactation.

2 Material and methods

Analyses were conducted based on productivity of primiparous cows in a herd of over 1000 cows. They were high-production Black and White Polish Holstein-Friesian (PHF). This study did not require a permit to carry out experiments on animals. No tests were performed on animals, and the data for publication have been obtained from a nationwide computer system (SYMLEK) collecting data from routine performance inspection of dairy cows. In 2014, in the herd of 1170 cows, the mean milk yield was 12 135 kg of milk per cow, with a mean of 3.66 % fat and 3.33 % protein. In view of these data this herd ranked second in Poland in the group of herds of over 500 cows (PFHBiPM Warszawa 2015). Cows, both during lactation and in the dry period, were kept in the free-stall system in individual cubicles (Fig. 1) bedded with separated cattle manure (SCM) coming from the farm (Fig. 2). Separated manure used as cubicle bedding contained 39–40 % solids (manure before separation contained 7–8 % solids). Cow manure was removed from the barn alleys with a grooved solid floor six times a day using a delta scraper (Fig. 3).

The herd was fed with total mixed ration (TMR) in identical amounts for each cow in the entire herd in lactation. In the study period the composition of the feed ration was as follows: maize silage, 18 kg; pressed sugar beet pulp, 10 kg; grass silage, 6 kg; molasses, 1 kg; wheat straw, 0.8 kg; and Lactasan concentrate, 9 kg. The nutritive value of the feed ration was as follows: energy content 7.2 MJ net energy lactation, at 16.5 % crude protein content. The feed ration was calculated for 36 kg of milk. TMR was administered twice daily and placed twice a day on the feeding table (Fig. 4).

Figure 1. A cubicle bedded with DMS.

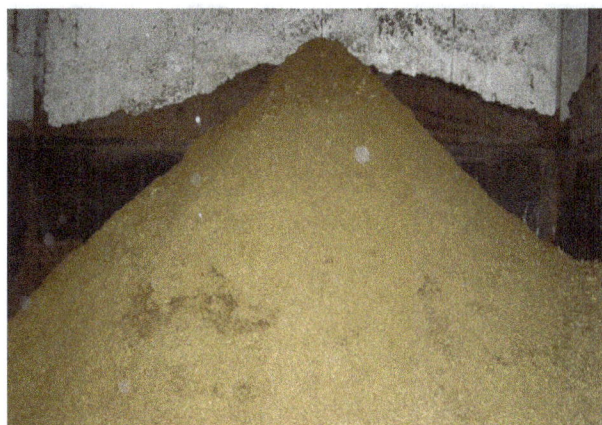

Figure 2. Separated manure (SM).

In the study period cows were milked three times a day in a 60-point carousel milking parlour with access to the udder outside of the carousel (Fig. 5). The farm is covered by the A4 milk recording system run by the Polish Federation of Cattle Breeders and Dairy Farmers (PFHBiPM).

Analyses were conducted on primiparous cows, which made it possible to standardise data and exclude several environmental factors. Analyses covered primiparous cows which calved between August 2013 and the end of February 2014.

Altogether data were collected for 242 primiparous cows during the period of a standard 305-day lactation. Analyses were conducted on milk recording data, i.e. quality expressed in SCC levels, daily milk yields and chemical composition in successive months of lactation. Statistical analyses were conducted using the SAS 9.2 (2009) statistical software package the MEAN, GLM and CORR procedures. Significance of differences between pairs of object means was tested using the least significant difference test.

Figure 3. Grooved floor in alleys.

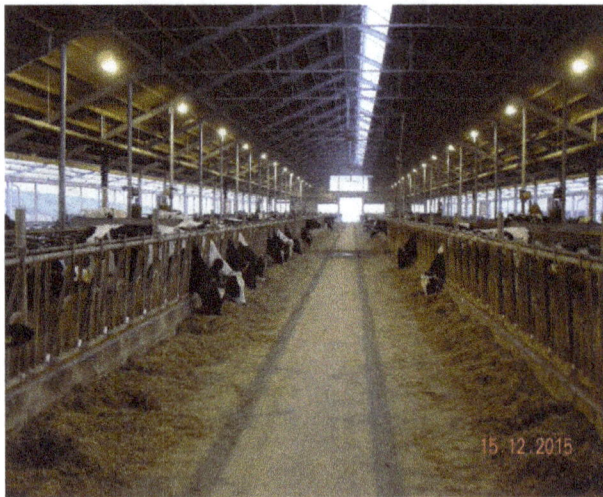

Figure 5. A 60-point carousel milking parlour.

Figure 4. Feeding table.

Table 1. Milk yields of primiparous cows in the course of lactation.

| Month of lactation | Mean | | Daily milk yield | |
	Mean $\overline{x}$ (kg)	% to max $\overline{x}$ day yields	Standard deviation (SD, kg)	Coefficient of variation (%)
1	29.2	75.9	7.35	25.2
2	35.9	93.4	6.69	18.6
3	38.0	98.9	5.80	15.3
4	**38.4**	**100.0**	**5.53**	**14.4**
5	37.6	97.8	5.89	15.7
6	36.0	93.9	6.08	16.9
7	34.6	90.2	4.94	14.3
8	34.2	89.0	5.44	15.9
9	33.2	86.5	5.55	16.7
10	32.1	83.6	5.96	18.6

3 Results

Mean daily milk yield in the herd in successive months of lactation was high and uniform (Table 1).

The lowest mean milk yield was recorded in the first month of lactation (29.2 kg); in the successive months it increased gradually, reaching the maximum in the fourth month (38.4 kg). Starting from the fifth month it decreased slightly. In the 10th month milk yield was 32.1 kg and it was 6.3 kg lower (by 16.4 %) than the maximum yield in the 4th month of lactation.

The coefficient of variation was highest in the first month, amounting to 25.2 %. In the other months of lactation it ranged from 14.4 to 18.6 % (Table 1).

Fat percentage content in milk was lowest at peak yield, in the 3rd and 4th months of lactation, after which it gradually increased to reach in 3.84 % in the 10th month (Table 2).

Mean protein content in milk was lowest in the 2nd month of lactation at 3.03 % and it gradually increased to reach 3.5 % in the 10th month.

Mean lactose content in milk was relatively stable in the course of lactation (Table 1). We need to stress here the low individual variation, as indicated by the low value of the coefficient of variation, generally in the range from 2 to 3 %.

Contents of milk solids changed in the course of lactation. Except for the first month, when colostrum was produced, the lowest content of solids was observed at the beginning and it gradually increased in the course of lactation (Table 2). Individual variation was small, as the coefficient of variation was generally around 5 %.

Mean urea content in milk in the months of lactation was similar, ranging from 210 to 243 mL (Table 2). In turn, considerable individual variation was found, as the coefficient of variation as a rule exceeded 15 % and in the second month of lactation it even exceeded 25 %.

Table 2. Changes in chemical composition in milk of primiparous cows in the course of lactation.

Month of lactation	Contents in milk									
	Fat (%)		Protein (%)		Lactose (%)		Solids (%)		Urea (mg mL^{-1})	
	$\bar{x}$*	SD**	$\bar{x}$	SD	$\bar{x}$	SD	$\bar{x}$	SD	$\bar{x}$	SD
1	4.37	0.70	3.24	0.29	4.89	0.17	13.12	0.90	220	48
2	3.44	0.57	3.03	0.19	4.98	0.11	11.97	0.67	210	54
3	3.33	0.48	3.07	0.19	4.96	0.10	11.89	0.61	216	45
4	3.37	0.48	3.16	0.21	4.94	0.11	12.00	0.63	227	45
5	3.44	0.47	3.19	0.20	4.92	0.11	12.09	0.59	227	43
6	3.56	0.48	3.25	0.22	4.87	0.13	12.25	0.62	238	36
7	3.69	0.45	3.33	0.23	4.87	0.12	12.48	0.64	243	34
8	3.73	0.45	3.36	0.23	4.85	0.14	12.57	0.59	235	40
9	3.72	0.44	3.41	0.27	4.86	0.15	12.64	0.64	242	39
10	3.84	0.45	3.50	0.22	4.88	0.13	12.86	0.62	241	33

* $\bar{x}$ – mean. ** SD – standard deviation.

Table 3 presents the distribution of milk samples in terms of milk quality expressed in SCC levels in the course of lactation. Percentage of individual quality classes changed in the successive months of lactation.

The primary direction of changes was found for the decrease in percentages of samples in classes 1 and 2 and the increase in classes 4 and 5 in successive months of lactation. The percentage of milk samples in the first SCC class up to mid-lactation exceeded 10 %. In turn, in the second half of lactation it gradually decreased, while in the 10th month of lactation it was as low as 3.5 %.

In SCC class 2 at the beginning of lactation there were 33–35 % milk samples, in mid-lactation it was 22–27 %, while at the end of lactation it was approximate 18 %.

In SCC class 3 the percentage of milk samples was relatively stable in the course of the entire lactation. It was also considerable, ranging from 28 to 34 %.

The share of milk samples in class 4 at the beginning of lactation was over 10 %, while at the end of lactation it increased considerably to 25 %. The share of milk samples in class 5 also grew considerably, from approximate 5 % at the beginning to 12 % at the end of lactation.

The percentage of milk samples in class 6 was relatively low, from 1.7 % in the second month to 5 % in the eight month, and only twice did it exceed 6 %. Relatively few milk samples were found in class 7, and in the third month there were no samples of that SCC class.

The modal value typically exceeded 30 % milk samples and it was found in SCC classes 2 and 3 – in the first 3 months in class 2 and from the 4th to the 10th month of lactation in class 3.

Statistical analysis to identify correlations between SCC class of milk and milk yield (Table 4) and its chemical composition was conducted (Tables 5 and 6).

The highest daily milk yield was recorded for cows producing milk classified to SCC classes 1 and 2. The difference between yields in these classes proved to be statistically non-significant. Differences in milk yield between classes from 3 to 7 also turned out to be statistically non-significant. In turn, differences between classes 1 and 2 and the others were statistically highly significant.

Mean percentage shares of fat, protein, lactose and solids as well as the weight content of urea in samples of milk classified to individual SCC classes varied for individual components.

For protein, lactose and urea, differences between classes proved to be statistically non-significant.

Statistical differences were observed for fat and milk solids (Table 5). Fat content gradually increased from 3.42 % in class 1, to 3.85 % in class 6, while only in class 7 did it decrease to 3.72 % (Table 5).

In the case of milk solids the lowest content was also reported in SCC class 1 at 12.11 %, while it was highest in class 2 at 12.77 % (Table 5). Differences between these classes with extreme values and the other classes were statistically highly significant.

4 Discussion

Milk yield of primiparous cows in the analysed herd was much greater both in comparison to the mean yield in the Wielkopolskie province, in which this herd is kept, and in comparison to the national average (Table 6). This refers to both the first 100 days of lactation (643 and 786 kg greater, respectively) and the entire period of 305-day lactation (1593 and 2130 kg greater, respectively).

We need to stress here two advantageous factors characterising the course of lactation: a long period of yield increase, to the 4th month of lactation, and the long-term persistent

Table 3. Distribution of milk samples in terms of SCC classes in months of lactation.

Milk SCC class	Somatic cell count (thousand mL)	Percentage of milk samples in months of lactation									
		1	2	3	4	5	6	7	8	9	10
1	≤ 25	8.5	12.7	11.9	14.3	11.1	6.9	6.1	6.2	4.7	3.5
2	≤ 50	**33.8***	**34.3**	**35.2**	27.3	25.3	22.5	19.2	18.6	18.3	18.9
3	≤ 100	26.9	31.4	28.0	**34.1**	**30.9**	**30.3**	**27.9**	**33.5**	**28.6**	**33.3**
4	≤ 200	16.9	12.7	17.4	12.6	16.6	22.1	24.0	21.5	27.7	25.0
5	≤ 400	7.5	5.1	4.7	7.1	10.6	10.8	11.4	11.6	11.3	12.7
6	≤ 800	4.0	1.7	2.8	2.1	4.1	4.8	6.6	5.0	6.1	4.8
7	> 800	2.4	2.1	0.0	2.5	1.4	2.6	4.8	3.6	3.3	1.8

* Bold – maximum number of observations in the SCC class.

Table 4. Distribution of milk samples in SCC classes and daily milk yield.

Milk SCC class	Somatic cell count (thousand mL)	Frequency of observations		Daily milk yield		
		n	%	Mean $\overline{x}$ (kg)	% to max $\overline{x}$ (daily yields)	Standard deviation (SD)
1	≤ 25	196	8.4	36.5[A]***	100.0	5.93
2	≤ 50	588	25.3	35.6[A]	97.3	6.56
3	≤ 100	709	30.6	34.6[B]	94.7	6.74
4	≤ 200	461	19.8	34.7[B]	94.9	6.82
5	≤ 400	213	9.2	33.7[B]	92.1	6.64
6	≤ 800	101	4.3	34.2[B]	93.7	6.29
7	> 800	56	2.4	33.1[B]	90.5	7.35

*** Means marked with identical letters are not significantly different. Capital letters ([A, B, C]) – highly statistically significant differences ($P \leq 0.01$).

milk production; in the 10th month mean daily yield was 32.1 kg, which accounts for 83.6 % in relation to maximum production (Table 1). The decisive cause for the very gentle decrease in the amount of obtained milk was connected with intensive feeding, particularly in the second part of lactation.

High variation in yield in the first month of lactation resulted from the conditions under which it was determined. On milk recording days for cows in the herd, the number of days which passed from the beginning of lactation varied for individual cows, and thus they were at different levels of milking. It also refers to the second month of lactation. In turn, a greater coefficient of variation in milk yield in the 10th month of lactation (18.6 %) was the effect of the final phase of lactation for several cows. An increase in fat contents in milk in the course of lactation is a natural physiological process. Such a character of changes was observed in both the analysed herd and nationwide (Table 6).

Protein content in milk indicates the degree of balance between yield and energy intake with feed (Ziemiński and Juszczak, 1997). At the state of balance the content of protein should fall between 3.2 and 3.6 %. It is relatively difficult

to provide energy with feed at a high yield at peak lactation. Thus, it is believed that a 3.1 % protein content is a minimum value.

In the analysed primiparous cows in the second and third months of lactation the mean value was lower, i.e. 3.03 and 3.07 %, respectively, thus indicating a negative energy balance of the diet in relation to yields (Table 2). In the fourth and fifth months of lactation the content of protein was also too low, and only starting from the 6th month did it meet the standard guidelines. In that period the daily yield of milk exceeded the energy value of the feed ration calculated for 36 kg milk. Starting from the sixth month of lactation, the daily yield was on average max. 36 kg. In that period the content of protein in milk was within the normal limits.

Among all milk components the content of lactose in milk from healthy cows was characterised by high stability. This results from its important role in providing electrolyte balance in the process of milk synthesis in the udder. At mastitis lactose content decreases at a simultaneous increase in the content of chlorine. Recorded data indicate that udders of analysed primiparous cows were generally healthy.

Table 5. Chemical composition of milk depending on SCC class.

Month of lactation		Contents in milk								
		Fat (%)		Protein (%)		Lactose (%)		Solids (%)		Urea (mg mL^{-1})	
		$\bar{x}$*	SD**	$\bar{x}$	SD	$\bar{x}$	SD	$\bar{x}$	SD	$\bar{x}$	SD
1	≤ 25	3.42^{A***}	0.60	3.18	0.22	4.98	0.11	12.11^{Aa}	0.76	227	48.4
2	≤ 50	3.57^{BC}	0.62	3.25	0.24	4.94	0.17	12.77^{BC}	0.77	227	43.8
3	≤ 100	3.67$^{B\,E}$	0.60	3.37	0.28	4.98	0.17	12.44^{BDE}	0.77	230	44.7
4	≤ 200	3.67$^{B\,E}$	0.54	3.29	0.26	4.87	0.13	12.44$^{BD\,g}$	0.71	234	42.7
5	≤ 400	3.75^{BD}	0.65	3.33	0.28	4.84	0.15	12.52^{BD}	0.78	232	41.9
6	≤ 800	3.85^{BDF}	0.54	3.35	0.25	4.82	0.16	12.64^{BDFhi}	0.68	232	35.9
7	> 800	3.72^{B}	0.54	3.26	0.23	4.77	0.13	12.36$^{bD\,j}$	0.59	232	44.1

* $\bar{x}$ – mean. ** SD – standard deviation. *** Means marked with different letters are significantly different in the pairs: $^{A-B}$, $^{C-D}$, $^{E-F}$, $^{G-H}$ and $^{I-J}$. Small letters ($^{a,\,b,\,c}$) describe statistically significant differences ($P \le 0.05$), while capital letters ($^{A,\,B,\,C}$) describe highly statistically significant differences ($P \le 0.01$).

Table 6. Yield of primiparous cows in the analysed herd in the province of Wielkopolskie and nationwide in 2014.

Primiparous cows – head	Number of animals	Yield (kg)			Content in milk (%)	
		milk	fat	protein	fat	protein
Analysed herd for						
- 100 days	302	3483	122	107	3.49	3.06
- 305 days	214	9520	362	317	3.80	3.33
Wlkp. province for						
- 100 days	44 464	2840	110	88	3.87	3.11
- 305 days	39 722	7927	310	262	3.91	3.31
Nationwide for						
- 100 days	181 837	2697	105	83	3.88	3.07
- 305 days	163 662	7390	293	242	3.96	3.27

An increase in solids content in successive months of lactation is a natural phenomenon observed due to the increase in the contents of fat and protein in milk.

Mean urea content in milk throughout lactation fell within the limits specified by Ziemiński and Juszczak (1997). Relatively uniform values in the course of lactation were determined by the stable composition of the feed ration.

The primary indicator for hygienic quality of milk is connected with the quality expressed by the somatic cell count (SCC). In Poland, similar to in many other countries, the upper SCC limit for liquid milk for human consumption is assumed to be 400 000 mL, which is a threshold value in Poland in milk buying (Regulation of the Minister of Agriculture and Rural Development, Journal of Laws no. 117, item 1011). Milk in classes 1–5 contains the number of somatic cells admissible for commercial sale. In the analysed herd there were a total of 93.3 % individual milk samples in these classes (Table 2) with relatively small deviations for individual months of lactation (Table 3). An elevated SCC value indicates the incidence of mastitis. An SCC level above 400 000 mL in-

dicates subclinical mastitis, while the level of over 800 000 shows clinical mastitis. In the analysed herd, mastitis was found in the course of entire lactation with a similar, low frequency in individual months of lactation (Table 3).

Mastitis in primiparous cows in the subclinical and clinical forms was already observed in the first month of lactation, at 4 and 2.4 %, respectively (Table 4). These cases may not have been caused by the use of DMS as bedding in cubicles, but rather an earlier effect of other factors. Parker et al. (2007) and Malinowski and Smulski (2007) indicate frequent mastitis cases already in primiparous cows. For this reason they recommend diagnostics and treatment of udders in heifers before calving. A comparable frequency of mastitis in primiparous cows in the analysed herd in the course of entire lactation indicates that DMS bedding in cubicles was not a source of mastitis hazard (Table 4).

Statistical analysis was conducted on differences in daily milk yields (Table 5) and milk chemical composition in SCC classes (Tables 6–8). The highest milk yield was obtained from cows from SCC classes 1 and 2, i.e. 36.5

and $35.6 \, \mathrm{kg \, day^{-1}}$ (Table 4). The difference between these classes proved to be statistically non-significant. The same statement was given in a report by the American National Mastitis Council (Laboratory and Field, 1987) and it was confirmed in a study by De Vliegher et al. (2005b). Differences in milk yield between SCC classes 1 and 2 and the others, i.e. from 3 to 7, proved to be statistically highly significantly. These differences are relatively slight; however, they indicate a deterioration of milk production functions of the udder.

The Laboratory and Field (1987) report showed that, starting from SCC class 3, milk yield decreases by 1.5 lb (approx. 0.7 kg) a day for each class. In our study no such uniform decrease was recorded. This incomplete consistency with data contained in the Laboratory and Field (1987) report results from the fact that this study concerns only one barn and for this reason it may not lead to generalisations.

A slightly different method was adopted by De Vliegher et al. (2005b), who determined SCC in primiparous cows between the 5th and 14th day of lactation. They stated that mastitis in that period causes a decrease in yield for the entire lactation.

The chemical composition of milk depends on nutrition and milk production functions of the udder. Protein content in milk is determined by the supply of energy and urea content from the supply of protein in feed (Ziemiński and Juszczak, 1997). Lactose content in milk is connected with the maintenance of osmotic pressure and it is regulated in the process of milk production. Fat content is determined by hereditary factors and nutrition, mainly fibre contents in feed.

In the analysed herd of primiparous cows contents of protein, lactose and urea were similar in SCC classes. Statistically significant differences were found only for contents of fat and solids (Table 5). The higher the SCC class, the greater the content of fat. There exists a relationship between milk yield and fat content – the higher the yield, the lower the fat percentage. In turn, for milk solids the lowest contents were found in SCC class 1, while it was highest for class 2. Such a situation is connected with the lowest contents of fat and protein in milk of cows in SCC class 1.

Changes in the chemical composition of milk, in classes from 1 to 6, were not the effect of SCC, but they were determined by dependencies between the daily yield of milk and contents of fat and protein. Recorded results indicate that at subclinical mastitis in SCC class 6 the milk production functions were not disturbed. Only at clinical mastitis in group 7 functional changes were found in milk synthesis. This is evidenced by lower contents of fat, protein and lactose in relation of group 6.

Recorded results show that the cubicle management of primiparous cows with high milk yields caused no deterioration of SCC quality of milk in the course of a standard 305-day lactation. Providing high comfort for cows when lying is the reason for a search for new bedding materials. Although they are many very good mats and mattresses for cubicles,

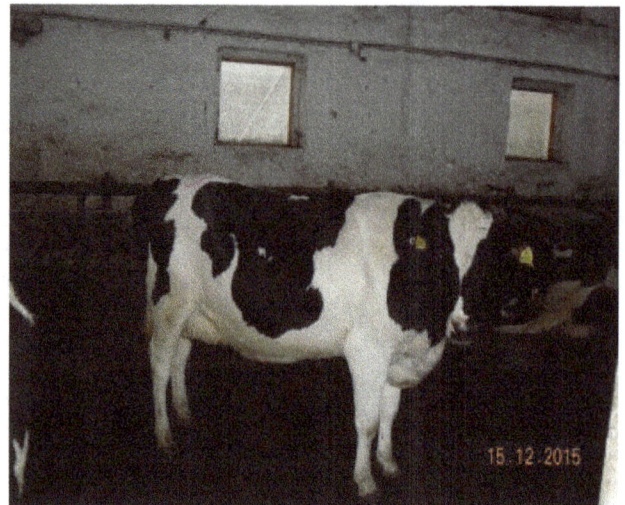

Figure 6. A clean cow standing on separated manure.

they are inferior to bedding, particularly in terms of protection of body surface (Schaub et al., 1999). In comparison to mattresses, DMS bedding is considerably more effective in prevention of skin lesions (Schrade et al., 2008). This was confirmed in the analysed herd. Cows are clean and no skin damage or lesions are observed (Fig. 6). Another advantage of DMS bedding is connected with its resilience and adaptability to the body contours of the lying animal. This may even be a certain problem, since humps are easily formed in the cubicle. For this reason manual levelling of bedding in the cubicle is required.

5 Conclusions

Analyses were conducted on high-production primiparous cows kept in cubicles bedded with separated cattle manure showed the following:

- DMS may successfully be used, with no deterioration of SCC milk quality, as bedding material in resting pens for dairy cows in well-managed (high-production) herds.

- The highest milk yield was recorded for cows with SCC levels below $50\,000 \, \mathrm{mL^{-1}}$. Their yields were statistically significantly greater in comparison to higher SCC classes.

6 Animal Welfare Statement

This study did not require a permit to carry out experiments on animals. No tests were performed on animals, and the data for publication have been obtained from a nationwide computer system SYMLEK collecting data from routine performance inspection of dairy cows.

Competing interests. The authors declare that they have no conflict of interest.

Edited by: M. Mielenz

References

De Vliegher, S., Barkema, H. W., Opsomer, G., de Kruif, A., and Duchatau, L.: Association Between Somatic Cell Count in Early Lactation and Culling of Dairy Heifers Using Cox Frailty Models, J. Dairy Sci., 88, 560–568, 2005a.

De Vliegher, S., Barkema, H. W., Stryhn, H., Opsomer, G., and de Kruif, A.: Impact of Early Lactation Somatic Cell Count in Heifers on Milk Yield Over the First Lactation, J. Dairy Sci., 88, 938–947, 2005b.

Engels, H.: To save money, DLG, Special Releases, 10–13, 2015.

Harrison, E., Bonhotal, J., Schwarz, M., and Fiesinger, T.: Using Manure Solid as Bedding, Final Raport, Cornell Waste Management Institute, 2008.

Hohenbrink, S., Boelhauve, M., Fiege, F., and Ickler, A. L.: Sprinkle grass in the pit – elephants, Top Agrar, 10, 14–17, 2013.

Laboratory and Field: Handbook on Bovine mastitis: National Mastitis Council USA, Library of Congress Catalog Number 87-081736, 1987.

Leifker, A.: „Gülle-Einstreu": Bleiben die Euter gesund?, Top Agrar, 3, 20–23, 2013.

Malinowski, E. and Kłossowska, A.: Diagnostic of infection and mastitis, Puławy National Veterinary Research Institute, 2002.

Malinowski, E. and Smulski, S.: Incidence and prevention of infections and mastitis in heifers, Veterinary Life, 82, 476–482, 2007.

Parker, K. J., Compton, C., Anniss, F. M., Weir, A., Heuer, C., and Mc Dougall, S.: Subclinical and Clinical Mastitis in Heifers Following the use of a Teat Sealant Precalving, J. Dairy Sci., 90, 207–218, 2007.

PFHBiPM: Polish Federation of Cattle Breeders and Dairy Farmers: Rating Dairy Cattle, Warsaw, 2015.

Regulation of the Minister of Agriculture and Rural Development of the 5th July 2002: on detailed special veterinary conditions required for obtaining, processing, storage and transport of milk and milk products, Journal of Laws, 117, 7710–7718, 2002.

SAS: Guide for Personal Computers, ver. 9.2. SAS Inst. Inc., Cary, NC, USA, 2009.

Schaub, J., Friedli, K., and Wechsler, B.: Soft mats for boxes for lying for dairy cows, Taenikon, No. 529, 1999.

Schlueter, D.: Seperated manure as a badding, Elite, 5, 50–53, 2012.

Schrade, S., Zaehner, M., and Schaeren, W.: Bedding for dairy cows in box for lying, Compost and solids from the separation of liquid manure as an alternative to straw-manure mattress, Taenikom, No. 699, 2008.

Schwarz, M., Bonhotal, J., and Staehr, A. E.: Use of Dried Manure Solids as Bedding for Dairy Cows, Cornell Waste Management Institute, 9, 2010.

Sontheimer, A.: (No) secretly sick, New Agriculture, 3, 75–76, 2011.

Zaehner, M., Schrade, S., Schaeren, W., and Schmidtko, J.: New materials as litter in resting pens of dairy barns, Science Conference on Organic Agriculture, Taenikon, Band 2, 50–53, 2009a.

Zaehner, M., Schmidtko, J., Schrade, S., Schaeren, W., and Otten, S.: Alternative bedding materials in cubicle/boxes for lying, Bautagung Raumberg, Gumpenstein, 33–38, 2009b.

Zieminski, R. and Juszczak, J.: Milk urea content as an indicator of protein:energy ratio in feed ration for dairy cows, Progress in Agricultural Sci., 3, 73–82, 1997.

15

Association of VLDLR haplotypes with abdominal fat trait in ducks

Shifeng Pan[1,2,*], **Cong Wang**[1,*], **Xuan Dong**[1], **Mingliang Chen**[1], **Hua Xing**[1,2], **and Tangjie Zhang**[1,2]

[1]College of Veterinary Medicine, Yangzhou University, Yangzhou, 225009, China
[2]Jiangsu Co-Innovation Center for the Prevention and Control of Important Animal Infectious Disease and
Zoonoses, Yangzhou, Jiangsu, 225009, China
[*]These authors contributed equally to this work.

Correspondence to: Tangjie Zhang (slx@yzu.edu.cn)

Abstract. This study aimed to determine the correlation among *VLDLR* (very low-density lipoprotein receptor) gene polymorphisms, body weight and abdominal fat deposition of Gaoyou ducks. A total of 267 Gaoyou ducks from one pure line was employed for testing. The polymorphisms of the *VLDLR* gene were screened by polymerase chain reaction and DNA sequencing. Four novel single nucleotide polymorphisms (SNPs) (g.151G > A, g.170C > T, g.206A > G and g.278–295del) were identified in the 5'-UTR and signal peptide region. Furthermore, eight haplotypes were identified based on the four SNPs. The H8 was the most common haplotype with a frequency of more than 31 %. The four SNPs and their haplotype combinations were shown to be significantly associated with body weight at 6–10 weeks of age ($P < 0.05$ or $P < 0.01$) and abdominal fat percentage (AFP) ($P < 0.05$ or $P < 0.01$). Remarkably, the H1H1 diplotype had an effect on increasing body weight and decreasing AFP from the 6th to the 10th weeks of age. However, increasing positive effects of the H5H8 diplotype were observed for both body weight and AFP. This study suggests that the *VLDLR* gene plays an important role in the regulation of body weight and fat-related traits and may serve as a potential marker for the marker-assisted selection program during duck breeding.

1 Introduction

Genetic markers closely linked to loci for economically important traits can be used to enhance the speed and effectiveness of progress in animal breeding. Once an association between DNA polymorphism and a trait was found, the DNA polymorphism could be considered as a candidate genetic marker for marker-assisted selection (MAS) programs.

Lipoprotein receptor is a member of the low-density lipoprotein receptor (LDLR) family and highly expressed in adipose tissue, heart and skeletal muscles, while it is absent in liver. VLDLR binds apolipoprotein E-triglyceride-rich lipoproteins and plays a critical role in lipid metabolism and the reelin signaling pathway. The *VLDLR* gene contains five functional domains (Willnow, 1998). VLDLR mediates the uptake of very low-density lipoprotein (VLDL) by peripheral tissues through lipoprotein lipase (LPL)-dependent lipolysis and participates in VLDL metabolism (Tacken et al., 2000;

Takahashi et al., 1995, 2004; Goudriaan et al., 2004). In addition, VLDLR has protective features against obesity, insulin resistance, premature heart disease, tumour growth, inflammation and angiogenesis (Nguyen et al., 2014; Yuan et al., 2011; Kyosseva et al., 2013). Studies with VLDLR knockout mice have linked VLDLR with obesity and VLDLR mutants exhibit modest reductions in body weight and adiposity (Frykman et al., 1995; Goudriaan et al., 2001; Eppig et al., 2015; Suwa et al., 2010). Patients with VLDLR mutations have abnormally lower body mass index when compared with control subjects (Boycott et al., 2005; Crawford et al., 2008), which is consistent with the results in mice. Furthermore, previous studies showed that VLDLR is related to body weight and adiposity in humans and mice (Brockmann et al., 1998; Kunej et al., 2013; Clemente-Postigo et al., 2011). Thus, VLDLR expression levels are likely associated with the phenotypic biomarkers for obesity (Kim et al., 2012). To date, most poultry VLDLR studies have focussed on reproduction because VLDLR can develop growing oocytes and deposit yolk lipoprotein (Shen et al., 1993; Wang et al., 2011; Wu et al., 2015), but the relationships between VLDLR polymorphism and fat deposition or body weight have not yet been investigated in poultry. Therefore, based on our previous results that the *VLDLR* gene is probably associated with duck abdominal fat deposition (Zhao et al., 2015), in the present study, we further investigate the association between the *VLDLR* gene and abdominal fat deposition in ducks by screening different polymorphic sites.

Growth rate and carcass lean content are two economically important traits in meat-producing animals. Therefore, a higher growth rate and lower body fat percentage are always to be preferred in breeding programs for commercial breeders. And this is why we screen the polymorphic sites in partial sequences of the *VLDLR* gene and correlate the VLDLR polymorphism with fat deposition and body weight traits. Although association studies cannot determine whether the gene markers are responsible for the variation in a trait or whether the variation is due to a closely linked locus that influences the trait, there is still evidence suggesting that the *VLDLR* gene would affect these traits. This study aimed to identify the polymorphism of the *VLDLR* gene and to analyse the associations among polymorphism, growth and main carcass traits (abdominal fat weight – AFW; abdominal fat percentage – AFP; carcass weight – CW) in Gaoyou ducks.

2 Materials and methods

2.1 Ethics statement

All animal experiments were approved by the Jiangsu Administrative Committee for Laboratory Animals (permission number SYXK-SU-2007-0005) and complied with the guidelines of Jiangsu laboratory animal welfare and the ethics of Jiangsu Administrative Committee of Laboratory Animals.

2.2 Sample collection and preparation

A total of 267 pure-line Gaoyou ducks were obtained from a high-quality Jiangsu Gaoyou duck farm in Jiangsu province that produces pure-bred animals without hybridisation. All ducks were raised in floor pens under the same standardised conditions of management and fed with commercial corn–soybean diets that met NRC nutrient requirements. Blood samples and phenotypic data on growth and carcass traits were collected from 267 individuals. They were euthanised at 10 weeks of age after 6 h with no access to food prior to euthanasia. CW was measured on the chilled carcass after the removal of feathers, heart, lungs, liver, kidneys, the gastrointestinal tract and abdominal fat. The ratio of these traits to BW10 (body weight at 10 weeks of age) was calculated as carcass percentage (CP) and AFP. Genomic DNA was ob-

tained by phenol and chloroform (1 : 1) extraction and stored at $-80\,°C$.

2.3 Primer design, PCR amplification and identification of gene polymorphism

The VLDLR genomic sequence (NM_001310401) was obtained from the National Center for Biotechnology Information (NCBI). One pair of primers (5′-ATTACACTGCCAAATGACC-3′ and 5′-CGGGAACTGGGATTCTTC-3′) was designed to amplify the signal peptide region of the duck *VLDLR* gene. The size of the product was 374 bp.

Polymerase chain reaction (PCR) was performed using 50 ng DNA templates, 10 pM of each primer, 0.20 mM dNTP, 2.5 mM $MgCl_2$ and 0.5 U Taq DNA polymerase. Thermal cycling began with an initial denaturation step of 94 °C for 10 min, followed by 35 cycles of 94 °C for 30 s, 53.5 °C annealing for 30 s, 72 °C for 30 s and an elongation step at 72 °C for 10 min. DNA sequencing was performed using an ABI 3130 genetic analyser (Applied Biosystems, USA). Sequencing variants were detected by visual examination of the sequencing map followed by alignment using DNAMAN.

2.4 Statistic analysis

The genotype and allelic frequencies, genotypic numbers, effective allele numbers (Ne), gene heterozygosity (He) and polymorphism information content (PIC) were calculated and the Hardy–Weinberg equilibrium was analysed using the χ^2 test of PopGene32 (version 1.31). SHEsis online version (http://analysis2.bio-x.cn/myAnalysis.php) was used to calculate the pairwise linkage disequilibrium. Haplotypes were obtained for each animal using the PHASE computer program, version 2.1. The association between VLDLR genotypes with growth and carcass traits, including the weight of birth, body weight at 2–10 weeks, CW, eviscerated weight, abdominal fat weight, dressing percentage and percentage of eviscerated weight, were evaluated according to the two-way analysis of the software SPSS (version 16.0), using the following model: $Y = \mu + G + L + G \times L + e$, where Y was the dependent variable (analysed traits), μ was the overall mean, G was the genotype of different variation for the *VLDLR* gene, L was the duck population, $G \times L$ was the interaction between genotype and duck population (it is a fixed effect), and e was the random error. The differences between genotypes were determined by least square analysis.

3 Results

3.1 Polymorphisms in the duck *VLDLR* gene

A pair of primers was used to amplify and screen single nucleotide polymorphisms (SNPs) in the entire signal peptide coding region of the duck *VLDLR* gene. PCR amplifi-

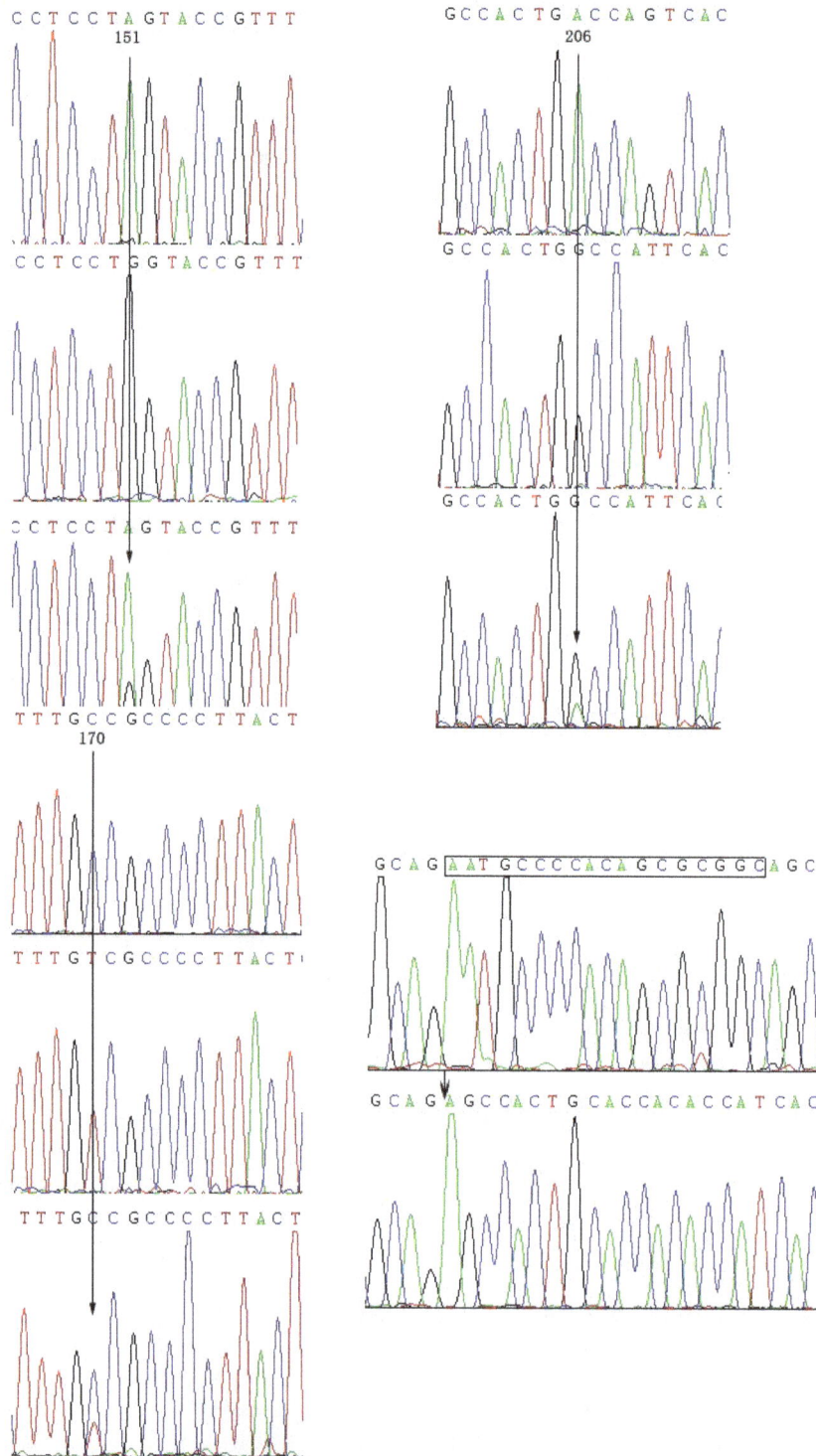

Figure 1. DNA sequencing maps from several DNA templates at four SNPs of duck *VLDLR*.

cation of one *VLDLR* gene fragment yielded a 374-bp fragment including the entire signal peptide coding region and partial exons. The polymorphism type and position were identified by direct DNA sequencing. Multiple sequence alignment showed that four SNPs (g.151G > A, g.170C > T,

g.206A > G, g.278–295del) were identified in the duck *VLDLR* gene, all of which were found in the 5'-UTR and the signal peptide coding region. In addition, the four SNPs were all deposited in the GenBank database (KU317918-KU317932). DNA sequencing maps were shown in Fig. 1.

Table 1. Population genetic indexes of four polymorphisms in signal peptide coding of VLDLR.

SNPs	Number of genotypes			Allele		He	Ne	PIC	χ^2	P
g.151G > A	GG (107)	GA (149)	AA (11)	G(0.68)	A(0.32)	0.44	1.77	0.34	21.20	N/A
g.170C > T	CC (133)	CT (59)	TT (75)	C(0.61)	T(0.39)	0.48	1.91	0.36	76.76	N/A
g.206A > G	AA (88)	AG (134)	GG (45)	A(0.58)	G(0.42)	0.49	1.95	0.37	0.25	0.62
g.278–295del	++ (174)	+− (29)	−− (64)	+ (0.70)	− (0.30)	0.42	1.71	0.33	145.56	N/A

Table 2. Haplotype and diplotype frequencies in the duck *VLDLR* gene.

Haplotype	SNP				Frequency (%)	Diplotype	Frequency (%)
	g.151G > A	g.170C > T	g.206A > G	g.278–295del			
H1	G	T	A	−	25.96	H1H1	27.18
H2	A	T	A	+	9.62	H2H8	16.51
H3	G	T	G	+	0.96	H4H8	28.16
H4	G	C	A	+	18.27	H5H8	6.80
H5	G	C	G	+	9.62	H4H4	1.94
H6	A	T	G	+	4.81	H5H5	2.91
H7	A	C	A	+	0.96	H4H5	3.88
H8	A	C	G	+	29.81	H2H5	1.94
						H3H5	0.97
						H2H2	0.97
						H4H7	2.91
						H3H8	0.97
						H6H8	0.97
						H7H8	0.97
						H8H8	2.91

The standardised measure of linkage disequilibrium (LD) denoted as r^2 was calculated for all pairs of the four SNPs (Fig. 1). If $r^2 > 0.33$, the linkage disequilibrium was considered strong (19) (Ardlie et al., 2002). Linkage disequilibrium was often observed between closely positioned loci (20) (Gibbs et al., 2003), as indicated among the four SNPs in the present study. In particular, there was strong linkage disequilibrium between g.206 and g.151, g.170 and g.278–295, and g.170 and g.206.

The g.151G > A, g.170C > T and g.206A > G mutations occurred in the 5'-UTR of the *VLDLR* gene. The g.278–295del lacked 18 bases, which encoding six amino acids of signal peptides.

3.2 Genetic variation in different populations

Minor allelic frequencies, Hardy–Weinberg equilibriums, He, Ne and PIC for each of the four SNPs in the *VLDLR* gene are shown in Table 1. The χ^2 test showed that the genotype distributions at loci g.151, g.17 and g.278–295 displayed deviation from the Hardy–Weinberg equilibrium in the Gaoyou duck population. The genotype distributions at g.206 was consistent with Hardy–Weinberg equilibrium.

3.3 Linkage disequilibrium and haplotype analysis

Haplotypes generally have more information than individual SNPs. Haplotypes were reconstructed with the four SNPs in all 267 ducks by employing the Phase computer program. All eight haplotypes, which accounted for 100 % of all the observations, were listed in Table 2. Among them, five hap-

lotypes, H1(GTA−), H2 (ATA+), H4 (GCA+), H5 (GCG+) and H8(ACG+), were prevalent and counted for 93.28 % of the observations. Fifteen diplotypes were obtained based on these eight haplotypes. Among them, the frequencies of four diplotypes were higher than 5.0 %. Two diplotypes, H1H1 and H4H8, accounted for 55.34 % of them.

The standardised measure of linkage disequilibrium (LD) denoted as r^2 was calculated for all pairs of the four SNPs (Fig. 2). If $r^2 > 0.33$, the linkage disequilibrium was considered strong (Ardlie et al., 2002). Linkage disequilibrium was often observed between closely positioned loci (Gibbs et al., 2003), as indicated among the four SNPs in the present study. Particularly, there was strong linkage disequilibrium between g.206 and g.151, g.170 and g.278–295, and g.170 and g.206.

3.4 Association of diplotypes with duck growth and abdominal fat deposition

Four diplotypes were reconstructed based on haplotypes. The frequencies of all these four diplotypes were higher than 5.0 %. The generalised linear model (GLM) analysis (Table 3) indicated the existence of associations between the

Table 3. Least square means (±SE) of the traits, by diplotype, of the duck *VLDLR* gene.

Traits	H1H1(73)	H5H8(34)	H2H8(52)	H4H8(83)
Weight (g) at birth	48.05 ± 0.79	51.00 ± 1.83	48.70 ± 1.13	48.25 ± 0.59
3 weeks	619.19 ± 17.26	652.00 ± 41.37	576.20 ± 33.42	594.10 ± 16.73
4 weeks	913.00 ± 33.24	948.83 ± 66.39	815.20 ± 44.75	869.95 ± 30.24
5 weeks	1170.71 ± 41.76	1141.17 ± 88.86	1077.10 ± 48.97	1150.95 ± 32.23
6 weeks	1543.14 ± 48.93^{a}	1517.33 ± 96.98^{a}	1365.70 ± 63.53^{b}	1477.30 ± 36.75^{a}
7 weeks	1968.67 ± 59.40^{Aa}	1890.83 ± 108.59^{Aa}	1715.20 ± 67.50^{Bb}	1810.15 ± 43.98^{Ab}
8 weeks	2088.05 ± 53.58^{Aa}	2025.83 ± 126.81^{Aa}	1832.60 ± 65.45^{Bb}	1936.55 ± 42.84^{Ab}
9 weeks	2326.71 ± 55.38^{Aa}	2232.67 ± 173.79^{Aa}	2013.70 ± 59.96^{Bb}	2112.10 ± 50.73^{Ab}
10 weeks	2537.71 ± 51.43^{Aa}	2402.14 ± 215.35^{Aa}	2274.00 ± 77.47^{Bb}	2384.98 ± 32.80^{Ab}
CP (%)	90.05 ± 0.60	90.63 ± 1.14	91.84 ± 0.44	89.74 ± 0.95
EWP (%)	74.36 ± 0.47	74.31 ± 0.63	75.12 ± 0.65	74.38 ± 0.79
AFP (%)	1.86 ± 0.15^{Bb}	2.79 ± 0.24^{Aa}	2.08 ± 0.12^{Bb}	2.22 ± 0.07^{ABb}

The data are expressed as least square means ± standard errors (mean ± SE).
Note that only significant associations are shown in this table. Values within a row without a common superscript letter differ, and values with superscript letters differ significantly ($P < 0.05$ and $P < 0.01$ each). Genotypes with a and b mean they differed significantly ($P < 0.05$), and genotypes with A and B mean they differed very significantly ($P < 0.01$).
EWP: eviscerated weight percentage.

diplotypes and the different traits (body weight (BW) and AFP). The results showed that the weight of ducks with the diplotype H1H1 and H5H8 was significantly higher than that of ducks with the other two diplotypes at 6, 7, 8, 9 and 10 ($P < 0.05$, $P < 0.01$) weeks old, although no significant differences for the weight at birth and 3-, 4- and 5-week weight were found from the least squares means of the four diplotypes. The weight of ducks the diplotype H2H8 was significantly lower than that of ducks with the other three diplotypes from the 6th to the 10th week ($P < 0.05$, $P < 0.01$). The diplotype H1H1 had an effect on increasing body weights, whereas H2H8 had an effect on decreasing body weights ($P < 0.05$).

Ducks with the diplotype H5H8 had a significantly higher abdominal fat percentage than those with the other three diplotypes ($P < 0.01$). Ducks with the diplotype H1H1 had a lower abdominal fat percentage than those with the other three diplotypes ($P < 0.01$). The results showed that the diplotype H5H8 had an increasing positive effect on abdominal fat deposition, and the diplotype H1H1 had an increasing negative effect on abdominal fat deposition. However, there was no significant correlation between the different four diplotypes and other traits

4 Discussion

Three of four SNPs in the Gaoyou VLDLR signal peptide coding region were in Hardy–Weinberg disequilibrium. The Hardy–Weinberg disequilibrium in the Gaoyou duck population may be due to the artificial selection of parents during long-term commercial breeding (e.g., growth, egg appraisal, carcass weight) since genotype frequency deviations from the Hardy-Weinberg equilibrium were expected to happen on loci under selection. The Gaoyou duck population had in-

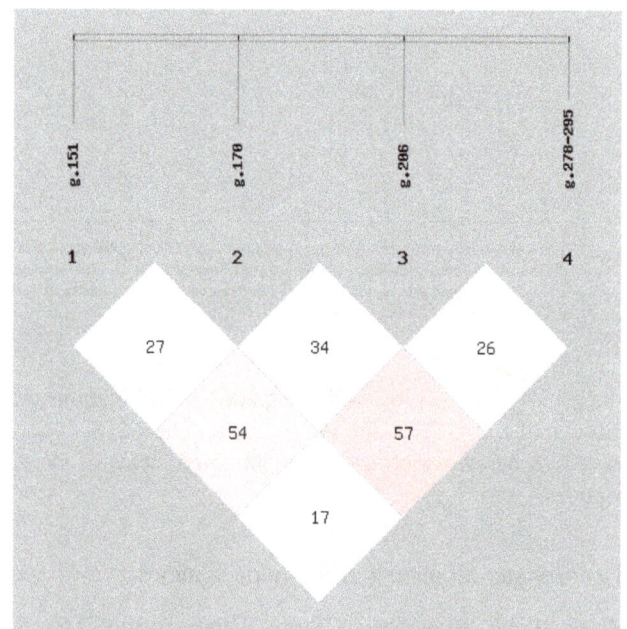

Figure 2. Linkage disequilibrium (LD) plot of the *VLDLR* gene in Gaoyou ducks. The colour scheme is according to SHEsis r^2 scheme. Numbers in each cell stand for the pairwise r^2 value (%) between the corresponding SNPs.

termediate levels of genetic diversity ($0.25 < \text{PIC} < 0.50 =$ intermediate polymorphism); therefore, there was sufficient genetic diversity for the effective selection on improving growth, egg and other traits during the breeding process.

Through the comparison of the fundamental frequency between different haplotypes, it was demonstrated that mutations are not directional; furthermore, the frequencies of H1 and H5 are not positively related to the complexity of their respective mutation. In addition, to evolve from haplotype H8, H1 needs a six-step mutation, while the H5 haplotype only needs a one-step mutation. However, the H5 frequency is much lower than that of H1, which reflected that selection also had a role in haplotype frequencies.

In the present study, the duck *VLDLR* gene was investigated as a possible genetic factor in determining the variance of abdominal fat content. Polymorphism-trait association studies cannot determine whether VLDLR is responsible for variation in a trait or whether a closely linked locus influences the trait. Many studies have revealed that obesity candidate genes were associated with genetics, nutritional disease, gastrointestinal and developmental disorders, and cancer (Castro et al., 2017; Ning et al., 2017). Peroxisome proliferator-activated receptor alpha (PPARα) and retinoid X receptor alpha (RXRα) were identified as central nodes, also called hub molecules (Kunej et al., 2013). Several studies have reported that adipose *VLDLR* gene expression can be regulated by PPARγ (Takazawa et al., 2009; Tao et al., 2010; Tao and Hajri, 2011; Gao et al., 2014). All these results suggested that VLDLR is important in regulating fat accumulation (Clemente-Postigo et al., 2011; Go and Mani, 2012; Rankinen et al., 2006).

Most proteins that are completely transported across the cytoplasmic membrane are synthesised with an amino-terminal signal peptide. Signal peptides directly deliver the protein to the proper organelle. In this study, the amplified fragment was 374 bp, containing the 243-bp region of 5'-UTR and the 131-bp region of the signal peptide, which encoded 37 amino acids of the signal peptide region. These mutations in 5'-UTR and signal peptide of the duck *VLDLR* gene implied that this region had effects on duck growth and fat deposition by gene translational efficiency and/or VLDLR synthesis, secretion and position. However, more duck populations and gene expression analysis are required to further confirm our results. Moreover, our results demonstrated that these polymorphisms have a functional effect on fat accumulation. In our previous study, eight SNPs were identified in the VLDLR epidermal growth factor (EGF) precursor homologous domain, and a significant association was revealed between the homologous domain of the VLDLR EGF precursor and the AFP in ducks (Zhao et al., 2015). In the present study, we obtained similar results through the polymorphism of signal peptide regions. Our research further demonstrated that VLDLR can be considered as a candidate gene for abdominal fat deposition. The growth pattern in ducklings shows high and rapid-growth rate during the initial raising period. The growth pattern in ducks approaches the Gompertz curve (Maruyama et al., 2001). The age at the inflection point, at which the curvature changes sign, is approximately 5 weeks for Gaoyou ducks. Ducks grow less rapidly after the inflection point. No significant difference ($P > 0.05$) in body weight was observed across different diplotypes from birth to the age of 5 weeks. A possible reason for this phenomenon was that Gaoyou ducks were in a rapid-growth phase, so body weight differences were not noticeable even though different diplotypes were evaluated.

Clearly, the *VLDLR* gene is worth studying further as a potential candidate gene involved in growth and fat deposition and for the genotyping of these alleles in other resource populations and in elite lines of commercial duck breeders.

In summary, as far as we know, this is the first study to investigate the associations of SNPs in the *VLDLR* gene 5'-UTR and signal peptide region with duck growth and abdominal fat deposition. The results suggest that the *VLDLR* gene plays an important role in the regulation of body weight and abdominal fat deposition and may be used as a potential marker in the molecular MAS program during duck breeding. Further investigations with different duck populations and larger sample sizes are needed to confirm this point.

Competing interests. The authors declare that they have no conflict of interest.

Acknowledgements. This work was supported by the Priority Academic Program Development of Jiangsu Higher Education Institutions and the Jiangsu Co-innovation Center for Prevention and Control of Important Animal Infectious Diseases and Zoonoses, the National Nature Science Foundation of China (201010766), the National Science Foundation of Jiangsu Province (202010188), the China Postdoctoral Science Foundation Funded Project and Postdoctoral Science Foundation Funded Project of Jiangsu Province (137070149), the Postgraduate Degree Innovation Projects of Jiangsu Province (SJLX15_0677), and the Students' Academic and Scientific Innovation Fund funded project of Yangzhou University (X2015713, X2015710).

Edited by: Steffen Maak

References

Ardlie, K. G., Kruglyak, L., and Seielstad, M.: Patterns of linkage disequilibrium in the human genome, Nat. Rev. Genet., 3, 299–309, 2002.

Boycott, K. M., Flavelle, S., Bureau, A., Glass, H. C., Fujiwara, T. M., Wirrell, E., Davey, K., Chudley, A. E., Scott, J. N., McLeod, D. R., and Parboosingh, J. S.: Homozygous deletion of the very low density lipoprotein receptor gene causes autosomal recessive cerebellar hypoplasia with cerebral gyral simplification, Am. J. Hum. Genet., 77, 477–483, 2005.

Brockmann, G. A., Haley, C. S., Renne, U., Knott, S. A., and Schwerin, M.: Quantitative trait loci affecting body weight and fatness from a mouse line selected for extreme high growth, Genetics, 150, 369–381, 1998.

Castro, N. P., Euclydes, V. V., Simões, F. A., Vaz-de-Lima, L. R., De Brito, C. A., Luzia, L. A., Devakumar, D., and Rondó, P. H.: The Relationship between Maternal Plasma Leptin and Adiponectin Concentrations and Newborn Adiposity, Nutrients, 9, E182, https://doi.org/10.3390/nu9030182, 2017.

Clemente-Postigo, M., Queipo-Ortuno, M. I., Fernandez-Garcia, D., Gomez-Huelgas, R., and Tinahones, F. J.:

Adipose tissue gene expression of factors related to lipid processing in obesity, PLoS ONE 6, e24783, https://doi.org/10.1371/journal.pone.0024783, 2011.

Crawford, D. C., Nord, A. S., Badzioch, M. D., Ranchalis, J., McKinstry, L. A., Ahearn, M., Bertucci, C., Shephard, C., Wong, M., Rieder, M. J., Schellenberg, G. D., Nickerson, D. A., Heagerty, P. J., Wijsman, E. M., and Jarvik, G. P.: A common VLDLR polymorphism interacts with APOE genotype in the prediction of carotid artery disease risk, J. Lipid Res., 49, 588–596, 2008.

Eppig, J. T., Blake, J. A., Bult, C. J., Kadin, J. A., and Richardson, J. E.: The Mouse Genome Database (MGD): facilitating mouse as a model for human biology and disease, Nucleic Acids Res., 43, 726–736, 2015.

Frykman, P. K., Brown, M. S., Yamamoto, T., Goldstein, J. L., and Herz, J.: Normal plasma lipoproteins and fertility in gene-targeted mice homozygous for a disruption in the gene encoding very low density lipo-protein receptor, P. Natl. Acad. Sci. USA, 92, 8453–8457, 1995.

Gao, Y., Shen, W., and Lu, B.: Upregulation of hepatic VLDLR via PPARα is required for the triglyceride-lowering effect of fenofibrate, J. Lipid Re., 55, 1622–1633, 2014.

Gibbs, R., Belmont, J., Hardenbol, P., Willis, T., Yu, F., and Yang, H.: The International HapMap Consortium, The International HapMap Project, Nature, 426, 789–796, 2003.

Go, G. W. and Mani, A.: Low-density lipoprotein receptor (LDLR) family orchestrates cholesterol homeostasis, Yale J. Biol. Med. Mar., 85, 19–28, 2012.

Goudriaan, J. R., Tacken, P. J., Dahlmans, V. E., Gijbels, M. J., van Dijk, K. W., Havekes, L. M., and Jong, M. C.: Protection from obesity in mice lacking the VLDL receptor, Arterioscl. Throm. Vas., 21, 1488–1493, 2001.

Goudriaan, J. R., Espirito Santo, S. M. S., Voshol, P. J., Teusink, B., van Dijk, K. W., and van Vlijmen, B. M. J.: The VLDL receptor plays a major role in chylomicron metabolism by enhancing LPL-mediated triglyceride hydrolysis, J. Lipid Res., 45, 1475–1481, 2004.

Gwang-Woong, G. and Arya, M.: Low-density lipoprotein receptor (LDLR) family orchestrates cholesterol homeostasis, Yale J. Biol. Med., 85, 19–28, 2012.

Kim, O. Y., Lee, S. M., Chung, J. H., Do, H. J., Moon, J., and Shin, M. J.: Arginase I and the very low-density lipoprotein receptor are associated with phenotypic biomarkers for obesity, Nutrition, 28, 635–639, 2012.

Kunej, T., Jevsinek Skok, D., Zorc, M., Ogrinc, A., Michal, J. J., Kovac, M., and Jiang, Z.: Obesity gene atlas in mammals, J. Genomics, 1, 45–55, 2012.

Kyosseva, S. V., Chen, L., Seal, S., and McGinnis, J. F.: Nanoceria inhibit expression of genes associated with inflammation and angiogenesis in the retina of Vldlr null mice, Exp. Eye Res., 116, 63–74, 2013.

Maruyama, K., Vinyard, B., and Akbar, M. K.: Growth curve analyses in selected duck lines, Brit. Poultry Sci., 42, 574–582, 2001.

Nguyen, A., Tao, H., Metrione, M., and Hajri, T.: Very low density lipoprotein receptor (VLDLR) expression is a determinant factor in adipose tissue inflammation and adipocyte-macrophage interaction, J. Biol. Chem., 289, 1688–1703, 2014.

Ning, T., Zou, Y., Yang, M., Lu, Q., Chen, M., Liu, W., Zhao, S., Sun, Y., Shi, J., Ma, Q., Hong, J., Liu, R., Wang, J., and Ning, G.: Genetic interaction of DGAT2 and FAAH in the development of

human obesity, Endocrine, https://doi.org/10.1007/s12020-017-1261-1, online first, 2017.

Rankinen, T., Zuberi, A., Chagnon, Y. C., Weisnagel, S. J., Argyropoulos, G., Walts, B., Perusse, L., and Bouchard, C.: The human obesity gene map: the 2005 update, Obes. Res., 14, 529–644, 2006.

Shen, X., Steyrer, E., Retzek, H., Sanders, E. J., and Schneider, W. J.: Chicken oocyte growth: receptor-mediated yolk deposition, Cell Tissue Res., 272, 459–471, 1993.

Suwa, A., Yoshino, M., Yamazaki, C., Naitou, M., Fujikawa, R., and Matsumoto, S. I.: Rmi1 deficiency in mice protects from diet and genetic-induced obesity, Febs Journal, 277, 677–686, 2010.

Tacken, P. J., Beer, F. D., Vark, L. C., Havekes, L. M., Hofker, M. H., and Willems Van Dijk, K.: Very-low-density lipoprotein binding to the apolipoprotein E receptor 2 is enhanced by lipoprotein lipase, and does not require apolipoprotein E, Biochem. J., 347, 357–361, 2000.

Takahashi, S., Suzuki, J., Kohno, M., Oida, K., Tamai, T., Miyabo, S., Yamamoto, T., and Nakai, T.: Enhancement of the binding of triglyceride-rich lipoproteins to the very low density lipoprotein receptor by apolipoprotein E and lipoprotein lipase, J. Biol. Chem., 270, 15747–15754, 1995.

Takahashi, S., Sakai, J., Fujino, T., Hattori, H., Zenimaru, Y., Suzuki, J., Miyamori, I., and Yamamoto T. T.: The very low-density lipoprotein (VLDL) receptor: characterization and functions as a peripheral lipoprotein receptor, J. Atheroscler. Thromb., 11, 200–208, 2004.

Takazawa, T., Yamauchi, T., Tsuchida, A., Takata, M., Hada, Y., Iwabu, M., Okada-Iwabu, M., Ueki, K., and Kadowaki, T.: Peroxisome proliferator-activated receptor gamma agonist rosiglitazone increases expression of very low density lipoprotein receptor gene in adipocytes, J. Biol. Chem., 284, 30049–30057, 2009.

Tao, H. and Hajri, T.: Very low density lipoprotein receptor promotes adipocyte differentiation and mediates the proadipogenic effect of peroxisome proliferator-activated receptor gamma agonists, Biochem. Pharmacol., 82, 1950–1962, 2010.

Tao, H., Aakula, S., Abumrad, N. N., and Hajri, T.: Peroxisome proliferator-activated receptor-gamma regulates the expression and function of very-low-density lipoprotein receptor, Am. J. Physiol. Endocrinol. Metab., 298, E68–E79, 2010.

Willnow, T. E.: The low-density lipoprotein receptor gene family: multiple roles in lipid metabolism, J. Mol. Med., 77, 306–315, 1998.

Wang, C., Li, S. J., Yu, W. H., Xin, Q. W., Li, C., Feng, Y. P., Peng, X. L., and Gong, Y. Z.: Cloning and Expression Profiling of the VLDLR Gene Associated with Egg Performance in Duck (Anas Platyrhynchos), Genet. Sel. Evol., 43, 144–144, 2011.

Wu, Y., Pi, J. S., Pan, A. L., Du, J. P., Shen, J., Pu, Y. J., and Liang, Z. H.: Two novel linkage SNPs of VLDLR gene intron 11 are associated with laying traits in two quail populations, Arch. Anim. Breed., 58, 1–6, https://doi.org/10.5194/aab-58-1-2015, 2015.

Yuan, G., Liu, Y., Sun, T., Xu, Y., Zhang, J., Yang, Y., Zhang, M., Cianflone, K., and Wang, D. W.: The therapeutic role of very low-density lipoprotein receptor gene in hyperlipidemia in type 2 diabetic rats, Hum. Gene Ther., 22, 302–312, 2011.

Zhao, N., Lin, S., Wang, Z., and Zhang, T.: VLDLR gene polymorphism associated with abdominal fat in Gaoyou domestic duck breed, Czech J. Anim. Sci., 60, 178–184, 2015.

16

SIRT1 gene polymorphisms associated with carcass traits in Luxi cattle

Guifen Liu[1,2,*], **Hongbo Zhao**[1,2,*], **Xiuwen Tan**[1,2], **Haijian Cheng**[1,2], **Wei You**[1,2], **Fachun Wan**[1,2], **Yifan Liu**[1,2], **Enliang Song**[1,2], **and Xiaomu Liu**[1,2]

[1]Shandong Key Lab of Animal Disease Control and Breeding, Sangyuan Road 8 Number, Ji'nan City, Shandong Province, 250100, China
[2]Institute of Animal Science and Veterinary Medicine, Shandong Academy of Agricultural Sciences, Sangyuan Road 8 Number, Ji'nan City, Shandong Province, 250100, China
*These authors contributed equally to this work.

Correspondence to: Guifen Liu (liuguifen126@126.com), Enliang Song (enliangs@126.com), and Xiaomu Liu (xmliu2002@163.com)

Abstract. *SIRT1* is the gene that codes for Sirtuin 1, an NAD (nicotinamide adenine dinucleotide)-dependent class III histone deacetylase. This gene plays a key role in adipose tissue and muscle development in animals. Chinese Luxi cattle ($n = 169$) were selected to identify *SIRT1* SNPs (single nucleotide polymorphisms) and investigate the relationship of these SNPs with carcass traits. Five SNPs (g.-382G > A, g.-274C > G, g.17324T > C, g.17379A > G, and g.17491G > A) were identified by direct sequencing. SNPs g.-382G > A and g.-274C > G were located within the promoter region of this gene. SNP g.-382G > A was significantly associated with dressing percentage, meat percentage, and striploin and ribeye weights, and the g.-274C > G polymorphism had a strong effect on carcass, tenderloin, and high rib weights in Luxi cattle. These findings will provide possible clues for the biological roles of *SIRT1* underlying beef cattle carcass traits.

1 Introduction

Sirtuin 1, also known as NAD (nicotinamide adenine dinucleotide)-dependent deacetylase Sirtuin 1, is an NAD+-dependent protein deacetylase; it has many established protein substrates and is thought to regulate an impressive list of biological functions (McBurney et al., 2013). Sirtuin 1 has an important function in endocrine signaling, specifically in glucose and fat metabolism in mammals (Zillikens et al., 2009; Picard et al., 2004). Increased hepatic *SIRT1* activity enhances gluconeogenesis and inhibits glycolysis (Rodgers et al., 2005; Zillikens et al., 2009). In the pancreatic β cells, *SIRT1* positively regulates insulin secretion in response to glucose (Bordone et al., 2006). It is also involved in cellular differentiation, apoptosis, metabolism, and aging (Shakibaei et al., 2012; Sasaki et al., 2014; Luna et al., 2013; Gueguen et al., 2014).

Previous studies have suggested that SNPs (single nucleotide polymorphisms) within *SIRT1* increase the risk of obesity (De Oliveira et al., 2012), type 2 diabetes, and Parkinson's disease (Schug and Li, 2011; Inamori et al., 2013; Shiota et al., 2012; Civelek et al., 2013; Dong et al., 2011; Rai et al., 2012; Figarska et al., 2013). In adipose tissue, Sirtuin 1 inhibits fat storage and increases lipolysis via the repression of peroxisome proliferator-activated receptor-γ (PPAR-γ). PPAR-γ is a key regulator in adipogenesis and fat storage, controlling the expression of many adipocyte-specific genes (Picard et al., 2004). These studies suggested that Sirtuin 1 is a key regulator of whole-body energy balance and plays a role in human health (Sasaki et al., 2014).

The bovine *SIRT1* gene, which includes nine exons on chromosome 28, is highly expressed in the liver and adipose tissue (Ghinis-Hozumi et al., 2011). *SIRT1* may play an important role in the development of bovine adipose tissue in vivo. Although *SIRT1*, forkhead box O1 (*FOXO1*), and

Table 1. PCR primers and conditions for identification of SNPs in *SIRT1* (NM_001192980).

SNPs	Primers (5′–3′)	Genotyping methods	Temp (°C)	Restriction enzyme	Genotype pattern (bp)
g.G-382A	F: GTTTAGCCTTAACGCCGTTCAGGAAATT* R: GTCTTTCAGAGTCTTCAAATCAGTGCCC	ACRS	56	Vsp I	166/136 + 30
g.C-274G	F: GTATAGTCCACGGGGTTACAG R: CCAAACTTGTCTTTCAGAGTC	PCR-RFLP	59	Sma I	273/235 + 38
g.T17324C	F (inner): GTTAGTAAACTTCAGAATTGCTTTgCT R (inner): TAATTTTTCCTACAAAACTAATATAAgGG F (outer): CTAGATGCTTTGAGATTGTCGTGTGTTG R (outer): ACTAAGCACACTATTTGAAACTTGAGTG	T-ARMS-PCR	58		550 bp (outer) 270 bp (alleteT) 330 bp (alleteC)
g.A17379G	F: TTCCAACCATCTCTTTGTCAC R: AATAATAAGGCTTAATCTGAATT*	ACRS	57	EcoR I	235/211 + 24
g.G17491A	F (inner): AAATACTGGCCTCAACTCTTAATTtTA R (inner): AAATCCAAATTAACATCTGACATTTtAC F (outer): TACTTCGCAACTATACTCAGAACATAGA R (outer): GTTTGATCTCTAGGTTAGGAAGATCCT	T-ARMS-PCR	58		477 bp (outer) 296 bp (alleteG) 236 bp (alleteA)

Note: * purposeful mismatch was introduced in the sequence to create a restriction site.

PPAR-γ expression appear to be nonlinear during the stages of preadipocyte differentiation, these genes play an important role during bovine adipocyte development in Lilu cattle (Liu et al., 2014). The study examined the variations of *SIRT1* in Luxi beef cattle by identified SNPs, and explored possible associations between *SIRT1* variants and carcass traits. These molecular markers will provide some theoretical basis for improving cattle carcass characteristics.

2 Materials and methods

2.1 Animals and genomic DNA isolation

In the Shandong province, 169 Chinese Luxi cattle were reared in same conditions. The animals were slaughtered at the age of 24 months according to Chinese national law (China Administration Rule of Laboratory Animal; Operating Procedure of Cattle Slaughtering GB/T 19477-2004). Carcass traits were recorded and blood samples were collected. Genomic DNA containing nucleotides from leukocytes was isolated from blood samples and stored at $-20\,°C$ following standard procedures (QIAamp DNA Blood Mini Kit, Qiagen, Germany).

2.2 SNP detection

We used the primer sequences from M. X. Li et al. (2013) (Table 1) to detect SNPs. The 30 μL reaction volume included 15 μL Taq $2 \times$ PCR MasterMix (QUANSHIJIN, Beijing, China), 3 μL DNA template ($20\,ng\,μL^{-1}$), 9.6 μL ddH$_2$O, and 1.2 μL of each primer ($10\,pmol\,μL^{-1}$).

The g.-382G > A and g.17379A > G polymorphisms were genotyped using the amplification-created restriction site (ACRS) method (Figarska et al., 2013). The tetra-primer amplification refractory mutation system PCR (T-ARMS-PCR) was carried out to genotype SNPs g.17324T > C and g.17491G > A (Haliassos et al., 1989). The PCR reactions were performed in a total volume of 10 μL, containing 10 pmol of each of the inner primers, 1 pmol of each of the outer primers, 200 mM of each dNTP, 2 mM of MgCl$_2$, $1 \times$ PCR buffer, 50 ng of DNA, and 0.2 U of Taq DNA polymerase (MBI, Fermentas, Waltham, MA, USA). To increase the specificity of the reaction, a touchdown profile was followed.

2.3 Statistical analysis

DNA sequences were assembled and aligned for mutation analysis with DNASTAR (DNAS Inc., Madison, WI, USA). Allele and genotype frequencies were directly calculated. Heterozygosity, effective number of alleles, and polymorphic information content (PIC) were estimated based on Botstein et al. (1980). A chi-square test assessed conformance with Hardy–Weinberg equilibrium (HWE). Association of genotype with performance traits was analyzed with the general linear model (GLM) procedure of SPSS 16.0.

3 Results

3.1 Identification of SNPs

Five SNPs were detected in the exons, flanking introns, and promoter sequences of *SIRT1*, including four transi-

Table 2. Genotypic and allelic frequencies (%), value of χ^2 test, and diversity parameters of the bovine *SIRT1* gene.

SNPS	Genotype	Number	GF	Allele	AF	χ^2 (HWE)	He	Hom	PIC	Ne
g.-382G > A	GG	96	0.568	G	0.722	9.230	0.402	0.598	0.321	1.671
	GA	52	0.308	A	0.278					
	AA	21	0.124							
g.-274C > G	CC	108	0.639	C	0.769	12.018	0.355	0.645	0.292	1.550
	CG	44	0.260	G	0.231					
	GG	17	0.101							
g.17324T > C	TT	113	0.669	T	0.805	3.031	0.314	0.686	0.265	1.458
	TC	46	0.272	C	0.195					
	CC	10	0.059							
g.17379A > G	AA	122	0.722	A	0.831	8.121	0.280	0.720	0.241	1.390
	AG	37	0.219	G	0.169					
	GG	10	0.059							
g.17491G > A	GG	139	0.823	G	0.888	20.434	0.200	0.800	0.180	1.249
	AG	22	0.130	A	0.112					
	AA	8	0.047							

Note: GF: genotypic frequency; AF: allelic frequency; χ^2 (HWE): Hardy–Weinberg equilibrium χ^2 value; χ^2 0.05 (df $= 2) = 5.99$, χ^2 0.01 (df $= 2) = 9.21$; He: gene heterozygosity; Hom: gene homozygosity; PIC: polymorphism information content; Ne: effective allele number.

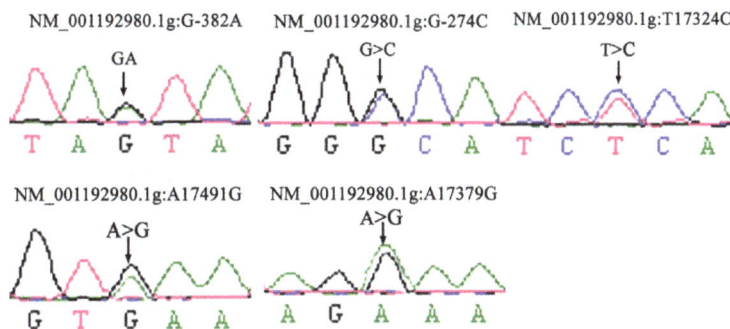

Figure 1. Schematic representation of the *SIRT1* gene with the localization of the five identified SNPs.

tions (G/A at g.-382G > A, T/C at g.17324T > C, A/G at g.17379A > G, and G/A at g.17491G > A) and one transversion (g.-274C > G) (Fig. 1). The nomenclature adopted for the SNPs was based on the convention described by the Human Genome Variation Society (Den Dunnen et al., 2016). No SNPs were found in the coding sequence from the set of animals used in this study. SNPs g.-382G > A and g.-274C > G were located in the promoter region and could cause disruption of several transcription factor binding sites, as predicted by MatInspector release 8.0 (Cakir et al., 2009). The other three SNPs were found in intron five. All five SNPs were successfully genotyped.

Genotypic and allelic frequencies, value of χ^2 test, and PIC of the bovine *SIRT1* gene have been shown in Table 2. The g.-382G > A, g.-274C > G, and g.17324T > C loci had moderate polymorphism and thus genetic diversity, which implies that these SNPs have a potential for selection. The

g.17379A > G and g.17491G > A loci had low genetic diversity and selection potential.

3.2 The relationship between SNPs and carcass traits

Significant differences between genotypes and carcass traits of beef cattle are shown in Table 3. In g.-382G > A, AA genotypes have a more significant difference ($P < 0.05$) in dressing percentage, meat percentage, and striploin than the GG and GA genotypes; however, there is no difference in ribeye. In g.-274C > G, AA genotypes have a more significant difference in carcass, tenderloin and high rib weight than GG and GC genotypes. However, no differences between SNPs and carcass traits were found when focusing on 17379A > G and g.17491G > A.

Based on these results, we predicted potential differential transcription factor (TF) binding sites according to the presence of different alleles using MatInspector Release 8.0. At

Table 3. Significant SNP, genotype, and carcass trait associations.

SNPs	Traits	Genotypes (mean ± SE)			P value
		GG	GA	AA	
g.-382	Dressing percentage/%	53.184 ± 1.258[a]	53.627 ± 0.951[a]	49.880 ± 1.165[b]	0.043
	Meat percentage/%	47.990 ± 1.632[a]	48.008 ± 1.234[a]	42.824 ± 1.511[b]	0.029
	Striploin/kg	8.113 ± 0.449[a]	6.898 ± 0.567[a]	6.367 ± 0.634[b]	0.041
	Ribeye/kg	8.618 ± 0.751	6.174 ± 0.312	5.029 ± 0.394	0.039
g.-274		GG	GC	CC	
	Carcass weight/kg	326.118 ± 12.910[a]	300.455 ± 16.049[a]	269.778 ± 17.743[b]	0.015
	Tenderloin/kg	4.47 ± 0.324[a]	4.213 ± 0.293[a]	3.608 ± 0.324[b]	0.039
	High rib/kg	10.605 ± 0.448[a]	9.579 ± 0.557[a]	8.85 ± 0.616[b]	0.027
g.17379		GG	GA	AA	
	Striploin/kg	7.703 ± 0.411	8.058 ± 0.232	6.968 ± 0.300	0.036
g.17324		TT	TC	CC	
	Bone weight/kg	18.269 ± 1.858	19.832 ± 2.220	12.248 ± 2.304	0.048

g.-382G > A, a myocyte-specific enhancer factor 2 (MEF2) binding site was generated on substitution to the A allele. At g.-274C > G, in the presence of the C allele, a binding site for a CDE (cell-cycle-dependent element) was generated, whereas the same binding site was abolished in the presence of the G allele.

4   Discussion

There are several variants associated with body mass index and risk of obesity in human *SIRT1* gene (Zillikens et al., 2009). Recent studies have found possibly useful SNPs in the *SIRT1* gene and explored the relationships between these SNPs and ultrasound-measured carcass traits in Qinchuan cattle (Gui et al., 2015). We identified five SNPs in bovine *SIRT1* and estimated the extent of associations between these SNPs and carcass traits in Chinese Luxi cattle.

Association analysis showed that SNP g.17379A > G was significantly associated with tenderloin, striploin, and ribeye and that polymorphisms with g.17324T > C had a strong effect on bone weight (these effects became non-significant following the Bonferroni correction). This SNP did not result in changes in amino acids. Such associations may be a result of linkage disequilibrium between *SIRT1* and other genes on the same chromosome that have a significant effect on these carcass traits. It is interesting to note that the SNP g.17379A > G was severely out of HWE. Subsequent sequencing showed that this was not due to technical error. We considered two possible explanations: (1) Luxi cattle have experienced high selection pressure. Artificial selection led to the loss of non-favored alleles. (2) The analyzed breed has an insufficiently large population size.

Five SNPs (g.-382G > A, g.-274C > G, g.17324T > C, g.17379A > G, and g.17491G > A) were identified in the Luxi cattle and are similar to previous research results (Ye et al., 2001; M. Li et al., 2013). The role of *SIRT1* as an inhibitor of adipogenesis and the recent demonstration of its involvement in white adipose tissue "browning" (M. X. Li et al., 2013) as well as the roles played by *SIRT1* in muscle metabolism (Qiang et al., 2012) have motivated us to further investigate the effects of the identified SNPs on beef cattle carcass traits. Our results showed that SNP g.-382G > A was significantly associated with dressing percentage, meat percentage, and striploin and ribeye weights, and g.-274C > G polymorphism had a strong effect on carcass, tenderloin, and high rib weights in Luxi cattle.

At g.-382G > A, a MEF2 binding site was generated on substitution to the A allele. At g.-274C > G, in the presence of the C allele, a binding site for a CDE was generated, whereas the same binding site was abolished in the presence of the G allele. These indicated that g.-382G > A and g.-274C > G polymorphisms might affect the binding affinity of the surrounding sequences with TF and further influence the activity of the *SIRT1* promoter that was associated with growth trait regulation.

Carcass traits are regulated by multiple genes and are influenced by interactions among them; thus, the effects of these SNPs should be further validated before they can be incorporated into beef cattle breeding practices.

Competing interests. The authors declare that they have no conflict of interest.

Acknowledgements. This study was supported by the Young Talents Training Program of Shandong Academy of Agricultural Science, National Natural Science Foundation of China (No. 31402098), the Sustentative Research Project of China Ministry of Science and Technology (2015BAD03B04), Breeding New Varieties Projects of Transgenic Organisms (2016ZX08007-002).

Edited by: S. Maak

References

Bordone, L., Motta, M. C., Picard, F., Robinson, A., Jhala, U. S., Apfeld, J., McDonagh, T., Lemieux, M., McBurney, M., Szilvasi, A., Easlon, E. J., Lin, S. J., and Guarente, L.: Sirt1 regulates insulin secretion by repressing UCP2 in pancreatic beta cells, PLoS Biol., 4, e1002346, doi:10.1371/journal.pbio.1002346, 2006.

Botstein, D., White, R. L., Skolnick, M., and Davis, R. W.: Construction of a genetic linkage map in man using restriction fragment length polymorphisms, Am. J. Hum. Genet., 32, 314–331, 1980.

Cakir, I., Perello, M., Lansari, O., Messier, N. J., Vaslet, C. A., and Nillni, E. A.: Hypothalamic Sirt1 Regulates Food Intake in a Rodent Model System, PloS One, 4, 8322, doi:10.1371/journal.pone.0008322, 2009.

Civelek, M., Hagopian, R., Pan, C., Che, N., Yang, W. P., Kayne, P. S., Saleem, N. K., Cederberg, H., Kuusisto, J., Gargalovic, P. S., Kirchgessner, T. G., Laakso, M., and Lusis, A. J.: Genetic regulation of human adipose microRNA expression and its consequences for metabolic traits, Hum. Molec. Gen., 22, 3023–3037, 2013.

Den Dunnen, J. T., Dalgleish, R., Maglott, D. R., Hart, R. K., Greenblatt, M. S., McGowan-Jordan, J., Roux, A. F., Smith, T., Antonarakis, S. E., and Taschner, P. E.: HGVS recommendations for the description of sequence variants: 2016 update, Hum. Mutat., 37, 564-569, 2016.

De Oliveira, R. M., Sarkander, J., Kazantsev, A. G., and Outeiro, T. F.: SIRT2 as a Therapeutic Target for Age-Related Disorders, Front. Pharmacol., 3, 82, doi:10.3389/fphar.2012.00082, 2012.

Dong, Y., Guo, T., Traurig, M., Mason, C. C., Kobes, S., Perez, J., Knowler, W. C., Bogardus, C., Hanson, R. L., and Baier, L. J.: SIRT1 is associated with a decrease in acute insulin secretion and a sex specific increase in risk for type 2 diabetes in Pima Indians, Molec. Gen. Metabol., 104, 661–665, 2011.

Figarska, S. M., Vonk, J. M., and Boezen, H. M.: SIRT1 Polymorphism, Long-Term Survival and Glucose Tolerance in the General Population, PloS One, 8, 58636, doi:10.1371/journal.pone.0058636, 2013.

Ghinis-Hozumi, Y., Gonzalez-Gallardo, A., Gonzalez-Davalos, L., Antaramian, A., Villarroya, F., Shimada, A., Varela-Echavarria, A., and Mora, O.: Bovine sirtuins: initial characterization and expression of sirtuins 1 and 3 in liver, muscle, and adipose tissue, J. Anim. Sci., 89, 2529–2536, 2011.

Gueguen, C., Palmier, B., Plotkine, M., Marchand-Leroux, C., and Bessson, V. C.: Neurological and histological consequences induced by in vivo cerebral oxidative stress: evidence for beneficial effects of SRT1720, a sirtuin 1 activator, and sirtuin 1-mediated neuroprotective effects of poly(ADP-ribose) polymerase inhi-

bition, PloS One, 9, 87367, doi:10.1371/journal.pone.0087367, 2014.

Gui, L., Hao, R., Zhang, Y., Zhao, X., and Zan, L.: Haplotype distribution in the class I sirtuin genes and their associations with ultrasound carcass traits in Qinchuan cattle (Bos taurus), Molec. Cell. Probes, 29, 102–107, 2015.

Haliassos, A., Chomel, J. C., Tesson, L., Baudis, M., Kruh, J., Kaplan, J. C., and Kitzis, A.: Modification of enzymatically amplified DNA for the detection of point mutations, Nucl. Acids Res., 17, 3606, 1989.

Inamori, T., Goda, T., Kasezawa, N., and Yamakawa-Kobayashi, K.: The combined effects of genetic variation in the SIRT1 gene and dietary intake of n-3 and n-6 polyunsaturated fatty acids on serum LDL-C and HDL-C levels: a population based study, Lipids Health Dis., 12, 1–8, 2013.

Li, M., Sun, X., Hua, L., Lai, X., Lan, X., Lei, C., Zhang, C., Qi, X., and Chen, H.: SIRT1 gene polymorphisms are associated with growth traits in Nanyang cattle, Molec. Cell. Probes, 27, 215–220, 2013.

Li, M. X., Sun, X. M., Zhang, L. Z., Wang, J., Huang, Y. Z., Sun, Y. J., Hu, S. R., Lan, X. Y., Lei, C. Z., and Chen, H.: A novel c.-274C > G polymorphism in bovine SIRT1 gene contributes to diminished promoter activity and is associated with increased body size, Animal Gen., 44, 584–587, 2013.

Liu, X., Liu, G., Tan, X., Zhao, H., Cheng, H., Wan, F., Wu, N., and Song, E.: Gene expression profiling of SIRT1, FoxO1, and PPARgamma in backfat tissues and subcutaneous adipocytes of Lilu bulls, Meat Science, 96, 704–711, 2014.

Luna, A., Aladjem, M. I., and Kohn, K. W.: SIRT1/PARP1 crosstalk: connecting DNA damage and metabolism, Genome Integrity, 4, 6, doi:10.1186/2041-9414-4-6, 2013.

McBurney, M. W., Clark-Knowles, K. V., Caron, A. Z., and Gray, D. A.: SIRT1 is a Highly Networked Protein That Mediates the Adaptation to Chronic Physiological Stress, Genes Cancer, 4, 125–134, 2013.

Picard, F., Kurtev, M., Chung, N., Topark-Ngarm, A., Senawong, T., Machado de oliveira, R., Leid, M., Mcburney, M. W., and Guarente, L.: Sirt1 promotes fat mobilization in white adipocytes by repressing PPAR-gamma, Nature, 429, 771–776, 2004.

Picard, F., Kurtev, M., Chung, N., Topark-Ngarm, A., Senawong, T., Machado De Oliveira, R., Leid, M., McBurney, M. W., and Guarente, L.: Sirt1 promotes fat mobilization in white adipocytes by repressing PPAR-gamma, Nature, 429, 771–776, 2004.

Qiang, L., Wang, L., Kon, N., Zhao, W., Lee, S., Zhang, Y., Rosenbaum, M., Zhao, Y., Gu, W., Farmer, S. R., and Accili, D.: Brown remodeling of white adipose tissue by SirT1-dependent deacetylation of Ppargamma, Cell, 150, 620–632, 2012.

Rai, E., Sharma, S., Kaul, S., Jain, K., Matharoo, K., Bhanwer, A. S., and Bamezai, R. N.: The interactive effect of SIRT1 promoter region polymorphism on type 2 diabetes susceptibility in the North Indian population, PloS One, 7, e48621, doi:10.1371/journal.pone.0048621, 2012.

Rodgers, J. T., Lerin, C., Haas, W., Gygi, S. P., Spiegelman, B. M., and Puigserver, P.: Nutrient control of glucose homeostasis through a complex of PGC-1alpha and SIRT1, Nature 434, 113–118, 2005.

Sasaki, T., Kikuchi, O., Shimpuku, M., Susanti, V. Y., Yokota-Hashimoto, H., Taguchi, R., Shibusawa, N., Sato, T., Tang, L., Amano, K., Kitazumi, T., Kuroko, M., Fujita, Y., Maruyama, J.,

Lee, Y. S., Kobayashi, M., Nakagawa, T., Minokoshi, Y., Harada, A., Yamada, M., and Kitamura, T.: Hypothalamic SIRT1 prevents age-associated weight gain by improving leptin sensitivity in mice, Diabetologia, 57, 819–831, 2014.

Schug, T. T. and Li, X.: Sirtuin 1 in lipid metabolism and obesity, Ann. Med., 43, 198–211, 2011.

Shakibaei, M., Shayan, P., Busch, F., Aldinger, C., Buhrmann, C., Lueders, C., and Mobasheri, A.: Resveratrol Mediated Modulation of Sirt-1/Runx2 Promotes Osteogenic Differentiation of Mesenchymal Stem Cells: Potential Role of Runx2 Deacetylation, PloS One, 7, 35712, doi:10.1371/journal.pone.0035712, 2012.

Shiota, A., Shimabukuro, M., Fukuda, D., Soeki, T., Sato, H., Uematsu, E., Hirata, Y., Rai, E., Sharma, S., Kaul, S., Jain, K., Matharoo, K., Bhanwer, A. S., and Bamezai, R. N.: The interactive effect of SIRT1 promoter region polymorphism on type 2 diabetes susceptibility in the North Indian population, PloS One, 7, 48621, doi:10.1371/journal.pone.0048621, 2012.

Ye, S., Dhillon, S., Ke, X., Collins, A. R., and Day, I. N.: An efficient procedure for genotyping single nucleotide polymorphisms, Nucl. Acids Res., 29, 88–96, 2001.

Zillikens, M. C., van Meurs, J. B., Rivadeneira, F., Amin, N., Hofman, A., Oostra, B. A., Sijbrands, E. J., Witteman, J. C., Pols, H. A., van Duijn, C. M., and Uitterlinden, A. G.: SIRT1 genetic variation is related to BMI and risk of obesity, Diabetes, 58, 2828–2834, 2009.

17

Nutritional modification of *SCD*, *ACACA* and *LPL* gene expressions in different ovine tissues

Katarzyna Ropka-Molik[1], **Jan Knapik**[2], **Marek Pieszka**[3], **Tomasz Szmatoła**[1], **and Katarzyna Piórkowska**[1]

[1]Department of Genomics and Animal Molecular Biology, National Research Institute of Animal Production, 32-083 Balice, Poland
[2]Department of Animal Genetics and Breeding, National Research Institute of Animal Production, 32-083 Balice, Poland
[3]Department of Animal Nutrition and Feed Science, National Research Institute of Animal Production, 32-083 Balice, Poland

Correspondence to: Katarzyna Ropka-Molik (katarzyna.ropka@izoo.krakow.pl)

Abstract. Fatty acid composition is one of the main factors affecting health benefits of food. Stearoyl-CoA desaturase 1 (*SCD*), acetyl-CoA carboxylase alpha (*ACACA*) and lipoprotein lipase (*LPL*) have been considered as the rate-limiting enzymes in the biosynthesis of different fatty acids critical in lipid metabolism. The aim of our study was the analysis of differences in expression profiles of three ovine genes related to lipid metabolism (*LPL*, *ACACA*, *SCD*) depending on feeding system and tissue type. The gene expression measurement was performed using a real-time PCR method on 60 old-type Polish Merino Sheep, which were divided into three feeding groups (I – complete pellet mixture, $n = 12$; II – complete mixture with addition of fresh grass, $n = 24$; III – complete mixture with addition of fresh red clover, $n = 24$). From all lambs, tissue samples – subcutaneous fat, perirenal fat and liver – were collected immediately after slaughter and *LPL*, *ACACA* and *SCD* expression was estimated based on two endogenous controls (*RPS2* – ribosomal protein S2; *ATP5G2* – H(+)-transporting ATP synthase). Our research indicated that supplementation of diet with an addition of fresh grass or red clover significantly ($P < 0.05$) decreased the expression of *SCD*, *ACACA* and *LPL* genes in fat tissue compared to standard complete pelleted mixture. On the other hand, the highest expression of *ACACA* was detected in liver tissue collected from sheep fed a diet with an addition of fresh red clover ($P < 0.05$). In turn, the highest expression of the *SCD* gene was detected in animals fed with grass supplementation ($P < 0.05$). Regardless of diet supplementation, the highest *SCD* transcript abundance was detected in perirenal fat, while *LPL* and *ACACA* expression was the highest in both perirenal and subcutaneous fat. The ability of nutrigenomic regulation of transcription of analyzed genes confirmed that these genes play a critical role in regulation of lipid metabolism processes in sheep and could be associated with fatty acid profiles in milk and meat.

1 Introduction

In sheep production, an increase and/or maintaining a satisfactory level of proper ratio of polyunsaturated (PUFAs) to saturated fatty acids (SFAs) in dairy products and meat (Wood et al., 2003) have become of the greatest importance. The modification of the fatty acid profile has been intended to increase the content of omega-3 fatty acids and conjugated linoleic acid (CLA) and is usually obtained by diet supple-

mentation. On the other hand, diet supplementation with expensive additions significantly increases costs of animal production, which can become unprofitable. It has been confirmed that modification of the expression of genes involved in lipid metabolism is associated with a change in lipid profiles in ovine tissues such as skeletal muscle, cardiac muscle and adipose tissue (Bonnet et al., 2000). Thus, due to the health-promoting properties of sheep products and relatively high costs of different diet supplements, research has been performed to establish genetic markers related to fatty acid composition.

The stearoyl-CoA desaturase 1 (*SCD*), acetyl-CoA carboxylase alpha (*ACACA*) and lipoprotein lipase (*LPL*) have been considered as the rate-limiting enzymes in the biosynthesis of different fatty acids critical in lipid metabolism. The SCD enzyme plays a key role in synthesis of mono-unsaturated fatty acids (MUFAs) from saturated fatty acids (SFAs; Ntambi et al., 2002), LPL participates in triglyceride-rich lipoprotein metabolism, and ACACA regulates biosynthesis of palmitic acids and long-chain fatty acids. All these enzymes are expressed ubiquitously throughout different tissues, but the highest expression has been detected in adipocyte tissue, liver and also in mammary glands during lactation (Jensen et al., 1991). Is has been established that the activity of the ACACA enzyme is associated with palmitic acid content in milk, which is considered as undesirable component increasing the risk of occurrence of cardiovascular diseases (Fattore and Fanelli, 2013). In turn, the *SCD* enzyme is responsible for biosynthesis of conjugated linoleic acid (CLA, isomer *cis*-9, *trans*-11), which is also secreted in milk and shows many beneficial effects on human health. Moreover, the ovine *SCD* gene is localized on chromosome 22 within a QTL (quantitative traits locus) region associated with the ratio of CLA to vaccenic acid in milk (Carta et al., 2008). The hydrolysis of the triacylglycerol (TAG) component of chylomicrons and the very low-density lipoprotein fraction of lipoproteins is catalyzed by LPL enzymes, which are also the primary source of long-chain fatty acids taken up by the mammary gland (Bernard et al., 2008). According to the function of LPL, ACACA and SCD enzymes, their genes are examined as candidate markers related to fatness traits in sheep.

Thus, the aim of our study was the analysis of differences in expression profiles of three genes related to lipid metabolism depending on feeding system and tissue type. Many reports show that the *LPL*, *ACACA* and *SCD* genes can be regulated by nutrition, but the exact mechanism of such modifications is still unknown. The obtained results will be useful to identify the molecular basis of lipid metabolism in ovine tissues and can be a base for further research focusing on determination of genes involved in fatty acid composition in milk and meat.

2   Material and method

2.1   Animals

The analysis was performed on 60 old-type male Polish Merino Sheep. All animals were maintained in the Experimental Station of the National Research Institute of Animal Production in Pawłowice under the same housing conditions. Lambs were divided into three feeding groups according to the system of feeding ad libitum with isoenergetic diets:

i. complete pelleted mixture according to INRA norm (IZ PIB-INRA, 2009, $n = 6$);

ii. complete mixture with addition of fresh pasture grass ($n = 12$);

iii. complete mixture with addition of fresh red clover ($n = 12$).

The detailed composition and nutritional value of complete mixture for lambs control group (diet I) and for experimental groups (diets II and III) were presented in Tables S1 and S2 in the Supplement. In diets II and III, the complete mixture was administered in an amount of 3 % of body weight, while supplementation of fresh grass/red clover was 3 kg per lamb administered after concentrates intake. To balance dietary fiber, animals received meadow hay as a supplement ($100\,\mathrm{g\,sheep^{-1}\,day^{-1}}$). The weight of refusals was daily controlled. The diets were formulated using a feed optimizing software tool (Zifo WIN v1.5; 2012). The composition and nutritional value of each diet and the exact feed conversion ratios (FCRs; $\mathrm{kg\,kg^{-1}}$) according to feeding system were presented in Table 1.

The feeding experiment was performed in a duplicate (the repetitions of experiment were performed within two years) in the same scheme and included 30 animals for each replicate. Animals were maintained in group pens (six lamb per pen) according to feeding system. Lambs were fattened from an average of 28 kg to a final body weight of 41.5–43 kg (day at slaughter was from 170 to 195, an average of 183 days), slaughter and dissected. From each animal, immediately after slaughter, tissue samples (liver, subcutaneous fat, perirenal fat) were collected into tubes with RNAlater® solution (Ambion, Thermo Fisher Scientific, USA) and stored at −20 °C. For fatty acids composition analysis, the longissimus dorsi muscle samples behind the last rib were collected and then stored at −20 °C. Muscle for expression measurements and fatty acid analysis was sampled from the same part of the longissimus dorsi muscle.

2.2   Genes expression measurement

Isolation of total RNA was carried out in total for 180 samples using a PureLink™ RNA Mini Kit (Ambion, Thermo Fisher Scientific, USA) according to the protocol. The quantity of the obtained RNA was estimated using a NanoDrop2000 spectrophotometer (Thermo Scientific, Wilmington, USA), while RNA degradation was checked on 2 % agarose gel. The cDNA was obtained from 250 ng of total RNA using a Maxima First Strand cDNA synthesis kit for RT-qPCR (Thermo Scientific) according to the attached protocol. The gene expression was measured on a 7500 real-time PCR system (Applied Biosystems, Thermo Fisher Scientific) using the TaqMan® Gene Expression Master Mix (Applied Biosystems, Thermo Fisher Scientific), primer sets and TaqMan probes. Primers and probes for target genes were de-

Table 1. The composition and nutritional value of each diet and the exact feed conversion ratios.

	Feeding groups		
	Diet I – complete mixture $n = 6$	Diet II – grass addition $n = 12$	Diet III – red clover addition $n = 12$
Composition of integrities, kg:			
complete mixture	1.65	0.92	0.95
meadow hay	0.08	–	–
meadow grass	–	2.37	–
red clover	–	–	2.44
Composition of integrities in the ration, kg:			
metabolizable energy, MJ	17.0	14.0	13.6
total protein, g	313.5	242.7	250.6
useful protein nBO, g	262.4	201.2	202.9
protein UDP, %	28.2	24.7	22.1
crude fat, g	43.2	39.4	37.9
ash content, g	133.3	109.8	113.4
crude fiber, g	144.4	133.4	121.1
calcium, g	19.3	15.9	18.8
phosphorus, g	8.4	6.9	6.7
sodium, g	6.4	2.7	2.9
magnesium, g	3.5	3.4	3.6
Fatty acid concentration (presented as % of total fatty acid content)			
linoleic acid (C18:2)	16.2	12.8	12.8
polyunsaturated fatty acid n-3 + n-6	18.6	16.0	15.7
feed conversion ratio (FCR) per 1 kg of gain	5.14	3.34	3.29
Feed unit for maintenance and meat production UFV	4.16	5.56	5.73

signed using Primer Express 3.0 software, while primers and probes for the two endogenous controls were synthesized by Primerdesign (Primerdesign, Southampton, UK) as a custom-designed real-time PCR assay with a double-dye probe and primer-limited concentration (Table S3). Relative quantification was performed in 45 cycles, for each sample in three replications, and multiplexed as follows: *LPL* and *ATP5G2*; *SCD1* and *RPS2*. For the *ACACA* gene, reactions were performed without multiplexing. Two genes were used as endogenous controls (*RPS2* – ribosomal protein S2; *ATP5G2* – H(+)-transporting ATP synthase; Ropka-Molik et al., 2016). The reaction temperature steps were as follows: UDG incubation (uracil-DNA glycosylase treatment) – 50 °C (2 min); AmpliTaq Gold, UP enzyme activation – 95 °C (10 min); and denaturation (95 °C; 15 s) and annealing/extending (60 °C; 1 min) in 45 PCR cycles.

For each analyzed gene, the PCR efficiency was estimated based on the standard curve method and the exact transcript abundance was calculated using the $\Delta\Delta$ Ct method $(1/E(\mathrm{Ct}))$, where E is efficiency ($10[-1/\text{slope}]$) and Ct is cycle determined by the threshold applied to the maximum am-

plification of the standard curve. The sample with the lowest expression level was used as a calibrator in each gene separately. The normality of the distribution of expression data was assessed with the use of the Shapiro–Wilk test (SAS v. 8.02), while the differences in genes expression levels between tissues were analyzed using the Kruskal–Wallis test.

2.3 Estimation of fatty acid content in longissimus dorsi muscle

The fatty acid profile of the meat samples (longissimus dorsi muscle) was estimated by gas chromatography according to the method of Folch et al. (1957). The analysis was performed on a basis of fat extraction mixture composed of methanol and chloroform in a 1 : 2 ratio and then the residue was saponified with NaOH. Then, the formed methyl esters of fatty acids were examined using gas chromatography on a Varian 3400 (Sugar Land, TX, USA) analyzer, an 8200 CX injector and a flame ionization detector (FID). Chromatographic separation was performed on a CP Wax 58 column (0.53 mm × 1 µm; Chrompack, USA) using different temperature programs ranging from 60 to 188 °C (4 °C change

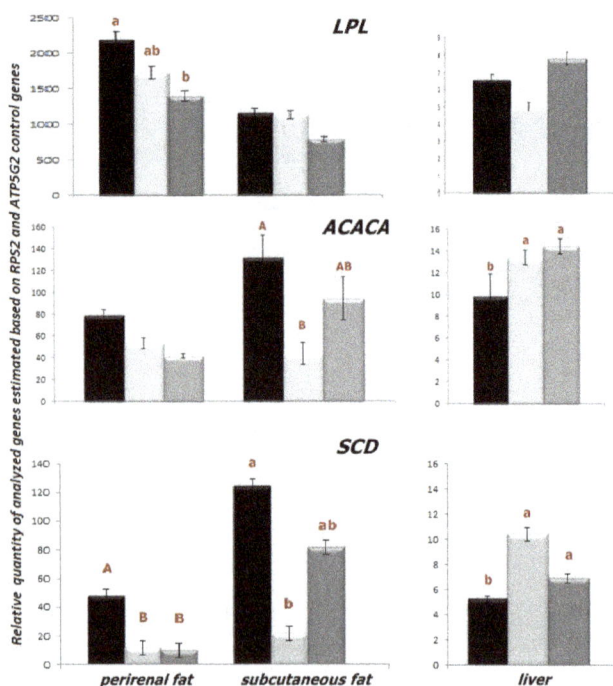

Figure 1. The transcript level for *LPL*, *ACACA* and *SCD* genes estimated in fat and liver tissues according to diet supplementation: complete pellet mixture (black column), grass addition (while column) and red clover addition (gray column). Data are presented as mean ± standard error; means with a, b and A, B differ significantly at $P < 0.05$ and $P < 0.01$, respectively.

per min) and then followed by temperature change from 60 to 220 °C (5 °C change per min). Injector and FID temperature was 200 and 260 °C, respectively. In addition, helium was used as a carrier gas (6 mL was added each minute). Hexane sample solutions (1 μL) were injected onto the column. The analyses were performed using standard solutions containing a mixture of standards (0.02–3.3 mg mL^{-1} in hexane) all purchased from Sigma-Aldrich (USA). Final results were adjusted for fatty acid content in a blank sample, which was prepared in a similar manner to the sample but without a weighed sample. Fatty acid content was expressed as percent of total fatty acids. The quantitative analysis of CLA isomers was performed by gas chromatography attached to Shimadzu GC-MS QP-2010 Plus mass spectrometry apparatus.

The significance of differences in fatty acid content between feeding groups were estimated using ANOVA with the Duncan post hoc test (SAS v. 8.02).

3 Results

3.1 Effect of the diet supplementation on transcript abundance of LPL, ACACA and SCD1

Our results showed that the expression of the *LPL* gene in perirenal fat tissue was modified by dietary supplement. Sig-

nificantly ($P < 0.05$) higher transcript abundances were detected in perirenal fat tissue of sheep fed with pelleted complete mixture without the addition of grass/clover as a supplement, while the lower expressions were observed in animals fed with fresh red clover and supplemental mixture (Fig. 1). A similar trend was observed in subcutaneous fat, but without statistical significance.

The supplementation of diet with an addition of grass or red clover significantly ($P < 0.05$) decreased the expression of the stearoyl-CoA desaturase 1 gene in both fat tissues compared to standard pelleted complete mixture. In subcutaneous fat, the *SCD1* expression in sheep fed with a complete mixture was 5.75-fold higher than in animals fed with grass supplementation and 1.52 than in animals fed with an addition of red clover (Fig. 1).

The present research also showed a lower transcript level of the *ACACA* gene in subcutaneous fat of sheep fed with grass compared to standard pellet mixture ($P < 0.01$), similar to the case of the *SCD1* gene (Fig. 1). Intermediate values were obtained for the subcutaneous fat of animals which had their diet supplemented with fresh red clover.

For the *LPL* gene, we did not observe any significant differences in liver tissue regardless of diet supplementation. For *ACACA* the highest expression was detected in liver tissue collected from sheep fed a diet with an addition of fresh red clover (Fig. 1; significant for the *ACACA* gene; $P < 0.05$). In the liver, the highest expression of the *SCD* gene was detected in animals fed with grass supplementation ($P < 0.05$).

3.2 The expression level of LPL, ACACA and SCD1 in different tissues

When comparing gene expression levels between different tissues in each feeding group, the highest amount of mRNA for all investigated genes was detected in fat, especially in perirenal fat. Regardless of diet supplementation, the highest *SCD* expression was detected in perirenal fat (Fig. 2), while *LPL* and *ACACA* transcript abundances were the highest in both perirenal and subcutaneous fat (Fig. 2). The lowest expression of all genes was detected in the liver.

3.3 Fatty acid composition following diet supplementation

Our results showed that meat (longissimus dorsi muscle) of lamb fed with fresh red clover and grass addition was characterized by significantly the highest SFA concentration as well as the highest total conjugated linoleic acid (CLA) content (Table 2). Furthermore, the highest concentration of c9-t11CLA was obtained for animals fed with red clover, while the lowest was found for lambs fed with the complete mixture ($p < 0.05$). For the rest of the isomers (CLA t10-c12, CLA c9-c11, CLA t9-t11) significant differences were not found. Moreover, the highest n-6 to n-3 polyunsaturated fatty acid ratio was detected in lambs fed with the standard com-

Table 2. The differences in selected fatty acid content estimated in lamb longissimus dorsi muscle depending on feeding system. Fatty acid contents are presented as g $100\,\text{g}^{-1}$ of total fatty acids in muscle samples.

Trait	Feeding groups		
	Complete mixture (diet I)	Grass addition (diet II)	Red clover addition (diet III)
Back fat of saddle (kg)	0.167 ± 0.05a	0.102 ± 0.03b	0.139 ± 0.05ab
SFA	46.07 ± 1.37b	49.54 ± 2.43a	48.39 ± 2.57a
MUFA	43.15 ± 5.59	41.45 ± 3.53	42.20 ± 1.44
PUFA	10.77 ± 4.87	8.99 ± 2.28	9.39 ± 1.77
n-6:n-3 PUFA	12.89 ± 4.22a	8.24 ± 2.16b	6.67 ± 1.06b
CLA	0.228 ± 0.05b	0.361 ± 0.10a	0.325 ± 0.09a
c9t11CLA	0.115 ± 0.03c	0.190 ± 0.06b	0.254 ± 0.05a
18:1cis 9	13.42 ± 1.37	14.39 ± 1.80	13.20 ± 0.95
18:2n 6	12.42 ± 1.37	13.40 ± 1.80	12.21 ± 0.95
18:3n 3	11.42 ± 1.37	12.40 ± 1.80	11.21 ± 0.96

SFA – saturated fatty acid; MUFA – monounsaturated fatty acid; PUFA – polyunsaturated fatty acid; CLA – conjugated linoleic acid; c9t11CLA – *cis*-9, *trans*-11 CLA isomer; a, b, c – $P < 0.05$.

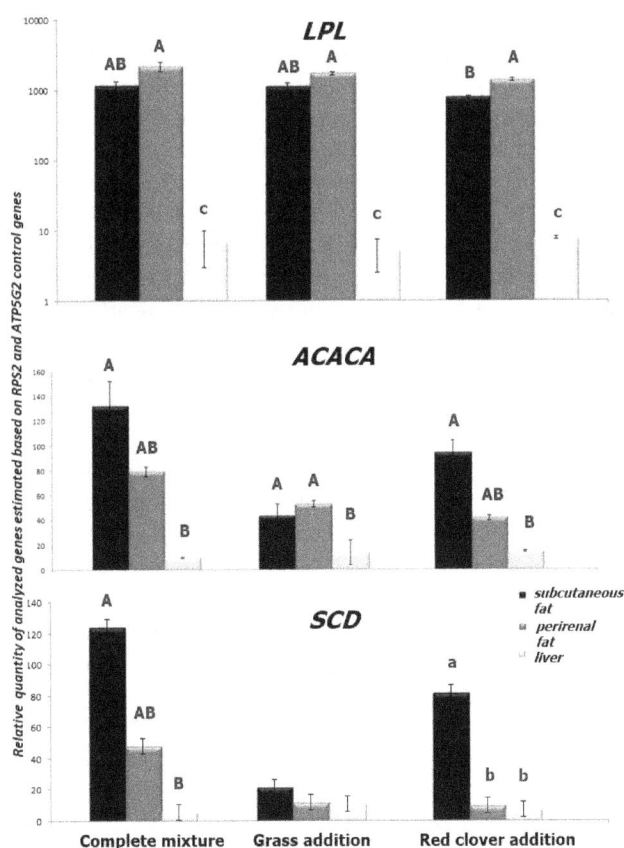

Figure 2. The differences in expression of *LPL*, *ACACA* and *SCD* genes between analyzed ovine tissues regardless of feeding supplementation. Data are presented as mean ± standard error; means with a,b and A,B differ significantly at $P < 0.05$ and $P < 0.01$, respectively.

plete mixture (which contains a higher concentration of carbohydrates compared to the other two feeding groups). The feeding system also affected the degree of fatness – lambs fed with complete mixture (diet I) had significantly higher back fat on the saddle compared to animals fed with fresh grass addition.

4   Discussion

Fatty acid composition is one of the main factors affecting health benefits of food. The ratio of polyunsaturated and monounsaturated to saturated fatty acid and the concentrations of omega-3 acids and conjugated linoleic acid (CLA) are considered as essential diet elements which may affect the prevalence of many diseases in society such cardiovascular disease, diabetes and cancer (Kritchevsky et al., 2000; Diniz et al., 2004; Liao et al., 2010). The main aim of the present research was to estimate the potential effect of diet supplementation on the transcript level of *LPL*, *ACACA* and *SCD* genes, which play a key role in lipid metabolism and adipogenesis process. Our studies are based on relatively affordable and accessible feeding supplements (grass and red clover addition to standard pellet mixture), which could potentially affect the expression of enzymes that are critical in fatty acid biosynthesis and/or hydrolysis of lipoprotein triacylglycerols. Previous research has confirmed that such nutritional regulation might result in modification of fatty acid composition, but the exact molecular mechanism of observed variations is still unknown (Murphy et al., 1999; Bonnet et al., 2000; Ntambi et al., 2004; Corazzin et al., 2013; Kęsek et al., 2014).

Our study showed that diet with an addition of fresh grass or red clover to standard pellet mixture significantly reduced the transcript levels of the *LPL* gene in perirenal fat, *ACACA* in subcutaneous fat and *SCD* in both fat tissues. In the rumen, unsaturated fatty acids have a short half-life according to their rapid hydrogenation to saturated fatty acids (Jenkins, 1993). It has been established that enrichment of the diet with pasture (grass or fresh red clover) results in greater fodder intake by animals due to the lower energy content in the diet. This leads to decreasing of residence time of forage in the rumen and results in reduction of hydrogenation process for some PUFAs. Moreover, the addition a red clover, which contains the polyphenol oxidase (PPO) enzyme, decreases proteolytic and lipolytic processes in the rumen and can reduce fatty acid biohydrogenation (Sullivan and Hatfield, 2006). Steinshamn and Thuen (2008) confirmed that the addition of red clover to standard diet leads to significant modifications of milk fatty acid profiles: an increase in C18:3n-3 and C18:2n-6 content and the reduction in n-6 / n-3 ratio. Our results indicated that feeding supplementation modified the expression of ovine *LPL*, *ACACA* and *SCD* genes in fat tissues as well as changed selected fatty acid profiles in lamb. Raes et al. (2004) showed that c9t11CLA content can be increased by feeding ruminants n-3-rich diets. In our study, it has been confirmed that diet with the high concentration of carbohydrates (complete mixture) led to the meat n-6 / n-3 ratio increasing by about 2 times compared to diet with red clover addition. Furthermore, diet supplementation with fresh grass or red clover increased total CLA content and c9t11CLA isomer concentration.

In the case of lipoprotein lipase, which is related to fat deposition in adipocyte tissue and synthesis of the low-density lipoprotein fraction, the addition of red clover reduced the *LPL* gene expression in perirenal fat. Intermediate values of transcript abundances were obtained for fat tissue of sheep fed with a grass addition. Research performed on chickens showed that a diet rich in n-3 and n-6 fatty acids reduced *LPL* expression measured in adipocytes compared to n-9 fatty acid addition (Montalto and Bensadoun, 1993; Sato and Akiba, 2002). On the other hand, in rats, a diet rich in n-3 acids increased LPL activity and transcript levels in adipocyte tissue (Chapman et al., 2000). Furthermore, the *LPL* expression can be up-regulated by refeeding, which was established in cattle (Bonnet et al., 1998) and sheep (Bonnet et al., 2000).

The present study shows that transcription of the *LPL* gene in sheep was most efficiently modified by diet supplemented with red clover, probably due to the highest concentration of polyphenol oxidase. In the rumen, the PPO enzyme regulates the lipolysis and is critical for deposition of PUFA in animal products (Lee, 2014). Similarly to research performed by Dervishi et al. (2011), we detected the highest *LPL* expression in sheep fed an intensive diet. Interestingly, in the liver, the highest *LPL* expression, but without statistical significance, was observed in sheep fed with red clover, in contrast to the perirenal fat tissue. This reverse trend in the modification of *LPL* expression by different diet additions might be related to tissue-specific ability of distribution and synthesis of fatty acids and/or secretion of triglyceride-rich lipoproteins (Pullen et al., 1990). The tissue-specific regulation of

LPL expression was also confirmed in rats, where a dietary content of n-3 PUFA affected *LPL* transcription in internal adipose tissue while lacking influence in subcutaneous adipose (Raclot et al., 1997).

Our study indicated that the transcript level of the *ACACA* gene might be regulated in a similar manner to the *LPL* gene. The feeding of sheep with grass or red clover significantly decreased expression of acetyl-CoA carboxylase alpha in subcutaneous fat ($P < 0.05$) and perirenal fat (without significance), while in the liver both supplements activated *ACACA* transcription ($P < 0.05$). Similar to the case of the *LPL* gene, the highest *ACACA* expression in the liver was detected in sheep fed with an addition of fresh red clover and grass. The ACACA catalyzed malonyl-CoA synthesis from acetyl-CoA and is mainly expressed in lipogenic tissues (Lopez-Casillas et al., 1991). Mao et al. (2003) showed that in, human transcription, the *ACACA* gene is regulated by three promoters and one of them is expressed in a tissue-specific manner. The adverse regulation of the ovine *ACACA* gene in the liver and fat in response to a different diet would suggest that, in sheep, this gene might have a different role in both tissues. Furthermore, in the liver and fat tissue different lipid metabolism processes occur that are related to modifications of distinct fatty acids which may also impose tissue-specific expression.

In turn, the results obtained showed that SCD expression in sheep fat tissue was down-regulated in response to a diet with an addition of grass and red clover. On the other hand, forage with fresh grass supplementation significantly increased SCD transcript levels in the liver. It has been proven that, in the most of tissues, a high-carbohydrate diet shows large stimulatory effects on the levels of SCD mRNA (Miyazaki et al., 2002). In mice, dietary intake of n-3 fatty acids, which is considered a strong SCD repressor, showed the inhibitory effect of stearoyl-CoA desaturase 1 expression in the liver, despite supplementation with carbohydrates (Ntambi, 1992, 1999). Our results indicated that an increase in intake of unsaturated fatty acid reduced *SCD* gene expression in ovine fat tissue but at the same time increased transcript level in the liver. Inversely, in both fat tissues of sheep fed with standard pellet mixture, which contains higher concentration of carbohydrates compared to two other forages, the detected expression of the *SCD* gene was higher, while in liver tissue it was lower. This suggests that in sheep transcription of the stearoyl-CoA desaturase 1 gene is modified in a tissue-specific manner, probably due to different metabolic activity of investigated tissues. It can also be related to fatty acid modification occurring in the rumen as well as diverse fatty acid composition and concentration which is transported to the liver and fat. The different regulatory mechanisms of the *SCD* gene that are dependent on tissue type have also been confirmed in porcine (Doran et al., 2006) and cattle (Hiller et al., 2011) muscles and adipose tissues.

5 Conclusions

In summary, our research showed that diet supplementation with fresh red clover and grass significantly modified expression of *LPL*, *ACACA* and *SCD* genes in fat and liver tissues, which can be indirectly related to fatty acid concentration in animal products. Furthermore, tissue-specific regulations of *SCD* and *ACACA* expression were identified and the opposite effect of diet additions was observed depending on tissue type. The ability of nutrigenomic regulation of the analyzed gene transcription confirmed that these genes play a critical role in regulation of lipid metabolism processes in sheep and could be associated with fatty acid profiles in milk and meat.

Competing interests. The authors declare that they have no conflict of interest.

Acknowledgements. The present study was supported by the National Research Institute of Animal Production statutory activity, research project no. 01-007.1.

Edited by: Steffen Maak

References

Bernard, L., Leroux, C., and Chilliard, Y.: Expression and nutritional regulation of lipogenic genes in the ruminant lactating mammary gland, Adv. Exp. Med. Biol., 606, 67–108, 2008.

Bonnet, M., Faulconnier, Y., Flechet, J., Hocquette, J. F., Leroux, C., Langin, D., Martin, P., and Chilliard, Y.: Messenger RNAs encoding lipoprotein lipase, fatty acid synthase and hormone-sensitive lipase in the adipose tissue of underfedrefed ewes and cows, Reprod. Nutr. Dev., 38, 297–307, 1998.

Bonnet, M., Leroux, C., Faulconnier, Y., Hocquette, J. F., Bocquier, F., Martin, P., and Chilliard, Y.: Lipoprotein lipase activity and mrna are up-regulated by refeeding in adipose tissue and cardiac muscle of sheep, J. Nutr., 130, 749–756, 2000.

Chapman, C., Morgan, L. M., and Murphy, M. C.: Maternal and early dietary fatty acid intake: changes in lipid metabolism and liver enzymes in adult rats, J. Nutr., 130, 146–151, 2000.

Carta, A., Casu S., Usai, M. G., Addis, M., Fiori, M., Fraghi, A., Miari, S., Mura, L., Piredda, G., Schibler, L., Sechi, T., Elsen, J. M., and Barillet, F.: Investigating the genetic component of fatty acid content in sheep milk, Small Ruminant Res., 79, 22–28, 2008.

Corazzin, M., Bovolenta, S., Saccà, E., Bianchi, G., and Piasentier, E.: Effect of linseed addition on the expression of some lipid metabolism genes in the adipose tissue of young Italian Simmental and Holstein bulls, J. Anim. Sci., 91, 405–12, 2013.

Dervishi, E., Serrano, C., Joy, M., Serrano, M., Rodellar, C., and Calvo, J. H.: The effect of feeding system in the expression of genes related with fat metabolism in semitendinous muscle in sheep, Meat Sci., 89, 91–97, 2011.

Diniz, Y. S., Cicogna, A. C., Padovani, C. R., Santana, L. S., Faine, L. A., and Novelli, E. L.: Diets rich in saturated and polyunsaturated fatty acids: metabolic shifting and cardiac health, Nutrition, 20, 230–234, 2004.

Doran, O., Moule, S. K., Teye, G. A., Whittington, F. M., Hallett, K. G., and Wood, J. D.: A reduced protein diet induces stearoyl-CoA desaturase protein expression in pig muscle but not in subcutaneous adipose tissue: relationship with intramuscular lipid formation, Br. J. Nutr., 95, 609–617, 2006.

Fattore, E. and Fanelli, R.: Palm oil and palmitic acid: a review on cardiovascular effects and carcinogenicity, Int. J. Food Sci. Nutr., 64, 648–59, 2013.

Folch, J., Lees, M., and Stanley, G. H.: A simple method for the isolation and purification of total lipides from animal tissues, J. Biol. Chem., 226, 497–509, 1957.

Hiller, B., Herdmann, A., and Nuernberg, K.: Dietary n-3 fatty acids significantly suppress lipogenesis in bovine muscle and adipose tissue: a functional genomics approach, Lipids, 46, 557–67, 2011.

IZ PIB-INRA: Ruminant Nutrition Standards. The nutritional value of French and national feed for ruminants, Ed. National Research Institution of Animal Production, 1–117, 2009.

Jensen, D. R., Bessesen, D. H., Etienne, J., Eckel, R. H., and Neville, M. C.: Distribution and source of lipoprotein lipase in mouse mammary gland, J. Lipid Res., 32, 733–742, 1991.

Kęsek, M., Szulc, T., and Zielak-Steciwko, A.: Genetic, physiological and nutritive factors affecting the fatty acid profile in cows' milk – a review, Anim. Sci. Pap. Rep., 32, 1–11, 2014.

Kritchevsky, D., Tepper, S. A., Wright, S., Tso, P., and Czarnecki, S. K.: Influence of conjugated linoleic acid (CLA) on establishment and progression of atherosclerosis in rabbits, J. Am. Coll. Nutr., 19, 472–477, 2000.

Lee, M. R. F.: Forage polyphenol oxidase and ruminant livestock nutrition, Front Plant Sci., 5, 1–9, 2014.

Liao, F. H., Liou, T. H., Shieh, M. J., and Chien, Y. W.: Effects of different ratios of monounsaturated and polyunsaturated fatty acids to saturated fatty acids on regulating body fat deposition in hamsters, Nutrition, 26, 811–817, 2010.

Lopez-Casillas, F., Ponce-Castaneda, M. V., and Kim, K. H.: In vivo regulation of the activity of the two promoters if the rat acetyl coenzyme-A carboxylase gene, Endocrinology, 129, 1049–1058, 1991.

Jenkins, T. C.: Advances in ruminant lipid metabolism, J. Dairy Sci., 76, 3851–3863, 1993.

Mao, J., Chirala, S. S., and Wakil, S. J.: Human acetyl-CoA carboxylase 1 gene: presence of three promoters and heterogeneity at the 5-prime-untranslated mRNA region, P. Natl. Acad. Sci. USA, 100, 7515–7520, 2003.

Miyazaki, M., Gomez, F. E., and Ntambi, J. M.: Lack of stearoyl-CoA desaturase-1 function induces a palmitoyl-CoA Delta6 desaturase and represses the stearoyl-CoA desaturase-3 gene in the preputial glands of the mouse, J. Lipid. Res., 43, 2146–2154, 2002.

Montalto, M. B. and Bensadoun A.: Lipoprotein lipase synthesis and secretion: Effects of concentration and type of fatty acids in adipocyte cell culture, J. Lipid Res., 34, 397–407, 1993.

Murphy, M. C., Brooks, C. N., Rockett, J. C., Chapman, C., Lovegrove, J. A., Gould, B. J., Wright, J. W., and Williams, C. M.: The quantitation of lipoprotein lipase mRNA in biopsies of human adipose tissue, using the polymerase chain reaction, and the effect of increased consumption of n-3 polyunsaturated fatty acids, Eur. J. Clin. Nutr., 53, 441–447, 1999.

Ntambi, J. M.: Dietary Regulation of Stearoyl-CoA Desaturase 1 Gene Expression in Mouse Liver, J. Biol. Chem., 267, 10925–10930, 1992.

Ntambi, J. M.: Regulation of stearoyl-CoA desaturase by polyunsaturated fatty acids and cholesterol, J. Lipid. Res., 40, 1549–1558, 1999.

Ntambi, J. M., Miyazaki, M., and Dobrzyn, A.: Regulation of stearoyl-CoA desaturase expression, Lipids, 39, 1061–1065, 2004.

Ntambi, J. M., Miyazaki, M., Stoehr, J. P., Lan H., Kendziorski, C. M., Yandell, B. S., Song, Y., Cohen, P., Friedman, J. M., and Attie, A. D.: Loss of stearoyl-CoA desaturase-1 function protects mice against adiposity, P. Natl. Acad. Sci. USA, 99, 11482–11486, 2002.

Pullen, D. L., Liesman, J. S., and Emery, R. S.: A species comparison of liver slice synthesis and secretion of triacylglycerol from nonesterified fatty acids in media, J. Anim. Sci., 68, 1395–1399, 1990.

Raclot, T., Groscolas, R., Langin, D., and Ferre, P.: Site-specific regulation of gene expression by n-3 polyunsaturated fatty acids in rat white adipose tissues, J. Lipid Res., 38, 1963–1972, 1997.

Raes, K., De Smet, S., and Demeyer, D.: Effect of dietary fatty acids on incorporation of long chain polyunsaturated fatty acids and conjugated linoleic acid in lamb, beef and pork meat: a review, Anim. Feed. Sci. Tech., 113, 199–221, 2004.

Ropka-Molik, K., Knapik, J., Pieszka, M., and Szmatoła, T.: The expression of the SCD1 gene and its correlation with fattening and carcass traits in sheep, Arch. Anim. Breed., 59, 37–43, 2016.

Sato, K. and Akiba, Y.: Lipoprotein lipase mRNA expression in abdominal adipose tissue is little modified by age and nutritional state in broiler chickens, Poult Sci. S., 81, 846–852, 2002.

Steinshamn, H. and Thuen, E.: White or red clover-grass silage in organic dairy milk production: Grassland productivity and milk production responses with different levels of concentrate, Livest. Sci., 119, 202–215, 2008.

Sullivan, M. L. and Hatfield, R. D.: Polyphenol xxidase and o-diphenols inhibit postharvest proteolysis in red clover and alfalfa, Crop Sci., 46, 662–670, 2006.

Wood, J. D., Richardson, R. I., Nute, G. R., Fisher, A. V., Campo, M. M., Kasapidou, E., Sheard, P. R., and Enser, M.: Effects of fatty acids on meat quality: a review, Meat Sci., 66, 21–32, 2003.

18

A new somatic cell count index to more accurately predict milk yield losses

Janez Jeretina[1], Dejan Škorjanc[2], and Drago Babnik[1]

[1]Agricultural Institute of Slovenia, Hacquetova ulica 17, 1000 Ljubljana, Slovenia
[2]University of Maribor, Faculty of Agriculture and Life Sciences, Pivola 10, 2311 Hoče, Slovenia

Correspondence to: Janez Jeretina (janez.jeretina@kis.si)

Abstract. Intramammary infection and clinical mastitis in dairy cows leads to considerable economic losses for farmers. The somatic cell concentration in cow's milk has been shown to be an excellent indicator for the prevalence of subclinical mastitis. In this study, a new somatic cell count index (SCCI) was proposed for the accurate prediction of milk yield losses caused by elevated somatic cell count (SCC). In all, 97 238 lactations (55 207 Holstein cows) from 2328 herds were recorded between 2010 and 2014 under different scenarios (high and low levels of SCC, four lactation stages, different milk yield intensities, and parities (1, 2, and ≥ 3)). The standard shape of the curve for SCC was determined using completed standard lactations of healthy cows. The SCCI was defined as the sum of the differences between the measured interpolated values of the natural logarithm of SCC (ln(SCC)) and the values for the standard shape of the curve for SCC for a particular period, divided by the total area enclosed by the standard curve and upper limit of ln(SCC) = 10 for SCC. The phenotypic potential of milk yield (305-day milk yield – MY305) was calculated using regression coefficients estimated from the linear regression model for parity and breeding values of cows for milk yield. The extent of daily milk yield loss caused by increased SCC was found to be mainly related to the early stage of lactation. Depending on the possible scenarios, the estimated milk yield loss from MY305 for primiparous cows was at least 0.8 to 0.9 kg day^{-1} and for multiparous cows it ranged from 1.3 to 4.3 kg day^{-1}. Thus, the SCCI was a suitable indicator for estimating daily milk yield losses associated with increased SCC and might provide farmers reliable information to take appropriate measures for ensuring good health of cows and reducing milk yield losses at the herd level.

1   Introduction

Development of intramammary infection (IMI) and occurrence of clinical mastitis in dairy cows leads to considerable economic losses for farmers (Nielsen et al., 2010; El-Awady et al., 2011), mainly owing to the reduction of milk production and lowering of milk technological traits (Bobbo et al., 2016). IMI has been shown to adversely affect fertility (Wolfenson et al., 2015) and reduce the longevity of dairy cows (Archer et al., 2013); it also increases the costs for implementation of veterinary services and extra labour.

The concentration of somatic cells in cow's milk has been shown to be an excellent and the main indicator for the estimation of the prevalence of subclinical mastitis. Cows with subclinical mastitis show no visible signs, but their somatic cell count (SCC, defined as the number of somatic cells per millilitre of milk) is elevated. Elevated SCC in milk suggests the presence of pathogens in the udder and is an indicator of IMI as well as a measure of the response to infection (Pyörälä, 2003; Heringstad et al., 2006). Thus, subclinical mastitis is considered as a hidden threat to healthy cows in a herd (Nyman et al., 2014).

For estimating the possible milk yield losses caused by subclinical mastitis, a definition of healthy or non-infected is essential. The threshold for a healthy udder was considered to be an SCC of $\leq 50\,000$ (Seegers et al., 2003) or approximately 70 000 (Djabri et al., 2002; Schukken et al., 2003). Some authors defined a healthy animal as having a slightly higher SCC, i.e., $\leq 100\,000$ (Hand et al., 2012). Cows with an SCC of less than 100 000 are considered to be non-infected, with no significant milk yield losses owing to subclinical mastitis. A new definition of subclinical mastitis assumed a new case if the SCC reached $> 100\,000$ after a test day when the SCC was $< 50\,000$ (Halasa et al., 2009). Therefore, the selection of an appropriate threshold for defining a non-infected udder depends on the purpose. At a lower threshold, more IMI (increased sensitivity and fewer false negatives) cases are identified, whereas the use of a higher threshold (increased specificity) might result in fewer false positives (Pantoja et al., 2009).

Together with high milk yield, the fat-to-protein ratio (FPR) can serve as an important risk factor for mastitis (Windig et al., 2005). Cows with mastitis are characterized by lower milk yield, elevated SCC, and a higher FPR (Jamrozik and Schaeffer, 2012). Thus, the FPR of milk was considered to be a suitable measure of the energy balance status of animals, especially during the initial and most metabolically stressful stage of lactation (Buttchereit et al., 2010).

A close association is known to exist between high milk yield and SCC. High-milk-yielding cows are more susceptible to mastitis (Jamrozik et al., 2010). The average SCCs calculated based on SCCs at different lactation stages are often used in mastitis control programs and in programs for the improvement of udder health. The drawback of using the lactation average of the SCCs is that it does not account for the SCC variability during lactation (De Hass et al., 2004). Variation in the shapes of the lactation curve during different lactation periods can be influenced by subclinical and/or clinical mastitis. Moreover, the types of pathogens associated with clinical mastitis occurrence can also differentially affect the lactation curve (De Hass et al., 2002). The early detection of elevated SCC during lactation is possible only by using test-day records. For detection of subclinical mastitis and possible IMI, comparing different test-day records of SCC is necessary. Timely detection and analysis of peaks in SCC during the different stages of lactation are important for the successful management of dairy farms.

However, considering the relationship between milk yield and SCC might lead to erroneous results since high milk yield might decrease the SCCs because of the dilution effect (Miller et al., 1993). The estimated SCCs of high-yielding dairy cows without IMI were found to be lower than those of low-yielding dairy cows (Green et al., 2006; Halasa et al., 2009; Boland et al., 2013). If the SCC concentration due to lower milk production in infected cows is neglected, the milk production loss might be overestimated. Overestimation of milk production loss can be avoided by using a dilution factor λ for the adjustment of the yield and category of SCC (Green et al., 2006). Nevertheless, few previous studies could not completely clarify the relation between dilution effect and estimated milk loss.

Several other effects are also related to subclinical mastitis development and elevated SCCs. Numerous studies have shown that different factors, such as stage of lactation, subsequent parity (PAR), milk yield (Nielsen et al., 2010; Boland et al., 2013), calving month (Rupp and Boichard, 2000) and

calving season, feeding and housing (Hortet and Seegers, 1998; Hagnestam-Nielsen et al., 2009), milking (Nyman et al., 2009), milk composition (Windig et al., 2005; Nyman et al., 2014), and test-day season, breed, pregnancy status, and health disorders, affect the SCC (Hagnestam-Nielsen et al., 2009). Subclinical mastitis is a very complex problem. Therefore, developing a simple, cost-effective, and efficient method for the estimation of the relationships between elevated SCC, subclinical mastitis, and potential milk yield loss in dairy cows is of great interest to the dairy sector.

Therefore, this study aimed to develop an index for excessive SCC, namely, the somatic cell count index (SCCI), for estimating the effect of subclinical mastitis on milk yield loss. Intervals of 30 days in milk test-day records were used to determine the relationships among SCC, calving year, calving season, age at the first calving, milk composition, stage of lactation, and milk yield; the effect of herd was also assigned for a more reliable prediction of milk yield losses.

2 Materials and methods

2.1 Data sources and analyses

In this study, 97 238 standard lactation records of 55 207 Holstein breed cows from 2328 herds were collected over 921 594 test-days between 2010 and 2014. The data were a part of the national milk recording from the Slovenian database (Jeretina et al., 1997) collected according to the International Committee for Animal recording (ICAR, 2016). The average herd size was 32 cows. The test-day records with clinical mastitis were discarded from the data set. Lactations with at least seven milk recordings were truncated at 305 days. Each record of later analyses included the number of test-day milk yields (TDMYs, kg), FPR, PARs, stage of lactation (days in milk, DIMs), the season of calving (S; 1: spring, 2: summer, 3: autumn, and 4: winter), age at first calving (AFC, days), breeding values of cows for milk yield (BVAs), and SCC ($\times 10^3$).

The model was developed using two steps by using the statistical application R (R Development Core Team, 2016) and the lme4 libraries (Bates et al., 2015). In the first step, we determined the standard shapes of the curves for the natural logarithm of SCC during lactation in healthy cows for PAR 1, 2, and ≥ 3 and estimated the phenotypic potential of milk yield for all cows to classify them into production groups. After that, we developed the SCCI for the estimation of SCC excess above the standard shape. In the second step, we analysed the effect of SCCI on milk yield loss for standard lactation according to the stage of lactation.

2.1.1 Somatic cell count index

We defined the SCCI for an individual lactation as the sum of the differences between the measured interpolated values of $\ln(\mathrm{SCC})$ (IP, Fig. 1) and the values of the standard shape of

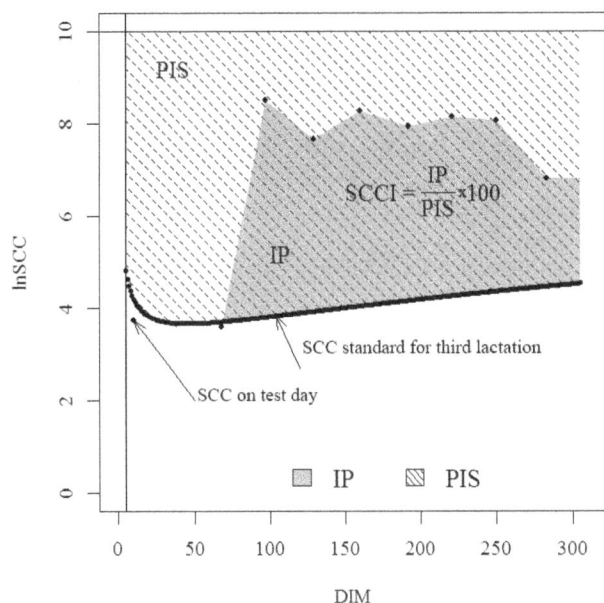

Figure 1. SCCI of a cow in the third lactation with somatic cell count (SCC) values on test days in relation to the standard shape of the curve for SCC. Jeretina et al. (2016).

the curve for SCC for a particular period, divided by the area above the standard shape of the curve for SCC (PIS, Fig. 1) (Eq. 1).

Therefore, the SCCI represents the area for IP in the percentage share of the total area for PIS above the standard shape of the curve for SCC (Fig. 1). By definition, the values of SCCI are between 0 and 100, wherein a value of 0 represents a small or inconsequential influence of SCC on milk yield for standard lactation, and 100 represents the maximal effect. When the SCCI was calculated, we considered the effect of dilution for high-milk-yielding cows without IMI. For cows with a daily milk yield above 10 kg and an SCC lower than 50 000, we performed a correction by using the factor −0.485 (Green et al., 2006). For calculation of the SCCI we used the following equation:

$$\mathrm{SCCI} = \frac{\sum\limits_{i=x_1}^{x_2}(\ln\mathrm{SCC}_i - \ln\mathrm{SCCM}_i)}{\sum\limits_{i=5}^{305}(10 - \ln\mathrm{SCCM}_i)} \times 100; \qquad (1)$$

$$\begin{cases} \ln\mathrm{SCC}_i > \ln\mathrm{SCCM}_i \\ x_1 \geq 5\,\mathrm{DIM} \\ x_2 \leq 305\,\mathrm{DIM} \end{cases},$$

where the SCCI is the SCC index excess calculated for an optional interval between x_1 DIM and x_2 DIM, lnSCCM is the natural logarithm of SCC for the standard shape of the curve at the ith DIM, and lnSCC is the natural logarithm of the measured SCC at the ith DIM.

To determine the standard shape of the curve for SCC, we included the completed standard lactations (305-day milk yield – MY305) of cows for which the average SCC for standard lactation (ASCC) did not exceed 100 000 SCC. This limit was set because in healthy cows, on the third day after calving, the SCC drops to 166 000 and by the 10th day of lactation, it reduces to 100 000 SCC (Barkema et al., 1999). In addition, we excluded the data for all cows in which the SCC between two consecutive milk recordings increased from less than 50 000 to more than 100 000. These numbers potentially indicate a suspected case of subclinical mastitis (Halasa et al., 2009). We also excluded data for which the sum of squared deviations at fitting of lactation curves through TDMY in MY305 according to the MilkBot model (Cole et al., 2012) was larger than 150 kg, which represents > 0.5 kg milk per production day.

We included fixed-effect PAR and the linear and quadratic regression effects of DIMs, which were used to explain the dependent variable ln(SCC/100), in the square regression model (Eq. 2).

$$\ln\mathrm{SCCM}_{ij} = \mu + \mathrm{PAR}_i + \beta_1 \times \mathrm{DIM} + \beta_2 \times \mathrm{DIM}^2 + e_{ij}, \quad (2)$$

where lnSCCM is the natural logarithm of SCC; PAR is the ith parity ($i = 1, 2, \geq 3$); $\beta_1 \times \mathrm{DIM}$ and $\beta_2 \times \mathrm{DIM}^2$ are the linear and quadratic regression effects of DIMs, respectively; and e_{ij} is the residual effect.

2.1.2 Predicting milk yield and classification in groups

We calculated the phenotypic potential of milk yield for cows by using regression coefficients estimated from the linear regression model as follows:

$$\mathrm{MY305}_{ij} = \mu + \mathrm{PAR}_i + \beta_1 \times \mathrm{BVA} + e_{ij}, \qquad (3)$$

where PAR is the fixed effect of parity ($i = 1, 2, \geq 3$), β_1 is the regression coefficient of BVA, and e_{ij} is the residual of the model.

Based on the phenotypic potential of milk yield, we classified the cows into the following four classes: cows with the lowest – milk quantity (MQ) = 1 – to the highest (MQ = 4) MY305. We determined the size classes based on MQ with regard to the average of the predicted milk yield of the studied cattle population within specific PARs and by considering the sizes of the standard deviations.

2.1.3 Estimation of the effect of SCCI on milk yield loss for standard lactation

We investigated the effect of SCCI on milk yield loss for standard lactation within a particular month of lactation (ΔMY30) and PAR for individual MQ by using multiple mixed regression models as follows:

$$\Delta\text{MY30}_{ijklm} = \mu + Y_i + S_j + \beta_1 \times \text{AFC} + \beta_2 \times \text{BVA} \quad (4)$$
$$+ \beta_3 \times \text{FPR} + \beta_4 \times \text{SCCI} + h_l + e_{ijklm},$$

where ΔMY30 is the estimated milk yield loss within a particular month of lactation, calculated as a difference between the predicted (Jeretina et al., 2013) and the estimated milk yields on the test day for an individual month; Y is the effect of calving year (2010–2014); S is the effect of calving season (1–4); β_1 is the linear regression coefficient of AFC; β_2 is the linear regression coefficient of the BVA; β_3 is the linear regression coefficient of the FPR; β_4 is the linear regression coefficient of SCCI; h is the random effect of l-herd; and e_{ijklm} is the residual of the model.

For milk yield loss on a daily basis (ΔMY), we used the linear model according to Ali and Schaeffer (AS; Ali and Schaeffer, 1987) to estimate the shape of the curve based on the SCCI points. Based on the estimated values of SCCI and DIMs within individual PARs and MQs, we used the following equation to predict ΔMY for ASCC with 200 000, 300 000, and 400 000 SCC:

$$\Delta\text{MY} = \sum_{\text{DIM}=x_1}^{\text{DIM}=x_2} \left(a + b \times \left(\frac{\text{DIM}}{305}\right) + c \times \left(\frac{\text{DIM}}{305}\right)^2 \right. \quad (5)$$
$$\left. + d \times \ln\left(\frac{305}{\text{DIM}}\right) + k \times \left(\ln\left(\frac{305}{\text{DIM}}\right)\right)^2 \right)$$
$$\times \text{SCCI}_{\text{DIM}}; x_1 \geq 5; x_2 \leq 305,$$

where ΔMY is the predicted milk yield loss for any day or period between x_1 and x_2; parameters a, b, c, d, and k are factors of the AS model; and SCCI_{DIM} is the SCC index excess for a particular day of DIMs.

2.1.4 Impact of somatic cell count level within specific PARs, lactation interval, and milk production level on milk yield loss

To determine the effect of IMI on ΔMY at different stages of lactation, we used the data for cows from the cattle population that exhibited average values of $200\,000 \pm 50\,000$, $300\,000 \pm 50\,000$, or $400\,000 \pm 50\,000$ ASCC for standard lactation. The lactation period was divided into four lactation intervals (LIs; LI1 = 0–80 DIMs, LI2 = 81–160 DIMs, LI3 = 161–240 DIMs, and LI4 = 241–305 DIMs), and ΔMY was calculated for each period. An individual LI with SCCI $\leq 5\,\%$ was considered as a period of lactation with a low increase in SCC; above this value, the lactation period was considered to show a high increase in SCC. Individual effects on ΔMY305 were estimated using the following multiple linear regression model (Eq. 6):

$$\Delta\text{MY305}_{ijklm} = \mu + \text{PAR}_i + \text{MQ}_j + \text{VAR}_k + \text{ASCC}_l \quad (6)$$
$$+ e_{ijklm},$$

where ΔMY305 is the ΔMY in standard lactation, PAR is the ith parity ($i = 1, 2,$ and ≥ 3), MQ ($j = 1$–4) is the level

of potential milk production (see Table 2), VAR ($k = 1$–8) represents eight different scenarios regarding the effect of a low or high level of SCCI within each of the four LIs (see Table 5), ASCC is the average SCC (200 000, 300 000, or 400 000 SCC), and e_{ijklm} is the residual of the model.

3 Results and discussion

The number of completed lactations with the average MY305, protein and fat contents in milk, and the average geometric mean of SCC per lactation are shown in Table 1. The average MY305 of primiparous cows, which was 40 % of all lactations in the analysis, was 6828 kg with 122 000 SCC, and that in the second lactation was 7622 kg with 158 000 SCC. In the later lactations, the milk yield was 7808 kg with 236 000 SCC.

To estimate the effect of SCCI on ΔMY, we used the linear AS model and assessed the value of the SCCI depending on DIMs. The parameters of the curve for individual PAR and MQ are shown in Table 2.

The corresponding parameter values (Table 2) were included in Eq. (5) and ΔMY values were calculated for different SCCs for any day or period. A representative estimation of the daily milk yield loss in kilogram per stage of lactation (DIM) and MQ for the consecutive PAR is shown in Fig. 2.

The results showed that in primiparous cows the regression influence of SCC on ΔMY was the lowest within 7 to 10 weeks after calving. It coincided with the peak of lactation in primiparous cows. Our results are in good agreement with those of a previous published study by Hagnestam-Nielsen et al. (2009). They set the limit between 3 and 8 weeks after calving for primiparous cows and between 3 and 16 weeks for later lactations. In the early stage of the second lactation, the smallest influence of SCC on ΔMY was noted between 8 and 10 weeks after calving. However, this was observed only in the case of high-milk-yielding cows (MQ = 3 and 4). Interestingly, this effect was also not found in the case of subsequent lactations. Unexpectedly, when the SCC was constant, its effect on milk yield was not the highest during the early stage of lactation. This can be attributed to the physiology and functionality of the udder gland. The mammary epithelial cells secrete the highest amount of milk during this period; a similar finding was noted for cell renewal capacity. During this period, the process of apoptosis is only initiated in the udder gland (Knight and Wilde, 1993). Therefore, IMI does not have a marked effect on milk yield loss. Notably, this is evident for primiparous cows, the mammary glands of which do not have a history of IMI. Another possibility is the high uncertainty related to the estimation of yield loss during early lactation because of the limited number of milk recordings. In this short period, the right type of model for the assessment of the shape of lactation curve after calving cannot be accurately selected, which might also undervalue

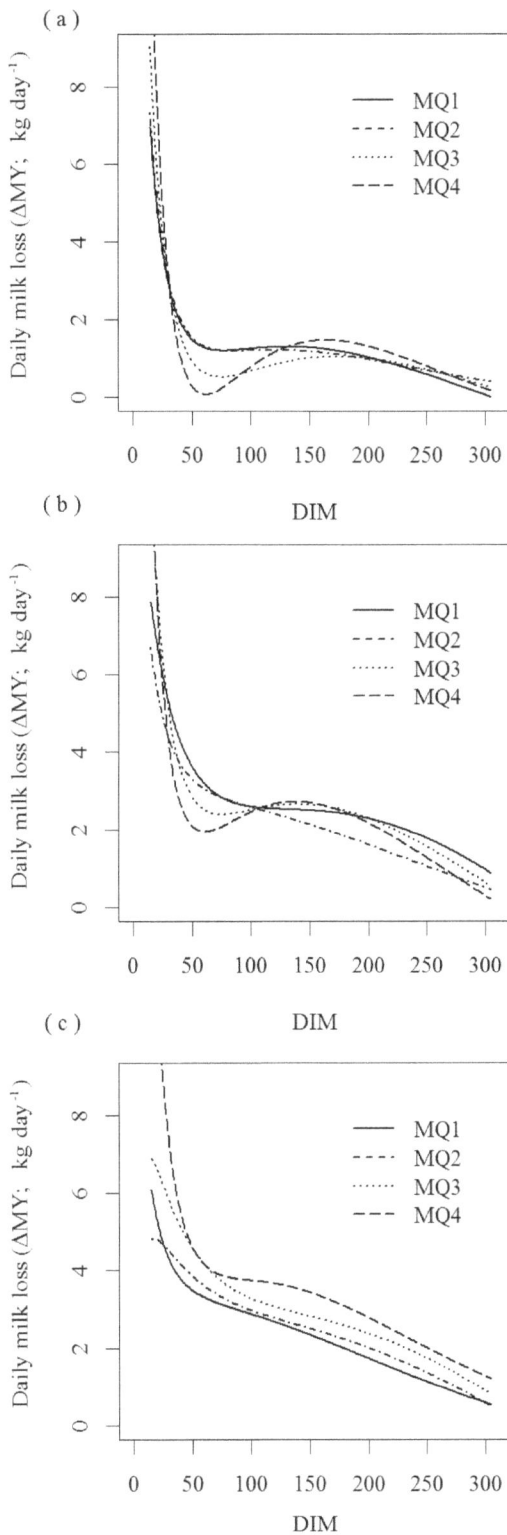

Figure 2. Estimated daily milk yield loss ($\mathrm{kg\,DIM^{-1}}$) for different milk production levels (MQ) and the specific value of 300 000 somatic cell count (SCC) in consecutive lactations – **(a)** first parity, **(b)** second parity, and **(c)** $\geq$ third parity. Jeretina et al. (2016).

the level and time of the peak lactation achieved (Elahi Torshizi et al., 2011).

About 77 % of the variance in milk yield loss caused by SCC could be explained by the effects such as subsequent PAR, MQ, LI, and ASCC (Tables 3 and 4).

For consecutive lactations, the model explained almost 44 % of the variability in milk yield loss ($P < 0.01$, Table 4). It is evident that milk loss is related to subsequent lactation. The estimated mean values for PAR2 and PAR3 represented 285.1 and 333.8 kg of additional loss of milk compared to PAR1 (Table 3). Although MQ explains only a small proportion of the variance in the model (2.2 %), the differences in milk yield loss between cows with different milk yields were statistically significant. The LI in the model explained 19.3 % of the variability in milk yield losses (ΔMY) by SCC. It is an important factor that influences the prediction of milk yield loss during lactation. The worst scenario (VAR5 to VAR7) regarding milk loss was when the SCC was high during the first interval of lactation (0–80 days, Table 3). In this case, reaching the normal lactation curve is impossible even in the case when the health of the udder is speculated to improve during later stages of lactation. In addition, the ASCC remarkably affected the explanation of variability. For standard lactation, the ASCC was often used for mutual comparison of cows with regard to the level of SCC that caused adverse effects and to predict ΔMY on a daily basis. The ASCC accounted for 11.6 % of the variability in milk yield loss caused by SCC.

Based on their literature review, Hortet and Seegers (1998) reported that in published studies 38 to 84 % of variability in milk yield loss at the test-day level was explained by regression models. Most of previously reported regression models can explain about 63 to 84 % of variability; thus, the 77 % variability explained in the present study is in good agreement with the findings of previous studies. In this study, in agreement with the findings of literature reports, a significant herd effect was noted on the investigated characteristics (data not presented). The herd effect was highly significant in milk composition and SCC even in studies that included low numbers of herds, i.e., with only two commercial herds with 149 and 106 Holstein cows per herd (Friggens and Rasmussen, 2001), 12 to 58 herds with $\geq$ 80 cows per herd of Swedish Holstein and Swedish Red breeds (Nyman et al., 2009), and 85 herds of Brown Swiss with a maximum of 15 cows per herd (Bobbo et al., 2016).

In the present study, the SCC was adjusted using a dilution factor, λ. A dilution factor of -0.485, according to Green et al. (2006), was applied for SCC only for those cows that had more than 10 kg of daily milk yield and an SCC of $< 50\,000$. Only a few studies proposed the existence of the dilution effect from increased milk yield on the SCC (at the level of 200 000) in cows without IMI. High-milk-yielding cows without IMI have been thought to exhibit lower SCC than low-milk-yielding cows (Miller et al., 1993; Green et al., 2006; Halasa et al., 2009). Dilution-adjusted SCC values fitted the data better and resulted in a slightly reduced milk loss

Table 1. Some descriptive statistics of data: number of lactations, average milk yield in standard lactation (MY305, kg), fat (% F), protein (% P), and geometric mean of the somatic cell count (SCC).

Parity	Number of lactations	MY305 $\bar{x}$	SD	% F $\bar{x}$	SD	% P $\bar{x}$	SD	SCC ($\times 10^3$) $\bar{x}$	SD
1	33 488	6828	1453	4.00	0.50	3.30	0.24	122	3.1
2	25 276	7622	1778	4.07	0.55	3.39	0.27	158	3.2
≥ 3	38 474	7808	1823	4.01	0.53	3.30	0.25	236	3.3

$\bar{x}$: average. SD: standard deviation.

Table 2. Regression coefficients calculated using Eq. (5) (Ali and Schaeffer, 1987) within parities (PARs) and the rank of milk production level (MQ) for the somatic cell count index estimation according to the stage of lactation.

PAR	MQ	Range of milk	Calculated regression coefficients				
		kg	a	b	c	d	k
1	1	≤ 4500	236.5	−259.7	23.3	−172.7	39.5
	2	4501–6500	286.7	−347.5	65.7	−201.9	44.6
	3	6501–8500	293.9	−322.0	31.1	−226.4	53.5
	4	> 8500	697.5	−844.2	148.7	−507.1	108.9
2	1	≤ 5500	−66.2	199.1	−121.2	26.9	5.5
	2	5501–7500	159.6	−166.1	12.5	−102	24.1
	3	7501–9500	338.7	−335.2	3.7	−252.8	60.1
	4	> 9500	600.5	−694.5	96.7	−425.6	90.9
≥ 3	1	≤ 6000	189.6	−207.7	26.3	−115.5	25.4
	2	6001–8000	−104.4	224.1	−112.1	73.3	−8.6
	3	8001–10 000	−148.3	305.2	−144.6	95.0	−9.1
	4	> 10 000	563.0	−654.1	109.6	−387.7	86.4

Table 3. Sources of variation for milk yield loss estimation included in the linear regression model with standard errors (SEs), t values, and P values.

Effects[a]	Mean	SEs	t value	P value
Intercept	156.4	4.1	37.81	< 0.01
PAR2	285.1	2.2	127.21	< 0.01
PAR3	333.8	2.4	136.73	< 0.01
MQ2	−7.9	3.2	−2.52	< 0.05
MQ3	55.3	3.2	17.24	< 0.01
MQ4	91.9	3.9	23.81	< 0.01
VAR2[b]	176.5	3.6	48.45	< 0.01
VAR3	46.6	4.3	10.87	< 0.01
VAR4	37.4	3.8	9.90	< 0.01
VAR5[b, c]	181.1	6.2	29.31	< 0.01
VAR6[b, c]	179.0	5.8	30.99	< 0.01
VAR7[c]	184.5	3.9	47.92	< 0.01
VAR8	145.2	3.9	37.55	< 0.01
ASCC300	138.2	2.5	55.37	< 0.01
ASCC400	248.6	3.2	78.97	< 0.01
Adj. R2	0.77			

[a] PAR: parity. MQ: milk production levels. VAR: scenarios. ASCC: average somatic cell count. [b, c] Mean values with the same letters indicate non-significance ($P > 0.05$).

Table 4. Analysis of variance[1] for milk yield loss by somatic cell count effect.

Effect	DF	SS	SS %	F value	P value
PAR	2	305 554 556	43.6	12 734.67	< 0.01
MQ	3	15 716 222	2.2	436.67	< 0.01
VAR	7	135 111 629	19.3	1608.88	< 0.01
ASCC	2	81 167 063	11.6	3382.82	< 0.01
Residuals	13 586	162 990 641	23.3		

[1] PAR: parities. MQ: milk production levels. VAR: scenarios. ASCC: average somatic cell count. DF: degree of freedom. SS: sum of squares. SS %: percentage of explained variability of the whole model.

compared to the unadjusted SCC (Green et al., 2006). Contrary to the findings of previous studies, Boland et al. (2013) found no dilution effect on the SCC in the Irish dairy cattle; this was more evident in the low-SCC category (< 200 000) in which adjustment for dilution yielded similar results as those for the unadjusted model.

Cows with an SCC of $\leq 100 000$ were considered as a reference group. Two steps were adapted: first, standard lactation curves were generated for specific milk yield (Jeretina et al., 2013); next, a standard curve of SCC in healthy cows with 1, 2, and ≥ 3 lactations was developed. In this study, an expected milk yield for standard lactation MY305$_{ij}$ was estimated for each cow. The expected milk yield of j-cows was equal to the population mean value corrected to the subsequent parity PAR_i and regression of BVA. This estimation was under the assumption that all fixed and random effects showed normal distribution with a mean value of 0 and were not under other environmental effects. From this curve, predictions for daily milk production were obtained. For each 30-day interval after calving, the test interval method (ICAR) was used to predict the milk yield, and this value was obtained from the area below the expected curve of milk yield.

The duration of subclinical mastitis was described using the SCCI, for which the area (IP) above the standard curve for SCC was used. The size of the area depended on the duration and intensity of the subclinical mastitis. In addition, all incidences for the increase in SCC were used for the prediction of the SCCI. Within each 30-day interval, the effect of the SCCI was estimated as the regression for 1, 2, and ≥ 3 lactations for four milk production levels (MQ). The mean values of regression of the SCCI (10 points in standard lactation) were used to generate the SCCI curve. The milk yield loss over a specific interval with increased SCC was computed. In this study, the prolonged effect of subclinical mastitis was not adjusted. Reduction of milk could be partly compensated for by other quarters within the udder (Holdaway, 1990), but it depended on the level of infection and stage of lactation, as well as the type of pathogen (Schepers et al., 1997; De Haas et al., 2002, 2004). In the present study, all milk samples with extremely changed characteristics were excluded from further analyses. In addition, each case of clinical mastitis on the test day was not included in the analysis. Because such information was not collected systematically in the regular milk recording system (AT method, according to the ICAR), in the period between two controls, some data regarding clinical mastitis presence might not have been collected. The possible carry-over effect of clinical mastitis from previous lactation was avoided by using the SCC measured for the previous milk control before dry-off. This effect was not significant and was later excluded from the model.

A specific cyclic nature of infection of mammary glands in the data collected was presumed form the estimated SCCI in relation to the eight different scenarios (high or low levels of SCC in four lactation stages for different parities) an equal average SCC was determined for the milk yield loss across these scenarios. The SCCI enables the investigation of different situations related to varying SCCs at different lactation stages, milk intensities, and subsequent PARs (Table 5). It has been demonstrated that milk yield loss is related to the lactation stage at which the elevated SCC appears. This means that milk yield loss cannot be reliably estimated on the basis of average SCC. Information on the course of events is needed. A too-long interval between test-day controls is a period with a lack of information. During this period, subclinical mastitis could heal spontaneously or progress to clinical mastitis and thus lead to an elevated SCC. Determining the number of days before or after the diagnosis of subclinical

mastitis is important for clinical mastitis detection. Such data can be accurately obtained on an experimental farm or by using a systematic collection of data on medical treatments by veterinarians and breeders, as in some countries (Scandinavian countries, Austria).

It is questionable whether management conditions on experimental farms (Lescourret and Coulon, 1994) are appropriate and comparable with larger numbers of commercial farms (Windig et al., 2005) to study applicable methods for the detection of subclinical mastitis and its effect on milk yield loss in field conditions. Hortet and Seegers (1998) reviewed selected literature on milk yield loss, milk composition, and elevated SCCs. Prediction of milk yield loss can be affected by several factors such as differences in the methodology used for preparing data sets and calculations; the specific design of studies regarding the number and interval between test-day controls; the number and size of investigated farms, herds, and types of farms (experimental vs. commercial); housing systems; number of PARs; duration of lactation; and threshold of SCCs for subclinical mastitis. Therefore, comparing our results (Table 5) regarding the effect of elevated SCC and lactation period, subsequent PAR, milk yield, and different scenarios within standard lactation on reduced milk production with those of literature findings is difficult.

A threshold of 200 000 SCC would suggest that a cow has IMI and is likely to be infected for at least an udder quarter. Across studies primiparous cows with 200 000 SCC showed, in kilograms per day, a milk loss of 0.13 (Boland et al., 2013), 0.23 to 1.76 (Rekik et al., 2008), 0.28 (Halasa et al., 2009), 0.61 (Hortet et al., 1999), 0.46 to 0.72 (Dürr et al., 2008), 0.35 to 0.80 (Hand et al., 2012). The results of the present study for thresholds of 200 000 and 400 000 SCC predicted higher milk yield losses in the range of 0.8 to 1.4 and 1.0 to 2.7 kg day^{-1}, respectively. The seriousness of the effect of SCC increase on milk yield loss from various studies is difficult to compare directly. This is due to the SCC range and model approach, different baseline that divides healthy from unhealthy cows, dilution adjustment, different stage of lactation in which IMI occurred, production practises, and differences due to geographic area in the relative importance of different mastitis pathogens. Depending on the possible scenario, VAR2 to VAR4 (Table 5), our estimated milk yield loss for primiparous cows was at least 0.8 to 0.9 kg day^{-1} at the level of 200 000 SCC. Higher milk yield loss was predicted at the same threshold of SCC but for other scenarios. High SCC during the first 80 days of lactation had a strong impact on the increase in milk yield loss, which was estimated to be 1.2 to 1.4 kg day^{-1} (VAR1 and VAR5 to VAR7). Even in the case when the SCC dropped later during lactation to a value lower than our estimation, milk yield reduction was at least 1.2 kg day^{-1}. These findings regarding the effect of elevated SCC early during lactation and its impact on milk yield loss for an extended period are in agreement with those reported previously. During the period when the SCC peaked,

Table 5. Daily milk yield loss (ΔMY, kg) in various scenarios (VAR) of somatic cell index (SCCI) (H – high, L – low) according to lactation intervals (LI1 = 0–80, LI2 = 81–160, LI3 = 161–240, LI4 = 241–305 days in milk) within parity, milk production level (MQ), and level of average SCC ($\times 10^3$) (ASCC).

ASCC	Schemes of increased SCCI (H, L) per LI	First parity				Second parity				≥ Third parity			
		MQ1	MQ2	MQ3	MQ4	MQ1	MQ2	MQ3	MQ4	MQ1	MQ2	MQ3	MQ4
					VAR1								
200		1.4	1.3	1.3	1.4	1.9	2.0	2.4	2.3	2.0	2.0	2.2	2.4
300		1.9	2.0	1.9	1.8	3.0	2.7	3.0	2.9	2.9	2.9	3.5	3.5
400		2.5	2.0	2.0	2.0	2.9	3.3	4.1	3.7	3.3	3.5	3.7	4.3
					VAR2								
200		0.9	0.8	0.8	0.9	1.5	1.5	1.8	1.8	1.8	1.8	2.0	2.1
300		1.1	1.1	1.0	0.9	2.2	2.0	2.4	2.4	2.4	2.4	2.5	2.8
400		1.0	1.2	1.2	1.3	2.4	2.4	3.0	3.0	2.5	2.8	3.2	3.3
					VAR3								
200		0.8	0.8	0.9	0.9	1.4	1.3	1.7	1.8	1.5	1.5	1.7	1.9
300		1.1	1.0	1.0	1.0	1.8	1.6	2.1	2.3	2.1	2.0	2.2	2.6
400		1.3	1.5	1.7	1.7	1.9	1.9	3.0	2.3	2.2	2.3	2.7	3.1
					VAR4								
200		0.9	0.8	0.8	0.9	1.5	1.5	1.8	1.9	1.9	1.8	2.0	2.1
300		1.1	1.0	1.0	1.1	1.9	1.9	2.3	2.3	2.3	2.3	2.5	2.8
400		1.3	1.2	1.1	1.3	2.3	2.2	2.7	2.8	2.7	2.7	2.9	3.2
					VAR5								
200		1.3	1.3	1.3	1.3	1.8	1.8	2.4	2.5	2.0	2.1	2.2	2.7
300		1.5	1.7	1.7	2.0	2.3	2.2	3.2	2.9	2.4	2.5	2.9	3.4
400		1.7	2.2	1.9	1.1	2.5	2.7	3.5	2.5	2.8	2.9	3.8	3.6
					VAR6								
200		1.2	1.2	1.2	1.4	1.9	2.0	2.2	2.3	2.2	2.0	2.3	2.5
300		1.7	1.8	1.6	1.4	2.1	2.3	2.9	3.2	2.6	2.6	3.0	3.4
400		1.6	1.8	1.9	2.7	2.9	2.8	3.2	3.8	2.9	3.1	3.3	3.9
					VAR7								
200		1.3	1.2	1.2	1.2	2.0	1.9	2.4	2.5	2.1	2.2	2.4	2.8
300		1.6	1.5	1.5	1.6	2.4	2.4	2.9	2.9	2.7	2.7	3.0	3.2
400		2.3	1.8	1.8	2.4	2.8	2.9	3.5	3.6	3.2	3.2	3.5	4.0
					VAR8								
200		1.2	1.2	1.2	1.2	2.0	1.9	2.3	2.4	2.2	2.1	2.2	2.6
300		1.5	1.4	1.5	1.5	2.3	2.3	2.8	2.9	2.6	2.6	2.8	3.2
400		1.8	1.6	1.7	1.7	2.6	2.7	3.2	3.4	3.0	3.0	3.3	3.6

the milk yield dropped and did not reach the pre-peak levels (Windig et al., 2005). During early lactation, mastitis development had a substantial effect on reduced milk yield when milk loss was noted for an extended period during the entire lactation (Lescourret and Coulon, 1994).

In the second and/or third and subsequent PARs, elevated SCC led to milk yield losses. Studies show, in kilograms per day, milk yield losses of 0.63 to 1.17 and from 0.60 to 1.85 for PAR2 and PAR $\geq$ 3 (Hortet et el., 1999), 0.50 for PAR $\geq$ 2 (Halasa et al., 2009), 0.6 for PAR $\geq$ 2 (Hortet and Seegers, 1998), 1.05 to 2.50 for PAR $\geq$ 2 (Dürr et al., 2008),

and 0.61 to 1.07 and 0.63 to 1.09 for PAR2 and PAR $\geq$ 3 (Hand et al., 2012). Our predicted values for milk yield loss in PAR2 and PAR $\geq$ 3 ranged from 1.3 to 2.5 and 1.5 to 2.8 kg day^{-1}, respectively. An elevated SCC of 400 000 in subsequent lactations was found to be significantly associated with the loss of milk yield in many studies. In studies, reported daily milk yield loss at an SCC of 400 000 ranged from 0.78 for PAR $\geq$ 2 (Halasa et al., 2009) to 1.0–3.0 for PAR $\geq$ 2, and from 1.58 to 3.74 for PAR $\geq$ 2 (Dürr al., 2008). Exceptions to the studies above are Dürr et al. (2008) and Hand et al. (2012) and our estimated milk yield loss are very close to theirs. Nevertheless, differences in milk yield loss existed even in all cases. One explanation for this discrepancy is the threshold, a lower limit at which SCC could affect the loss of milk yield owing to IMI. Dürr et al. (2008) set the limit for subclinical mastitis at more than 7400 SCC, Hand et al. (2012) at 200 000, and in our study more than 100 000 SCC with additional specific limitations. Various lower limits of SCC over a broad range have been proposed in previous studies: 51 000 to more than 1 000 000 in PAR of one to four (Boland et al., 2013), 31 000 to more than 400 000 (Green et al., 2006), less than 200 000 (Rekik et al., 2008), 100 000 to 600 000 (Hortet et al., 1999), and 500 000 for primiparous and multiparous cows (Hagnestam-Nielsen et al., 2009). In some higher classes, the ranges were from less than 200 000 to 1 000 000 SCC with milk yield loss of 0.387 to 2.351 kg day^{-1} (Guo et al., 2010), 403 000 to 1 900 000 SCC with 0.6 to 3.8 kg estimated milk yield loss (Yalçin et al., 2000), and 200 000 to 2 000 000 SCC with a reduction of milk production of 0.35–1.09 kg day^{-1} to 1.49–4.70 kg day^{-1} (Hand et al., 2012). Another possible problem for comparison is the average milk yield in the population of the investigated cows. In this study, lower milk yield for standard lactation and higher variability for this trait was observed, unlike that in previous studies. Furthermore, not only genetic differences but also a probable lower level of herd management could be responsible for the variation in the predicted milk yield loss.

In the present study, elevated SCC had a higher effect on milk yield loss in multiparous than in primiparous cows. Multiparous cows in late lactation might be responsible for the majority of milk production loss at the herd level caused by elevated SCC (Hagnestam-Nielsen et al., 2009). Moreover, if milk losses owing to subclinical mastitis were not estimated appropriately, i.e., by using average loss per lactation, milk loss could be overestimated in the beginning of lactation, thereby remarkably underestimating losses toward the end of lactation (Dürr et al., 2008). Therefore, the SCCI developed in the present study allows corrections for the estimation of milk loss in the population of Holstein cows. Furthermore, it is applicable to other cow populations, but the standard curves and breeding value prediction methods for specific cow populations need to be determined.

4 Conclusions

Improving herd management requires the recognition of the dynamics and peaks of elevated somatic cell count with relation to daily milk loss during lactation. A standard of the average somatic cell count as a criterion for comparing cows with regard to the health status of their udder glands does not allow identification of time-related consequences of IMI for cow and herd management. The newly introduced somatic cell count index might enable the mutual comparison of milk yield loss across cows in relation to the level of SCC, effect

of consecutive parity, stage of lactations, and milk yield intensity. The SCCI has been proposed as an indicator of IMI to provide farmers reliable information to apply appropriate measures regarding cow health management and overall economical cow milk production.

Author contributions. JJ and DB designed the experiment. JJ analysed the data, and DŠ drafted the paper.

Competing interests. The authors declare that they have no conflicts of interest.

Acknowledgements. The comments and suggestions of the three anonymous referees are greatly appreciated. The authors would like to thank the Slovenian Research Agency for the financial support (P1-0164, P4-0133).

Edited by: Nina Melzer

References

Ali, T. E. and Schaeffer, L. R.: Accounting for covariances among test day milk yields in dairy cows, J. Anim. Sci., 67, 637–644, 1987.

Archer, S. C., Mc Coy, F., Wapenaar, W., and Green, M. J.: Association between somatic cell count early in the first lactation and the lifetime milk yield of cows in Irish dairy herds, J. Dairy Sci., 96, 2951–2959, 2013.

Barkema, H. W., Deluyker, H. A., Schukken, Y. H., and Lam, T.: Quarter-milk somatic cell count at calving and at the first six milkings after calving, Prev. Vet. Med., 38, 1–9, 1999.

Bates, D., Maechler, M., Bolker, B., Walker, S.: Fitting linear mixed-effects models using lme4, J. Stat. Softw., 67, 1–48, 2015.

Bobbo, T., Cipolat-Gotet, C., Bittante, G., and Cecchinato, A.: The nonlinear effect of somatic cell count on milk composition, coagulation properties, curd firmness modelling, cheese yield, and curd nutrient recovery, J. Dairy Sci., 99, 5104–5119, 2016.

Boland, F., O'Grady, L., and More, S. J.: Investigating a dilution effect between somatic cell count and milk yield and estimating milk production losses in Irish dairy cattle, J. Dairy Sci., 96, 1477–1484, 2013.

Buttchereit, N., Stamer, E., Junge, W., and Thaller, G.: Evaluation of five lactation curve models fitted for fat : protein ratio of milk and daily energy balance, J. Dairy Sci., 93, 1702–1712, 2010.

Cole, J. B., Ehrlich, J. L., and Null, D. J.: Short communication: projecting milk yield using best prediction and the MilkBot lactation model, J. Dairy Sci., 95, 4041–4044, 2012.

De Haas, Y., Veerkamp, R. F., Barkema, H. W., Gröhn, Y. T., and Schukken, Y. H.: Associations between pathogen-specific cases of clinical mastitis and somatic cell count patterns, J. Dairy Sci., 87, 95–105, 2004.

De Hass, Y., Barkema, H. W., and Veerkamp, R. F.: The effect of pathogen-specific clinical mastitis on the lactation curve for somatic cell count, J. Dairy Sci., 85, 1314–1323, 2002.

Djabri, B., Bareille, N., Beaudeau, F., and Seegers, H.: Quarter milk somatic cell count in infected dairy cows: a meta-analysis, Vet. Res., 33, 335–357, 2002.

Dürr, J. W., Cue, R. I., Monardes, H. G., Moro-Méndez, J., and Wade, K. M.: Milk loss associated with somatic cell counts per breed, parity and stage of lactation in Canadian dairy cattle, Livest. Sci., 117, 225–232, 2008.

Elahi Torshizi, M., Aslamenejad, A. A., Nassiri, M. R., and Farhangfar, H.: Comparison and evaluation of mathematical lactation curve functions of Iranian primiparous Holsteins, S. Afr. J. Anim. Sci., 41, 104–115, 2011.

El-Awady, H. G. and Oudah, E. Z. M.: Genetic and economic analysis for the relationship between udder health and milk production traits in Friesian cows, Asian-Aust. J. Anim. Sci., 24, 1514–1524, 2011.

Friggens, N. C. and Rasmussen, M. D.: Milk quality assessment in automatic milking systems: accounting for the effects of variable intervals between milkings on milk composition, Livest. Prod. Sci., 73, 45–54, 2001.

Green, L. E., Schukken, Y. H., and Green, M. J.: On distinguishing cause and consequence: do high somatic cell counts lead to lower milk yield or does high milk yield lead to lower somatic cell count?, Prev. Vet. Med., 76, 74–89, 2006.

Guo, J., Liu, X., Xu, A., and Xia, Z.: Relationship of somatic cell count with milk yield and composition in Chinese Holstein population, Agric. Sci. China, 9, 1491–1496, 2010.

Hagnestam-Nielsen, C., Emanuelson, U., Berglund, B., and Strandberg, E.: Relationship between somatic cell count and milk yield in different stages of lactation, J. Dairy Sci., 92, 3124–3133, 2009.

Halasa, T., Nielen, M., De Roos, A. P. W., Van Hoorne, R., De Jong, G., Lam, T. J. G. M., Van Werven, T., and Hogeveen, H.: Production loss due to new subclinical mastitis in Dutch dairy cows estimated with a test-day model, J. Dairy Sci., 92, 599–606, 2009.

Hand, K. J., Godkin, A., and Kelton, D. F.: Milk production and somatic cell counts: a cow-level analysis, J. Dairy Sci., 95, 1358–1362, 2012.

Heringstad, B., Gianola, D., Chang, Y. M., Ødegård, J., and Klemetsdal, G.: Genetic associations between clinical mastitis and somatic cell score in early first-lactation cows, J. Dairy Sci., 89, 2236–2244, 2006.

Holdaway, R. J.: A comparison of methods for the diagnosis of bovine subclinical mastitis within New Zealand dairy herds: a thesis presented in partial fulfilment of the requirements for the degree of Doctor of Philosophy in veterinary clinic sccince at Massey University, Massey University, Massey, New Zealand, p. 413, 1990.

Hortet, P. and Seegers, H.: Calculated milk production losses associated with elevated somatic cell counts in dairy cows: review and critical discussion, Vet. Res., 29, 497–510, 1998.

Hortet, P., Beaudeau, F., Seegers, H., and Fourichon, C.: Reduction in milk yield associated with somatic cell counts up to

600.000 cells mL^{-1} in French Holstein cows without clinical mastitis, Livest. Prod. Sci., 61, 33–34, 1999.

ICAR: ICAR International agreement of recording practices, available at: http://www.icar.org/wp-content/uploads/2016/03/Guidelines-Edition-2016.pdf (last access: 15 February 2017), 2016.

Jamrozik, J. and Schaeffer, L. R.: Test-day somatic cell score, fat-to-protein ratio, milk yield, as indicators traits for sub-clinical mastitis in dairy cows, J. Anim. Breed. Genet., 129, 11–19, 2012.

Jamrozik, J., Bohmanova, J., and Schaeffer, L. R.: Relationships between milk yield and somatic score in Canadian Holsteins from simultaneous and recursive random regression models, J. Dairy Sci., 93, 1216–1233, 2010.

Jeretina, J., Ivanovič, B., Podgoršek, P., Perpar, T., Logar, B., Sadar, M., Jenko, J., Glad, J., Božič, A., Žabjek, A., Babnik, D., and Verbič, J.: Centralno podatkovna zbirka Govedo, Kmetijski inštitut Slovenije, Ljubljana, Slovenija, available at: http://www.govedo.si (last access: 10 January 2017), 1997.

Jeretina, J., Babnik, D., and Škorjanc, D.: Modeling lactation curve standards for test-day milk yield in Holstein, Brown Swiss and Simmental cows, J. Anim. Plant Sci., 23, 754–762, 2013.

Knight, C. H. and Wilde, C. J.: Mammary cell changes during pregnancy and lactation, Livest. Prod. Sci., 35, 3–19, 1993.

Lescourret, F. and Coulon, J. B.: Modeling the impact of mastitis on milk production by dairy cows, J. Dairy Sci., 77, 2289–2301, 1994.

Miller, R. H., Paape, M. J., Fulton, L. A., and Schutz, M. M.: The relationship of milk somatic cell count to milk yields for Holstein heifers after first calving, J. Dairy Sci., 76, 728–733, 1993.

Nielsen, C., Østergaard, S., Emanuelson, U., Andresson, H., Berglund, B., and Strandberg, E.: Economic consequences of mastitis and withdrawal of milk with high somatic cell count in Swedish dairy herds, Animal, 4, 1758–1770, 2010.

Nyman, A.-K., Emanuleson, U., Gustafsson, A. H., and Persson Waller, K.: Management practices associated with udder health of first-parity dairy cows in early lactation, Prev. Vet. Med., 88, 138–149, 2009.

Nyman, A.-K., Persson Waller, K., Bennedsgaard, T. W., Larsen, T., and Emanuelson, U.: Association of udder-health indicators with cow factors and with intramammary infection in dairy cows, J. Dairy Sci., 97, 5459–5473, 2014.

Pantoja, J. C. F., Hulland, C., and Ruegg, P. L.: Dynamics of somatic cell counts and intrammamary infections across the dry period, Prev. Vet. Med., 90, 43–54, 2009.

Pyörälä, S: Indicators of inflammation in the diagnosis of mastitis, Vet. Res., 34, 565–578, 2003.

R Core Team, R: A language and environment for statistical computing, R Foundation for statistical computing, Vienna, Austria, available at: http://www.R-project.org, last access: 15 July 2016.

Rekik, B., Ajili, N., Belhani, H., Ben Gara, A., and Rouissi, H.: Effect of somatic cell count on milk and protein yields and female fertility in Tunisian Holstein dairy cows, Livest. Sci., 116, 309–317, 2008.

Rupp, R. and Boichard, D.: Relationship of early lactation somatic cell count with risk of subsequent first clinical mastitis, Livest. Prod. Sci., 62, 169–180, 2000.

Schepers, A. J., Lam, T. J. G. M., Schunkken, Y. H., Wilmink, J. B. M., and Hanekamp, W. J. A.: Estimation of variance components for somatic cell counts to determine thresholds for uninfected quarters, J. Dairy Sci., 80, 1833–1840, 1997.

Schukken, Y. H., Wilson, D. J., Welcome, F., Garrison-Tikofsky, L., and Gonzales, R. N.: Monitoring udder health and milk quality using somatic cell counts, Vet. Res., 34, 579–596, 2003.

Seegers, H., Fourichon, C., and Beaudeau, F.: Production effects related to mastitis and mastitis economics in dairy cattle herds, Vet. Res., 34, 475–491, 2003.

Windig, J. J., Calus, M. P. L., De Jong, G., and Veerkamp, R. F.: The association between somatic cell count patterns and milk production prior to mastitis, Livest. Prod. Sci., 96, 291–299, 2005.

Wolfenson, D., Leitner, G., and Lavon, Y.: The disruptive effects of mastitis on reproduction and fertility in dairy cows, Ital. J. Anim. Sci., 14, 650–654, 2015.

Yalçin, C., Cevger, Y., Türkyilmaz K., and Uysal, G.: Estimation of milk yield losses from subclinical mastitis in dairy cows, Turk. J. Vet. Anim. Sci., 24, 599–604, 2000.

19

Invited review: Further progress is needed in procedures for the biological evaluation of dietary protein quality in pig and poultry feeds

Frank Liebert

Chair of Animal Nutrition, University of Göttingen, Kellnerweg 6, 37077 Göttingen, Germany

Correspondence to: Frank Liebert (flieber@gwdg.de)

Abstract. Recently, biological procedures for feed protein evaluation in pig and poultry diets have been based on the amino acid composition of feed ingredients considering the animal's losses during processes of digestion or total protein utilization in a different manner. Such a development towards individual amino acids (AAs) was inevitable according to the disadvantage of traditional protein quality measures, like biological value (BV) or net protein utilization (NPU), to be non-additive in complex animal diets. In consequence, such measures are generally not suitable for predicting the final protein quality of protein mixtures from the individual protein value of feed ingredients. Otherwise, recent measures of AA disappearance from the small intestine up to the end of the ileum (ileal AA digestibility) also do not provide a true reflection of the biological availability of individual feed AAs independent of the extent of taking into account endogenous AA losses during digestion processes. Sophisticated procedures for protein evaluation are needed considering the AA losses, both during absorption and utilization after absorption. Advantages and limitations of important developments in procedures are discussed. Accordingly, the development of an exponential modelling approach is described (the "Göttingen approach"), which overcomes some of the traditional disadvantages by measuring the individual AA efficiency. Connecting feed protein evaluation, the modelling of quantitative AA requirements, and improved ideal protein concepts offers different fields of application. In addition, as demonstrated by example, the modelling of nitrogen losses per unit protein deposition and the minimizing of this parameter yields a further interesting tool for lowering the nitrogen burden from protein utilization processes. Finally, it is pointed out that traditional laboratory procedures also need to be updated, adapted to current knowledge, and validated according to the increasing hurdles for animal studies from the viewpoint of animal welfare. Modelling is a procedure with the potential to reduce the number of experimental animals significantly. This development needs more attention, higher acceptance, and wider application in the future of protein evaluation.

1 Introduction

Today, the importance of valid protein evaluation systems in animal nutrition is not a point of dispute. However, the procedures underlay a continuous development over more than 75 years. The implications for the sustainable use of feed protein resources in animal nutrition, which are partly in concurrency with human needs, are clear. Environmental aspects also increase the pressure to further lower the dietary protein supply in animal diets without a decline in the animal's performance data. Consequently, an extended number of in-

dispensable amino acids (AA) have become more interesting as a feed additive to compensate for the suboptimal dietary supply of individual AAs. This process yields lower nitrogen (N) loads which have to be eliminated from the animal's metabolism by urea or uric acid synthesis. In summary, all these factors are a driving force for ongoing research on protein evaluation in animal nutrition. The current review aims to summarize the important steps in the development of this important area of nutritional research over decades and also to discuss the advantages and limitations of approaches and draw some conclusions for focusing further research work.

In contrast to earlier reviews (Bock, 1975; Bergner, 1994), an extended focus on biological protein evaluation (Hackler, 1977) will overcome the excessive view on digestibility-related processes in the digestive tract. This current procedure is supported by the general view of Fuller (2012) who pointed out that digestibility is not the only determinant of nutrient bioavailability; an integration of factors is needed for factors that limit the extent of absorption and the availability of AA for metabolism. Despite the well-known difficulties of such an integrated procedure, it will provide the most validated information from a nutritional point of view and consequently also the guideline for the present review.

2   Important developments

The starting point of intensified protein research would be expected in the middle of the 19th century, but it seems speculative to name the first scientist who recognized the nutritional importance of N-containing substances in feed or food. An excellent review by Block and Mitchell (1946) indicates that up to the beginning of the 20th century it was believed that only intact proteins were of nutritional value for the consumer. However, several studies applying hydrolyzed proteins provided the experimental background for the current view that protein nutrition is in fact an AA nutrition, and last but not least, that it is necessary to decide between dispensable and indispensable AAs depending on species and age (see Block and Mitchell, 1946).

Up to now, it has been a fundamental principle of biological protein evaluation to relate the effect of a given protein intake to the animal's response as measured by different, but mostly growth-related, criteria. Osborne et al. (1919) have set the starting point by creating the protein efficiency ratio (PER) in experiments with the laboratory rat to define the maximum PER for individual protein sources based on experiments with graded dietary protein supply.

PER(protein efficiency ratio) = Body weight gain : Protein intake

In fact, the observed maximum PER of individual feed proteins differs depending on the dietary protein quality, but the maximum PER is achieved with a different dietary protein supply. In consequence, several later PER applications have modified the original approach through a standardization of protein intake. Block and Mitchell (1946) discussed these procedures in detail and mostly under critical view. However, based on the rather easy way of measuring both the protein intake and the gain response in experimental animals, the PER approach is also currently in use as a complex measure of dietary protein quality, mainly for protein substitution studies in fish nutrition (e.g. Peres and Olivia-Teles, 2005; Slawski et al., 2011; Piccolo et al., 2017) or as response criteria in requirement studies (e.g. Ahmed and Khan, 2004). Several limitations in the procedure are mostly overlooked. In spite of the uncomplicated measure of body weight gain,

the age-dependent variation in body nutrient composition is not taken into account. However, the response of this influence factor to derived dietary protein quality is lower in standardized rat growth trials, but not in agricultural animals.

In addition, Eggum et al. (1971, cited by Bock, 1975) proposed a nitrogen efficiency ratio (NER) to eliminate effects resulting from the transfer factor (6.25) for crude protein calculation from analysed nitrogen content on protein quality assessment. In consequence, a more precise distinction between different feed proteins was expected.

$$\text{NER(nitrogen efficiency ratio)} = \frac{\text{Body weight gain}}{\text{Nitrogen intake}}$$

This assumption was not validated in general, and consequently the modification was not widely introduced in animal nutrition.

The recommendation of Mitchell (1924) to evaluate feed proteins based on the biological value (BV) became much more precise by taking into account more physiologically based data, like N deposition (ND), N maintenance requirement (NMR), and the true digestibility of the feed protein:

$$\text{BV(biological value)} = \frac{\text{ND} + \text{NMR}}{\text{N intake as truly digested}} \times 100.$$

In addition to the observed N deposition (ND) data as response criteria provided by N-balance studies, information about the quantity of endogenous N losses was required. The nitrogen maintenance requirement (NMR) is a reflection of the N quantity needed to replace the metabolic (endogenous) N losses via faeces and urine, respectively. Finally, data about BV were achieved by relating the sum of N deposition (ND) and NMR to the uptake of truly digested feed protein as a measure of N utilization following the process of absorption. Over decades, this procedure dominated the field of feed and food protein evaluation for single-bowl animals. In consequence, based on N-balance studies, fundamental concepts were developed to provide comparable information about the complex protein value of individual feedstuffs or diets, in spite of the fact that the knowledge about protein metabolism and functional properties of individual AAs increased (Lintzel, 1939).

Mitchell and Carman (1924) created a net protein value taking into account protein content, protein digestibility, and BV as the three important factors for the dietary protein value. The net protein value of an individual protein source was achieved by multiplying these data (Mitchell et al., 1945). Later on, multiplying the coefficient of true protein digestibility and BV provided a useful measure of total utilization or net utilization of a dietary protein (Block and Mitchell, 1946). Accordingly, Bender and Miller (1953a) defined the net protein value (NPV) based on results of the traditional N-balance technique. However, due to an elevated number of N analyses and a time-consuming procedure, a short rat assay estimating the N content in the body from a strong correlation between body water and whole body nitrogen content (Bender and Miller, 1953b) was later recommended (Miller and Bender, 1955) for assessing net protein utilization (NPU):

$$\text{NPU(net protein utilization)} = \frac{B - (Bk - Ik)}{I},$$

where B is the total body N of the rats on the test protein, Bk is the total body N of the rats on a non-protein diet, I is the N intake of the test protein group, and Ik is the N intake of the non-protein group.

Expressed as a percentage, the NPU reflects the efficacy of net protein utilization (Miller and Bender, 1955; Bender and Doell, 1957). From the current point of view, the term "net" indicates that a separate non-protein group of rats was utilized to create a measure for metabolic N losses, which need to be replaced by the dietary protein supply.

Summarizing the expressiveness of both true N digestibility and BV, Lintzel (1941) proposed the term "Physiologischer Nutzwert":

$$\text{Physiologischer Nutzwert} = \frac{\text{True N digestibility} \times \text{biological value}}{100}.$$

A new understanding about protein metabolism led to acceptance that a mixture of absorbed exogenous and endogenous AA from protein catabolism can be utilized to replace the endogenous metabolic losses. Lintzel and Rechenberger (1940) and Gebhardt (1966) established the PNu as a benchmark for evaluating dietary protein quality:

$$\text{PNu(physiological value of protein)} = \frac{\text{ND} + \text{NMR}}{\text{NI}} \times 100.$$

In fact, the application of this formula yields equal results with Lintzel (1941). Additionally, all experimental data were related to the metabolic body weight ($\text{BW}_{\text{kg}}^{0.67}$). The sum of ND and NMR was described as N retention (NR) and needs to be distinguished from ND in terminology.

This was the initial situation when Gebhardt (1966) developed the new basic concept of an exponential N utilization model. N balance experiments with the laboratory rat and general agreement about the importance of replacing endogenous N losses in future protein evaluation systems provided the platform. An exponential function conforming to the biological laws of growth (von Bertalanffy, 1951) provided a physiologically well-founded response curve of body N deposition depending on both the quantity and quality of feed protein intake. A significant driving force for this research was the observed restriction for the application of traditional procedures for complex protein evaluation of individual feedstuffs or mixed diets. Unfortunately, traditional measures, like PER, BV, and NPU, were not independent of the actual level of dietary protein intake (Block and Mitchell, 1946). Each of these parameters was modulated with characteristic course when the dietary protein supply of the same protein was increased or lowered.

In Germany, a special working group on protein evaluation was established to discuss fundamental problems of BV and NPU during the 12th annual meeting of the Society for Nutrition Physiology (Gebhardt and Brune, 1960). This was indeed the starting point to improve the reliability of feed protein evaluation. Accordingly, the new concept of Gebhardt (1966) was at first focused on standardization to improve the comparability of protein quality measures. Consequently, the exponential model was developed as a tool to make dietary protein quality parameters independent of N intake. Due to the common principle of several procedures taking into consideration the cost of N maintenance metabolism, the common term NPU is subsequently applied for protein quality measures making use of the relation between NR and N intake (NI):

$$\text{NPU(net protein utilization)} = \frac{\text{NR}}{\text{NI}} \times 100.$$

This application is valid independent of different methods and different adequacy to reflect the real quantitative N costs for maintenance metabolism. In this context, no distinction is made between N-balance data and the results of comparative slaughter techniques with whole body analyses to quantify ND in the animal. This type of model application is still in use for evaluating the complex protein value of mixed feeds. However, in the meantime the application field of the approach was significantly extended and adapted to recent expectations for protein nutrition research in food-producing animals.

3   Current applications

Several reports provide the details of current developments and applications of the basic concept as initiated by Gebhardt (1966) and further developed by Liebert and Gebhardt (1988). Today, the procedure is called the "Göttingen approach" due to further developments over 2 decades at the University of Göttingen (Liebert, 2015; Dorigam et al., 2017; Samadi et al., 2017). However, it will not be possible to outline in detail how the different issues of the current procedure differ from other approaches recently in use. Model-specific parameters as utilized in current applications were justified in earlier and recent publications (e.g. Liebert et al., 2000; Thong and Liebert, 2004a–c; Samadi and Liebert, 2006a, b, 2007a, b, 2008; Liebert and Benkendorff, 2007a, b; Liebert, 2008, 2009, 2015; Liebert and Wecke, 2008; Samadi et al., 2017; Wecke and Liebert, 2009, 2010, 2013; Wecke et al., 2016; Dorigam et al., 2017) and can be condensed as follows:

$$\text{NR} = \text{NR}_{\max} T \left(1 - e^{-\text{NI} \cdot b}\right)$$

$$\text{ND} = \text{NR}_{\max} T \left(1 - e^{-\text{NI} \cdot b}\right) - \text{NMR}, \tag{1}$$

where NR is daily N retention (ND + NMR) $[\text{mg} \, (\text{BW}_{\text{kg}}^{0.67})^{-1}]$, ND is daily N deposition or N bal-

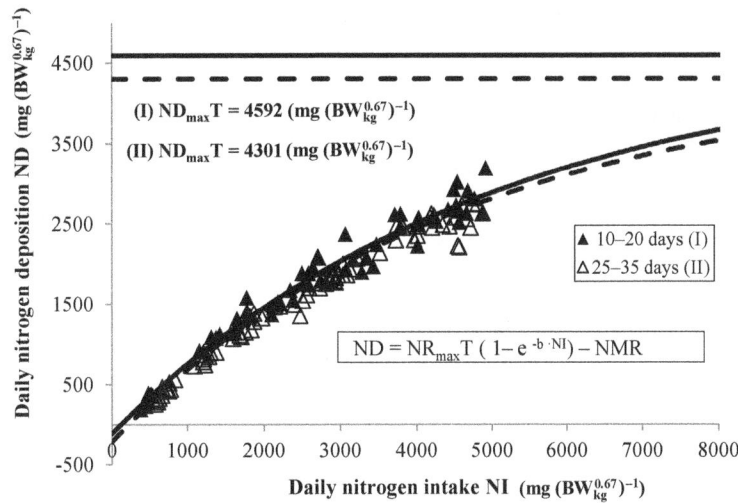

Figure 1. Results of N rise experiments with growing chickens (Ross 308) depending on age period for estimating the threshold value $ND_{max}T$ (Pastor et al., 2013).

ance $(NI - NEX \ [mg \, (BW_{kg}^{0.67})^{-1}])$, NI is daily N intake $[mg \, (BW_{kg}^{0.67})^{-1}]$, NEX is daily N excretion $[mg \, (BW_{kg}^{0.67})^{-1}]$, NMR is daily N maintenance requirement $[mg \, (BW_{kg}^{0.67})^{-1}]$, $NR_{max}T$ is the theoretical maximum for daily NR $[mg \, (BW_{kg}^{0.67})^{-1}]$, $ND_{max}T = NR_{max}T - NMR$ is the theoretical maximum for daily ND $[mg \, (BW_{kg}^{0.67})^{-1}]$, b is the slope of the NR curve indicating the dietary protein quality (the slope of the curve for a given protein quality is independent on NI), and e is the basic number of the natural logarithm (ln).

The attribute "theoretical" suggests that the threshold values ($ND_{max}T$ or $NR_{max}T$) are generally not in the scope of practical growth performance data but yield an estimate of the genetic potential when each of the limiting factors for maximum growth are eliminated (Samadi and Liebert, 2006a). This is a theoretical situation indeed, but at least not a limiting factor to derive practical data from modelling. Figure 1 gives an example for this application from current studies.

The genetic potential is defined as an unreachable theoretical threshold value of the exponential function and cannot be realized even with an optimized feeding strategy or in ideal environmental conditions. If the ranking of such a threshold value is clear, no problem exists for further model applications. Accordingly, individual amino acid (AA) requirement data are derived for daily protein deposition data in line with practical growth data. The threshold value ($ND_{max}T$ resp. $NR_{max}T$) is used only as a model parameter to relate the real rate of deposition to the estimated genetic potential.

A validation of the model parameter b as a measure of dietary protein quality, which is independent of the actual level of protein intake, has been reported in several pig and poultry studies (e.g. Thong and Liebert, 2004a, b, c; Wecke

and Liebert, 2009; Farke, 2011; Pastor et al., 2013; Pastor, 2014). According to the basic concept of standardizing NI for valid feed protein evaluation, the model is also currently applied for assessing the dietary protein quality in mixed diets with alternative protein sources making use of a standardized value of NPU (Brede et al., 2016; Dietz et al., 2016; Dietz and Liebert, 2017; Neumann et al., 2017). More diversified applications of such an important tool could help to overcome misleading conclusions about the reality of distinctions in feed protein value between protein sources (Neumann et al., 2017).

3.1 Characterization of developing the genetic potential

As already discussed, the estimation of $ND_{max}T$ is required as a threshold value for basic applications of the exponential model, but as a given percentage of the theoretical threshold value $ND_{max}T$ real performance data are utilized to derive AA requirement data depending on graded aimed animal performance (e.g. Wecke et al., 2016; Samadi et al., 2017). However, also from the viewpoint of animal breeding, the observed $ND_{max}T$ data are of interest because they provide additional information about breeding success. An example for this application with growing chickens is demonstrated in Fig. 1.

The threshold value of the exponential function ($ND_{max}T$) is estimated by statistical application of the Levenberg–Marquardt algorithm (Marquardt, 1963) as reported elsewhere (e.g. Samadi and Liebert, 2008; Wecke and Liebert, 2009; Pastor et al., 2013). The applicability of the procedure was also demonstrated in fish nutrition (Liebert et al., 2006) and utilized for AA requirement studies in *Oreochromis niloticus* (Liebert and Benkendorff, 2007a, b; Liebert, 2009).

Table 1. Age-dependent $ND_{max}T$ $[mgN\,(BW_{kg}^{0.67})^{-1}\,day^{-1}]$ of fattening pigs with different genders and years as derived from N-balance studies with graded dietary protein supply and approximated functions for $NR_{max}T$ depending on body weight (BW).

Average BW	Estimated $ND_{max}T$		
(kg)	Boars[1]	Female pigs[2]	Boars[3]
30	1740	2515	3800
40	1538	2020	3348
50	1395	1696	2998
60	1287	1466	2716
70	1201	1293	2479
80	1130	1157	2276
90	1071	1046	2098
100	1020	955	1941
110	975	877	1798

[1] Gebhardt (1973); $NR_{max}T = 6995.7 \times BW_{kg}^{-0.3635}$; NMR = 292.
[2] Liebert and Gebhardt (1988); $NR_{max}T = 29038 \times BW_{kg}^{-0.6776}$; NMR = 283 (Nörenberg, 1987). [3] Wecke and Liebert (2009); $NR_{max}T = -1619.3 \times lnBW_{kg} + 9733.6$; NMR as a function of BW.

Table 2. Age-dependent $ND_{max}T$ $[mgN\,(BW_{kg}^{0.67})^{-1}\,day^{-1}]$ of male meat-type chicken genotypes.

	Estimated $ND_{max}T$			
Genotype	Ross 308[1]	Ross 308[2]	Na/na[3]	Na/Na[3]
	Age period (days)			
10–20	3412	4592	3741	3501
25–35	2713	4301	3166	3056

[1] Farke (2011); [2] Pastor et al. (2013); [3] Khan et al. (2015) (Na/na heterozygous; Na/Na homozygous naked neck meat-type chicken).

Table 3. Age-dependent $ND_{max}T$ $[mgN\,(BW_{kg}^{0.67})^{-1}\,day^{-1}]$ of older male meat-type chicken genotypes (earlier data as summarized by Liebert, 2008).

	Estimated $ND_{max}T$			
Genotype	Cobb 500	Ross 308	I 657*	Red JA*
	Age period (days)			
10–25	3634	3663	2807	2789
30–45	2783	2751	2723	2688
50–65	1783	1936	1486	1419
70–85	1386	nd	1191	1043

* Hubbard ISA extensive genotypes of meat-type chicken (Samadi and Liebert, 2007a).

The summarized results of a series of experiments, both earlier and current, are given in Tables 1–3. In consequence, the estimated $ND_{max}T$ data give an indication of the influencing factors, like age period, gender, and breeding progress. As demonstrated in Table 1, parameter $ND_{max}T$ declines with increasing age, but the course of the threshold value is also dependent on the gender. In addition, the breeding progress in the modern genotype is clear.

The age and genotype effect is also valid in growing meat-type chickens (Tables 2 and 3).

It has to be repeated that $ND_{max}T$ data are not real data, but theoretical values resulting from a statistical estimation of threshold values of the N-rise curve dependent on N intake. This cannot be seen as a disadvantage of the approach because modelling quantitative AA requirements makes use of real ND data.

3.2 Amino acid requirements based on dietary amino acid efficiency

In addition to the validated evaluation of dietary protein quality (model parameter b or standardized NPU), the "Göttingen approach" may also be applied to AA requirement studies making use of the principles from the diet dilution technique (Gous and Morris, 1985). Generally, a defined limiting AA (LAA) in the diet under study is a prerequisite for these applications because protein deposition in the animal is strictly limited by the dietary supply of this AA.

In this case, the shape of the NR curve is not only a function of NI, but also of the daily intake of the LAA (LAAI) as a part of the feed protein fraction. For that important appli-

cation, the basic function (1) is logarithmically transformed (natural logarithm, ln) and provides Eqs. (2) and (3):

$$NI = [\ln NR_{max}T - \ln(NR_{max}T - NR)] : b, \qquad (2)$$
$$b = [\ln NR_{max}T - \ln(NR_{max}T - NR)] : NI. \qquad (3)$$

The derived NI by Eq. (2) gives the daily quantity of dietary protein (N × 6.25) which is needed to yield the intended level of growth performance (in terms of NR) at a given or observed dietary protein quality (in terms of model parameter b). In addition, the model parameter b is derived by Eq. (3). Equations (1)–(3) have demonstrated earlier model applications for which the main focus was on questions of complex protein evaluation and the AA composition of the feed protein was not of top priority. However, since the review by Block and Mitchell (1946), the importance of feed protein AA composition as the most important factor in dietary protein value is well known. When the emphasis of the model changes to AA-based applications, a further important transformation is required: the function needs to be adapted because the independent variable determining the resultant dietary protein quality (b) is the concentration (c) of the LAA in the dietary protein. This fundamental connection needs to be "translated" into the traditional model applications. As reported in detail earlier (e.g. Liebert and Gebhardt, 1988; Liebert, 1995, 2008; Samadi and Liebert, 2006a, b, 2007a, b; Liebert and Wecke, 2008; Liebert, 2015), Eq. (2)

Table 4. Example for modelling lysine (Lys) requirement data during the starter and grower periods of male meat-type chickens (Ross 308) depending on graded daily protein deposition, different in-feed efficiency of Lys, and predicted daily feed intake (Wecke et al., 2016).

Starter period (d10–20, mean BW 600 g)									
PD (g day^{-1})		9			10			11	
BWG (g day^{-1})		55			61			67	
	(1)	(2)	(3)	(1)	(2)	(3)	(1)	(2)	(3)
bc_{Lys}^{-1}	53.1	50.4	47.8	53.1	50.4	47.8	53.1	50.4	47.8
Lys required									
(mg (BW$_{kg}^{0.67}$)$^{-1}$ day^{-1})	901	948	1001	1044	1099	1160	1206	1270	1340
(mg day^{-1})	640	673	711	741	780	823	857	902	952
Lys content needed in the starter diet (%)									
FI (g day^{-1})									
70	0.91	0.96	1.02	1.06	1.12	1.18	1.22	1.29	1.36
80	0.80	0.84	0.89	0.93	0.98	1.03	1.07	1.13	1.19
90	0.71	0.75	0.79	0.82	0.87	0.92	0.95	1.00	1.06
Grower period (d25–35, mean BW 1800 g)									
PD (g d^{-1})		15			16.5			18	
BWG (g day^{-1})		91			100			109	
	(1)	(2)	(3)	(1)	(2)	(3)	(1)	(2)	(3)
bc_{Lys}^{-1}	64.5	61.3	58.1	64.5	61.3	58.1	64.5	61.3	58.1
Lys required									
(mg (BW$_{kg}^{0.67}$)$^{-1}$ day^{-1})	753	793	837	858	903	953	975	1026	1083
(mg day^{-1})	1117	1175	1241	1272	1339	1413	1446	1522	1606
Lys content needed in the grower diet (%)									
FI (g day^{-1})									
150	0.74	0.78	0.83	0.85	0.89	0.94	0.96	1.02	1.07
170	0.66	0.69	0.73	0.75	0.79	0.83	0.85	0.90	0.95
190	0.59	0.62	0.65	0.67	0.70	0.74	0.76	0.80	0.84

PD is daily protein deposition (N deposition × 6.25), BWG is daily body weight gain (crude protein content in BWG 16.5 %), bc_{Lys}^{-1} is lysine efficiency: (1) as observed, (2) 5 % lower as observed, (3) 10 % lower as observed. Lys supply required is the lysine requirement for targeted PD. FI is daily feed intake.

can be transformed into Eq. (4) when taking into account the concentration of LAA in the feed protein:

$$LAAI = [\ln NR_{max} T - \ln (NR_{max} T - NR)] : 16bc^{-1}, \quad (4)$$

where LAAI is the daily intake of the LAA [mg (BW$_{kg}^{0.67}$)$^{-1}$], c is the concentration of the LAA in the feed protein [g 16 gN^{-1}], and bc^{-1} is the observed dietary efficiency of the LAA.

Equation (4) is widely applied for assessing quantitative AA requirement data in both earlier (Liebert et al., 1987; Liebert and Gebhardt, 1988; Thong and Liebert, 2004a–c; Samadi and Liebert, 2006a, b, 2007a, b; Liebert, 2009; Wecke and Liebert, 2009, 2010) and recent studies (Pastor et al., 2013; Wecke and Liebert, 2013; Khan et al., 2015; Dorigam et al., 2017; Samadi et al., 2017). An important precondition for validated conclusions is that experimental

data are available which describe the NR or ND response to a defined LAAI at a specific level of dietary efficiency of the LAA, as reflected by the model parameter (bc^{-1}). The existing relationship between the aimed daily ND, graded dietary efficiency of the AA under study, and required LAAI in context with the expected level of feed intake is demonstrated in Table 4.

It is shown by example that the finally recommended in-feed concentration of lysine is under the influence of both animal factors and feed factors, which need to be taken into account for the validity of the recommended in-feed AA concentrations. The real feed intake depends on age, gender, and genotype, but environmental variables, like climate, are also generally underestimated influence factors. More attention has to be given to the modulating effects of such zootechnical factors. If not, it cannot be expected that requirement studies

Table 5. Optimal dietary ratios for individual amino acids as related to lysine; results of a meta-analysis (Wecke and Liebert, 2013).

	N^*	Average	SD
Lysine	26	100	0
Methionine	22	40	4
Methionine + cysteine	24	74	2
Threonine	24	66	3
Tryptophan	22	16	1
Arginine	25	105	4
Histidine	12	34	4
Isoleucine	24	69	4
Valine	21	80	4
Leucine	12	110	6
Phenylalanine	8	66	3
Phenylalanine + tyrosine	9	120	7

* Number of references involved.

under controlled conditions will yield generalizable requirement data. These factors are also important when traditional dose–response experiments are applied in AA requirement studies, but they are insufficiently taken into account as currently demonstrated by Samadi et al. (2017).

Dose–response experiments are widely applied when the efficacy of supplemented AAs is under study. However, misleading efficacy for L- and DL-methionine isomers was concluded (Shen et al., 2014) when both the basic preconditions for the application of statistical procedures and factors as discussed above are ignored. In contrast, applications of the "Göttingen approach" yielded similar methionine efficiency for both of the isomers in chicken studies (Liebert et al., 2015) in agreement with recent reports (e.g. Htoo and Morales, 2016). This example underlines the importance of a verified experimental design and validated physiologically based statistical procedures for generalized conclusions about the efficacy of supplemented feed AAs.

3.3 Improvements on the ideal protein concept

One approach to realize the high efficiency of protein utilization in agricultural animals is the earlier concept to recommend an ideal dietary protein composition for diet formulation (Almquist and Grau, 1944; Oser, 1951; Dean and Scott, 1965). Later on, the dietary ideal amino acid ratio (IAAR) was introduced by Cole et al. (1980) and taken over by the British ARC (1981) for pig nutrition.

Currently, the IAAR concept is widely accepted in pig and poultry nutrition (e.g. Baker, 2003; Wecke and Liebert, 2013; Wecke et al., 2016). The individual indispensable AAs have to be related to a reference AA, mostly lysine (Lys), which is almost exclusively utilized for body protein deposition in growing animals. In addition to the quantitative AA requirement data (Table 4), applications of the "Göttingen approach" may also contribute to improving the IAAR both indirectly via individual AA requirements and directly by relating the observed AA efficiency data (model parameter bc^{-1}) as reported recently (Samadi and Liebert, 2008; Pastor et al., 2013; Wecke and Liebert, 2013; Khan et al., 2015; Liebert, 2015). According to Samadi and Liebert (2008), the reciprocal relationship between Lys efficiency (as reference AA) and the observed efficiency of the individual LAA under study is utilized to derive optimal dietary AA ratios (Eq. 5):

Table 6. IAAR of growing meat-type chickens as derived by directly relating observed amino acid efficiency data according to Eq. (5) (Wecke and Liebert, 2013).

Amino acid	Starter	Grower
Lysine	100	100
Threonine	60	62
Tryptophan	19	17
Arginine	105	105
Isoleucine	55	65
Valine	63	79

$$\text{IAAR} = bc_{\text{LYS}}^{-1} : bc_{\text{LAA}}^{-1}. \tag{5}$$

As already mentioned, model parameter b linearly depends on LAA concentration (c) in the protein, and the slope (bc^{-1}) is an expression of AA efficiency by summarizing both digestibility and post-absorptive utilization of the LAA in general agreement with Lintzel (1941). In addition, the order of observed AA efficiency data from the individual AA under study is indirectly related to the specific physiological AA requirement per unit of protein deposition. From this point of view, both feed factors and animal factors are involved when comparisons are made at the level of observed AA efficiency data. As pointed out by Wecke et al. (2016), the reliability of measured AA efficiency data for the reference AA Lys is a fundamental precondition for such applications. The summarized results of a meta-analysis are given in Table 5.

Actually, the complete information about the IAAR of indispensable AAs with the "Göttingen approach" is not available. A summary of current results based on applications of Eq. (5) is given in Table 6.

According to the fact that both feed and animal factors modulate the observed AA efficiency data, further studies have to enlighten their individual quantitative importance. The sulfur-containing AAs methionine and cysteine are the focus of ongoing experiments.

4 Future applications

Eggum and Christensen (1974) basically demonstrated the additivity of the protein digestibility data in a mixture in relation to the protein digestibility of individual ingredients. However, the missing additivity of traditional protein quality parameters for individual feed proteins, as discussed above, is the main limitation to making use of these parameters in optimizing animal feeds. Consequently, the further development of protein quality evaluation systems had to be founded on evaluation of individual AAs. At least the specific contribution of the individual feed proteins is added, and in summary it yields the AA content of the final diet.

Over many years, only the chemically analysed total AA content was utilized in feed formulation for monogastric agricultural animals. A next step to come closer to the utilization process in the animal was focused on AA digestibility as measured at the end of the digestive tract (digestible AA). However, increasing knowledge about the significance of microbial processes in the digestive tract, namely in the post-ileal sections of the intestine, led to procedures for measuring the individual AA digestibility up to the end of the small intestine (e.g. Low, 1980; Sauer and Ozimek, 1986; Van Leeuwen et al., 1987; Lemme et al., 2004; Stein et al., 2007). Since Low (1980), it is generally accepted that ileal measurement is preferred to the faecal method in simple-stomached animals when the digestion and absorption of AAs is to be evaluated. However, ileal digestibility may be expressed as apparent, standardized, or true digestibility. Endogenous losses are separated into basal and specific losses, and specific losses are induced by feed ingredient characteristics, like fiber content, type of fiber, and anti-nutritional factors (Stein et al., 2007). In consequence, a high modulation of endogenous AA losses can be expected but is sufficiently taken into account only in part. Currently, only basal AA losses are estimated depending on feed intake and providing a standardized ileal digestibility. In consequence, a database for standardized AA digestibility in pig and poultry was created (e.g. Evonik, 2016). The advantage is that standardized AA digestibility data are more likely to be taken into account in mixed diets compared with apparent ones (Stein et al., 2005). In this context, it is important to note again that standardized ileal AA digestibility only means that basal endogenous AA losses are considered. In addition, several proposals were made to standardize the experimental procedures as a whole, namely the section of the small intestine taken for chyme sampling in poultry studies (e.g. Kluth and Rodehutscord, 2006, 2009). Generally, for an improved validity of the observed AA digestibility data, a standard type of experiment is required taking into account more than the procedure of chyme sampling (Ravindran et al., 2017).

However, according to Stein et al. (2007) all measures of AA digestibility are generally based on the disappearance of AA from the digestive tract only. These measures do not reflect the net breakdown or synthesis of AA in the intestinal lumen and the absorption of chemical forms, like Maillard reaction products (Maillard, 1912) with Lys, which are precluded from metabolic utilization for protein synthesis. The ε-amino group of Lys is the primary target for an attack by reducing carbohydrates, and up to 70 % of the Lys residues of a protein are reactive and can be damaged depending on the factors time and temperature (Finot et al., 1977). Previous work with growing pigs has demonstrated that the ileal digestibility assay overestimates the availability of Lys, but also threonine, methionine, and tryptophan in heat-processed proteins (Batterham et al., 1990; Batterham, 1992). It appears that a considerable portion of these amino acids is absorbed but inefficiently utilized. In the case of isoleucine, it was indicated that ileal digestibility more closely reflected the proportion of the AA that can be utilized by the pig (Batterham and Andersen, 1994). Consequently, in the case of heat-processed feed proteins it cannot be expected that measures of the ileal AA digestibility are generally a valid indicator of the available AA supply in pigs. According to Carpenter (1973), reactive amino groups can also be provided by arginine and histidine, indicating that Lys represents not the only problem but the most important one.

In addition, microbial fermentation in the small intestine may also contribute to the synthesis and catabolism of AA, and in consequence to discrepancies between ileal AA digestibility data and AA bioavailability, which include AA utilization following the absorption process (Fuller, 2003).

Summarizing these aspects with a focus on future developments in feed protein evaluation, it cannot be accepted to commit only to ileal AA digestibility. In addition, strengthened animal protection laws are limiting surgery techniques to make use of fistulated pigs or caecectomized birds. In consequence, it remains doubtful whether the needed database update can be sufficiently ensured by in vivo studies. The applications of traditional procedures, like feeding experiments and digestibility and balance studies, are also relevant from the viewpoint of animal welfare when metabolism cages restrict activities, movement, and inter-individual contact. Consequently, the demand from the viewpoint of animal science needs to be stated for further scientific development (Committee for Requirement Standards of the Society of Nutrition Physiology, 2017).

Unfortunately, measures of AA bioavailability based on the response of growth parameters or body protein deposition, which can sort below the maximum permissible load from the viewpoint of animal protection, are generally restricted to investigating the LAA under study. In consequence, both the procedure AA efficiency ("Göttingen approach") and each of the other techniques to measure AA bioavailability cannot provide an enlarged database usable for feed protein evaluation systems. The only way out for routine protein evaluation is to create more in vitro techniques as proposed earlier (e.g. Savoie and Gauthier, 1986; Galibois et al., 1989; Huang et al., 2000; Van Kempen and Bodin, 1998; Boisen, 2000).

In addition, analytical procedures for the evaluation of AA bioavailability, extensively starting with Carpenter (1960, 1973) and Booth (1971), may yield improved information when they are further developed (e.g. Hurrell et al., 1979; Nordheim and Coon, 1984). The use of the rat as a model animal for growing pigs was discussed by Rutherford and Moughan (2003). The potential for such alternative procedures can be seen when they are adapted to current knowledge and validated in vivo. However, systemic developments in this field are unfortunately missing.

The further potential of the modelling procedure as presented consists of estimating N losses during protein conversion processes in the animal, depending on both feed factors and animal factors (Dänicke and Liebert, 1992; Liebert, 1996; Liebert and Wecke, 2010, 2012). Such a tool has the potential to be developed into a physiologically based estimate for N excretion (NEX) per unit ND (NEX : ND) deposition depending on the aimed animal's performance (ND) and the available feed protein in terms of quantity and quality. An example for this application is given in Fig. 2.

Clearly, the lowest ratio NEX : ND in a 50 kg growing pig was achieved at approximately 2500 mg NI per $BW_{kg}^{0.67}$, corresponding to 215 g of daily crude protein intake and providing 115 g of daily protein deposition. It is indicated that both a lower and higher protein supply create a higher ratio NEX : ND. However, the course of the response curve is also dependent on the age period and the dietary protein quality. In consequence, the better the protein quality, the lower the required protein supply, and the ratio NEX : ND will further decline. In addition, requirement recommendations for individual AAs can be derived for an optimal level to make use of the $ND_{max}T$ depending on genotype and corresponding to a minimized NEX : ND. Such a sophisticated application of the modelling procedure needs an enlarged database for model parameter $ND_{max}T$ (e.g. Nörenberg, 1987; Farke, 2011; Wecke and Liebert, 2009; Khan et al., 2015) and observed individual AA efficiency data in mixed diets with and ingredient composition near practical feeding conditions (e.g. Liebert, 2008; Samadi and Liebert, 2008; Wecke and Liebert, 2009, 2010, 2013; Wecke et al., 2016; Pastor et al., 2013; Samadi et al., 2017), which may reflect the real variation in this model parameter in common feedstuffs.

Finally, modelling protein metabolism with the physiologically based "Göttingen approach" lays the foundations for the most important applications in the field of current protein evaluation for simple-stomached growing animals:

- defining the genotype in terms of the theoretical potential for N deposition ($ND_{max}T$);

- assessing feed protein value based on observed efficiency of the limiting AA;

- concluding AA requirements taking into account graded dietary AA efficiency;

- modelling AA requirements depending on the aimed level of performance (percent of $ND_{max}T$);

- evaluating the efficacy of supplemented AAs as related to protein-bound AAs or different isomers or analogues of the added-feed AAs;

- and modelling the N losses from the N utilization process in terms of minimized N excretion per unit ND.

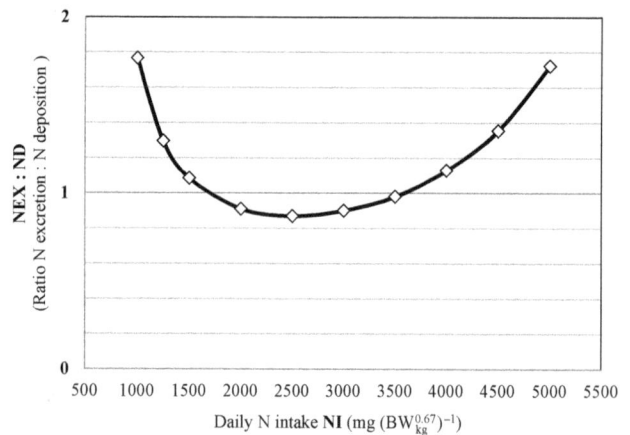

Figure 2. Course of N excretion per unit N deposition (NEX : ND) as derived from N balance data and exponential function of NEX dependent on N intake in growing pigs of 50 kg BW (Wecke and Liebert, 2009).

Greater acceptance by both scientific societies and applied research groups is needed to make use of each type of complex modelling procedure. It would be desirable to compensate for the upcoming limitation on in vivo studies due to increasing standards for animal welfare and animal protection through the extended application of physiologically based modelling, also in the field of protein evaluation for pig and poultry diets.

Competing interests. The author declares no conflict of interest.

Edited by: Manfred Mielenz

References

Ahmed, I. and Khan, M. A.: Dietary lysine requirement of fingerling Indian major carp, *Cirrhinus mrigala* (Hamilton), Aquaculture, 235, 499–511, 2004.

Almquist, H. J. and Grau, C. R.: The Amino Acid Requirement of the Chick, J. Nutr., 28, 325–331, 1944.

ARC – Agricultural Research Council: The Nutrient Requirement of Pigs, Commonwealth Agricultural Bureaux, Slough, UK, 1981.

Baker, D. H.: Ideal Amino Acid Patterns for Broiler Chicks, in: Amino Acids in Animal Nutrition, 2nd Edn., edited by: D'Mello, J. P. F., CAB International, Wallingford, Oxon, UK, 223–235, 2003.

Batterham, E. S.: Availability and utilization of amino acids for growing pigs, Nutr. Res. Rev., 5, 1–18, 1992.

Batterham, E. S. and Andersen, L. M.: Utilization of ileal digestible amino acids by growing pigs: isoleucine, Br. J. Nutr., 71, 531–541, 1994.

Batterham, E. S., Andersen, L. M., Baigent, D. R., Beech, S. A., and Elliot, R.: Utilization of ileal digestible amino acids by pigs: lysine, Br. J. Nutr., 64, 679–690, 1990.

Bender, A. E. and Doell, B. H.: Note on the determination of net protein utilization by carcass analysis, Br. J. Nutr., 11, 138–139, 1957.

Bender, A. E. and Miller, D. S.: A new brief method of estimating net protein value, Biochem. J., 53, vii, 1953a.

Bender, A. E. and Miller, D. S.: Constancy of the N / H$_2$0 ratio of the rat and its use in the determination of the net protein value, Biochem. J., 53, vii–viii, 1953b.

Bergner, H.: Ermittlung der Eiweißqualität von Nahrungs- und Futtermitteln, Arch. Anim. Nutr., 45, 293–332, 1994.

Block, R. J. and Mitchell, H. H.: The correlation of the amino acid composition of proteins with their nutritive value, Nutr. Abstr. Rev., 16, 249–278, 1946.

Bock, H.-D.: Methoden zur Beurteilung der Proteinqualität von Nahrungs- und Futtermitteln, Fortschrittsberichte für die Landwirtschaft und Nahrungsgüterwirtschaft, 13, 1–67, 1975.

Boisen, S.: In vitro digestibility methods: history and specific approaches, in: New developments in feed evaluation, edited by: Moughan, P. J., Verstegen, M. W. A., and Visscher, M., CABI, Wageningen, 153–168, 2000.

Booth, V. H.: Problems in the determination of FDNB-available lysine, J. Sci. Food Agric., 22, 658–666, 1971.

Brede, A., Neumann, C., Velten, S., and Liebert, F.: Evaluation of *Hermetia illucens* and *Spirulina platensis* proteins in semisynthetic diets for the laboratory rat, Proc. Soc. Nutr. Physiol., 25, 31, 2016.

Carpenter, K. J.: The estimation of the available lysine in animal-protein foods, Biochem. J., 77, 604–610, 1960.

Carpenter, K. J.: Damage to lysine in food processing: its measurement and its significance, Nutr. Abstr. Rev., 43, 424–451, 1973.

Cole, D. J. A., Yen, H. T., and Lewis, D.: The Lysine Requirements of Growing and Finishing Pigs – The Concept of an Ideal Protein, in: Proceedings of the 3rd International Symposium on Protein Metabolism and Nutrition, Vol. II, edited by: Oslage, H. J. and Rohr, K., EAAP, Braunschweig, Germany, 658–668, 1980.

Committee for Requirement Standards of the Society of Nutrition Physiology: Stellungnahme zur Unerlässlichkeit von Tierversuchen und zur Eignung von Ersatzmethoden in der Tierernährungsforschung, Proc. Soc. Nutr. Physiol., 26, 218–224, 2017.

Dänicke, S. und Liebert, F.: Modellierung des Wachstums und der N-Ausscheidung wachsender Masthähnchen aufder Grundlage des Konzeptes wirksamer Aminosäure, Proc. Int. Tagung Schweine- und Geflügelernährung, Halle, 17–24, 1992.

Dean, W. F. and Scott, H. M.: The Development of an Amino Acid Reference Diet for the Early Growth of Chicks, Poult. Sci., 44, 803–808, 1965.

Dietz, C. and Liebert, F.: Insect protein in aquafeed – effect of substituting soy protein on protein quality of Tilapia feed, Proc. Soc. Nutr. Physiol., 26, 147, 2017.

Dietz, C., Liebert, F. and Winter, B.: Protein hydrolysates from animal by-products as sustainable protein source. I: Effect of thermal hydrolysis and origin of basic material on protein quality, Proc. Soc. Nutr. Physiol., 25, 124, 2016.

Dorigam, J., Sakomura, N., and Liebert, F.: Modelling of lysine requirement in broiler breeder hens based on daily nitrogen reten-

tion and efficiency of dietary lysine utilization, Anim. Feed Sci. Technol., 226, 29–38, 2017.

Eggum, B. O. and Christensen, K. D.: Protein digestibility of a feed mixture in relation to the protein digestibility of individual protein components, Br. J. Nutr., 31, 213–218, 1974.

Evonik: AMINODat$^{®}$ 5.0 – Animal nutritionist's information edge, in: book I and II, Evonik Nutrition & Care GmbH, Hanau, Wolfgang, 2016.

Farke, J.: Studien zur Aminosäurenwirksamkeit beim Mastgeflügel unter spezifischer Betrachtung der schwefelhaltigen Aminosäuren, Diss. Univ. Göttingen, 199 pp., http://webdoc.sub.gwdg.de/diss/2011/farke/farke.pdf (last access: Augst 2017), 2011.

Finot, P. A., Bujard, E., and Arnaud, M.: in: Protein cross-linking – Nutritional and medical consequences, edited by: Friedman, M., Plenum Press, New York, London, 51–71, 1977.

Fuller, M.: AA bioavailability – A brief history, in: Vol. 1, Proc. 9th Int. Symp. on Digestive Physiology in Pigs, edited by: Ball, R. O., Univ. Alberta, Canada, 183–198, 2003.

Fuller, M.: Determination of protein and amino acid digestibility in foods including implications of gut microbial amino acid synthesis, Br. J. Nutr., 108, S238–S248, 2012.

Galibois, I., Savoie, L., Simoes Nunes, C., and Rerat, A.: Relation between in vitro and in vivo assessment of amino acid availability, Reprod. Nutr. Dev., 29, 495–507, 1989.

Gebhardt, G.: Die Bewertung der Eiweißqualität von Nahrungs- und Futtermitteln mit Hilfe des N-Bilanzversuches, in: Vergleichende Ernährungslehre des Menschen und seiner Haustiere, edited by: Hock, A., Fischer Verl., Jena, 323–348, 1966.

Gebhardt, G.: Parameter des N-Stoffwechsels und Wachstumsgesetzmäßigkeiten, Math.-Nat. wiss. R. 22, Wiss. Z. Karl-Marx-Universität Leipzig, Leipzig, 201–213, 1973.

Gebhardt, G. and Brune, H.: 12. Tagung der Gesellschaft für Ernährungsphysiologie der Haustiere, Arbeitskreis für Eiweißbewertung, Z. Tierphysiol. Tierernähr. Futtermittelkd., 15, 308–320, 1960.

Gous, R. M. and Morris, T. R.: Evaluation of a diet dilution technique for measuring the response of broiler chickens to increasing concentrations of lysine, Br. Poult. Sci., 26, 147–161, 1985.

Hackler, L. R.: Methods of measuring protein quality: A review of bioassay procedures, Cereal Chem., 54, 984–995, 1977.

Htoo, J. K. and Morales, J.: Bioavailability of l-methionine relative to dl-methionine as a methionine source for weaned pigs, J. Anim. Sci., 94, 249–252, 2016.

Huang, R.-L., Tan, Z.-L., Xing, T.-X., Pan, Y.-F., and Li, T.-J.: An in vitro method for the estimation of ileal crude protein and amino acid digestibility using the dialysis tubing for pig feedstuffs, Anim. Feed Sci. Technol., 88, 79–89, 2000.

Hurrel, R. F., Lerman, P., and Carpenter, K. J.: Reactive lysine in foodstuffs as measured by a rapid dye-binding procedure, J. Food Sci., 44, 1221–1231, 1979.

Khan, D. R., Wecke, C., Sharifi, A. R., and Liebert, F.: Evaluating the age dependent potential for protein deposition in naked neck meat type chicken, Animals, 5, 56–70, 2015.

Kluth, H. and Rodehutscord, M.: Comparison of amino acid digestibility in broiler chickens, turkeys and Pekin ducks, Poult. Sci., 85, 1953–1960, 2006.

Kluth, H. and Rodehutscord, M.: Standardisierte Futterbewertung auf der Basis der Aminosäurenverdaulichkeit beim Geflügel, Übers. Tierernährg., 37, 1–26, 2009.

Lemme, A., Ravindran, V., and Bryden, W. L.: Ileal digestibility of amino acids in feed ingredients for broilers, World's Poult. Sci. J., 60, 423–437, 2004.

Liebert, F.: Methodische Untersuchungen zur Beurteilung von Lysinverwertungskennzahlen von Schweinen nach extremen Veränderungen von Proteinmenge und -zusammensetzung, Arch. Anim. Nutr., 48, 319–327, 1995.

Liebert, F.: Level of protein deposition and N-excretion: N-deposition ratio in the growing pig and chicken, in: Proc. 7th Int. Symp. Protein metabolism and nutrition, EAAP publ. no. 81, edited by: Nunes, A. F., Portugal, A. V., Costa, J. P., and Ribeiro, J. R., Santarèm, Portugal, p. 397, 1996.

Liebert, F.: Modelling of protein metabolism yields amino acid requirements dependent on dietary amino acid efficiency, growth response, genotype and age of growing chicken, Avian Biol. Res., 1, 101–110, 2008.

Liebert, F.: Amino acid requirement studies in *Orechromis niloticus* by application of principles of the diet dilution technique, J. Anim. Physiol. Anim. Nutr., 93, 787–793, 2009.

Liebert, F.: Basics and applications of an exponential nitrogen utilization model ("Goettingen approach") for assessing amino acid requirements in growing pigs and meat type chickens based on dietary amino acid efficiency, in: Nutritional Modelling for Pigs and Poultry, edited by: Sakomura, N. K., Gous, R. M., Kyriasakis, I., and Hauschild, L., CABI Publishing, Wallingford, 73–87, 2015.

Liebert, F. and Benkendorff, K.: Modeling lysine requirements of *Oreochromis niloticus* due to principles of the diet dilution technique, Aquaculture, 267, 100–110, 2007a.

Liebert, F. and Benkendorff, K.: Modelling of threonine and methionine requirements of *Oreochromis niloticus* due to principles of the diet dilution technique, Aquacult. Nutr., 13, 397–406, 2007b.

Liebert, F. and Gebhardt, G.: Ergebnisse zur Wirksamkeit und zum Bedarf an ausgewählten Aminosäuren beim wachsenden weiblichen Schwein. 6. Mitt.: Zusammenfassende Diskussion und Wertung sowie Anwendungsempfehlungen zur vorgestellten Methode, Arch. Anim. Nutr., 38, 453–462, 1988.

Liebert, F. and Wecke, C.: Models for further developing the evaluation of protein and amino acids as well as for predicting performance from energy and amino acids intake, in: Recommendations for the Supply of Energy and Nutrients to Pigs, edited by: Staudacher, W., DLG-Verlag, Frankfurt am Main, 219–230, 2008.

Liebert, F. and Wecke, C.: Nitrogen losses per unit of nitrogen deposition as derived from modelling of protein utilization depending on dietary protein quality parameters and age of growing barrows, in: Proc. 3rd Int. Symp. on Energy and Protein Metabolism and Nutrition, Parma, Italy, EAAP publ. no. 127, edited by: Matteo Crovetto, G., Acadademic Publishers, Wageningen, 443–444, 2010.

Liebert, F. und Wecke, C.: Zur Modellierung von N-Stoffwechselparametern als Basis für die Bewertung der Nachhaltigkeit von Ausschöpfungsstrategien für Wachstumspotenziale bei Masthähnchen und Mastschweinen, VDLUFA-Schriftenreihe 68, VDLUFA-Verlag Darmstadt, 749–755, 2012.

Liebert, F., Le Khac, H., and Gebhardt, G.: Ergebnisse zur Wirksamkeit und zum Bedarf an ausgewählten Aminosäuren beim wachsenden weiblichen Schwein. 4. Mitt.: Kombinationen von Proteinträgern mit Lysin-, Methionin/Zystin- bzw. Threoninlimitanz. Arch. Anim. Nutr., 37, 559–568, 1987.

Liebert, F., Rimbach, M., and Peisker, M.: Model for estimation of amino acid utilization and its requirement in growing animals, Proc. Aust. Poult. Sci. Symp., 12, 88–92, 2000.

Liebert, F., Sünder, A., and Mohamed, K.: Assessment of nitrogen maintenance requirement and potential for protein deposition in juvenile Tilapia genotypes by application of an exponential nitrogen utilization model, Aquaculture, 261, 1346–1355, 2006.

Liebert, F., Wecke, C., and Sünder, A.: Besteht Korrekturbedarf bei der optimalen Versorgung von Masthähnchen mit schwefelhaltigen Aminosäuren? (Invited review), in: Proc. 13. Tag. Schweine- und Geflügelernährung, Wittenberg, 32–39, 2015.

Lintzel, W.: Über einige neue Gesichtspunkte und Möglichkeiten der Erforschung des Eiweißstoffwechsels der landwirtschaftlichen Nutztiere, Z. Tierernähr. Futtermittelkd., 2, 32–44, 1939.

Lintzel, W.: Über den Nährwert des Eiweißes der Speisepilze, Biochem. Z., 308, 413–419, 1941.

Lintzel, W. and Rechenberger, J.: Experimentelle Studien zur Theorie des Eiweißstoffwechsels. III. Mitt. Die eiweißsparende Wirkung unvollständiger Eiweiße (Zein und Gelatine) beim Menschen, Biochem. Z., 304, 214–222, 1940.

Low, A. G.: Nutrient absorption in pigs, J. Sci. Food Agric., 31, 1087–1130, 1980.

Maillard, L. C.: Action of amino acids on sugars. Formation of melanoidins in a methodical way, Compt. Rend., 154, 66–68, 1912.

Marquardt, D. W.: An algorithm for least squares estimation of nonlinear parameters, J. Soc. Ind. Appl. Math., 11, 431–441, 1963.

Miller, D. S. and Bender, A. E.: The determination of the net utilization of proteins by a shortened method, Br. J. Nutr., 9, 382–388, 1955.

Mitchell, H. H.: A method of determining the biological value of protein, J. Biol. Chem., 58, 873–903, 1924.

Mitchell, H. H. and Carman, G. G.: The biological value for maintenance and growth of the proteins of whole wheat, eggs, and pork, J. Biol. Chem., 60, 613–620, 1924.

Mitchell, H. H., Hamilton, T. S., and Beadles, J. R.: The importance of commercial processing for the protein value of food products, J. Nutr., 29, 13–25, 1945.

Neumann, C., Velten, S., Kubitza, D., and Liebert, F.: Protein quality of chicken diets with complete substitution of soybean meal by insect meal (*Hermetia illucens*) or algae meal (*Spirulina platensis*) and graded fortification of dietary amino acid supply, Proc. Soc. Nutr. Physiol., 26, 79, 2017.

Nordheim, J. P. and Coon, C. N.: A comparison of four methods for determining available lysine in animal protein meals, Poult. Sci., 63, 1040–1051, 1984.

Nörenberg, P.: Untersuchungen zum maximalen Stickstoffretentionsvermögen wachsender weiblicher Schweine, Diss. Univ. Leipzig, Leipzig, 1987.

Osborne, T. B., Mendel, L. B., and Ferry, E. L.: A method of expressing numerically the growth promoting value of proteins, J. Biol. Chem., 37, 223–229, 1919.

Oser, B. L.: Method of Integrating Essential Amino Acid Content in the Nutritional Evaluation of Protein, J. Am. Diet. Assoc., 27, 396–402, 1951.

Pastor, A.: Aminosäurenwirksamkeit beim Mastgeflügel unter spezifischer Betrachtung der verzweigtkettigen Aminosäuren, Diss. Univ. Göttingen, 321 pp., http://hdl.handle.net/11858/ 00-1735-0000-0022-5E64F (last access: August 2017), 2014.

Pastor, A., Wecke, C., and Liebert, F.: Assessing the age dependent optimal dietary branched-chain amino acid ratio in growing chicken by application of a non-linear modeling procedure, Poult. Sci., 92, 3184–3195, 2013.

Peres, H. and Olivia-Teles, A.: The effect of dietary protein replacement by crystalline amino acid on growth and nitrogen utilization of turbot Scophthalmus maximus juveniles, Aquaculture, 250, 755–764, 2005.

Piccolo, G., Iaconisi, V., Marono, S., Gasco, L., Loponte, R., Nizza, S., Bovera, F., and Parisi, G.: Effect of Tenebrio molitor larvae meal on growth performance, in vivo nutrients digestibility, somatic and marketable indexes of gilthead sea bream (Sparus aurata), Anim. Feed Sci. Technol., 226, 12–20, 2017.

Ravindran, V., Adeola, O., Rodehutscord, M., Kluth, H., van der Klis, J. D., van Eerden, E., and Helmbrecht, A.: Determination of ileal digestibility of amino acids in raw materials for broiler chickens – Results of collaborative studies and assay recommendations, Anim. Feed Sci. Technol., 225, 62–72, 2017.

Rutherford, M. S. and Moughan, P. J.: The rat as a model animal for the growing pig in determining ileal amino acid digestibility in soya and milk proteins, J. Anim. Physiol. Anim. Nutr., 87, 292–300, 2003.

Samadi and Liebert, F.: Modelling of threonine requirement in fast growing chickens depending on age, sex, protein deposition and dietary threonine efficiency, Poult. Sci., 85, 1961–1968, 2006a.

Samadi and Liebert, F.: Estimation of nitrogen maintenance requirements and potential for nitrogen deposition in fast-growing chickens depending on age and sex, Poult. Sci., 85, 1421–1429, 2006b.

Samadi and Liebert, F.: Threonine requirement of slow-growing male chickens depends on age and dietary efficiency of threonine utilization, Poult. Sci., 86, 1140–1148, 2007a.

Samadi and Liebert, F.: Lysine requirement of fast growing chickens – Effects of age, sex, level of protein deposition and dietary lysine efficiency, J. Poult. Sci., 44, 63–72, 2007b.

Samadi and Liebert, F.: Modelling the optimal lysine to threonine ratio in growing chickens depending on age and efficiency of dietary amino acid utilisation, Br. Poult. Sci., 49, 45–54, 2008.

Samadi, Wecke, C., Pastor, A., and Liebert, F.: Assessing lysine requirement of growing chicken by direct comparison between supplementation technique and "Goettingen approach", Open J. Anim. Sci., 7, 56–69, 2017.

Sauer, W. C. and Ozimek, L.: Digestibility of amino acids in swine: Results and their practical applications. A review, Livest. Prod. Sci., 15, 367–388, 1986.

Savoie, L. and Gauthier, S. F.: Dialysis cell for the in vitro measurement of protein digestibility, J. Food Sci., 51, 494–498, 1986.

Shen, Y. B., Weaver, A. C., and Kim, S. W.: Effect of feed grade l-methionine on growth performance and gut health in nursery pigs compared with conventional dl-methionine, J. Anim. Sci., 92, 5530–5539, 2014.

Slawski, H., Adem, H., Tressel, R.-P., Wysujack, K., Koops, U., Wuertz, S., and Schulz, C.: Replacement of fishmeal with rapeseed protein concentrate in diets fed to wels catfish (Silurus glanis L.), Aquacult. Nutr., 17, 605–612, 2011.

Stein, H. H., Pedersen, C., Wirt, A. R., and Bohlke, R. A.: Additivity of values for apparent and standardized ileal digestibility of AA in mixed diets fed to growing pigs, J. Anim. Sci., 83, 2387–2395, 2005.

Stein, H. H., Seve, B., Fuller, M. F., Moughan, P. J. and de Lange, C. F. M.: Invited review: Amino acid bioavailability and digestibility in pig feed ingredients: Terminology and application, J. Anim. Sci., 85, 172–180, 2007.

Thong, H. T. and Liebert, F.: Potential for protein deposition and threonine requirement of modern genotype barrows fed graded levels of protein threonine as the limiting amino acid, J. Anim. Physiol. Anim. Nutr., 88, 196–203, 2004a.

Thong, H. T. and Liebert, F.: Amino acid requirement of growing pigs depending on efficiency of amino acid utilisation and level of protein deposition. 1. Lysine, Arch. Anim. Nutr., 58, 69–88, 2004b.

Thong, H. T. and Liebert, F.: Amino acid requirement of growing pigs depending on efficiency of amino acid utilisation and level of protein deposition. 2. Threonine, Arch. Anim. Nutr., 58, 157–168, 2004c.

Van Kempen, T. and Bodin, J.: Near-infrared reflectance spectroscopy (NIRS) appears to be superior to nitrogen-based regression as a rapid tool in predicting the poultry digestible amino acid content of commonly used feedstuffs, Anim. Feed Sci. Technol., 76, 139–147, 1998.

Van Leeuwen, P., Sauer, W. C., Huisman, J., van Weerden, E. J., van Kleef, D., and den Hartog, L. A.: Methodological aspects for the determination of amino acid digestibilities in pigs fitted with ileocecal re-entrant cannulas, J. Anim. Physiol. Anim. Nutr., 58, 122–133, 1987.

von Bertalanffy, L.: Theoretische Biologie, in: 2. Bd., Stoffwechsel, Wachstum, A. Francke AG Verlag, Bern, 1951.

Wecke, C. and Liebert, F.: Lysine requirement studies in modern genotype barrows dependent on age, protein deposition and dietary lysine efficiency, J. Anim. Physiol. Anim. Nutr., 93, 295–304, 2009.

Wecke, C. and Liebert, F.: Optimal dietary lysine to threonine ratio in pigs (30–110 kg BW) derived from observed dietary amino acid efficiency, J. Anim. Physiol. Anim. Nutr., 94, E277–E285, 2010.

Wecke, C. and Liebert, F.: Improving the reliability of optimal in-feed amino acid ratios based on individual amino acid efficiency data from N balance studies in growing chicken, Animals, 3, 558–573, 2013.

Wecke, C., Pastor, A., and Liebert, F.: Validation of the lysine requirement as reference amino acid for ideal in-feed amino acid ratios in modern fast growing meat-type chickens, Open J. Anim. Sci., 6, 185–194, 2016.

20

Evaluation of novel SNPs and haplotypes within the *ATBF1* gene and their effects on economically important production traits in cattle

Han Xu, Sihuan Zhang, Xiaoyan Zhang, Ruihua Dang, Chuzhao Lei, Hong Chen, and Xianyong Lan

College of Animal Science and Technology, Northwest A&F University, Shaanxi Key Laboratory of Molecular Biology for Agriculture, Yangling, Shaanxi 712100, China

Correspondence to: Xianyong Lan (lanxianyong79@126.com) and Hong Chen (chenhong1212@126.com)

Abstract. AT motif binding factor 1 (*ATBF1*) gene can promote the expression level of the growth hormone 1 (*GH1*) gene by binding to the enhancers of the *POU1F1* and *PROP1* genes; thus, it affects the growth and development of livestock. Considering that the *ATBF1* gene also has a close relationship with the Janus kinase–signal transductor and activator of transcription (JAK–STAT) pathway, the objective of this work was to identify novel single-nucleotide polymorphism (SNP) variations and their association with growth traits in native Chinese cattle breeds. Five novel SNPs within the *ATBF1* gene were found in 644 Qinchuan and Jinnan cattle for first time using 25 pairs of screening and genotyping primers. The five novel SNPs were named as AC_000175:g.140344C>G (SNP1), g.146573T>C (SNP2), g.205468C>T (SNP3), g.205575A>G (SNP4) and g.297690C<T (SNP5). Among them, SNP1 and SNP2 were synonymous coding SNPs, while SNP5 was a missense coding SNP, and the other SNPs were intronic. Haplotype analysis found 18 haplotypes in the two breeds, and three and five closely linked loci were revealed in Qinchuan and Jinnan breeds, respectively. Association analysis revealed that SNP1 was significantly associated with the height across the hip in Qinchuan cattle. SNP2 was found to be significantly related to chest circumference and body side length traits in Jinnan cattle. SNP3 was found to have significant associations with four growth traits in Qinchuan cattle. Moreover, the different combined genotypes, SNP1–SNP3, SNP1–SNP4 and SNP2–SNP5 were significantly associated with the growth traits in cattle. These findings indicated that the bovine *ATBF1* gene had marked effects on growth traits, and the growth-trait-related loci can be used as DNA markers for maker-assisted selection (MAS) breeding programs in cattle.

1 Introduction

With the fast improvement in living standards in developing countries, especially China, the demand for beef consumption has increased quickly. Although cattle breed resources are very abundant in China, poor quality and low growth rate of many breeds are still barriers to an increase in cattle production. It is difficult to meet our needs and improve the breeding speed by using traditional methods; thus, the efficient genetical methods, such as DNA marker-assisted selection (MAS), should be used to improve the efficiency of production and lay the foundation for breeding new breeds (Pedersen et al., 2009). As the most practical and economic

method, the MAS strategy relies on the numerous single-nucleotide polymorphisms (SNPs) associated with production traits. Therefore, more functional SNPs, which could be applied in MAS breeding of domestic livestock, should be discovered. An example is the single A-to-G substitution near the ovine *CLPG* gene, which has been used for double-muscle livestock breeding (Cockett et al., 1994).

The AT motif binding factor 1 (*ATBF1*) gene encodes a transcription factor with multiple homeodomains and zinc finger motifs; thus, it is also named zinc finger homeobox 3 (*ZFHX3*). *ATBF1* was first isolated as an AT (adenine and thymine)-binding factor of human α-fetoprotein (AFP) (Morinaga et al., 1991). It was reported to function as a tumor suppressor in several cancers (Kawaguchi et al., 2016; Sun et al., 2014, 2015). More importantly, it plays an important role in regulating myogenesis, adipose tissue development and transactivating the cell cycle inhibitor (Jung et al., 2005; Postigo and Dean, 1997, 1999; Richard and Stephens, 2014).

Furthermore, *ATBF1* could promote the expression level of the growth hormone 1 (*GH1*) gene by binding to the enhancers of the *POU1F1* and *PROP1* genes (Araujo et al., 2013; Qi et al., 2008), which are the key genes in mammalian growth, development and the lactation-related hypothalamic–pituitary–adrenal (HPA) axis pathway (*PITX2/PITX1 – HESX1 – LHX3/LHX4 – PROP1 – POU1F1*) (Davis et al., 2010; Ma et al., 2017). In addition, another key gene in the HPA axis pathway, *PITX2*, has a positive regulation relationship with *ATBF1* under the participation of miR-1 (Huang et al., 2015).

Meanwhile, *ATBF1* has close relationships with STAT family genes, which are growth-related genes. *STAT3* and *STAT5A* are two key genes in Janus kinase–signal transducer and activator of transcription pathway (JAK–STAT), and JAK–STAT is responsible for promoting the secretion of a variety of cytokines, growth factors and *GH1* (Herrington et al., 2000; Liongue and Ward, 2013; Trovato et al., 2012). *ATBF1* could enhance the suppression of STAT3 signaling by interaction with *PIAS3*, which is a protein inhibitor of the activated STAT family (Lao et al., 2016; Nojiri et al., 2004; S. F. Yang et al., 2016). Thus, it was surmised that the *ATBF1* gene plays an important role in regulating mammalian growth and development.

SNP research is a crucial step for the application of the genome project in human and MAS breeding in mammals. The genes mentioned above, *CLPG*, *STAT3*, *STAT5A*, *POU1F1* and *PROP1*, all have SNPs associated with growth traits, and some SNP genotypes were found to be significantly associated with mRNA expression levels (Jia et al., 2015; Lan et al., 2007, 2009; Wu et al., 2014; Zhao et al., 2013). In humans, functional SNPs that were associated with disease were found in the *ATBF1* gene (Liu et al., 2014; Tsai et al., 2015). Moreover, four SNPs that were significantly associated with goat growth traits were identified in the goat *ATBF1* gene (Zhang et al., 2015b).

Considering the important roles of the *ATBF1* gene in the HPA axis and JAK–STAT pathways, which are related to mammalian growth and development, and the significance of SNP in biological process and livestock breeding, the purpose of this study was to identify crucial SNP variations within the *ATBF1* gene in native Chinese cattle breeds. This will also help to promote the understanding of *ATBF1* gene function and better apply the excellent local cattle germplasm resources in cattle MAS breeding.

2 Materials and methods

2.1 Animal samples and data collection

Experimental animal samples used in this study were approved by the Faculty Animal Policy and Welfare Committee of Northwest A&F University under contract. The care and use of experimental animals fully complied with local animal welfare laws, guidelines and policies.

A total of 644 blood samples were collected from healthy and unrelated adult cattle belonging to two well-known Chinese native cattle breeds, Qinchuan cattle (459) and Jinnan cattle (185). All Qinchuan individuals were reared in a native breeding farm in Fufeng County, Shaanxi Province. Jinnan individuals were reared on a Yuncheng cattle farm in Shanxi Province.

The growth trait data of the Qinchuan cattle were collected from the Qinchuan breeding farm, including body weight, body height, body length, chest circumference, hucklebone width, height across the hip, chest width, chest depth, rump length and hip width. The growth trait data of the Jinnan cattle, including body height, height across the hip, chest circumference, rump length and body side length, were collected from the Jinnan cattle farm. All growth trait data were measured as described by Zhang et al. (2015a).

2.2 DNA isolation and genomic DNA pool construction

Genomic DNA samples were extracted from the leukocytes of the blood samples as described by Dang et al. (2014). All genomic DNA samples were diluted to the working concentration $50\,\mathrm{ng}\,\mathrm{\mu L}^{-1}$ for the DNA pool construction and polymerase chain reaction (PCR) amplification. To construct DNA pools, 30 DNA samples were randomly and selected from the Qinchuan and Jinnan cattle. The two DNA pools were used as templates for PCR amplification, and the product of amplification was used to sequence and explore genetic variations in the *ATBF1* gene.

2.3 Primer design and PCR amplification for SNP screening

To expose novel SNPs in the bovine *ATBF1* gene, 20 pairs of primer were designed using the Primer Premier 5 software based on the bovine *ATBF1* gene DNA sequence (NCBI reference sequence: AC_000175.1) (Table 1). The $25\,\mathrm{\mu L}$ PCR reaction volume includes $50\,\mathrm{ng}$ of genomic DNA from the DNA pool, $0.5\,\mathrm{\mu M}$ of each primer and $12.5\,\mathrm{\mu L}$ $2 \times$ EcoTaq PCR SuperMix (+dye) (Beijing TransGen Biotech Co., Ltd., Beijing, China). The touchdown PCR program was executed as follows: pre-denaturation at $95\,^{\circ}\mathrm{C}$ for 4 min, followed by 18 cycles of denaturation at $94\,^{\circ}\mathrm{C}$ for 30 s, annealing at $68\,^{\circ}\mathrm{C}$ for 30 s (decreased by $1\,^{\circ}\mathrm{C}$ per cycle) and extending at $72\,^{\circ}\mathrm{C}$ for $1\,\mathrm{kb}\,\mathrm{min}^{-1}$, then another 22 cycles at $94\,^{\circ}\mathrm{C}$ for 30 s, $50\,^{\circ}\mathrm{C}$ for 30 s and $72\,^{\circ}\mathrm{C}$ for $1\,\mathrm{kb}\,\mathrm{min}^{-1}$, finally extending

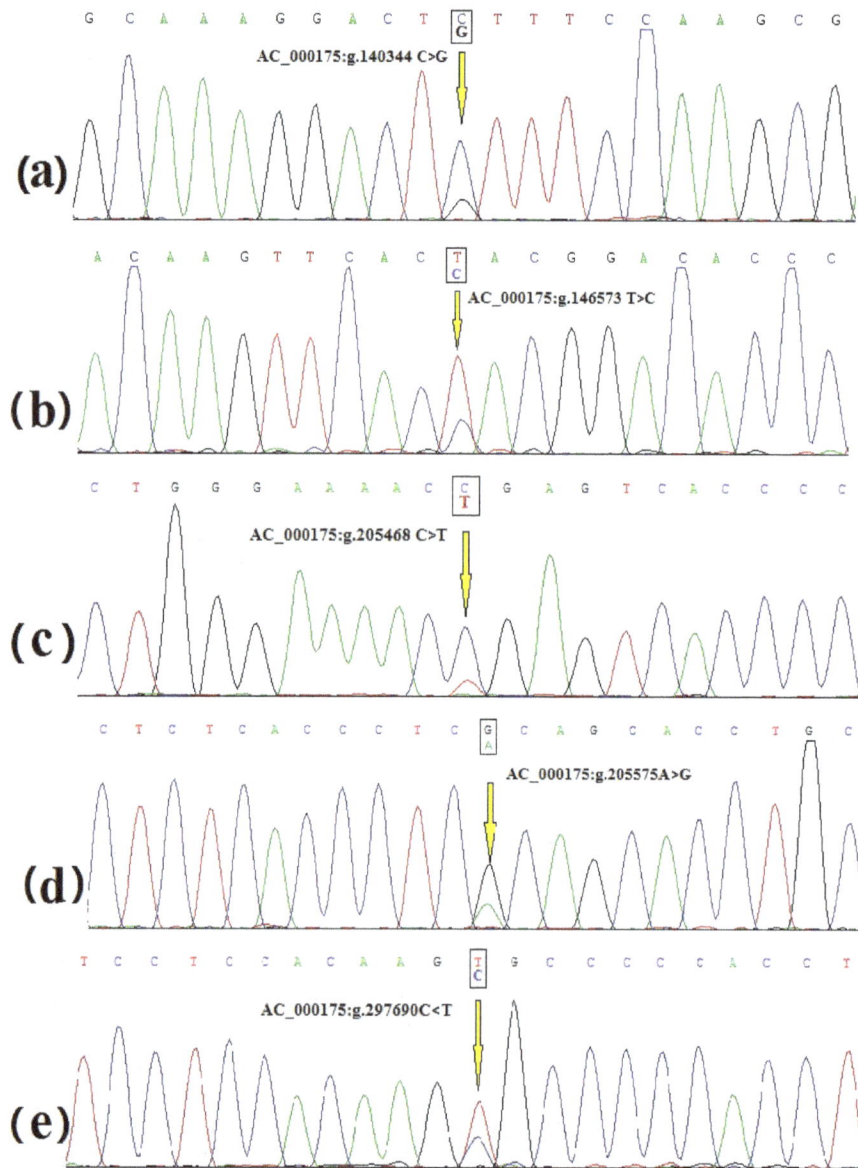

Figure 1. Sequence chromas of five novel SNP loci in the bovine *ATBF1* gene. Note: panels **(a)**, **(b)**, **(c)**, **(d)** and **(e)** represent the pooling sequence chromas of SNP1, SNP2, SNP3, SNP4 and SNP5, respectively.

at 72 °C for 10 min. The products of PCR amplification were sequenced to screen the SNP loci.

2.4 Primer design and genotyping by PCR-RFLP, forced PCR-RFLP and T-ARMS-PCR methods

DNA pool sequencing and sequence analysis identified five novel SNPs within the Qinchuan and Jinnan bovine *ATBF1* gene, namely, AC_000175:g.140344C>G (SNP1), g.146573T>C (SNP2), g.205468C>T (SNP3), g.205575A>G (SNP4) and g.297690C<T (SNP5) (Fig. 1). According to the sequencing results, PCR restriction fragment length polymorphism (PCR-RFLP), forced PCR-RFLP

and tetra-primer amplification refractory mutation system PCR (T-ARMS-PCR) methods were used to detect genotypes of Qinchuan and Jinnan individuals (Table 2). The primers of PCR-RFLP and forced PCR-RFLP were designed by the Primer Premier 5 software, and T-ARMS-PCR primers were designed on the Primer1 website (http://primer1.soton.ac.uk/primer1.html) (Collins, 2012; Ye et al., 2001). The genotyping methods used on different SNP loci were introduced as below:

SNPs genotyped with the PCR-RFLP and forced PCR-RFLP methods: SNP3 and SNP5 were detected using the PCR-RFLP method, and the PCR amplification products were digested with DdeI and HhaI restriction enzymes. SNP2

Table 1. Amplification PCR primers for screening the novel SNPs within the bovine *ATBF1* gene; bp: base pair.

Names	Primer sequences (5′ → 3′)	Product sizes (bp)	Location
P1	F: GAAAGGGCTTCTCCTGACG R: GATACCGCACCCATTGTCC	367	Intron 1
P2	F: CCTGACTCTAACGCTGTGCT R: GGATGGGCTTCCTCTTGC	1264	Exon 2
P3	F: CTTTCCACATAGCCTCATCCTT R: TTTATTGGCACTTTCATCAGCA	1202	Exon 2
P4	F: TGCTGATGAAAGTGCCAATA R: GAGCATCCAGTCGTCCCTT	1116	Exon 2
P5	F: GTGTCAGGTGTCCCATAGCC R: AATGCCAGTCCCTCCAGTTA	1153	Exon 3
P6	F: GATTATTGTGCCAGGAAGCC R: GATCTGAACCCAAAGACTGAA	714	Exon 4
P7	F: GCTCAGGCACCACGAAG R: CAGGACACCAGGGATACAAA	1080	Exon 5
P8	F: GACTCTTACCCAGCACGTACCCT R: TAACAGAAACCCACCATCCACAA	1461	Exon (6+7)
P9	F: CAGGACACCCTCTGGGCTAC R: ATGGAGACATCATAAGGGAG	1454	Exon 8
P10	F: CATTGGGCTTGATTTCTAT R: GGTGGCATTCCTACACTTT	1138	Exon 9
P11	F: CACCTTTACCACCACCAAC R: TACGAGGCCGCTTATTCT	1442	Exon 9
P12	F: GCCCATCTTCTCGCCACT R: ATCCTGCCCTTCCTCGTC	802	Exon 9
P13	F: CAGGATGACAGCCAGAATG R: CTTGCCAGCAGTGGGTTA	680	Exon 9
P14	F: TACAGCATCCTCTGCGTTCT R: CCGTGCCTTCCACCTTGA	1235	Exon 9
P15	F: GATGGCAATGTCTGAGTATGA R: ACCCTGGTCTGTGCTGAA	1031	Exon 9
P16	F: AACCGTCCTCAGCATCGC R: CGTGTCAGACTCCTCCGAAT	1457	Exon 10
P17	F: CGCTCACTCAAACGACAG R: AATCTACTCAACACCGAAAA	1221	Exon 10
P18	F: TTCTCAGGTCAATCGCTCAC R: CACCGCTCAGACTGCCTA	1261	Exon 10
P19	F: TGTTAGGCAGTCTGAGCG R: TTCTGGGTTAATGTGGAG	1409	Exon 10
P20	F: TTCTCCACATTAACCCAG R: TCAGTCAGCTCCATCACC	1314	Exon 10

was genotyped using the forced PCR-RFLP method, and the C nucleic acid on g.146575 was changed to T to make a locus that can be recognized by the EcoT14I (StyI) restriction enzyme. The PCR reaction volume for the two methods was 13 μL, including 25 ng of genomic DNA, 0.2 μM of each primer and 6.5 μL of 2 × EcoTaq PCR SuperMix (+dye). The amplification system was as follows: pre-denaturation at 95 °C for 4 min, followed by 35 cycles of denaturation at 94 °C for 30 s, optimal annealing temperature for 30 s and extending at 72 °C for 1 kb min^{-1}, finally extending at 72 °C for 10 min. Then the amplification products were digested with a special restriction enzyme and special temperature for

12 to 16 h. The volume of digestion contains 2 U (U is unit of restriction enzyme) restriction enzyme, 2 μL of 10 × buffer, 10 μL of PCR product and 6 μL of distillation H$_2$O. Then the enzyme-digested products were genotyped using agarose gel electrophoresis. The electrophoretic band size and genotyping information are shown in Table 2.

SNPs genotyped with the T-ARMS-PCR method: SNP1 and SNP4 were genotyped using the T-ARMS-PCR method for failing to search for a suitable restriction enzyme. The special primers and genotyping information are exhibited in Table 2. The PCR reaction volume was 13 μL. The touchdown PCR program (from 68 to 50 °C; decreased by 1 °C

Table 2. PCR primer sequences for *ATBF1* gene genotyping in cattle; bp: base pairs.

Loci	Primer sequence (5′ → 3′)	AT* (°C)	Size (bp)	Detection method
P21-SNP1 (g.140344C>G)	F inner: AAGAGGAGGAGGAGGGCTGCAAAGGAGTC R inner: AGCTCGTCGTCCAGCTCGCTTGGATAC F outer:GGGGCAGCAGAAGGAGAGAAGCAAGAAG par R outer: TCGACAGGGTCTGGAGCACATTAGGCAT	Touchdown PCR	180/200/325	T-ARMS-PCR C allele: 180 bp; G allele: 200 bp; Product size of two outer primers: 325 bp
P22-SNP2 (g.146573T>C)	F: GGGCAGTGCCTCAGGTAGGA R: CAGCAGGTCCAGGGTGTCCAT	61.7	231	Forced PCR-RFLP (EcoT14 I (StyI)) (TT=231 bp; TC=231+209+22 bp; CC=209+22 bp)
P23-SNP3 (g.205468C>T)	F: GATTATTGTGCCAGGAAGCC R: GATCTGAACCCAAAGACTGAA	60	714	PCR-RFLP (DdeI) (CC=589 bp; CT=589+440 bp; TT=440 bp)
P24-SNP4 (g.205575A>G)	F-inner: AGGGCACGTCCCTCTCTCTCACCCGCA R-inner: CACTCTCGTGCTGCTGCAGGTGCGGC F-outer: GGAAGGGCCCCCTGGGAAACCGAGTCAC R-outer: TCCTCGTCCTCCTCGGGGAGGCCCTTCT	Touch down PCR	216/153/316	T-ARMS-PCR A allele: 216 bp G allele: 153 bp Product size of two outer primers: 316 bp
P25-SNP5 (g.297690C<T)	F: TACAGCATCCTCTGCGTTCT R: CCGTGCCTTCCACCTTGA	60	1235	PCR-RFLP (HhaI) (TT=594 bp; CT=594+564 bp; CC=564 bp)

Note: the single nucleic acid that is underlined is different from the reference sequence, and the change is required for forced PCR-RFLP and T-ARMS-PCR primer designing.
* AT: annealing temperature.

per cycle) was executed for PCR amplification. Then the products were genotyped using agarose gel electrophoresis directly.

2.5 Statistical analysis

Genotypic frequencies and allelic frequencies were calculated according to Botstein's method (Botstein et al., 1980). Population genetic diversity index, homozygosity (Ho), effective allele number (Ne) and polymorphism information content (PIC) were calculated successively on the MSRcall website (http://www.msrcall.com/). Hardy–Weinberg equilibrium (HWE), linkage disequilibrium (LD) structure and haplotypes of the five SNP loci in Qinchuan and Jinnan breeds were calculated using the SHEsis program (http://analysis.bio-x.cn) (Li et al., 2009; Q. Yang et al., 2016).

The relationship between genotypes, haplotypes and the growth traits in Qinchuan and Jinnan populations were analyzed using the SPSS software (version 17.0) (IBM Corp., Armonk, NY, USA). Since all cattle were adult females and each breed was fed the same nutritional diet on their respective farms, the basic linear model $Y = \mu + G + e$ was used to determine the relationship between genotypes, haplotypes and growth traits for each breed. In the formulate, Y denotes the trait data of each animal, μ the overall mean for each trait, G the effect of genotype and e the random error (Dang et al., 2014; Zhang et al., 2015a).

3 Results

3.1 Novel SNP identification and genotyping of the bovine *ATBF1* gene

According to the sequence chromas, five novel SNPs (SNP1 to SNP5) were identified within the Qinchuan and Jinnan cattle *ATBF1* gene (Fig. 1). Among them, SNP1 and SNP2 were synonymous mutations, which were located at exon 2 and exon 3, respectively. SNP1 and SNP2 loci code the 503th leucine and the 963th threonine of the cattle ATBF1 protein, respectively. SNP3 and SNP4 were located at intron 3, and SNP4 was close to the exon 4 splicing site (four base distances). SNP5 was a missense coding SNP at exon 9, resulting in the 2488th amino acid valine to alanine. The genotyping results can be seen from the agarose gel electrophoresis photos, which shows that SNP1–SNP5 were successfully genotyped by their own methods (Fig. 2).

3.2 Genetic diversity analysis of the bovine *ATBF1* gene

Genotype frequency and allelic frequency were calculated according to the genotyping results. At the SNP1 locus, the frequency of allele C was distinctly higher than allele G in the two breeds. Genotype CC is the most prevalent. At the SNP2 locus, the frequency of TT was significantly higher than the other genotypes. At the SNP3 locus, the frequency of allele T is higher than C. At the SNP4 locus, there was no

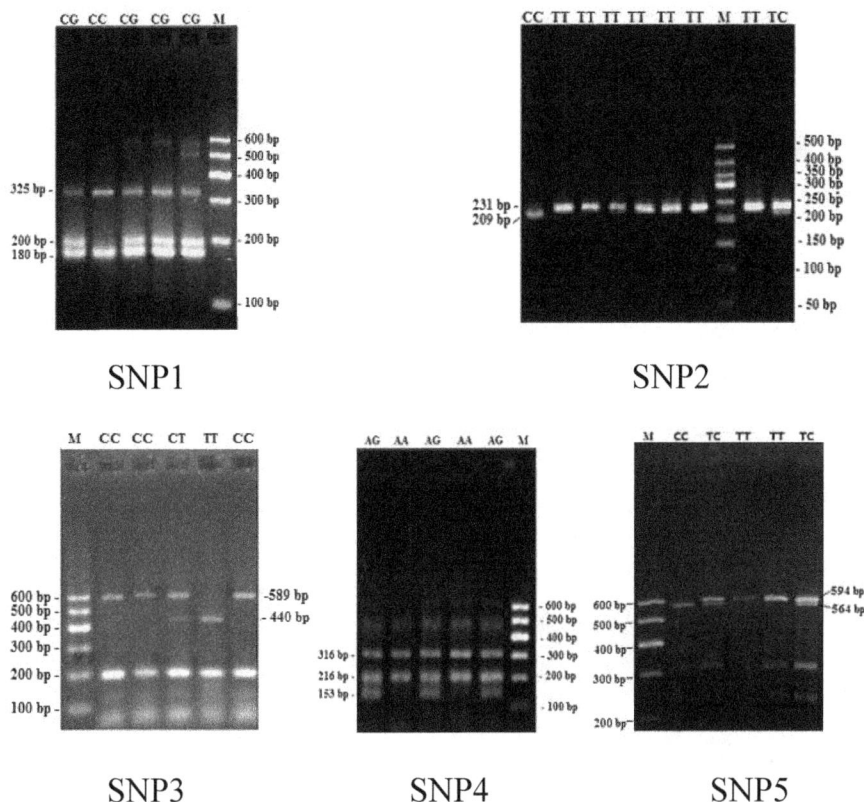

Figure 2. Agarose gel electrophoresis patterns of five novel SNPs of the bovine *ATBF1* gene. Note that the letter "M" above the figure represents the DNA marker.

GG genotype, and the frequency of AG was higher than AA. At the SNP5 locus, genotype frequency of the heterozygote was higher than the other genotypes, and the frequency of allele T was higher than C (Table 3).

The genetic diversity parameters Ho, Ne and PIC of the five loci of the Qinchuan and Jinnan populations were calculated and are shown in Table 3. These results suggest that these loci are polymorphic in these two cattle breeds. However, the values of PIC suggest that these loci are low polymorphisms ($0 < \text{PIC} < 0.25$) and intermediate polymorphisms ($0.25 < \text{PIC} < 0.5$) (Table 3) (Pan et al., 2013). The Hardy–Weinberg equilibrium P value shows that some loci were at Hardy–Weinberg equilibrium ($P > 0.05$) and some were in disequilibrium ($P < 0.05$).

3.3 Haplotype and linkage disequilibrium analysis of the five SNP loci

Haplotype pairwise linkage disequilibrium analysis indicated that there were a total of 18 haplotypes in Qinchuan and Jinnan cattle. Among these haplotypes, seven were shared by these two populations. Nine haplotypes were unique to Qinchuan cattle and two haplotypes were unique to Jinnan cattle. The frequencies of the haplotypes showed that Hap 7 (CTCGT) and Hap 5 (CTCAT) were the main haplotypes in

the Qinchuan and Jinnan cattle populations, respectively (Table 4). Based on the D' and r^2 values, three closely linked loci were revealed in the Qinchuan breed and five were revealed in the Jinnan breed (Fig. 3). The D' values were 0.756 (SNP1 and SNP3), 0.608 (SNP1 and SNP4) and 0.624 (SNP2 and SNP5) in the Qinchuan cattle breed. In the Jinnan cattle breed, the D' values were 0.640 (SNP1 and SNP3), 0.999 (SNP1 and SNP4), 0.997 (SNP2 and SNP4), 1.000 (SNP3 and SNP4) and 0.696 (SNP4 and SNP5) (Fig. 3). Thus, we further analyzed the effects of the combined genotypes above and growth traits in cattle.

3.4 Relationships between the genetic variations and growth-related traits

Association analysis found that different genotypes of the SNP1 locus were similar, with a significant association with the height across the hip in Qinchuan cattle ($P = 0.05$), and the heterozygote carriers had the highest value (Table 5). At the SNP2 locus, the different genotypes were significantly associated with chest circumference and body side length traits in Jinnan cattle, and the CC genotype carriers had the highest growth trait index (Table 5). For SNP3, the different genotypes were found to have a significant association with chest width, chest depth and hucklebone width growth traits,

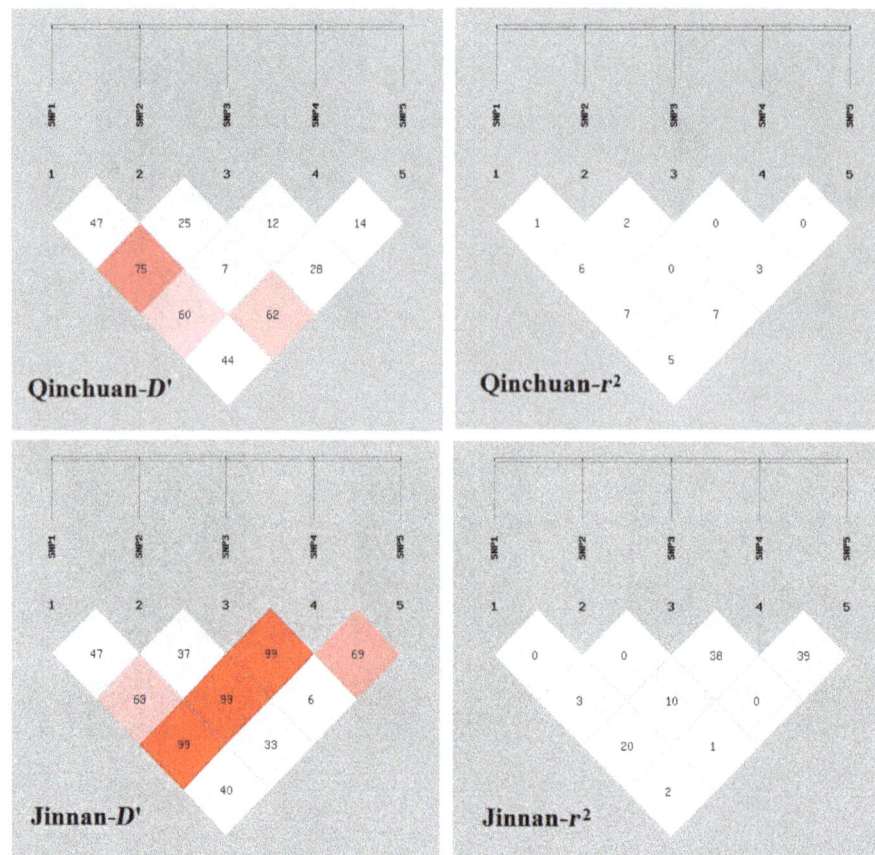

Figure 3. Linkage disequilibrium (LD) plot (D' and r^2) of five novel SNP loci within the *ATBF1* gene in Qinchuan and Jinnan cattle.

and CC carriers had the best growth trait index in Qinchuan cattle. Moreover, the genotypes had a similar significant association with body height ($P = 0.05$), and the heterozygote carriers had the best value in Qinchuan cattle (Table 5). However, no significant association between different genotypes of SNP4 and SNP5 loci and growth-related traits were found.

The association analysis found that three and one combined genotypes were associated with growth traits in Qinchuan and Jinnan breeds, respectively. At the SNP1–SNP3 loci, the combined genotype CG–CC carriers had significantly wider chests than the CC–TT carriers in the Qinchuan breed ($P < 0.05$) (Table 6). At the SNP1–SNP4 loci, the CG–AG carriers had a higher height across the hip than CC–AA carriers in the Qinchuan breed ($P = 0.05$) (Table 6). For the SNP2–SNP5 loci in Qinchuan cattle, TC–CC had the smallest chest circumference, chest width and hucklebone width values among all the combined genotypes ($P < 0.05$) (Table 6). For the SNP1–SNP3 loci in Jinnan cattle, CG–CT carriers had the smallest body height, height across the hip and body side length, and CG-CC had the largest chest circumference and rump length, among all the combined genotypes (Table 6). In addition, no significant association was found between the other combined genotypes and growth-related traits.

4 Discussion

Due to the important roles of *ATBF1* in regulating myogenesis and adipose tissue development and its close relationship with the HPA axis and JAK–STAT pathways in livestock science (Huang et al., 2015; Jiang et al., 2014; S. F. Yang et al., 2016; Zhao et al., 2016), *ATBF1* was chosen as the candidate gene. In this study, we recovered five SNPs in the bovine *ATBF1* gene for the first time. Three novel SNPs were exonic, while the other two novel SNPs were intronic. For individual genotype identification, three methods, namely PCR-RFLP, forced PCR-RFLP and T-ARMS-PCR, were applied. For the SNP locus, where the nucleotide sequence could be recognized using restriction enzymes, the PCR-RFLP method can be used to identify individual genotypes (Wang et al., 2013). For the SNP locus without the sequence that could be recognized using restriction enzymes, the forced PCR-RFLP primers were designed. This method needs to change one or two nucleotides, which are close to the SNP locus, to make this sequence recognizable using restriction enzymes and genotyping (Huang et al., 2014). For the SNP locus, which is difficult to genotype using the two methods above, T-ARMS-PCR is available (Li et al., 2014; Wang et al., 2014; Zhang et al., 2015a).

Table 3. Genotype, allelic distribution and genetic diversity of five SNP loci of the bovine *ATBF1* gene.

Locus/ Breed	Genotype frequency (Number of animals)			Allelic frequency		*P* value (HWE)[a]	Diversity parameters[b]		
							Ho	Ne	PIC
SNP1	CC (*n*)	CG (*n*)	GG (*n*)	C	G				
Qinchuan	0.556 (60)	0.407 (44)	0.037 (4)	0.759	0.241	$P>0.05$	0.634	1.576	0.299
Jinnan	0.721 (111)	0.221 (34)	0.058 (9)	0.831	0.169	$P<0.05$	0.719	1.390	0.241
SNP2	TT (*n*)	TC (*n*)	CC (*n*)	T	C				
Qinchuan	0.794 (247)	0.196 (61)	0.010 (3)	0.892	0.108	$P>0.05$	0.808	1.238	0.174
Jinnan	0.836 (153)	0.148 (27)	0.016 (3)	0.910	0.090	$P>0.05$	0.836	1.196	0.151
SNP3	CC (*n*)	CT (*n*)	TT (*n*)	C	T				
Qinchuan	0.528 (171)	0.457 (148)	0.015 (5)	0.756	0.244	$P<0.05$	0.631	1.584	0.301
Jinnan	0.486 (90)	0.459 (85)	0.054 (10)	0.716	0.284	$P>0.05$	0.595	1.680	0.323
SNP4	AA (*n*)	AG (*n*)	GG (*n*)	A	G				
Qinchuan	0.420 (21)	0.580 (29)	0 (0)	0.710	0.290	$P<0.05$	0.588	1.700	0.327
Jinnan	0.227 (40)	0.773 (136)	0 (0)	0.614	0.386	$P<0.05$	0.526	1.902	0.362
SNP5	CC (*n*)	TC (*n*)	TT (*n*)	C	T				
Qinchuan	0.202 (58)	0.453 (130)	0.345 (99)	0.429	0.571	$P>0.05$	0.510	1.960	0.370
Jinnan	0.227 (42)	0.459 (85)	0.314 (58)	0.457	0.543	$P>0.05$	0.504	1.986	0.373

Note: [a] *P* value (HWE): Hardy–Weinberg equilibrium *P* value.
[b] Diversity parameters: Ho: gene homozygosity; Ne: effective allele numbers; PIC: polymorphism information content.

Table 4. Haplotype frequency within the *ATBF1* gene in Qinchuan and Jinnan cattle.

Haplotype name	SNP1	SNP2	SNP3	SNP4	SNP5	Haplotype	Haplotype frequencies	
							Qinchuan cattle	Jinnan cattle
Hap 1	C	C	C	A	C	CCCAC	0.052	0
Hap 2	C	C	T	A	T	CCTAT	0.030	0
Hap 3	C	C	T	G	C	CCTGC	0.040	0.050
Hap 4	C	T	C	A	C	CTCAC	0.129	0.059
Hap 5	C	T	C	A	T	CTCAT	0.129	0.391
Hap 6	C	T	C	G	C	CTCGC	0.049	0.100
Hap 7	C	T	C	G	T	CTCGT	0.163	0
Hap 8	C	T	T	A	C	CTTAC	0.111	0
Hap 9	C	T	T	G	C	CTTGC	0.001	0.100
Hap 10	G	C	C	A	T	GCCAT	0.022	0.150
Hap 11	G	C	C	G	C	GCCGC	0.014	0
Hap 12	G	C	T	A	T	GCTAT	0.018	0
Hap 13	G	T	C	A	C	GTCAC	0.048	0.050
Hap 14	G	T	C	A	T	GTCAT	0.135	0
Hap 15	G	T	C	G	C	GTCGC	0.044	0
Hap 16	G	T	T	A	T	GTTAT	0.017	0
Hap 17	G	T	T	G	C	GTTGC	0	0.041
Hap 18	G	T	T	G	T	GTTGT	0	0.059

The preferred methods are PCR-RFLP and forced PCR-RFLP because they are more accurate and mature than T-ARMS-PCR (Cai et al., 2013; Sun et al., 2013). T-ARMS-PCR is easy to operate, using less time and money, but the accuracy is lower than the other two methods. This study performed both PCR-RFLP and T-ARMS-PCR methods to calculate the accuracy of them (Li et al., 2014). The result showed that the consistency of these two methods is 98.8 %, 40 % of inconsistency was caused by PCR-RFLP and 60 % of inconsistency was caused by the T-ARMS-PCR method (Li

Table 5. Association of *ATBF1* gene SNP3 genotypes and cattle growth traits.

Locus/breed	Growth trait	Observed genotypes (LSM ± SE)[*]			*P* value
SNP1		CC	CG	GG	
Qinchuan	height across the hip	124.75[ab] ± 1.07	127.34[a] ± 1.13	122.67[b] ± 1.45	$P = 0.05$
SNP2		TT	TC	CC	
Jinnan	chest circumference	183.77[b] ± 1.26	185.04[ab] ± 2.84	201.33[a] ± 5.55	$P < 0.05$
Jinnan	body side length	151.50[ab] ± 0.99	150.88[b] ± 1.63	158.00[a] ± 2.00	$P < 0.05$
SNP3		CC	CT	TT	
Qinchuan	body height	128.88[ab] ± 0.51	129.27[a] ± 0.64	124.00[b] ± 2.72	$P = 0.05$
Qinchuan	chest width	38.25[a] ± 0.44	38.37[a] ± 0.44	33.90[b] ± 0.95	$P < 0.05$
Qinchuan	chest depth	64.20[A] ± 0.51	64.21[A] ± 0.60	59.60[B] ± 0.93	$P < 0.01$
Qinchuan	hucklebone width	43.21[a] ± 0.43	41.96[b] ± 0.48	41.80[ab] ± 1.83	$P = 0.05$

Note: [*](LSM ± SE), LSM: least squares mean; SE: standard error. The LSM values with different superscripts within the same row differ significantly at $P < 0.05$ for a and b and $P < 0.01$ for A and B.

Table 6. Associations between combined genotypes and growth traits in Qinchuan and Jinnan cattle.

Loci/breed	Growth traits	Combined genotypes (number)/observed genotypes (LSM ± SE)*						*P* value
SNP1–SNP3		CC–CC (24)	CC–CT (24)	CC–TT (4)	CG–CC (21)	CG–CT (13)		
Qinchuan	chest width	39.69[ab] ± 1.22	38.00[ab] ± 1.26	34.38[b] ± 1.07	39.83[a] ± 1.18	38.54[ab] ± 1.29		$P < 0.05$
SNP1–SNP4		CC–AA (6)	CC–AG (12)		CG–AA (6)	CG–AG (11)		
Qinchuan	height across the hip	122.67[b] ± 2.32	124.25[ab] ± 1.99		125.00[ab] ± 2.45	129.09[a] ± 1.90		$P = 0.05$
SNP2–SNP5		TT–CC (26)	TT–CT (76)	TT–TT (70)	TC–CC (17)	TC–CT (26)	TC–TT (7)	
Qinchuan	chest circumference	178.40[A] ± 2.99	177.86[A] ± 3.06	177.41[A] ± 3.11	151.76[B] ± 14.62	180.23[A] ± 3.07	181.43[A] ± 5.22	$P < 0.01$
Qinchuan	chest width	37.94[ab] ± 0.91	38.07[ab] ± 0.74	38.43[ab] ± 0.59	35.82[b] ± 1.43	39.31[a] ± 1.04	38.29[ab] ± 1.25	$P < 0.05$
Qinchuan	hucklebone width	23.90[a] ± 0.91	23.68[a] ± 0.74	23.10[ab] ± 0.59	21.53[b] ± 1.43	23.87[a] ± 1.04	22.79[ab] ± 1.25	$P < 0.05$
SNP1–SNP3		CC–CC (40)	CC–CT (56)	CC–TT (7)	CG–CC (17)	CG–CT (12)	GG–CC (6)	
Jinnan	body height	127.70[ab] ± 0.93	129.41[ab] ± 0.85	129.57[ab] ± 1.76	131.18[a] ± 2.02	125.83[b] ± 1.63	127.33[ab] ± 1.45	$P < 0.05$
Jinnan	height across the hip	130.43[ab] ± 1.23	132.04[ab] ± 0.94	134.00[ab] ± 2.06	133.76[a] ± 2.37	127.58[b] ± 1.89	128.50[ab] ± 2.60	$P < 0.05$
Jinnan	body side length	186.00[b] ± 1.97	151.68[ab] ± 1.59	151.71[ab] ± 4.83	155.47[a] ± 2.35	144.67[b] ± 3.60	148.33[ab] ± 4.29	$P < 0.05$
Jinnan	chest circumference	152.60[ab] ± 2.67	185.50[ab] ± 1.86	184.29[ab] ± 5.59	188.59[a] ± 4.13	177.08[b] ± 4.09	177.83[ab] ± 5.24	$P < 0.05$
Jinnan	rump length	47.86[ab] ± 0.90	47.83[b] ± 0.63	46.43[ab] ± 1.82	50.71[a] ± 0.99	45.36[b] ± 0.89	47.00[ab] ± 3.61	$P < 0.05$

Note: *(LSM ± SE), LSM: least squares mean; SE: standard error. The LSM values with different superscripts within the same row differ significantly at $P < 0.05$ for a and b and $P < 0.01$ for A and B.

et al., 2014). Our previous study also identified that the accuracy of T-ARMS-PCR and PCR-RFLP reached 99.07 and 99.69 %, respectively, based on the sequencing result (Zhang et al., 2015a).

Genetic diversity analysis found that three loci were not at Hardy–Weinberg equilibrium. The disequilibrium of SNP4 in the two breeds may be caused by the deficiency of genotype GG. A possible explanation for the disequilibrium is that artificial selection promotes the mutation of these loci, and these mutations only happened a few generations ago.

The association analysis found that the synonymous coding SNPs, SNP1 and SNP2, were associated with three growth-related traits. There are studies that have the same results as this study, which are that coding SNPs are associated with economically important production traits. For example, the synonymous mutation AC_000163:g.18161C>G SNP in

the goat *PITX2* gene is associated with milk density in the Guanzhong dairy goat (Zhao et al., 2013). The synonymous mutation might produce codon usage bias, thereby influencing the production traits (Lan et al., 2007).

Furthermore, the different genotypes of intronic variation SNP3 were significantly associated with four growth-related traits in Qinchuan cattle. The combined genotypes containing the SNP3 locus were significantly associated with chest width in the Qinchuan breed and five other traits in the Jinnan breed. Studies showed that the intronic mutation G3072A in sheep *IGF2* was associated with skeletal muscle development (Cockett et al., 1994) and the intronic mutation AC_000163:g.18353TNC in the goat *PITX2* gene was associated with more than 10 milk production traits (Zhao et al., 2013). The intronic mutation might affect the binding of the DNA sequence and DNA binding factors, such as transcription factors and splicing factors. Moreover, intronic mutation might influence the transcriptional efficiency as well as stability of mRNA (Zhao et al., 2013).

Furthermore, association analysis of genotypes of single SNP loci and growth-related traits is an important way to evaluate the effects of a gene in animal breeding. However, the association analysis between combined genotypes and growth related traits will be more reliable and efficient for evaluating the effects of genetic variations in a gene (Akey et al., 2001; Schaid, 2004). Thus, we analyzed the association between the combined genotypes with higher D' value and growth straits. A total of three different combined genotypes were found to have effects on four and five different growth traits in Qinchuan and Jinnan cattle, respectively. These results demonstrated the important roles of *ATBF1* single-nucleotide variations in cattle.

5 Conclusions

In the present study, three novel SNPs (SNP1, SNP2 and SNP3) and three combined genotypes (SNP1–SNP3, SNP1–SNP4 and SNP2–SNP5) in the *ATBF1* gene were significantly associated with growth-related traits in cattle. SNP1 was similarly significantly associated with the height across the hip in Qinchuan cattle ($P = 0.05$), and the heterozygote carriers had the highest value. SNP2 was significantly associated with chest circumference ($P < 0.05$) and body side length traits ($P < 0.05$) in Jinnan cattle, and the CC genotype carriers had the highest growth trait index. For SNP3, associations between *ATBF1* genotypes and body height ($P = 0.05$), chest width ($P < 0.05$), chest depth ($P < 0.01$) and hucklebone width ($P = 0.05$) of Qinchuan cattle were found, and CC carriers had the best growth trait indexes for the first three traits. Moreover, a total of three different combined genotypes (SNP1–SNP3, SNP1–SNP4 and SNP2–SNP5) were found to have effects on four and five different growth traits in Qinchuan and Jinnan cattle, respectively. Thus, SNP1, SNP2 and SNP3 have the potential to be useful DNA markers for the improvement of growth-related traits in cattle.

Competing interests. The authors declare that they have no conflict of interest.

Acknowledgements. This work was supported by the National Natural Science Foundation of China (no. 31672400), Science and Technology Coordinator Innovative engineering projects of Shaanxi Province (2014KTZB02-02-02-02) and the Program of National Beef Cattle and Yak Industrial Technology System (no. CARS-37).

Edited by: Steffen Maak

References

Akey, J., Jin, L., and Xiong, M.: Haplotypes vs single marker linkage disequilibrium tests: what do we gain?, Eur. J. Hum. Genet., 9, 291–300, 2001.

Araujo, R. V., Chang, C. V., Cescato, V. A., Fragoso, M. C., Bronstein, M. D., Mendonca, B. B., Arnhold, I. J., and Carvalho, L. R.: PROP1 overexpression in corticotrophinomas: evidence for the role of PROP1 in the maintenance of cells committed to corticotrophic differentiation, Clinics., 68, 887–891, 2013.

Botstein, D., White, R. L., Skolnick, M., and Davis, R. W.: Construction of a genetic linkage map in man using restriction fragment length polymorphisms, Am. J. Hum. Genet., 32, 314–331, 1980.

Cai, H., Lan, X., Li, A., Zhou, Y., Sun, J., Lei, C., Zhang, C., and Chen, H.: SNPs of bovine HGF gene and their association with growth traits in Nanyang cattle, Res. Vet. Sci., 95, 483–488, 2013.

Cockett, N. E., Jackson, S. P., Shay, T. L., Nielsen, D., Moore, S. S., Steele, M. R., Barendse, W., Green, R. D., and Georges, M.: Chromosomal localization of the callipyge gene in sheep (Ovis aries) using bovine DNA markers, P. Natl. Acad. Sci. USA, 91, 3019–3023, 1994.

Collins, A.: Primer1: primer design web service for tetra-primer ARMS-PCR, Open Bioinformatics Journal, 6, 55–58, 2012.

Dang, Y., Li, M., Yang, M., Cao, X., Lan, X., Lei, C., Zhang, C., Lin, Q., and Chen, H.: Identification of bovine NPC1 gene cSNPs and their effects on body size traits of Qinchuan cattle, Gene., 540, 153–160, 2014.

Davis, S. W., Castinetti, F., Carvalho, L. R., Ellsworth, B. S., Potok, M. A., Lyons, R. H., Brinkmeier, M. L., Raetzman, L. T., Carninci, P., Mortensen, A. H., Hayashizaki, Y., Arnhold, I. J., Mendonca, B. B., Brue, T., and Camper, S. A.: Molecular mechanisms of pituitary organogenesis: In search of novel regulatory genes, Mol Cell Endocrinol., 323, 4–19, 2010.

Herrington, J., Smit, L. S., Schwartz, J., and Carter-Su, C.: The role of STAT proteins in growth hormone signaling, Oncogene., 19, 2585–2597, 2000.

Huang, Y. Z., Zhan, Z. Y., Li, X. Y., Wu, S. R., Sun, Y. J., Xue, J., Lan, X. Y., Lei, C. Z., Zhang, C. L., Jia, Y. T., and Chen, H.: SNP and haplotype analysis reveal IGF2 variants associated with growth traits in Chinese Qinchuan cattle, Mol. Biol. Rep., 41, 591–598, 2014.

Huang, Y., Wang, C., Yao, Y., Zuo, X., Chen, S., Xu, C., Zhang, H., Lu, Q., Chang, L., Wang, F., Wang, P., Zhang, R., Hu, Z., Song, Q., Yang, X., Li, C., Li, S., Zhao, Y., Yang, Q., Yin, D., Wang, X., Si, W., Li, X., Xiong, X., Wang, D., Huang, Y., Luo, C., Li, J., Wang, J., Chen, J., Wang, L., Wang, L., Han, M., Ye, J., Chen, F., Liu, J., Liu, Y., Wu, G., Yang, B., Cheng, X., Liao, Y., Wu, Y., Ke, T., Chen, Q., Tu, X., Elston, R., Rao, S., Yang, Y., Xia, Y., and Wang, Q. K.: Molecular basis of gene-gene interaction: cyclic cross-regulation of gene expression and post-GWAS gene-gene interaction involved in atrial fibrillation, PLoS Genet., 11, e1005393, https://doi.org/10.1371/journal.pgen.1005393, 2015.

Jia, W., Wu, X., Li, X., Xia, T., Lei, C., Chen, H., Pan, C., and Lan, X.: Novel genetic variants associated with mRNA expression of signal transducer and activator of transcription 3 (STAT3) gene significantly affected goat growth traits, Small Ruminant Research, 129, 25–36, 2015.

Jiang, Q., Ni, B., Shi, J., Han, Z., Qi, R., Xu, W., Wang, D., Wang, D. W., and Chen, M.: Down-regulation of ATBF1 activates STAT3 signaling via PIAS3 in pacing-induced HL-1 atrial myocytes, Biochem. Biophys. Res. Commun., 449, 278–283, 2014.

Jung, C. G., Kim, H. J., Kawaguchi, M., Khanna, K. K., Hida, H., Asai, K., Nishino, H., and Miura, Y.: Homeotic factor ATBF1 induces the cell cycle arrest associated with neuronal differentiation, Development, 132, 5137–5145, 2005.

Kawaguchi, M., Hara, N., Bilim, V., Koike, H., Suzuki, M., Kim, T. S., Gao, N., Dong, Y., Zhang, S., Fujinawa, Y., Yamamoto, O., Ito, H., Tomita, Y., Naruse, Y., Sakamaki, A., Ishii, Y., Tsuneyam, K., Inoue, M., Itoh, J., Yasuda, M., Sakata, N., Jung, C. G., Kanazawa, S., Akatsu, H., Minato, H., Nojima, T., Asai, K., and Miura, Y.: A diagnostic marker for superficial urothelial bladder carcinoma: lack of nuclear ATBF1 (ZFHX3) by immunohistochemistry suggests malignant progression, BMC Cancer., 16, 805, 2016.

Lan, X. Y., Pan, C. Y., Chen, H., Zhang, C. L., Li, J. Y., Zhao, M., Lei, C. Z., Zhang, A. L., and Zhang, L.: An AluI PCR-RFLP detecting a silent allele at the goat POU1F1 locus and its association with production traits, Small Ruminant Research, 73, 8–12, 2007.

Lan, X. Y., Pan, C. Y., Li, J. Y., Guo, Y. W., Hu, S., Wang, J., Liu, Y. B., Hu, S. R., Lei, C. Z., and Chen, H.: Twelve novel SNPs of the goat POU1F1 gene and their associations with cashmere traits, Small Ruminant Research, 85, 116–121, 2009.

Lao, M., Shi, M., Zou, Y., Huang, M., Ye, Y., Qiu, Q., Xiao, Y., Zeng, S., Liang, L., Yang, X., and Xu, H.: Protein Inhibitor of Activated STAT3 Regulates Migration, Invasion, and Activation of Fibroblast-like Synoviocytes in Rheumatoid Arthritis, J. Immunol., 196, 596–606, 2016.

Li, M. X., Sun, X. M., Jiang, J., Sun, Y. J., Lan, X. Y., Lei, C. Z., and Chen, H.: Tetra-primer ARMS-PCR is an efficient SNP genotyping method: an example from SIRT2, Anal. Methods., 6, 1835–1840, 2014.

Li, Z., Zhang, Z., He, Z., Tang, W., Li, T., Zeng, Z., He, L., and Shi, Y.: Apartition-ligation-combination-subdivision EMalgorithm for haplotype inference with multiallelic markers: update of the SHEsis, Cell Res., 19, 519–523, 2009.

Liongue, C. and Ward, A. C.: Evolution of the JAK-STAT pathway, JAKSTAT., 2, e22756, https://doi.org/10.4161/jkst.22756, 2013.

Liu, Y., Ni, B., Lin, Y., Chen, X. G., Fang, Z., Zhao, L., Hu, Z., and Zhang, F.: Genetic polymorphisms in ZFHX3 are associated with atrial fibrillation in a Chinese Han population, PLoS One, 9, e101318, https://doi.org/10.1371/journal.pone.0101318, 2014.

Ma, L., Qin, Q. M., Yang, Q., Zhang, M., Zhao, H. Y., Pan, C. Y., Lei, C. Z., Chen, H., and Lan, X. Y.: Associations of six SNPs of POU1F1-PROP1-PITX1-SIX3 pathway genes with growth traits in two Chinese indigenous goat breeds, Ann. Anim. Sci., 17, 399–411, 2017.

Morinaga, T., Yasuda, H., Hashimoto, T., Higashio, K., and Tamaoki, T.: A human alpha-fetoprotein enhancer-binding protein, ATBF1, contains four homeodomains and seventeen zinc fingers, Mol. Cell Biol., 11, 6041–6049, 1991.

Nojiri, S., Joh, T., Miura, Y., Sakata, N., Nomura, T., Nakao, H., Sobue, S., Ohara, H., Asai, K., and Ito, M.: ATBF1 enhances the suppression of STAT3 signaling by interaction with PIAS3, Biochem. Biophys. Res. Commun., 314, 97–103, 2004.

Pan, C., Wu, C., Jia, W., Xu, Y., Lei, C., Hu, S., Lan, X., and Chen, H.: A critical functional missense mutation (H173R) in the bovine PROP1 gene significantly affects growth traits in cattle, Gene., 531, 398–402, 2013.

Pedersen, L. D., Sørensen, A. C., and Berg, P.: Marker-assisted selection can reduce true as well as pedigree-estimated inbreeding, J. Dairy Sci., 92, 2214–2223, 2009.

Postigo, A. A. and Dean, D. C.: ZEB, a vertebrate homolog of Drosophila Zfh-1, is a negative regulator of muscle differentiation, EMBO J., 16, 3935–3943, 1997.

Postigo, A. A. and Dean, D. C.: Independent repressor domains in ZEB regulate muscle and T-cell differentiation, Mol. Cell Biol., 19, 7961–7971, 1999.

Qi, Y., Ranish, J. A., Zhu, X., Krones, A., Zhang, J., Aebersold, R., Rose, D. W., Rosenfeld, M. G., and Carriere, C.: Atbf1 is required for the Pit1 gene early activation, P. Natl. Acad. Sci. USA, 105, 2481–2486, 2008.

Richard, A. J. and Stephens, J. M.: The role of JAK-STAT signaling in adipose tissue function, Biochim. Biophys. Acta, 1842, 431–439, 2014.

Schaid, D. J.: Evaluating associations of haplotypes with traits, Genet. Epidemiol., 27, 348–364, 2004.

Sun, X. M., Li. M. X., Li, A. M., Lan, X. Y., Lei, C. Z., Ma, W., Hua, L. S., Wang, J., Hu, S. R., and Chen, H.: Two novel intronic polymorphisms of bovine FGF21 gene are associated with body weight at 18 months in Chinese cattle, Livestock Science., 155, 23–29, 2013.

Sun, X., Fu, X., Li, J., Xing, C., Frierson, H. F., Wu, H., Ding, X., Ju, T., Cummings, R. D., and Dong, J. T.: Deletion of Atbf1/Zfhx3 in mouse prostate causes neoplastic lesions, likely by attenuation of membrane and secretory proteins and multiple signaling pathways, Neoplasia, 16, 377–389, 2014.

Sun, X., Xing, C., Fu, X., Li, J., Zhang, B., Frierson Jr., H. F., and Dong, J. T.: Additive Effect of Zfhx3/Atbf1 and Pten Deletion on Mouse Prostatic Tumorigenesis, J. Genet. Genomics., 42, 373–382, 2015.

Trovato, L., Riccomagno, S., Prodam, F., Genoni, G., Walker, G. E., Moia, S., Bellone, S., and Bona, G.: Isolated GHD: investigation and implication of JAK/STAT related genes before and after rhGH treatment, Pituitary, 15, 482–489, 2012.

Tsai, C. T., Hsieh, C. S., Chang, S. N., Chuang, E. Y., Juang, J. M., Lin, L. Y., Lai, L. P., Hwang, J. J., Chiang, F. T., and Lin, J. L.: Next-generation sequencing of nine atrial fibrillation candidate genes identified novel de novo mutations in patients with extreme trait of atrial fibrillation, J. Med. Genet., 52, 28–36, 2015.

Wang, H. L., Li, Z. X., Wang, L. J., He, H., Yang, J., Chen, L., Niu, F. B., Liu, Y., Guo, J. Z., and Liu, X. L.: Polymorphism in PGLYRP-1 gene by PCR-RFLP and its association with somatic cell score in Chinese Holstein, Res. Vet. Sci., 95, 508–514, 2013.

Wang, Z. N., Li, M. J., Lan, X. Y., Li, M. X., Lei, Z. C., and Chen, H.: Tetra-primer ARMS-PCR identifies the novel genetic variations of bovine HNF-4α gene associating with growth traits, Gene., 546, 206–213, 2014.

Wu, X., Jia, W., Zhang, J., Li, X., Pan, C., Lei, C., Chen, H., Dang, R., and Lan X.: Determination of the novel genetic variants of goat STAT5A gene and their effects on body measurement traits in two Chinese native breeds, Small Ruminant Research, 121, 232–243, 2014.

Yang, Q., Zhang, S., Liu, L., Cao, X., Lei, C., Qi, X., Lin, F., Qu, W., Qi, X., Liu, J., Wang, R., Chen, H., and Lan, X.: Application of mathematical expectation (ME) strategy for detecting low frequency mutations: An example for evaluating 14-bp insertion/deletion (indel) within the bovine PRNP gene, Prion., 10, 409–419, 2016.

Yang, S. F., Hou, M. F., Chen, F. M., Ou-Yang, F., Wu, Y. C., Chai, C. Y., and Yeh, Y. T.: Prognostic value of protein inhibitor of activated STAT3 in breast cancer patients receiving hormone therapy, BMC Cancer, 16, 20, 2016.

Ye, S., Dhillon, S., Ke, X., Collins, A. R., and Day, I. N.: An efficient procedure for genotyping single nucleotide polymorphisms, Nucleic Acids Res., 29, E88, https://doi.org/10.1093/nar/29.17.e88, 2001.

Zhang, S., Dang, Y., Zhang, Q., Qin, Q., Lei, C., Chen, H., and Lan, X.: Tetra-primer amplification refractory mutation system PCR (T-ARMS-PCR) rapidly identified a critical missense mutation (P236T) of bovine ACADVL gene affecting growth traits, Gene., 559, 184–188, 2015a.

Zhang, X., Wu, X., Jia, W., Pan, C., Li, X., Lei, C., Chen, H., and Lan, X.: Novel Nucleotide Variations, Haplotypes Structure and Associations with Growth Related Traits of Goat AT Motif-Binding Factor (ATBF1) Gene, Asian-Australas J. Anim. Sci., 28, 1394–1406, 2015b.

Zhao, H., Wu, X., Cai, H., Pan, C., Lei, C., Chen, H., and Lan, X.: Genetic variants and effects on milk traits of the caprine paired-like homeodomain transcription factor 2 (PITX2) gene in dairy goats, Gene., 532, 203–210, 2013.

Zhao, D., Ma, G., Zhang, X., He, Y., Li, M., Han, X., Fu, L., Dong, X. Y., Nagy, T., Zhao, Q., Fu, L., and Dong, J. T.: Zinc Finger Homeodomain Factor Zfhx3 Is Essential for Mammary Lactogenic Differentiation by Maintaining Prolactin Signaling Activity, J. Biol. Chem., 291, 12809–12820, 2016.

21

A combined genotype of three SNPs in the bovine *PPARD* gene is related to growth performance in Chinese cattle

Jieping Huang[1,2]**, Qiuzhi Zheng**[1]**, Shuzhe Wang**[1]**, Qiongqiong Zhang**[1]**, Lijun Jiang**[1,2]**, Ruijie Hao**[1]**, Fen Li**[1,2]**, and Yun Ma**[1,2]

[1]College of Life Sciences, Xinyang Normal University, Xinyang, Henan, China
[2]Institute for Conservation and Utilization of Agro-Bioresources in Dabie Mountains, Xinyang, Henan, China

Correspondence to: Yun Ma (mayun_666@126.com)

Abstract. *PPARD* is involved in multiple biological processes, especially for those associated with energy metabolism. *PPARD* regulates lipid metabolism through up-regulate expression of genes associating with adipogenesis. This makes *PPARD* a significant candidate gene for production traits of livestock animals. Association studies between *PPARD* polymorphisms and production traits have been reported in pigs but are limited for other animals, including cattle. Here, we investigated the expression profile and polymorphism of bovine *PPARD* as well as their association with growth traits in Chinese cattle. Our results showed that the highest expression of *PPARD* was detected in kidney, following by adipose, which is consistent with its involvement in energy metabolism. Three SNPs of *PPARD* were detected and used to undergo selection pressure according the result of Hardy–Weinberg equilibrium analysis ($P < 0.05$). Moreover, all of these SNPs showed moderate diversity ($0.25 < \text{PIC} < 0.5$), indicating their relatively high selection potential. Association analysis suggested that individuals with the GAAGTT combined genotype of three SNPs detected showed optimal values in all of the growth traits analyzed. These results revealed that the GAAGTT combined genotype of three SNPs detected in the bovine *PPARD* gene was a significant potential genetic marker for marker-assisted selection in Chinese cattle. However, this should be further verified in larger populations before being applied to breeding.

1 Introduction

Peroxisome-proliferator-activated receptors (PPARs) are a group of transcription factors belonging to the nuclear hormone receptor superfamily (Evans et al., 2004). Many studies have revealed PPARs take part in numerous biological processes, including lipid metabolism, the insulin signaling pathway, glucose metabolism, and adipocyte differentiation (Youssef and Badr, 2013). To date, three subtypes of PPARs have been discovered: *PPARA*, *PPARD*, and *PPARG*. Among these, *PPARD* is widely expressed in various tissues, including kidney, liver, heart, intestine, and adipose (Abbott, 2009). *PPARD* regulates lipid metabolism through up-regulate expression of genes involved in the adipogenesis process (Vedhachalam et al., 2007). Recently, studies have suggested that *PPARD* is essential for adipogenesis as well (Garbacz et al., 2015; Barroso et al., 2015; Palomer et al., 2016).

Genetic variation in *PPARD* is proved to be associated with human diseases. The *PPARD* rs2016520 polymorphism was reported to affect repaglinide response in Han Chinese patients with type 2 diabetes mellitus (Song et al., 2015). Furthermore, this mutation was shown to be associated with brain diseases (Huang et al., 2015) and colorectal cancer (Rosalesreynoso et al., 2017). For its vital role in various biological processes, *PPARD* is a potential gene affecting production traits of livestock animals. Polymorphisms of *PPARD* were shown to affect ear size (Ren et al., 2011) and litter size (Spötter et al., 2010) in pigs. Recently, functional SNPs in the 5′ regulatory region of the porcine *PPARD* gene have been reported to be significantly associated with fat deposition traits (Zhang et al., 2015). However, association studies between *PPARD* and production traits in other animals are limited, including cattle.

Marker-assisted selection (MAS) has been widely used as a breeding strategy in livestock (Margawati, 2012). To detect functional SNPs of *PPARD* for MAS in cattle, we (i) analyzed the expression profile of *PPARD* in different tissues, (ii) detected SNPs in the bovine *PPARD* gene by direct sequencing using 514 Chinese cattle, and (iii) assessed the relationship between detected SNPs and growth traits in partial cattle.

2 Material and methods

2.1 Samples

Samples used in this study were shown in a previous study (Huang et al., 2017). Briefly, seven tissues were collected at the slaughter house for reverse transcription polymerase chain reaction (RT-PCR), including heart, liver, spleen, lung, kidney, muscle, and adipose of three Jiaxian cattle (bullock, 30 months). A total of 514 individuals from six Chinese cattle breeds – including 141 Jiaxian cattle, 139 Nanyang cattle, 114 Luxi cattle, 30 Qinchuan cattle, 30 Bohai, cattle and 60 Gaoyuan yak – were used for SNP genotyping. Birth weight and six growth traits (body weight, body height, body length, heart girth, hip width, and average daily gain) of Jiaxian and Nanyang cattle at 6, 12, 18, and 24 months as well as nine traits (body height, body length, heart girth, abdominal circumference, hip width, sciatic width, height at hip cross, body weight, and beef performance index) at around 28–30 months of age in 300 Henan cattle (100 Jiaxian, 100 Nanyang, and 100 Luxi) were recorded for association analysis.

2.2 Expression analysis of *PPARD*

In order to understand the potential biological effect of *PPARD* on cattle, expression levels of *PPARD* in seven tissues were investigated by RT-PCR. Details of the method are shown in a previous study (Huang et al., 2017). Total RNA was reversely transcribed into cDNA using a PrimeScript-sRT reagent kit with gDNA Eraser (TaKaRa, Japan). RT-PCR was performed using SYBR Green I with two-step reactions. Primers of *PPARD* (NM_001083636.1) and reference genes (*TUBA1A*, NM_001166505.1; β-actin, NM_173979.3) for RT-PCR are shown in Table S1 (Supplement). The relative expression level of each tissue was presented as mean $\pm$SD.

2.3 SNP detection and genotyping

In order to investigate polymorphism of *PPARD* (AC_000180.1) in Chinese cattle, nine pairs of primers covering CDS and partial upstream regions were synthesized (Table S1). The methods for SNP detection and genotyping were as in a previous study (Huang et al., 2017). Pooled DNA samples were used as a PCR template for SNP detection. For SNPs detected (Table 1), three pairs of specific

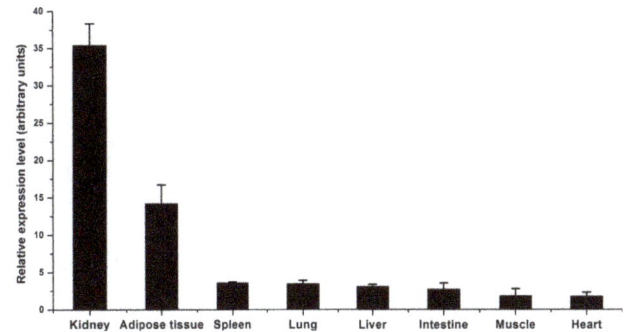

Figure 1. Tissue distribution of bovine *PPARD* mRNA assessed by RT-PCR. Values shown in this figure are averages of three independent experiments. Error bars represent SD ($n = 3$) of relative mRNA levels. Expression data were normalized using geometric mean of mRNA levels for two control genes (*TUBA1A* and β-actin).

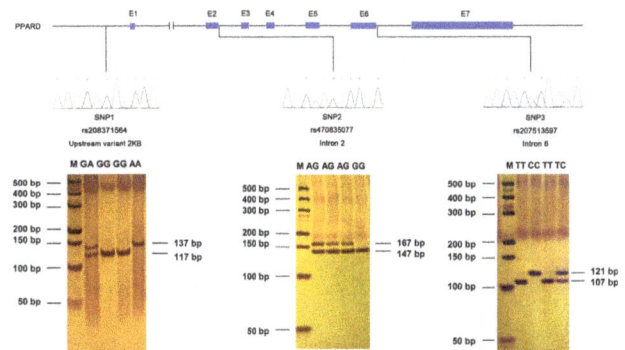

Figure 2. Schematic characteristic of SNPs identified in bovine *PPARD* and genotyping. From top to bottom: structure of bovine *PPARD*, mutant peaks of sequencing, details of SNPs identified, and electrophoretogram of genotyping.

primers were designed (Table 2). The traditional PCR-RFLP method was used for SNP genotyping in 514 Chinese cattle from six breeds. It should be noted that PCR production of PPARD-MluI primers contained two recognition sites of MluI, one of which was native and the other was introduced from primers for genotyping.

2.4 Statistical analysis

The genetic characteristics of each mutation were investigated after genotyping, including allele frequencies, Hardy–Weinberg equilibrium (HWE), heterozygosity (H_e), effective allele numbers (N_e), and polymorphism information content (PIC). To evaluate the potential relationship between the *PPARD* gene and development of cattle, an association study was performed based on the genotyping results and growth traits in Nanyang, Jiaxian, and Luxi cattle. Significant analysis was performed by SPSS 19.0 using general linear model. Results were presented as means $\pm$SE. Other details can be found in a previous study (Huang et al., 2017).

Table 1. Details of SNPs detected in the bovine *PPARD* gene.

Label	Position	Alleles	rs number	Functional consequence
SNP1	AC_000180.1: 9268142	G > A	rs208371564	upstream variant 2KB
SNP2	AC_000180.1: 9341130	A > G	rs470835077	intron 2
SNP3	AC_000180.1: 9352706	T > C	rs207513597	intron 6

Table 2. Details of primers and restriction enzymes used for genotyping.

Name	Primer sequence (5′-3′)	T_m (°)	Size (bp)	Used for	Restriction enzyme	Main fragments
PPARD -Hha I	F:GCAGGATATAGTTCCCAGC	55	137	SNP1	Hha I	GG: 117 bp
	R:GACTTGTCATCCCAACCTT					GA: 137 bp, 117 bp
						AA: 137 bp
PPARD -Pvu II	F:TCCTTCCAGCAGCTACACAG	57.5	195	SNP2	Pvu II	GG: 147 bp
	CT					AG: 167 bp, 147 bp
	R:GGGAGACAACTCGCCCAAG					
	A					
PPARD -Mlu I	F:ATGGCAGTGGGACACGCG	63	121	SNP3	Mlu I	TT: 107 bp
	R:CCACCAGAAATAACCCCCAT					TC: 121 bp, 107 bp
	C					CC: 121 bp

Note: letters underlined in the primer sequence are the introduced mutant for genotyping.

3   Results and discussion

3.1   Expression profile of *PPARD*

The expression profile of *PPARD* has been widely investigated in rodent and human development, but it is limited in cattle. In order to understand the potential biological effect of *PPARD* on cattle, expression levels of *PPARD* were investigated (Fig. 1). Consistent with previous studies, *PPARD* was widely expressed in main tissues, suggesting that it was involved in multiple biological processes. The highest level of expression was detected in kidney, followed by adipose tissue (Fig. 1), indicating its significant biological role in kidney and adipose tissues. Expression levels of the *PPARD* gene in the other six tissues were nearly the same, with relatively low values. In fact, the expression pattern of *PPARD* was found to be variable in different studies. *PPARD* was expressed in kidney with a high level in adult rats (Braissant and Wahli, 1996), adult mice (Girroir et al., 2008), and adult human (Auboeuf et al., 1997). PPARs were identified as the genetic sensor responsive to fatty acid ligands (Feige et

al., 2006) and involved in lipid metabolism and the insulin signaling pathway (Youssef and Badr, 2013). In addition, chronic kidney disease was attributed to metabolic disorders mainly through the mechanisms of insulin resistance and resultant hyperinsulinemia (Perlstein et al., 2007). In fat tissue, only a moderate level was detected in adult rats and humans (Braissant and Wahli, 1996; Auboeuf et al., 1997). These results were consistent with our study. However, a moderate to high level of expression in liver, heart, and lung was detected in adult rodents and humans (Girroir et al., 2008; Tugwood et al., 1996; Mukherjee et al., 1997), which was nearly contradictory with our result. Regardless, all of these results underline multiple functions of *PPARD* in the development of mammals. Thus, *PPARD* should be necessary for cattle development.

3.2   SNP detection and genetic characteristics of *PPARD* in Chinese cattle

In total, three SNPs were detected (Table 1 and Fig. 2), including AC_000180.1:g.9268142 G > A in the upstream

Table 3. Association analysis between combined genotypes of three SNPs in *PPARD* and growth traits of 300 adult cattle.

Growth traits	Combined genotypes								
	GAAGCC (10)	GAAGTC (66)	GAAGTT (35)	GAGGTC (45)	GAGGTT (21)	GGAGTC (29)	GGAGTT (14)	GGGGTC (41)	GGGGTT (20)
BH (cm)	129.450 ± 3.028	129.106 ± 1.179	130.843 ± 1.618	130.033 ± 1.427	129.286 ± 2.089	129.448 ± 1.778	127.286 ± 2.559	131.598 ± 1.495	127.450 ± 2.141
BL (cm)	145.900 ± 4.034	144.068 ± 1.570	147.743 ± 2.156	143.500 ± 1.901	143.619 ± 2.783	143.017 ± 2.369	146.643 ± 3.409	145.415 ± 1.992	141.750 ± 2.852
HG (cm)	171.400 ± 5.195	174.538 ± 2.022	176.414 ± 2.777	174.156 ± 2.449	168.214 ± 3.585	169.414 ± 3.051	171.357 ± 4.391	175.634 ± 2.566	168.750 ± 3.673
AC (cm)	200.100 ± 6.774	203.008 ± 2.637^{ab}	205.514 ± 3.621^{a}	204.156 ± 3.193^{a}	192.476 ± 4.674^{b}	199.069 ± 3.978	200.143 ± 5.725	201.415 ± 3.345	190.700 ± 4.790^{bc}
HW (cm)	45.200 ± 1.897^{a}	45.023 ± 0.738^{a}	45.471 ± 1.014^{a}	45.344 ± 0.894^{A}	43.310 ± 1.309	44.190 ± 1.114	45.286 ± 1.603^{a}	43.341 ± 0.937	41.100 ± 1.341^{Bb}
SW (cm)	28.500 ± 1.456^{A}	28.659 ± 0.567^{A}	28.586 ± 0.778^{A}	27.489 ± 0.687^{A}	27.548 ± 1.005^{A}	26.914 ± 0.855^{A}	28.107 ± 1.231^{A}	25.549 ± 0.719^{a}	22.600 ± 1.030^{Bb}
HHC (cm)	129.300 ± 2.691	129.394 ± 1.048	131.357 ± 1.438	129.122 ± 1.269	129.238 ± 1.857	131.621 ± 1.580	128.214 ± 2.274	132.232 ± 1.329	128.575 ± 1.903
BW (kg)	391.453 ± 32.883	388.720 ± 12.800	413.462 ± 17.577^{a}	399.380 ± 15.501	385.273 ± 2.691	391.784 ± 19.310	378.964 ± 27.791	391.686 ± 16.240	350.251 ± 3.252^{b}
BPI (kg cm^{-1})	3.017 ± 0.194	3.001 ± 0.076	3.113 ± 0.104^{a}	3.045 ± 0.092	2.906 ± 0.134	2.987 ± 0.114	2.962 ± 0.164	2.956 ± 0.096	2.738 ± 0.137^{b}

BH: body height; BL: body length; HG: heart girth; AC: abdominal circumference; HW: hip width; SW: sciatic width; HHC: height at hip cross; BW: body weight; BPI: beef performance index.
Lowercase letters mean difference of the value at $P < 0.05$; uppercase letters mean difference of the value at $P < 0.01$.

region (SNP1, rs208371564), AC_000180.1:g.9341130 A > G in intron 2 (SNP2, rs470835077), and AC_000180.1:g.9352706 T > C in intron 6 (SNP3, rs207513597). A total of 3801 SNPs of the bovine *PPARD* gene can be searched in the SNP database of NCBI (https://www.ncbi.nlm.nih.gov/snp/), including 710 detected by cluster and 3091 with no information. No more SNPs could be found from other studies. Thus, the three SNPs detected in this study were not further analyzed although they had been detected by cluster previously.

Then, genetic characteristics of SNPs were investigated based on the genotyping result (Table S2). We noted that the AA genotype of SNP2 was absent in all of the populations in this study. At the same time, the A allele was not rare in Chinese cattle. Therefore, we speculated that individuals with the AA genotype of SNP2 died during the embryonic stage or were culled because of disease at an early age. Amazingly, approximately half of the breeds were not in agreement with the HWE ($P < 0.05$) at each of these SNP loci, suggesting that they might undergo selection pressure. All of the three SNP loci showed moderate diversity ($0.25 < PIC < 0.5$), indicating their relatively high selection potential. Further selec-

tion could be implied if a positive effect were found among these SNPs in cattle breeds investigated.

3.3 Association study between *PPARD* and growth traits

Potential genomic mutations of the *PPARD* gene might be related to growth traits of cattle. First, relationship between *PPARD* and growth traits were investigated in 173 Henan cattle based on a single SNP locus (Table S3). Several significant differences were identified without regularity. Generally, the phenotypic value should change along with the variation in genotype (in the order of wild type, heterozygous type, and homozygous mutant type) with a specific trend. Moreover, this trend should be the same among different breeds and ages. However, significant differences detected in Table S3 did not conform to such trends and showed disorder. This might be due to the low sample size, or else multiple loci affect the same traits with different weight. Therefore, it was hard to estimate the real association between these SNPs and growth traits in cattle.

We speculated that coordination among multiple SNPs loci might contribute to development or be linked with growth traits of cattle. Based on such a hypothesis, association analysis between combined genotypes of these three SNPs and growth traits of 300 adult cattle was performed. Combined genotypes with less than 10 individuals were removed. In total, nine combined genotypes were used for analysis (Table 3). Interestingly, all traits showed the highest values in the GAAGTT combined genotype. Among these, abdominal circumference, hip width, sciatic width, body weight, and beef performance index showed significant differences. Obviously, individuals with the GAAGTT combined genotype of these three SNPs showed optimal growth performance. In fact, association analysis based on combined genotype has been widely used in studies on the relationship between genetic variation and diseases in humans (Kamitani et al., 1995; Boulet et al., 2008; Stelma et al., 2016) and traits in livestock animals (Garaulet et al., 2012). However, results from this analytical method need further verification from multiple points. The bovine *PPARD* gene is identified in chromosome 23 (9.27–9.36 Mb). By searching the quantitative trait locus (QTL) database of cattle (http://www.animalgenome.org/cgi-bin/QTLdb/BT/browse) for those QTLs associated with growth trait, four QTLs were obtained, including a QTL (5.9–16.3 Mb) for body weight of adult cattle (McClure et al., 2010), a QTL (0.6–17.5 Mb) for body weight before slaughter (Elo et al., 1999), a QTL (7.2–21.1 Mb) for body weight at weaning (McClure et al., 2010), and a QTL (7.2–21.1 Mb) for body weight at 12 months (McClure et al., 2010). These QTLs further suggested that *PPARD* was a potential significant candidate gene for production traits of cattle. Three SNPs identified in this study were in the non-coding region. In recent years, transcripts (non-coding RNAs) from non-coding region have been shown to regulate the transcription of the origin genes and then affect the biological function of the origin genes. However, the non-coding region might provide the binding site for some enzymes relating to transcription. Thus, mutations in the non-coding region could play a role in the regulation mechanism. However, SNPs may only be markers associated with production traits and do not affect any biological process.

4 Conclusions

The bovine *PPARD* gene is expressed widely in the main tissues of adult cattle. Three SNPs of *PPARD* were identified in Chinese cattle. The GAAGTT combined genotype of these three SNPs showed optimal growth performance, which could be a potential marker for MAS of cattle. However, further identification should be performed in larger populations before being applied to breeding of cattle.

Author contributions. YM and FL designed experiments and collected samples. QZ, SW, QZ, LJ and RH carried out the experiment; JH analyzed the data and wrote the manuscript.

Competing interests. The authors declare that they have no conflict of interest.

Acknowledgements. This study was supported by the National Natural Science Foundation of China (no. 31672403), the Chinese National High Technology Research and Development Program (no. 2013AA102505-4), the Technology Innovation Teams in Universities of Henan Province (no. 14IRTSTHN012), and the Nanhu Scholars Program for Young Scholars of XYNU.

Edited by: Steffen Maak

References

Abbott, B. D.: Review of the expression of peroxisome proliferator-activated receptors alpha (PPAR alpha), beta (PPAR beta), and gamma (PPAR gamma) in rodent and human development, Reprod. Toxicol., 27, 246–257, 2009.

Auboeuf, D., Rieusset, J., Fajas, L., Vallier, P., Frering, V., Riou, J.P., Staels, B., Auwerx, J., Laville, M., and Vidal, H.: Tissue distribution and quantification of the expression of mRNAs of peroxisome proliferator–activated receptors and liver X receptor-α in humans: no alteration in adipose tissue of obese and NIDDM patients, Diabetes, 46, 1319–1327, 1997.

Barroso, E., Rodríguez-Rodríguez, R., Chacón, M.R., Maymó-Masip, E., Ferrer, L., Salvadó, L., Salmerón, E., Wabistch, M., Palomer, X., Vendrell, J., Wahli, W., and Vázquez-Carrera, M.: PPARβ/δ ameliorates fructose-induced insulin resistance in adipocytes by preventing Nrf2 activation, Biochim. Biophys. Acta., 1852, 1049–1058, 2015.

Boulet, S., Kleyman, M., Kim, J. Y., Kamya, P., Sharafi, S., Simic, N., Bruneau, J., Routy, J. P., Tsoukas, C. M., and Bernard, N. F.: A combined genotype of KIR3DL1 high expressing alleles and HLA-B* 57 is associated with a reduced risk of HIV infection, AIDS, 22, 1487–1491, 2008.

Braissant, O., Foufelle, F., Scotto, C., Dauça, M., and Wahli, W.: Differential expression of peroxisome proliferator-activated receptors (PPARs): tissue distribution of PPAR-alpha, -beta, and -gamma in the adult rat, Endocrinology, 137, 354–366, 1996.

Elo, K. T., Vilkki, J., de Koning, D. J., and Mäki-Tanila, A. V.: A quantitative trait locus for live weight maps to bovine chromosome 23, Mamm. Genome, 10, 831–835, 1999.

Evans, R. M., Barish, G. D., and Wang, Y. X.: PPARs and the complex journey to obesity, Nat. Med., 10, 355–361, 2004.

Feige, J. N., Gelman, L., Michalik, L., Desvergne, B., and Wahli, W.: From molecular action to physiological outputs: peroxisome proliferator-activated receptors are nuclear receptors at the crossroads of key cellular functions, Prog. Lipid Res., 45, 120–159, 2006.

Garaulet, M., Tardido, A. E., Lee, Y. C., Smith, C. E., Parnell, L. D., and Ordovás, J. M.: SIRT1 and CLOCK 3111T>C combined genotype is associated with evening preference and weight loss resistance in a behavioral therapy treatment for obesity, Int. J. Obes., 36, 1436–1441, 2012.

Garbacz, W. G., Huang, J. T., Higgins, L. G., Wahli, W., and Palmer, C. A. N.: PPARα is required for PPARδ action in regulation of body weight and hepatic steatosis in mice, PPAR Res., 1, 927057, doi:10.1155/2015/927057, 2015.

Girroir, E. E., Hollingshead, H. E., He, P., Zhu, B., Perdew, G. H., and Peters, J. M.: Quantitative expression patterns of peroxisome proliferator-activated receptor-beta/delta (pparbeta/delta) protein in mice, Biochem. Biophys. Res. Commun., 371, 456–461, 2008.

Huang, J., Chen, N., Li, X., An, S., Zhao, M., Sun, T., Hao, R., and Ma, Y.: Two novel SNPs of *PPARγ* significantly affect weaning growth traits of Nanyang cattle, Animal Biotechnol., 1–7, doi:10.1080/10495398.2017.1304950, 2017.

Huang, Y., Nie, S., Zhou, S., Li, K., Sun, J., Zhao, J., Fei, B., Wang, Z., Ye, H., Hong, Q., Gao, X., and Duan, S.: PPARD rs2016520 polymorphism and circulating lipid levels connect with brain diseases in Han Chinese and suggest sex-dependent effects, Biomed. Pharmacother., 70, 7–11, 2015.

Kamitani, A., Rakugi, H., Higaki, J., Ohishi, M., Shi, S. J., Takami, S., Nakata, Y., Higashino, Y., Fujii, K., and Mikami, H.: Enhanced predictability of myocardial infarction in Japanese by combined genotype analysis, Hypertension, 25, 950–953, 1995.

Margawati, E.: A global strategy of using molecular genetic information to improve genetics in livestock, Reprod. Domest. Anim., 47, 7–9, 2012.

McClure, M. C., Morsci, N. S., Schnabel, R. D., Kim, J. M., Yao, P., Rolf, M. M., McKay, S. D., Gregg, S. J., Chapple, R. H., Northcutt, S. L., and Taylor, J. F.: A genome scan for quantitative trait loci influencing carcass, post-natal growth and reproductive traits in commercial Angus cattle, Anim. Genet., 41, 597–607, 2010.

Mukherjee, R., Jow, L., Croston, G. E., and Paterniti, J. R.: Identification, characterization, and tissue distribution of human peroxisome proliferator-activated receptor (PPAR) isoforms PPARγ2 versus PPARγ1 and activation with retinoid X receptor agonists and antagonists, J. Biol. Chem., 272, 8071–8076, 1997.

Palomer, X., Barroso, E., Zarei, M., Botteri, G., and Vázquez-Carrera, M.: PPARβ/δ and lipid metabolism in the heart, Biochim. Biophys. Acta., 1861, 1–10, 2016.

Perlstein, T. S., Gerhard-Herman, M., Hollenberg, N. K., Williams, G. H., and Thomas, A: Insulin induces renal vasodilation, increases plasma renin activity, and sensitizes the renal vasculature to angiotensin receptor blockade in healthy subjects[J], J. Am. Soc. Nephrol., 18, 944–951, 2007.

Ren, J., Duan, Y., Qiao, R., Yao, F., Zhang, Z., Yang, B., Guo, Y., Xiao, S., Wei, R., Ouyang, Z., Ding, N., Ai, H., and Huang, L.: A missense mutation in PPARD causes a major QTL effect on ear size in pigs, PloS Genet., 7, e1002043, doi:10.1371/journal.pgen.1002043, 2011.

Rosalesreynoso, M. A., Wencechavez, L. I., Arredondovaldez, A. R., Dumois-Petersen, S., Barros-Núñez, P., Gallegos-Arreola, M. P., Flores-Martínez, S. E., and Sánchez-Corona, J.: Protective role of +294 T/C (rs2016520) polymorphism of PPARD in Mexican patients with colorectal cancer, Genet. Mol. Res., 16, doi:10.4238/gmr16019324, 2017.

Song, J. F., Wang, T., Zhu, J., Zhou, X. Y., Lu, Q., Guo, H., Zhang, F., Wang, Y., Li, W., Wang, D. D., Cui, Y. W., Lv, D. M., and Yin, X. X.: PPARD rs2016520 polymorphism affects repaglinide response in Chinese Han patients with type 2 diabetes mellitus, Clin. Exp. Pharmacol. Physiol., 42, 27–32, 2015.

Spötter, A., Hamann, H., Müller, S., and Distl, O.: Effect of polymorphisms in four candidate genes for fertility on litter size in a German pig line, Reprod. Domest. Anim., 45, 579–584, 2010.

Stelma, F., Jansen, L., Sinnige, M. J., van Dort, K. A., Takkenberg, R. B., Janssen, H. L., Reesink, H. W., and Kootstra, N. A.: HLA-C and KIR combined genotype as new response marker for HBeAg-positive chronic hepatitis B patients treated with interferon-based combination therapy, J. Viral. Hepat., 23, 652–659, https://doi.org/10.1111/jvh.12525, 2016.

Tugwood, J. D., Aldridge, T. C., Lambe, K. G., MacDonald, N., and Woodyatt, N. J.: Peroxisome proliferator-activated receptors: structures and function, Ann. N. Y. Acad. Sci., 804, 252–265, 1996.

Vedhachalam, C., Duong, P. T., Nickel, M., Nguyen, D., Dhanasekaran, P., Saito, H., Rothblat, G. H., Lund-Katz, S., and Phillips, M. C.: Mechanism of ATP-binding cassette transporter A1-mediated cellular lipid efflux to apolipoprotein AI and formation of high density lipoprotein particles, J. Biol. Chem. 282, 25123–25130, 2007.

Youssef, J. and Badr, M. Z.: PPARs: history and advances, Methods Mol. Biol., 952, 1–6, 2013.

Zhang, Y., Gao, T., Hu, S., Lin, B., Yan, D., Xu, Z., Zhang, Z., Mao, Y., Mao, H., Wang, L., Wang, G., Xiong, Y., and Zuo, B.: The functional SNPs in the 5'regulatory region of the porcine PPARD gene have significant association with fat deposition traits, PloS One, 10, e0143734, doi:10.1371/journal.pone.0143734, 2015.

22

Carcass and meat quality characteristics of lambs reared in different seasons

Hulya Yalcintan, Bulent Ekiz, Omur Kocak, Nursen Dogan, P. Dilara Akin, and Alper Yilmaz

Istanbul University Veterinary Faculty, Department of Animal Breeding and Husbandry, Avcilar, 34320 Istanbul, Turkey

Correspondence to: Hulya Yalcintan (hyalcint@istanbul.edu.tr)

Abstract. Thirty-six Kivircik male lambs were used to determine the effects of rearing season (winter rearing – WR; spring–summer rearing – SSR; and autumn rearing – AR) on carcass and meat quality characteristics. Average daily gain in the period 0–134 days, final weight, cold carcass weight and real dressing percentage were higher in WR lambs than lambs from SSR and AR groups. Furthermore, SSR and AR lambs did not show significant differences for these traits ($P > 0.05$). WR lambs had the highest values in terms of back fat thickness, subjective carcass fatness score and fat percentage in pelvic limb, which gives information about the carcass fatness. Final meat pH, expressed juice and meat lightness 1 h after cutting were higher for SSR lambs than for WR and AR lambs. SSR lambs had the lowest scores in terms of flavour intensity, flavour quality and overall acceptability in the sensory evaluation panel. In conclusion, WR lambs yielded better carcass quality than SSR and AR lambs. When the rearing season is to be decided, the higher carcass quality of WR lambs and the lesser appreciation of meat of SSR lambs by consumers should be considered.

1 Introduction

Sheep breeding in Turkey is usually performed by the use of native breeds and also using traditional methods – which are similar to many countries in the Middle East. Sheep in the traditional production system proliferate according to natural breeding period and lambs suckle their dams until the slaughter age (Ekiz et al., 2012). Generally, lambs are slaughtered at 4–5 months of age or breeders apply a proper fattening programme after weaning to increase slaughter live weight of lambs depending on market demand. Therefore, lambs can usually be supplied to the market at predetermined times of the year.

The meat quality is influenced by consumer demand. Red meat consumers tend to prefer meat of grazed lambs, considering that such lamb meat is much healthier, tastier and more natural than meat from concentrate-based production systems (Font i Furnols and Guerrero, 2014). Butchers and restaurants which cater for high-income consumers demand high-quality lamb meat throughout the year. For this reason, some farmers apply synchronization programmes in order to supply meat as demanded by the market year round. According to the natural breeding programme, lambs are usually born in winter and are kept with their dams in sheepfold until spring as the pasture conditions are not suitable for lamb grazing in this period. But pasture conditions change with the season. Thus, farmers who aim to obtain lambs in other seasons take advantage of pasture and graze their lambs to provide for their nutritional needs. Therefore, the feeding system of lambs changes according to the rearing season.

The feeding system and husbandry conditions which change with rearing season may affect lamb growth performance (Santos-Silva et al., 2002; Joy et al., 2008; Ekiz et al., 2013) and influence the carcass and meat quality characteristics of lambs (Priolo et al., 2002; Joy et al., 2008; Carrasco et al., 2009; Ekiz et al., 2012). Previous studies have indicated that lambs fed with concentrate diet generally have higher carcass weight (Díaz et al., 2002; Priolo et al., 2002), dressing percentage and fatness level (Díaz et al., 2002; Ekiz et al., 2012, 2013) than lambs grazed on pasture. Studies also show that meat from lambs grazed on pasture is tougher and darker (Priolo et al., 2002; Cañeque et al., 2003; Ekiz et al., 2012) than meat from concentrate-fed lambs.

Table 1. Means $\pm$ SE (standard errors) for birth weight, final live weight and average daily gain in lambs reared in different seasons.

Characteristics	WR*	SSR*	AR*	P value
Birth weight, kg	4.29 ± 0.27	4.19 ± 0.19	4.01 ± 0.12	0.631
Final weight, kg	$28.00^a \pm 1.39$	$21.66^b \pm 0.61$	$21.85^b \pm 0.99$	< 0.001
ADG^d, g	$176.99^a \pm 10.27$	$130.42^b \pm 4.27$	$133.09^b \pm 7.22$	< 0.001
Slaughter age	134.67 ± 1.05	134.00 ± 1.19	133.75 ± 0.51	0.787

[a,b] Means in the same line with different superscripts are significantly different. [d] ADG, average daily weight gain from birth to 134 days of age. * WR: winter rearing; SSR: spring–summer rearing; AR: autumn rearing.

The aim of this study is to determine the carcass and meat quality of Kivircik lambs reared in autumn, winter and spring–summer seasons in relation to the feed sources of these seasons.

2 Material and methods

The Ethics Committee of Istanbul University approved the experimental protocol of the current study (Approval number: 2015/05).

2.1 Animals and experimental design

The study was conducted at the sheep farm of Istanbul University (during the 2015–2016 breeding season). The herd was separated into three groups at the beginning of the breeding season and Eazi-Breed^{TM} CIDR (controlled integral drug release) were used to synchronize ewes in each mating period (May, February, October). A total of 36 Kivircik male lambs raised in different seasons (autumn, 12 lambs; spring–summer, 12 lambs; winter, 12 lambs) were used in the research. The rearing seasons of lambs investigated in the current study are as follows:

a. *Spring–summer-reared lambs (SSR)*: lambs were kept with their mothers in the sheepfold for the 2 months following birth. In addition to mother's milk, lambs were fed with 500 g concentrate per day (87.7 % dry matter, 17.15 % crude protein, 4.8 % crude fat, 6.4 % crude cellulose, 7.96 % crude ash, 10.15 % neutral detergent fibre (NDF), 24.56 % acid detergent fibre (ADF), 0.82 % calcium, 0.5 % phosphorus and 2650 kcal kg^{-1} metabolizable energy (ME)) per lamb and ad libitum access to alfalfa hay (87.8 % dry matter, 12.88 % crude protein, 2.33 % crude fat, 37.3 % crude cellulose, 9.85 % crude ash, 43.89 % NDF, 58.79 % ADF and 1843 kcal kg^{-1} ME) after 2 weeks of age in a different pen which lambs can pass but mothers cannot, in the sheepfold. Lambs were grazed on natural pasture together with their dams during the daytime (09:00–16:00) from the beginning of June to slaughter age.

b. *Autumn-reared lambs (AR)*: the feeding system of AR lambs was similar to the SSR group; however, these lambs were grazed on natural pasture with their dams in the daytime from the beginning of October to slaughter age.

c. *Winter-reared lambs (WR)*: these lambs were born in middle November and kept with their dams until slaughter age in the sheepfold. In addition to mother's milk, lambs were fed with 500 g concentrate per day per lamb and ad libitum access to alfalfa hay after 2 weeks of age in a different pen which lambs can pass but mothers cannot, in the sheepfold.

The pasture was natural and composed of (on dry matter basis) 52 % Gramineae (*Festuca* spp. and *Lolium* spp.), 22 % Leguminosae (mainly *Trifolium* spp., *Medicago* spp. and *Vicia* spp.) and 26 % other families (mainly *Conium* spp., *Geranium* spp., *Viola* spp., *Rumex* spp. and *Plantago* spp.) (Ekiz et al., 2013). Lambs in all experimental groups were weighed on a fortnightly basis and were slaughtered at the average age of 134 days (Table 1).

2.2 Slaughter procedure, carcass characteristics and dissections

At the end of the each rearing season (March, August, December) 12 Kivircik male lambs were brought to the experimental abattoir of Istanbul University Veterinary Faculty, and were kept in the lairage pen (4×5 m) for 2 h. After lairage, lambs were weighed and then slaughtered after electrical stunning. After the removal of non-carcass components, lamb carcasses were chilled at 4 °C for 24 h and then carcass weights were recorded. Commercial dressing percentage was calculated based on pre-slaughter live weight, while real dressing percentage was calculated based on empty body weight (EBW). Subjective fatness and conformation scores were given by using the European system (EUROP) as described by Johansen et al. (2006). Subcutaneous fat colour was measured from tail root with a Minolta CR 400 colorimeter (Minolta Camera Co., Osaka, Japan) using CIELAB colour space (L^*, a^*, b^*) as described by Ekiz et al. (2012). Carcasses were split into two halves and the left side of each carcass was separated into five joints according to the methodology described by Colomer-Rocher et al. (1987). Longissimus dorsi (LD) muscle section area and back fat thickness were measured as described by Boggs and

Merkel (1993). In order to estimate carcass composition the left pelvic limb was used. After carcass jointing, pelvic limbs were vacuum-packed and then put into deep freeze ($-18\,^{\circ}$C) until the day of dissection. The day before dissection, pelvic limbs were thawed at $4\,^{\circ}$C for 24 h. Each pelvic joint was dissected into muscle, bone, subcutaneous fat, intermuscular fat and other tissues by the dissection method described by Fisher and De Boer (1994).

2.3 Meat instrumental quality

In order to be used in meat quality analyses LD muscle was removed from left side of the carcasses at 24 h post-slaughter and were packed. Meat samples were taken after 72 h ageing process at $4\,^{\circ}$C in refrigerator. For sensory evaluation, meat samples were packed under vacuum, frozen and kept at $-18\,^{\circ}$C until panel evaluation.

The pH of longissimus dorsi was measured immediately after dressing (pH_0) and at 24 h after dressing (pH_{final}) from the muscle between the 12th and 13th thoracic vertebrae using a digital pH meter (Testo 205, Testo AG, Lenzkirch, Germany). Drip loss was estimated by the ratio of weight loss to initial weight after meat samples were stored for 24 h at $4\,^{\circ}$C in refrigerator (Honikel, 1998). Expressed juice was determined by modified Grau and Hamm pressure method (Beriain et al., 2000). To determine cooking loss, the meat samples were put into heat-resistant polyethylene bags and were cooked in a water bath at $80\,^{\circ}$C for 45 min and then cooled under running water. Cooking loss was estimated by means of percentage of weight loss of the cooked samples to initial weight (Honikel, 1998). Six sub-samples cut parallel to the muscle fibres with a cross section of 1×1 cm were prepared from each cooked sample to measure tenderness (Ekiz et al., 2012) with an Instron Universal Testing Machine (model 3343, Instron Corp., Norwood, MA, USA). Meat colour was determined on LD muscle at 1 and 24 h after cutting on cut surface with a Minolta CR 400 colorimeter (Minolta Camera Co., Osaka, Japan) using CIE (1976) standards as described by Ekiz et al. (2012).

2.4 Meat sensory evaluation

Meat samples, which were served to the panellists, were prepared according to the methodology described by Ekiz et al. (2012). Panel evaluation was carried out in four sessions and in each session three sub-samples from each group were served to the panellist. Sensory characteristics of cooked samples were assessed by seven panellists using an eight-point category scale described by Sañudo et al. (1998). The panellists assessed lamb odour intensity, tenderness, juiciness, flavour intensity, flavour quality and overall acceptability (scale $1 =$ no odour, extremely tough, extremely dry, no flavour, dislike flavour extremely and dislike extremely; scale $8 =$ very strong lamb odour, extremely tender, ex-

tremely juicy, very strong flavour, like flavour extremely and like extremely) (Sañudo et al., 1998).

2.5 Statistical analyses

In order to determine the effect of rearing season on carcass and meat quality characteristics, one-way ANOVA was performed using the SPSS 13 statistical package (SPSS Inc, Chicago, Illinois, USA). Data for meat colour parameters were re-analysed by adding the slaughter weight as a covariate into the statistical model. General linear model analyses were performed to analyse the data for sensory characteristics. The model used to analyse data for sensory characteristics included the fixed effects of feeding system, session, panellist and also significant two-way interactions between these main effects. Pearson correlation analysis in the SPSS 13 statistical package was applied in order to obtain coefficients of correlation among sensory characteristics.

3 Results

The effects of rearing season on lamb growth performance are presented in Table 1. Lambs' slaughter ages in all groups in the research were similar as a result of the experimental plan. The average daily weight gain from birth to 134 days and final live weight were significantly higher in WR lambs than in AR and SSR lambs, while AR and SSR lambs had similar values for these traits.

The effects of the rearing season on carcass quality characteristics and fat colour parameters are shown in Table 2. WR lambs presented higher empty body weight, carcass weight, real dressing percentage and back fat thickness than their SSR and AR counterparts. However, there were no significant differences between SSR and AR groups in terms of these traits. Moreover, WR lambs had the highest score in subjective carcass fatness while SSR lambs had the lowest score for the same trait. Rearing season had no significant influence on carcass fat colour parameters except fat lightness (L^* value) and hue angle. The lowest lightness and hue values were observed in the SSR lambs, while there were no significant differences between WR and AR lambs with regard to these traits. Mean values of pelvic limb composition are presented in Table 3. Rearing season had a significant effect on lean, bone and fat percentages in different levels. SSR lambs presented higher lean and bone percentages and lower subcutaneous and intermuscular fat percentages than their WR and AR counterparts. The lean / fat ratio was significantly higher in SSR lamb carcasses than other experimental groups.

Mean values of meat pH, drip loss, cooking loss, expressed juice and Warner–Bratzler (WB) shear force values are shown in Table 4. Rearing season had no significant effect on pH_0, drip loss, cooking loss and WB shear force. While meat from SSR lambs had higher pH_{final} and expressed juice values than those of other groups, there were no significant

Table 2. Means $\pm$ SE for certain carcass quality characteristics and fat colour parameters of lambs reared in different seasons.

Characteristics	WR	SSR	AR	P value
Empty body weight, kg	$23.38^a \pm 1.01$	$15.55^b \pm 0.50$	$17.61^b \pm 0.89$	< 0.001
Carcass weight, kg	$12.62^a \pm 0.64$	$7.51^b \pm 0.28$	$8.63^b \pm 0.56$	< 0.001
Commercial dressing, %[d]	$45.14^a \pm 0.81$	$36.27^c \pm 0.64$	$39.41^b \pm 0.81$	< 0.001
Real dressing, %[f]	$53.77^a \pm 0.57$	$48.18^b \pm 0.44$	$48.63^b \pm 0.74$	< 0.001
Shrinkage loss, %	$2.51^c \pm 0.10$	$4.13^a \pm 0.10$	$2.84^b \pm 0.13$	< 0.001
LD muscle section area, mm^2	$9.97^a \pm 0.46$	$8.90^{a,b} \pm 0.43$	$8.16^b \pm 0.49$	0.030
Back fat thickness, mm	$0.85^a \pm 0.15$	$0.24^b \pm 0.03$	$0.26^b \pm 0.14$	0.001
Conformation score (1–15)	$5.42^a \pm 0.29$	$4.67^{a,b} \pm 0.19$	$4.25^b \pm 0.39$	0.031
Fatness score (1–15)	$5.50^a \pm 0.15$	$2.33^c \pm 0.23$	$3.17^b \pm 0.32$	< 0.001
	Fat colour parameters			
Fat lightness (L^*)	$72.38^a \pm 0.56$	$66.96^b \pm 1.29$	$73.49^a \pm 0.62$	< 0.001
Fat redness (a^*)	4.21 ± 0.33	4.61 ± 0.40	4.37 ± 0.30	0.718
Fat yellowness (b^*)	5.02 ± 0.46	4.23 ± 1.22	5.45 ± 0.48	0.558
Chroma (C^*)	6.59 ± 0.52	6.88 ± 0.95	7.06 ± 0.46	0.883
Hue angle (H^*)	$49.76^a \pm 2.01$	$29.16^b \pm 9.49$	$50.33^a \pm 2.48$	0.021

[a,b,c] Means in the same line with different superscripts are significantly different. [d] Commercial dressing: cold carcass dressing based on pre-slaughter live weight. [f] Real dressing: cold carcass dressing based on empty body weight.

Table 3. Means $\pm$ SE for pelvic limb composition of lambs reared in different seasons.

Characteristics	WR	SSR	AR	P value
Pelvic limb weight, kg	$2.09^a \pm 0.10$	$1.28^b \pm 0.05$	$1.47^b \pm 0.09$	< 0.001
	Composition of pelvic limb			
Lean, %	$58.83^b \pm 0.80$	$66.00^a \pm 0.62$	$60.07^b \pm 0.53$	< 0.001
Bone, %	$19.75^c \pm 0.30$	$23.48^a \pm 0.51$	$21.38^b \pm 0.70$	< 0.001
Subcutaneous fat, %	$11.11^a \pm 1.02$	$1.23^c \pm 0.38$	$7.40^b \pm 1.08$	< 0.001
Intermuscular fat, %	$5.12^a \pm 0.42$	$2.81^b \pm 0.45$	$5.25^a \pm 0.57$	0.001
Total fat, %	$16.23^a \pm 0.90$	$4.04^c \pm 0.81$	$12.65^b \pm 1.17$	< 0.001
Other tissues[d], %	$3.14^b \pm 0.30$	$4.43^a \pm 0.19$	$3.30^b \pm 0.21$	0.001
Evaporation losses, %	2.05 ± 0.10	2.05 ± 0.15	2.58 ± 0.29	0.101
Lean / bone ratio	2.98 ± 0.05	2.83 ± 0.06	2.84 ± 0.09	0.234
Lean / fat ratio	$3.79^b \pm 0.29$	$25.33^a \pm 4.75$	$5.35^b \pm 0.65$	< 0.001

[a,b,c] Means in the same line with different superscripts are significantly different. [d] Major blood vessels, tendons, larger nerves and lymph nodes.

differences between WR and AR lambs considering these traits. Results for meat colour parameters of LD muscle at 1 and 24 h after cutting are presented in Table 5. The differences among rearing season in terms of meat lightness (L^*) and redness (a^*) were significant at 1 and 24 h after cutting, while feeding system had no significant effect of yellowness values. SSR lambs had higher lightness values than those of WR and AR lambs at 1 h and that of WR lambs at 24 h. However, WR lambs had higher redness values in LD muscle at 1 and 24 h after cutting. Meat obtained from WR lambs had greater Chroma values than those of SSR and AR lambs at 1 and 24 h, while it had lower hue angle values than the other two groups at 24 h after cutting. The effects of

slaughter weight, as a covariate, on meat lightness (L^*) and redness (a^*) values were significant ($P < 0.05$ and $P < 0.01$ respectively). On the other hand, the effect of the rearing season becomes insignificant when slaughter weight is added the statistical model as a covariate, except for $(a^*)_{24h}$ values.

The effects of rearing season on meat sensory characteristics are presented in Table 6. There were no significant differences between treatments in terms of odour intensity, tenderness and juiciness in meat samples ($P > 0.05$). But the lowest scores were given to meat samples from SSR lambs for flavour intensity, flavour quality and overall acceptability by panellists in the sensory panel. Moreover, there was positive and significant correlation between flavour quality and over-

Table 4. Means ± SE for pH, drip loss, cooking loss, expressed juice and Warner–Bratzler (WB) shear force values of longissimus dorsi muscle of lambs reared in different seasons.

Characteristics	WR	SSR	AR	P value
pH_0	6.52 ± 0.07	6.48 ± 0.04	6.35 ± 0.04	0.072
pH_{final}	$5.78^b \pm 0.04$	$5.97^a \pm 0.03$	$5.70^b \pm 0.03$	< 0.001
pH^d_{0-24h}	$0.74^a \pm 0.08$	$0.51^b \pm 0.05$	$0.65^{a,b} \pm 0.05$	0.043
Drip loss, %	2.78 ± 0.40	3.08 ± 0.32	2.61 ± 0.21	0.575
Cooking loss, %	24.70 ± 0.70	27.40 ± 1.00	24.09 ± 1.16	0.050
Expressed juice, %	$9.64^b \pm 0.80$	$13.28^a \pm 0.83$	$8.12^b \pm 0.34$	< 0.001
WB shear force, kg	5.79 ± 0.53	4.66 ± 0.25	4.79 ± 0.65	0.239

[a,b] Means in the same line with different superscripts are significantly different. [d] pH decline between 0 min and 24 h post mortem ($pH_0 - pH_{24h}$).

Table 5. Means ± for meat colour parameters of longissimus dorsi muscle of lambs reared in different seasons.

Characteristics	WR	SSR	AR	P value
	Colour parameters at 1 h			
$(L^*)_{1h}$	$37.38^b \pm 0.69$	$41.65^a \pm 0.76$	$39.32^b \pm 0.59$	< 0.001
$(a^*)_{1h}$	$18.44^a \pm 0.36$	$16.49^b \pm 0.32$	$16.22^b \pm 0.46$	< 0.001
$(b^*)_{1h}$	4.12 ± 0.45	3.47 ± 0.40	3.12 ± 0.15	0.157
$Chroma_{1h}$	$18.94^a \pm 0.42$	$16.90^b \pm 0.35$	$16.53^b \pm 0.46$	< 0.001
Hue angle$_{1h}$	12.36 ± 1.23	11.79 ± 1.24	10.93 ± 0.52	0.631
	Colour parameters at 24 h			
$(L^*)_{24h}$	$40.55^b \pm 0.69$	$42.85^a \pm 0.56$	$42.12^{a,b} \pm 0.42$	0.023
$(a^*)_{24h}$	$17.89^a \pm 0.49$	$16.20^b \pm 0.27$	$14.47^c \pm 0.28$	< 0.001
$(b^*)_{24h}$	6.22 ± 0.36	6.63 ± 0.26	6.66 ± 0.16	0.437
$Chroma_{24h}$	$18.96^a \pm 0.54$	$17.52^b \pm 0.32$	$15.94^c \pm 0.26$	< 0.001
Hue angle$_{24h}$	$19.04^c \pm 0.89$	$22.20^b \pm 060$	$24.78^a \pm 0.68$	< 0.001

[a,b,c] Means in the same line with different superscripts are significantly different.

all acceptability ($P < 0.01$) and flavour intensity and overall acceptability ($P < 0.01$) according to the Pearson correlation results (Table 7).

4 Discussion

4.1 Lamb growth performance

WR lambs were slaughtered at higher slaughter weight than their SSR and AR counterparts although all lambs were at similar ages in the research. These differences in final live weight might be explained by the nutrient sources depending on the season. Supporting the current results, various studies (Karim et al., 2007; Joy et al., 2008; Ekiz et al., 2013) showed that lambs fed concentrate diet in sheepfold had higher growth rate than lambs fed on pasture. Growth rate increases with rising proportion of concentrate in the diet (Santos-Silva et al., 2002). Also, great energy deprivation might occur because of increasing physical activities and basal metabolism due to grass consumption in their diet for pasture-fed lambs (Díaz et al., 2002). In the current study,

energy needs of SSR and AR lambs increased as they were grazed together with their dams.

4.2 Carcass quality

WR lambs had higher empty body weight, dressing percentage and back fat thickness. The increased dressing percentages of WR lambs might be explained by the higher fatness in their carcasses. Supporting the current study, various authors (Díaz et al., 2002; Cañeque et al., 2003; Karim et al., 2007; Ekiz et al., 2012) reported higher fatness level in lambs fed concentrate in sheepfold than lambs fed on pasture and concomitant dressing percentage increase. Peña et al. (2005) also reported an increase in fatness level of carcass with increasing slaughter weight of lambs. In this study there are differences between rearing groups in terms of slaughter weight of lambs which were slaughtered at the same age – WR lambs had the highest slaughter weight compared with other groups. Back fat thickness of concentrate-fed lambs' carcasses were also higher than that of pasture-fed lambs. Feeding with high-energy nutrition and less physical activity

Table 6. Means $\pm$ SE for meat sensory characteristics of lambs reared in different seasons.

Characteristics	WR	SSR	AR	P value
Lamb odour intensity	4.52 ± 0.14	4.24 ± 0.14	4.60 ± 0.14	0.168
Tenderness	4.93 ± 0.18	4.67 ± 0.18	5.18 ± 0.18	0.139
Juiciness	4.41 ± 0.13	4.43 ± 0.13	4.63 ± 0.13	0.421
Flavour intensity	$4.67^a \pm 0.13$	$4.29^b \pm 0.13$	$4.80^a \pm 0.13$	0.015
Flavour quality	$5.10^a \pm 0.13$	$4.62^b \pm 0.13$	$5.04^a \pm 0.13$	0.020
Overall acceptability	$5.07^a \pm 0.15$	$4.56^b \pm 0.15$	$5.07^a \pm 0.15$	0.024

[a,b] Means in the same line with different superscripts are significantly different.

Table 7. Coefficient of correlation among for meat sensory characteristics.

Characteristics	Tenderness	Juiciness	Flavour intensity	Flavour quality	Overall acceptability
Odour intensity	0.27	0.147	0.518**	0.292	0.130
Tenderness	–	0.631**	0.597**	0.784**	0.825**
Juiciness		–	0.397*	0.634**	0.693**
Flavour intensity			–	0.628**	0.576**
Flavour quality				–	0.946**

$* \; P < 0.05$ and $** \; P < 0.01$.

might be the cause of the stored fat in carcasses of WR lambs which were fed with concentrate in addition to their mothers' milk in the sheepfold.

WR lambs had highest mean values in terms of subjective fatness score, subcutaneous fat percentage and total fat percentage in pelvic limb which provide information about carcass fatness. Several authors who have investigated the effects of feeding/management system on carcass quality (Díaz et al., 2002; Priolo et al., 2002; Ekiz et al., 2012) also reported less developed fat depots for pasture-raised lambs when compared with sheepfold lambs as a result of increasing physical activity or lower energy intake in grazing lambs. Díaz et al. (2002) noted that increased physical activity in pasture-raised lambs bring about less carcass fatness of lambs as a result of the increased mobilization of reserve lipids to form muscle tissue. Similarly, Carrasco et al. (2009) observed lower total carcass fat in grazing lambs caused by lower proportion of subcutaneous fat than in lambs which had free access to concentrate.

Increasing mean values of the fat b^* value represent the rising yellowness of carcass. Yellow fat is not generally appreciated by consumers worldwide (Priolo et al., 2002). The differences in subcutaneous fat colour might be due to carotenoids which are stored in fat depots in pasture lambs (Joy et al., 2008; Carrasco et al., 2009). However, the feeding system did not affect the subcutaneous fat b^* values (between 4.23 and 5.45) in the current study. Priolo et al. (2002) observed that carcasses which had subcutaneous fat b^* values above 22 were easily distinguished from the other carcasses by the naked eye. According to this fact, yellowness values

for subcutaneous fat in the current study were not as high as to be rejected by consumers.

Weights of carcass cuts rise with increasing slaughter weight (Majdoub-Mathlouthi et al., 2013). Higher pelvic limb weights in WR lambs were the result of the higher final weights in the current study. Furthermore, increased slaughter weight causes a change in carcass composition, which appears as a decrease in lean ratio and an increase in the fat ratio, especially subcutaneous fat, in the carcass (Díaz et al., 2002; Santos-Silva et al., 2002; Majdoub-Mathlouthi et al., 2013). On the other hand, differences between two grazing groups (SSR and AR) in terms of lean and bone percentages in pelvic limb were probably caused by different fatness levels due to grazing conditions.

4.3 Meat instrumental quality

Final pH value is the main indicator of meat quality commercially and has effects on instrumental meat quality characteristics such as texture, meat colour and water holding capacity. It also affects meat sensory characteristics (Miller, 2002). Werdi Pratiwi et al. (2007) reported that pH values higher than 6.0 are related to lower meat quality. In the current study, final pH values obtained for lambs from treatment groups varied between 5.70 and 5.97, which might be considered within the desired range for quality meat.

Meat juiciness was detected measuring water-holding capacity by using expressed juiciness, cooking loss and drip loss analyses. In these analyses, meat from SSR lambs had lower water-holding capacity as it lost more water than meat from WR and AR lambs. Similarly, several authors (Cañeque et al., 2003; Santos-Silva et al., 2002) who compared different management/feeding systems reported lower water-holding capacity for pasture lambs.

Hopkins et al. (2006) reported that tenderness (shear force values) and intramuscular fat level were significant predictors of the consumer sensory traits. Meat tenderness can be affected by the structure of the connective tissue, carcass fatness and collagen levels of meat (Sañudo et al., 2000; Díaz et al., 2002). Rearing season did not affect the meat tenderness, which was assessed by WB shear force in the current study. The results of shear force analysis suggest that meat from SSR lambs was as tender as the meat from the other groups.

Meat colour is not an important eating characteristic, but it is a major factor for the initial selection of meat by consumers (Beriain et al., 2000). Consumers mostly prefer light or pink lamb meat in Turkey, as is also the case in other Mediterranean countries (Ekiz et al., 2012). Priolo et al. (2001) noted that meat colour may be influenced by various factors such as carcass fatness, final pH, physical activity, slaughter age, carcass weight and intramuscular fat content.

Numerous authors (Díaz et al., 2002; Priolo et al., 2002; Cañeque et al., 2003) reported darker meat from pasture-fed lambs compared to concentrate-fed counterparts and they explained this difference by the physical activity and greater blood pigment concentration of animals raised on pasture. Ekiz et al. (2013) also reported darker meat colour for lambs raised on pasture compared to lambs raised on concentrate-based system and they explained this result as a combined effect of carcass fatness, slaughter age and final pH. Contrary to previous reports, WR lambs fed on a concentrate-based diet had darker meat colour at 24 h after cutting than SSR and AR counterparts which were fed a pasture-based diet in the current study. These differences could be explained by the higher pre-slaughter live weight of WR lambs. Meat colour became darker (decreasing L^* value) and had a lower yellowness value (decreasing b^* value) with increasing slaughter weight (Santos-Silva et al., 2002). Furthermore, Beriain et al. (2000) noted that darker meat colour is connected with increasing muscle myoglobin concentration which rises with slaughter weight or age. Indeed, the effect of slaughter weight on meat colour parameters was significant according to the results of covariance analyses. Moreover, if slaughter weight was included in the statistical model, rearing season had no significant influence on meat colour which indicates that effects of slaughter weight and rearing season overlap for determination of meat colour.

4.4   Meat sensory evaluation

Lamb odour intensity and tenderness perception of panellists were not influenced by rearing season in the present study. The lack of differences among treatment groups in terms of tenderness score is consistent with the result of WB shear force analysis.

Meat preferences of consumers are associated with socio-economic factors, ethics or religious beliefs and tradition (Font i Furnols and Guerrero, 2014). For instance, a highly preferred meat flavour in one culture, region or country may be perceived as less preferable or unacceptable in another (Schreurs et al., 2008). Red meat consumers generally tend to think that meat from pasture raised lambs is more tasty and healthy as these lambs grow in a natural environment and consume less concentrate feed when compared with lamb meat from sheepfold raised lambs (Font i Furnols and Guerrero, 2014). Yet panellists gave the highest flavour and overall acceptability scores to meats from WR and AR lambs in the present study. Supporting the current results, Font i Furnols et al. (2009) observed that EU consumers preferred lamb meat from concentrate or mixed systems (concentrate and pasture) to meat of lambs fed only on pasture. Furthermore, Sañudo et al. (2000) and Muela et al. (2010) reported that meat from fattier carcasses had higher flavour intensity.

Overall acceptability scores given to lamb could be reflection of the meat tenderness, flavour intensity and quality perception of panellists (Ekiz et al., 2012). In particular, flavour which can be the determining feature in acceptance or rejection of the meat, is an important aspect for consumer preferences (Schreurs et al., 2008). The highest scores in terms of flavour intensity and quality were given to meat from WR and AR lambs, while there were no significant differences in tenderness scores among whole groups. This is also supported by Pearson correlation results which are positive and significant between flavour quality and overall acceptability and flavour intensity and overall acceptability in agreement with the results reported by Ekiz et al. (2012) and Font i Furnols et al. (2009).

5   Conclusions

We found that growth rate, empty body weight, carcass weight, dressing percentages and pelvic limb weight decrease in the pasture-raised lambs in spring–summer and autumn. On the other hand, WR and AR lambs had higher values in terms of certain carcass fatness parameters (subjective fatness score, intermuscular, subcutaneous and total fat percentage in pelvic limb) than SSR lambs. Furthermore, results of sensory analyses indicate that meat from WR lambs had similar flavour intensity, flavour quality and overall acceptability scores as meat from pasture-raised lambs in autumn (AR). The results of the current study indicate that WR lambs yield better carcass quality than SSR and AR lambs, and meat of SSR lambs is least appreciated by consumers.

Author contributions. HY and BE conceived and designed the experiment; HY coordinated the research project; HY, BE, OK, ND, PDA and AY performed the experiments; HY and BE analysed the data; ND and PDA helped in drafting of the tables; and HY wrote the paper.

Competing interests. The authors declare that they have no conflict of interest.

Acknowledgements. The present study was supported by the Scientific and Technological Research Council of Turkey (TÜBİTAK/project no. 115O840).

Edited by: Steffen Maak

References

Beriain, M. J., Horcada, A., Purroy, A., Lizaso, G., Chasco, J., and Mendizabal, J. A.: Characteristics of Lacha and Rasa Aragonesa lambs slaughtered at three weights, J. Anim. Sci., 78, 3070–3077, 2000.

Boggs, D. L. and Merkel, R. A.: Live Animal Carcass Evaluation and Selection Manual, Kendall/Hunt Publishing Company, Iowa, 1993.

Cañeque, V., Velasco, S., Diaz, M. T., Huidobro, F. R., Perez, C., and Lauzurica, S.: Use of whole barley with a protein supplement to fatten lambs under different management systems and its effect on meat and carcass quality, Anim. Res., 52, 271–285, 2003.

Carrasco, S., Ripoll, G., Sanz, A., Alvarez-Rodriguez, J., Panea, B., Revilla, R., and Joy, M.: Effect of feeding system on growth and carcass characteristics of Churra Tensina light lambs, Livest. Sci., 121, 56–63, 2009.

Colomer-Rocher, F., Morand-Fehr, P., and Kirton, A. H.: Standard methods and procedures for goat carcass evaluation, jointing and tissue separation, Livest. Prod. Sci., 17, 149–159, 1987.

Díaz, M. T., Velasco, S., Caneque, V., Lauzurica, S., Ruiz de Huidobro, F., Perez, C., Gonzalez, J., and Manzanares, C.: Use of concentrate or pasture for fattening lambs and its effect on carcass and meat quality, Small Rum. Res., 43, 257–268, 2002.

Ekiz, B., Yilmaz, A., Ozcan, M., and Kocak, O.: Effect of production system on carcass measurement and meat quality of Kivircik lambs, Meat Sci., 90, 465–471, 2012.

Ekiz, B., Demirel, G., Yilmaz, A., Ozcan, M., Yalcintan, H., Kocak, O., and Altinel, A.: Slaughter characteristics, carcass quality and fatty acid composition of lambs under four different production systems, Small Rum. Res., 114, 26–34, 2013.

Fisher, A. V. and De Boer H.: The EAAP standard method of sheep carcass assessment. Carcas measurement and dissection procedures. Report of the EAAP working group on carcass evaluation, in cooperation with the CIHEAM Instituto Agronomico Mediterraneo of Zaragoza and the CEC Directorate General for Agriculture in Brussels, Livest. Prod. Sci., 38, 149–159, 1994.

Font i Furnols, M. and Guerrero, L.: Consumer preference, behaviour and perception about meat products: An overview, Meat Sci., 98, 361–371, 2014.

Font i Furnols, M., Realini, C. E., Guerrero, L., Oliver, M. A., Sañudo, C., Campo, M. M., Nute, G. R., Cañeque, V., Álvarez, I., San Julián, R., Luzardo, S., Brito, G., and Montossi, F.: Acceptability of lamb fed on pasture, concentrate or combinations of both systems by European consumers, Meat Sci., 81, 196–202, 2009.

Honikel, K. O.: Reference methods for the assessment of physical characteristics of meat, Meat Sci., 49, 447–457, 1998.

Hopkins, D. L., Hegarty, R. S., Walker, P. J., and Pethick, D. W.: Relationship between animal age, intramuscular fat, cooking loss, pH, shear force and eating quality of aged meat from sheep, Aust. J. Exp. Agr., 46, 879–884, 2006.

Johansen, J., Aastveit, A., Egelandsdal, B., Kvaa, K., and Røe, M.: Validation of the EUROP system for lamb classification in Norway, repeatability and accuracy of visual assessment and prediction of lamb carcass composition, Meat Sci., 74, 497–509, 2006.

Joy, M., Alvarez-Rodriguez, J., Revilla, R., Delfa, R., and Ripoll, G.: Ewe metabolic performance and lamb carcass traits in pasture and concentrate-based production systems in Churra Tensine breed, Small Rum. Res., 75, 24–35, 2008.

Karim, S. A., Porwal, K., Kumar, S., and Singh, V. K.: Carcass traits of Kheri lambs maintained on different system of feeding management, Meat Sci., 76, 395–401, 2007.

Majdoub-Mathlouthi, L., Saïd, B., Say, A., and Kraiem, K.: Effect of concentrate level and slaughter body weight on growth performances, carcass traits and meat quality of Barbarine lambs fed oat hay based diet, Meat Sci., 93, 557–563, 2013.

Miller, R. K.: Factors affecting the quality of raw meat, in: Meat processing: improving quality, edited by: Kerry, J. and Ledward, D., CRC Press LLC and Woodhead Publishing Ltd., Cambridge, 27–63, 2002.

Muela, E., Sañudo, C., Campo, M. M., Medel, I., and Beltrán, J. A.: Effects of cooling temperature and hot carcass weight on the quality of lamb, Meat Sci., 84, 101–107, 2010.

Peña, F., Cano, T., Domenech, V., Alcalde Ma., J., Martos, J., García-Martinez, A., Herrera, M., and Rodero, E.: Influence of sex, slaughter weight and carcass weight on "non-carcass" and carcass quality in segureñ lambs, Small Rum. Res., 60, 247–254, 2005.

Priolo, A., Micol, D., and Agabriel, J.: Effects of grass feeding systems on ruminant meat colour and flavour. A review, Anim. Res., 50, 185–200, 2001.

Priolo, A., Micol, D., Agabriel, J., Prache, S., and Dransfield, E.: Effect of grass or concentrate feeding systems on lamb carcass and meat quality, Meat Sci., 62, 179–185, 2002.

Santos-Silva, J., Mendes, I. A., and Bessa, R. J. B.: The effect of genotype, feeding system and slaughter weight on the quality of light lambs 1. Growth, carcass composition and meat quality, Livest. Prod. Sci., 76, 17–25, 2002.

Sañudo, C., Nute, G. R., Campo, M. M., Maria, G., Baker, A., Sierra, I., Enser, M. E., and Wood, J. D.: Assessment of commercial lamb meat quality by British and Spanish taste panels, Meat Sci., 48, 91–100, 1998.

Sañudo, C., Enser, M. E., Campo, M. M., Nute, G. R., Maria, G., Sierra, I., and Wood, J. D.: Fatty acid composition and sensory characteristics of lamb carcasses from Britain and Spain, Meat Sci., 54, 339–346, 2000.

Schreurs, N. M., Lane, G. A., Tavendale, M. H., Barry, T. N., and McNabb, W. C.: Pastoral flavour in meat products from ruminants fed fresh forages and its amelioration by forage condensed tannins, Anim. Feed Sci. Technol., 146, 193–221, 2008.

Werdi Pratiwi, N. M., Murray, P. J., and Taylor, D. G.: Feral goats in Australia: a study on the quality and nutritive value of their meat, Meat Sci., 75, 168–177, 2007.

23

Morphological assessment of the Zebu Bororo (Wodaabé) cattle of Niger in the West African zebu framework

M. Maaouia A. Moussa[1], Moumouni Issa[1], Amadou Traoré[2], Moustapha Grema[1], Marichatou Hamani[1], Iván Fernández[3], Albert Soudré[4], Isabel Álvarez[3], Moumouni Sanou[2], Hamidou H. Tamboura[2], Yenikoye Alhassane[1], and Félix Goyache[3]

[1]Université Abdou Moumouni de Niamey, Faculté des sciences et techniques, BP 10960 Niamey, Niger
[2]INERA, Laboratoire de Biologie et santé animals, 04 BP 8645 Ouagadougou 04, Burkina Faso
[3]SERIDA-Deva, Camino de Rioseco 1225, 33394 Deva-Gijón (Asturias), Spain
[4]Université de Koudougou, BP 376 Koudougou, Burkina Faso

Correspondence to: Félix Goyache (fgoyache@serida.org) and Amadou Traoré (traore_pa@yahoo.fr)

Abstract. A total of 357 adult cows and 29 sires belonging to the long-horned Niger Zebu Bororo cattle population were assessed for 13 body measurements and 11 qualitative traits. Data were jointly analysed with 311 cows and 64 sires belonging to other four West African zebu cattle populations, sampled in Burkina Faso and Benin, representative of both the short-horned and the long-horned West African zebu groups using multivariate statistical methods. Besides the other long-horned zebu breed analysed (Zebu Mbororo of Burkina Faso), Zebu Bororo cattle tended to have the highest mean values for all body measurements. Mahalanobis distance matrices further informed that pairs involving Zebu Bororo cattle had the higher differentiation of the dataset. However, contour plots constructed using eigenvalues computed via principal component analysis (PCA) illustrated a lack of differentiation among West African zebu cattle populations at the body measurements level. Correspondence analysis carried out on the 11 qualitative traits recorded allowed for ascertaining a clear differentiation between the Zebu Bororo and the other zebu cattle populations analysed which, in turn, did not show a clear differentiation at the qualitative type traits level. In our data, Zebu Bororo cattle had in high frequency qualitative features such as dropped ears, lyre-shaped horns and red-pied coat colour that are not frequently present in the other West African zebu populations analysed. A directional selection due to a rough consensus of the stock-keepers may be hypothesised. Performance of further analyses to assess the degree in which such breeding differences may be related to genetic or production differences are advised.

1 Introduction

Although it is accepted that zebu cattle moved and spread into Africa in different historical waves (Payne, 1970; Hanotte et al., 2002) there is archaeological and genetic evidence suggesting that West African zebu cattle is the present-day representative of an ancient introgression of zebu cattle into Africa (Magnavita, 2006; Pérez-Pardal et al., 2010). Together with their historical and genetic importance, West African zebu cattle breeds are a major source of meat, milk and draught power for large human populations. Therefore, their

characterisation is the first step for future implementation of effective improvement programmes accounting for their current low levels of productivity (Ibeagha-Awemu and Erhardt, 2006).

Traditionally, West African zebu is classified into two main groups (Rege and Tawah, 1999): the long-horned Fulani zebu cattle and the short-horned Gudali zebu. The long-horned zebu includes two subgroups according to horn shape: the lyre-horned subgroup, with the Gobra and White Fulani cattle as the main representatives, and the long-lyre-horned subgroup consisting of the Red Fulani cattle, which

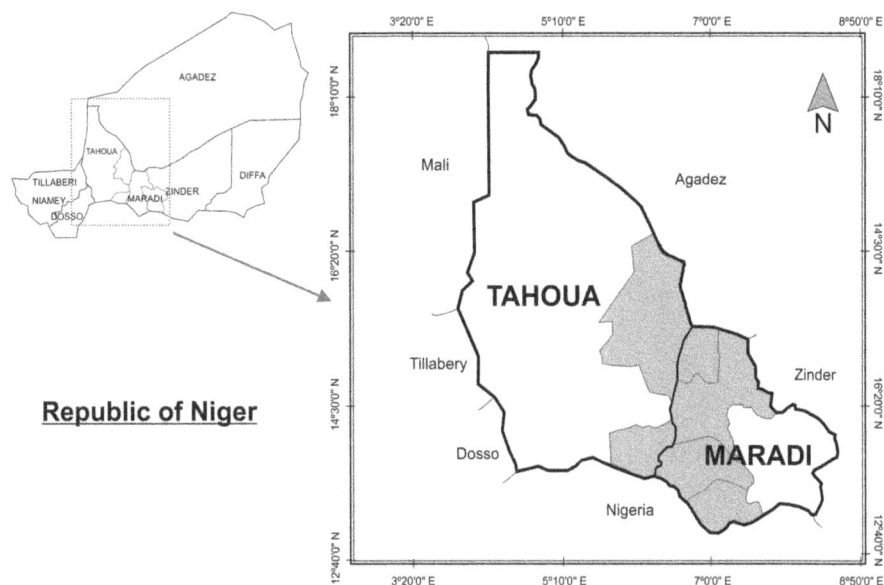

Figure 1. Map (own work of Iván Fernández and Maaouia A. Moussa) illustrating the limits of the provinces of Tahoua and Maradi in the Republic of Niger. Areas in which sampling was carried out are in light grey.

have many local names such as Fellata, Bororo or Mbororo. In any case, definition of breeds within African livestock groups does not follow strict criteria and is usually carried out considering either the geographic areas of spreading or the main ethnic groups acting as stockbreeders (Simon, 1999; Traoré et al., 2008a, b). West African zebu is mainly managed by nomadic stockbreeders such as the Fulani (Peul) people, all across the territory known as the Sahel, the eco-agricultural zone between the Sahara and the coastal rain forests, covering from Sudan in the east to Senegal in the west. Although no selection policies exist and the existence of a strong east-to-west gene flow is accepted (Hanotte et al., 2002), there is consensus on the existence of local populations within West African zebu that can be considered different breeds. Some of them, such as the long-horned Red Bororo (Ibeagha-Awemu and Erhardt, 2006) and White Fulani (Ibeagha-Awemu and Erhardt, 2006; Tawah and Rege, 1996a; Yakubu et al., 2009) or the short-horned zebu Gudali (Tawah and Rege, 1996b), have received some attention to document their main phenotypic and genetic characteristics. However, large morphological and genetic differences among local subpopulations within a breed are likely to exist (Rege and Tawah, 1999), and therefore accumulating information on a different local population is still a challenge in the characterisation of native African livestock.

Recently, nine West African cattle breeds have been jointly analysed for body measurements and qualitative type traits using multifactorial analyses (Traoré et al., 2015, 2016). These analyses included individuals of the Burkina Faso populations of Zebu Azawak (also known as Touareg) and Zebu Mbororo (Red Fulani), as well as the Benin and Burkina Faso populations of the short-horned Zebu Peul. The Zebu Bororo breed, basically owned by the Wodaabé (Bororo) ethnical group, is included in international databases of farm animal genetic resources together with other long-horned representatives of the West African zebu population. However, the amount of information available on this breed is extremely scant (see as an example the DAGRIS database at http://dagris.ilri.cgiar.org/display.asp?ID=142; Rege et al., 2007).

The aim of the current research was to document the phenotypic variation on body measurements and qualitative type traits in a representative sample of Niger Bororo cattle using the methodology reported in Traoré et al. (2015, 2016). Morphological variation was compared with that of four West African zebu cattle populations of Burkina Faso and Benin considering jointly all measured morphological variables using multifactorial techniques to assess within- and between-populations variation.

2 Materials and methods

2.1 Data

A total of 357 adult cows (age ranging from 4 to 16 years old) and 29 sires (from 4 to 5 years old) belonging to the Niger Zebu Bororo cattle population were assessed for 13 body measurements and 11 qualitative traits. Assessment was carried out in 12 different sites of the provinces of Tahoua (5 sites, 75 cows and 3 sires) and Maradi (7 sites, 282 cows and 26 sires) of the Republic of Niger (Fig. 1). Although the northern parts of these two provinces belong to the Sahara, sampled sites were located within the Sahel area. Sites sampled in the Tahoua province belonged to

Table 1. Least squares means (in cm) and their standard errors (in brackets) for 13 body measurements assessed in cows belonging to five West African zebu cattle breeds. Results and sample size (N) are given per breed. Eigenvectors computed for two factors identified via principal component analysis (eigenvalues > 1) are also given. Eigenvectors higher than |0.492| are in bold. Factor 1 explains 56.43 % of the total variability. Factor 2 explains 9.32 % of the total variability. Different superscripts denote significantly different trait raw means at $p < 0.05$. "(BF)" and "(B)" mean Burkina Faso and Benin, respectively.

	Zebu Bororo	Zebu Azawak	Zebu Mbororo	Zebu Peul (BF)	Zebu Peul (B)	Factor 1	Factor 2
N	357	29	64	266	128		
Facial length	51.0^b (0.3)	50.3^c (0.6)	52.5^a (0.5)	49.3^c (0.4)	43.5^d (0.4)	**0.843**	0.244
Muzzle circumference	38.6^a (0.3)	37.5^b (0.6)	$36.1^{b,c}$ (0.5)	35.9^c (0.3)	37.1^b (0.3)	0.358	**0.668**
Horn length	71.2^a (1.3)	17.7^e (2.4)	50.9^b (1.9)	33.0^c (1.3)	26.9^d (1.3)	**0.712**	0.451
Ear length	23.8^a (0.3)	18.4^c (0.5)	20.9^b (0.4)	18.7^c (0.3)	15.9^d (0.3)	**0.775**	0.375
Height at withers	120.5^a (1.0)	115.2^b (1.9)	117.8^a (1.5)	112.6^b (1.0)	101.2^c (1.0)	**0.850**	0.296
Heart girth	156.7^a (1.3)	$145.9^{b,c}$ (2.4)	148.0^b (1.9)	142.9^c (1.3)	133.7^d (1.3)	**0.754**	0.464
Height at hips	127.0^a (1.0)	122.5^b (2.0)	125.8^a (1.5)	119.9^b (1.1)	105.7^c (1.1)	**0.885**	0.257
Body length	134.8^a (1.1)	126.3^b (2.0)	131.8^a (1.6)	126.8^b (1.1)	118.7^c (1.1)	**0.817**	0.233
Thorax depth	67.1^a (0.8)	55.1^c (1.4)	57.0^b (1.1)	$53.3^{c,d}$ (0.8)	52.1^d (0.8)	**0.702**	0.324
Shoulder width	23.9^a (0.3)	21.4^c (0.6)	22.5^b (0.5)	$21.7^{c,d}$ (0.4)	19.0^d (0.4)	0.227	**0.677**
Pelvic width	43.3^a (0.4)	$31.6^{b,c}$ (0.8)	32.6^b (0.6)	30.6^c (0.4)	$31.9^{b,c}$ (0.4)	0.479	**0.678**
Ischium width	14.0^a (0.3)	13.1^b (0.5)	13.6^a (0.4)	12.9^b (0.3)	14.3^a (0.3)	0.188	**0.659**
Rump length	36.3^c (1.1)	$41.0^{a,b}$ (2.1)	42.9^a (1.6)	$40.3^{a,b}$ (1.1)	$38.7^{b,c}$ (1.1)	0.070	0.036

Table 2. Least squares means (in cm) and their standard errors (in brackets) for 13 body measurements assessed in sires belonging to five West African zebu cattle breeds. Results and sample size (N) are given per breed. Eigenvectors computed for two factors identified via principal component analysis (eigenvalues > 1) are also given. Eigenvectors higher than |0.510| are in bold. Factor 1 explains 61.88 % of the total variability. Factor 2 explains 8.21 % of the total variability. Different superscripts denote significantly different trait raw means at $p < 0.05$. "(BF)" and "(B)" mean Burkina Faso and Benin, respectively.

	Zebu Bororo	Zebu Azawak	Zebu Mbororo	Zebu Peul (BF)	Zebu Peul (B)	Factor 1	Factor 2
N	29	8	14	24	17		
Facial length	53.6^a (1.0)	51.5^b (1.7)	$52.9^{a,b}$ (1.4)	50.8^b (1.1)	42.6^c (1.0)	**0.764**	**0.568**
Muzzle circumference	41.3^a (1.3)	39.1^b (2.0)	38.7^b (1.7)	$39.2^{a,b}$ (1.4)	37.3^b (1.2)	**0.612**	0.324
Horn length	65.9^a (3.5)	12.0^d (5.6)	46.1^b (4.6)	27.6^c (3.8)	24.4^c (3.2)	**0.867**	0.033
Ear length	24.6^a (0.7)	20.4^c (1.2)	22.3^b (1.0)	$20.3^{b,c}$ (0.8)	15.6^d (0.7)	**0.845**	0.270
Height at withers	128.4^a (3.1)	124.2^a (5.0)	125.6^a (4.1)	121.5^a (3.4)	101.3^b (2.9)	**0.708**	**0.603**
Heart girth	160.3^a (4.5)	$155.4^{a,b}$ (7.2)	149.4^b (5.9)	$151.6^{a,b}$ (4.9)	128.5^c (4.1)	**0.731**	0.483
Height at hips	136.4^a (3.2)	$131.5^{a,b}$ (5.1)	$136.2^{a,b}$ (4.1)	128.4^b (3.4)	105.8^c (2.9)	**0.703**	**0.615**
Body length	138.1^a (3.2)	$137.0^{a,b}$ (5.1)	131.2^b (4.2)	132.2^b (3.5)	115.5^c (2.9)	**0.737**	0.503
Thorax depth	62.2^a (2.5)	57.4^b (4.0)	58.8^b (3.3)	56.8^b (2.8)	51.3^b (2.3)	**0.705**	0.307
Shoulder width	25.0^a (1.7)	25.8^a (2.6)	$22.9^{a,b}$ (2.2)	23.9^a (1.8)	18.5^b (1.5)	**0.525**	0.154
Pelvic width	43.2^a (1.4)	31.1^b (2.3)	29.3^b (1.8)	31.5^b (1.5)	30.9^b (1.3)	**0.831**	−0.057
Ischium width	13.2^a (1.3)	12.6^a (2.0)	12.9^a (1.7)	14.0^a (1.4)	13.4^a (1.2)	0.104	0.143
Rump length	$36.6^{a,b}$ (1.3)	$41.0^{a,b}$ (2.1)	42.1^a (1.7)	$40.7^{a,b}$ (1.4)	36.5^b (1.2)	0.035	**0.867**

the municipalities of Boundou (14°50′ N, 6°56′ E), Goulbi (13°27′ N, 7°02′ E), Bangui (13°40′ N, 06°16′ E) and Djangangari (14°55′ N, 06°34′ E) while sites sampled in the Maradi province belonged to the municipalities of Chigrenne (15°03′ N, 06°36′ E), Bermo (15°16′ N, 06°44′ E) and Birni Lalley (14°26′ N, 06°45′ E). Each sampling location was georeferenced using GPS Garmin-50 devices (Garmin Ltd., Olathe, KS, USA). Assessments were carried out in 2015, during the rainy season (August–September).

The original dataset included records of 21 quantitative traits. However, only 15 of them coincided with those obtained by Traoré et al. (2015, 2016) in nine West African cattle populations. Moreover, definition of two of the traits recorded (tail length and facial width) differed between the current research and the project of Traoré et al. (2015, 2016). Therefore, after editing, only 13 quantitative traits were used for analyses (Tables 1 and 2): facial length (from orbital fossa to upper lip), muzzle circumference, horn length (greater curvature), ear length, height at withers, heart girth, height at hips (tuber coxae), body length (from lateral tuberosity of the humerus to tuber ischii), thorax depth, shoulder width (between lateral tuberosities of the humerus), pelvic width (between tuber ischii), ischium width (between tuber ischii) and rump length (from tuber coxae to tuber ischii).

Eleven qualitative traits were scored with the same within-trait levels, codes and definitions used in Traoré et al. (2015, 2016): cephalic profile, ear shape, muzzle pigmentation, eyelid pigmentation, hoof pigmentation, horn colour, dewlap size, backline profile, horn shape, spotting pattern and coat colour pattern (Table 3).

Age of the individuals was approximated by examining dentition and direct enquiries to owners. For statistical purposes, the age of the individuals was grouped as follows: 4 years old (5 cows and 6 sires), from 5 to 10 years old (252 cows and 23 sires) and older than 10 years (100 cows). Body measurements were carried out, with the animals standing stationary on a flat floor, using Lydthin stick, tape measure and Vernier caliper. No ethics statement was required for data collection. Body measurements and trait scores were obtained from different technicians visiting farms with the permission of the owners. Animals were managed by the owners.

Data were jointly analysed with 375 individuals belonging to the 4 West African zebu cattle populations previously analysed by Traoré et al. (2015, 2016): Zebu Azawak (29 cows and 8 sires), Zebu Mbororo (64 cows and 14 sires), Zebu Peul of Burkina Faso (266 cows and 24 sires) and Zebu Peul of Benin (128 cows and 17 sires). Zebu Mbororo of Burkina Faso and Zebu Bororo of Niger can be considered local representatives of the long-horned West African zebu cattle group (Rege and Tawah, 1999). However, in the current analysis both samples were considered as belonging to different populations. Consequently, the name given by Traoré et al. (2015, 2016) to the long-horned zebu of Burkina Faso (Zebu Mbororo) was used throughout the text to point out the different geographical origin of data. For consistency, Zebu Peul of Burkina Faso and Benin were treated as belonging to different populations. Traoré et al. (2015, 2016) did not find clear differentiation at the qualitative traits level among West African zebu populations. Therefore, when necessary for descriptive purposes, frequencies of qualitative traits were pooled for West African zebu cattle other than Zebu Bororo.

2.2 Statistical analyses

Body traits were analysed separately for cows and sires to avoid bias due to sexual dimorphism (see Traoré et al., 2016, for a review on this issue). Preliminary analyses showed that neither country of origin (Benin, Burkina Faso and Niger) nor type of West African zebu (long-horned: Zebu Bororo and Zebu Mbororo; short-horned: Zebu Peul and Zebu Azawak) had statistically significant influence on data. Therefore, a very simple model, including the effect of the breed

Table 3. Between-breeds Mahalanobis distance matrices computed within sex using 13 body measurements. The distance matrix computed on the cows' dataset are below diagonal. The distance values corresponding to the sires' dataset are above diagonal. All distance pairs are statistically significant for $p < 0.0001$. "(BF)" and "(B)" mean Burkina Faso and Benin, respectively.

	Zebu Bororo	Zebu Azawak	Zebu Mbororo	Zebu Peul (BF)	Zebu Peul (B)
Zebu Bororo		47.49	22.48	29.66	30.62
Zebu Azawak	41.65		12.74	3.83	21.37
Zebu Mbororo	20.12	11.22		4.83	17.69
Zebu Peul (BF)	27.26	4.05	2.89		10.98
Zebu Peul (B)	28.23	18.05	19.57	12.33	

(five levels) and the age of the individual (three levels for cows and two levels for sires), was fitted using PROC GLM of the SAS/STAT™ package (SAS Institute Inc., Cary, NC) to estimate least squares means, and their corresponding standard errors, for each level of the breed effect. Additionally, Duncan's multiple-range test was performed on the breed effect means. Furthermore, the between-breeds Mahalanobis distance matrix was computed on body measurements using the CANDISC procedure of SAS/STAT.

Relationships among body measurements were summarised via principal component analysis (PCA), using the PROC FACTOR of SAS/STAT, to determine the number of independent traits that account for most of the phenotypic variation in body measurements. This analysis was computed from the correlation matrix among measurements to ensure that all traits were treated as equally important, giving the same weight to the variables regardless their own variance. A VARIMAX rotation was applied to the retained components in order to obtain factor pattern coefficients considerably less correlated than the original body measurements. Only factors accounting for more variation than any individual type trait (eigenvalue > 1) were retained.

Frequencies of each level of the qualitative traits analysed were computed using the PROC FREQ of SAS/STAT™. Statistical significance of the differences in the frequencies observed among the Zebu Bororo breed and the other West African zebu cattle was assessed pooling both sexes via chi-squared Mantel–Haenszel test. Relationships between qualitative traits were assessed via correspondence analysis using the PROC CORRESP of SAS/STAT™. Two canonical dimensions and their eigenvectors were computed to account for association between the levels of the traits scored. Following Parés-Casanova and Jordana (1999) and Grema et al. (2017) each score of the qualitative traits assessed was consider arbitrarily as polymorphism of the trait and used to compute the between-breeds Reynolds distance matrix using the program MolKin (Gutiérrez et al., 2005). Statistical confidence on the distance values was assessed via bootstrapping using 1000 replicates.

Eigenvectors computed for each individual via PCA and correspondence analyses were used to construct contour plots illustrating 75 % confidence region (per breed) of the relationships among individuals using the library ggplot2 of R (http://CRAN.R-project.org/). Eigenvectors computed for each individual via PCA were regressed on latitude (in decimal format) of the sampling site using the Proc REG of SAS/STAT to assess variation related to geography.

3 Results

3.1 Continuous traits

Least-squares means for the body measurements assessed by breed are given in Table 1 (cows) and Table 2 (sires). Data corresponding to the Zebu Azawak, Zebu Mbororo and Zebu Peul populations were previously analysed in Traoré et al. (2015, 2016). However, least-squares estimates may vary due to differences in both the datasets analysed (including the split of Zebu Peul into two different geographical populations) and the models fitted for analyses. For both males and females, Zebu Bororo cattle tended to have the higher estimates of the dataset and the Zebu Peul of Benin the lower. In any case, differences in estimates for body measurements between Zebu Bororo and the other long-horned zebu population analysed (Zebu Mbororo) were low. Even in the case of facial length, Zebu Bororo cows had lower estimates than Zebu Mbororo (51.0 ± 0.3 vs. 52.5 ± 0.5, respectively). Furthermore, these two breeds had very similar values for key traits such as height at hips and ischium width for both cows and sires. Actually, the higher differences between the Niger Zebu Bororo and the Burkina Faso Zebu Mbororo populations were assessed in measurement traits such as ear length and horn length closely related with key qualitative features such as ear shape and horn shape.

Both in males and in females, PCA identified two factors with eigenvalues higher than 1. In the females dataset factor 1 (eigenvalue $= 7.34$) accounted for 56.43 % of the total variation and factor 2 (eigenvalue $= 1.21$) explained 9.32 %. In the sires dataset factor 1 (eigenvalue $= 8.04$) accounted for 61.88 % of the total variation and factor 2 (eigenvalue $= 1.07$) explained 8.21 %. In females (Table 1) factor 1 clearly summarised the general size of the individuals (height and length) while factor 2 summarised the body width. In sires (Table 2), with a relatively limited sample size, factor 2 was less informative on body trait variation. The eigenvalues computed via PCA for each individual were plotted in a two-dimensional space to illustrate the between-breeds relationships for body measurements (Fig. 2). No differentiation among West African zebu populations was found. The 75 % confidence regions computed for Zebu Bororo, Zebu Azawak, Zebu Mbororo, Zebu Peul of Burkina Faso and Zebu Peul of Benin were intermingled, particularly in the case of

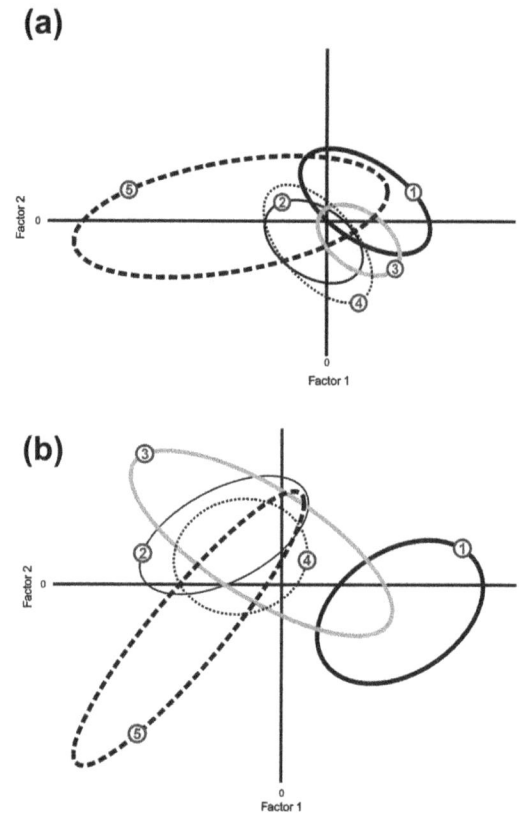

Figure 2. Contour plots summarising, per breed, the information provided by the 13 body traits analysed via principal component analysis. Contours show the 75 % confidence region of the within-breed relationships among individuals. Factor 1 is on the x axis and factor 2 on the y axis. Panel (**a**) corresponds to the female subset and (**b**) corresponds to the sires subset. Numbers on contours mean the following: (1) Zebu Bororo; (2) Zebu Azawak; (3) Zebu Mbororo; (4:) Zebu Peul of Burkina Faso; and (5) Zebu Peul of Benin.

females in which sample size was not a limitation (Fig. 2a). The eigenvalues computed for each individual corresponding to the factor 1 identified either the cows' or sires' dataset were also regressed on latitude to ascertain the existence of a geographical pattern of variation on body measurements in West African zebu cattle. Both in females ($R^2 = 0.598$) and in males ($R^2 = 0.370$) positive and significant regression coefficients were computed (0.34467 in cows and 0.24049 in sires) showing that overall body size increased with latitude).

The between-breeds Mahalanobis distance matrix is given in Table 3. All pairwise distances were statistically significant ($p < 0.001$). The hypothesis that the breeds' means are equal in the populations analysed was also tested using Wilks' lambda. This parameter took a significant value ($p < 0.0001$) for both the cows ($\lambda = 0.04662812$; $F = 75.37$; degrees of freedom $= 52$) and the sires ($\lambda = 0.02895334$; $F = 8.64$; degrees of freedom $= 52$) datasets. Therefore, differences found were statistically different from zero. For both sires and cows, the largest distances were found between the

Zebu Bororo cattle and the other breeds. In both sexes the higher differentiation was found for the pair Zebu Bororo–Zebu Azawak (41.65 for cows and 47.49 for sires). Zebu Bororo had the lowest differentiation with Zebu Mbororo.

3.2 Qualitative traits

The frequencies (in percentage) of each level of the 11 qualitative traits recorded for Zebu Bororo cows and sires are given in Table 4. Data corresponding to the Zebu Azawak, Zebu Mbororo, Zebu Peul of Burkina Faso and Zebu Peul of Benin populations were pooled with no sex separation to be used as a reference for Zebu Bororo. A chi-squared Mantel–Haenszel test showed that incidence of all the analysed traits varied significantly between Zebu Bororo cows and sires and the other West African zebu cattle for $p < 0.001$ except for cephalic profile and backline profile (Table 4). For these two traits zebu cattle were majority straight, regardless of the population to which the individuals belonged. However, Zebu Bororo cattle had large differences for most qualitative traits with the other West African zebu cattle. The "most frequent" Zebu Bororo individuals had dropped ears, non-pigmented muzzle, grey-coloured lyre-shaped horns and red-pied coat colour pattern. This general appearance substantially departs from that of the other West African zebu cattle populations analysed in which horizontal ears and pigmented muzzle are the rule and it is possible to find a high variation in horn shape and coat colour (Table 4).

A correspondence analysis was carried out on the 11 qualitative traits recorded. Two correspondence dimensions identified accounted for 36.77 and 21.60 %, respectively, in the cows' dataset and 32.43 and 26.57 % of the total variation, respectively, in the sires' dataset. The solutions provided for each individual by the correspondence analysis were plotted in a two-dimensional space. Figure 3 shows that the 75 % confidence region computed for the Zebu Bororo breed is clearly separated from those of the other zebu populations at both the cows and the sires level which. In turn, the non-Niger West African zebu populations showed highly intermingled 75 % confidence regions.

Between-breeds Reynolds' distance matrices were computed for each sex to quantify differentiation due to qualitative type traits (Table 5). The pairs involving Zebu Bororo cattle had the higher distance values.

4 Discussion

Although analyses focusing on body measurements are relatively frequent in East African zebu (Mwacharo et al., 2006), zoometric studies on West African zebu cattle were scant before Traoré et al. (2015, 2016). Many times, zoometrical studies cannot be compared straightforwardly due to differences in both the definition of the traits assessed and the heterogeneity of the samples used. Hall (1991) using 10 Sokoto Gudali and 29 White Fulani humped Nigerian cattle re-

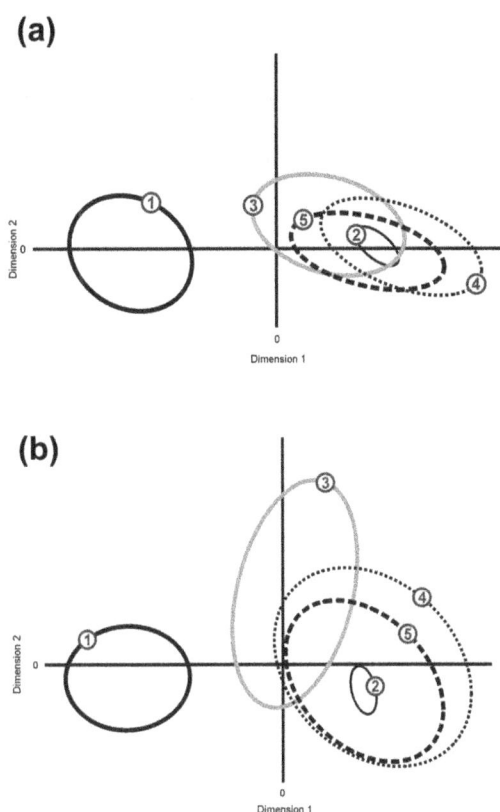

Figure 3. Contour plots summarising, per breed, the information provided by the 11 qualitative traits analysed via correspondence analyses. Contours show the 75 % confidence region of the within-breed relationships among individuals. Correspondence dimension 1 is on the x axis and correspondence dimension 2 on the y axis. Panel **(a)** corresponds to the female subset and **(b)** corresponds to the sires subset. Numbers on contours mean the following: (1) Zebu Bororo; (2) Zebu Azawak; (3) Zebu Mbororo; (4) Zebu Peul of Burkina Faso; and (5) Zebu Peul of Benin.

ported mean values for body length and height at withers of 123.6 ± 2.3 and 130.9 ± 2.5 cm, respectively, for the former and 117.1 ± 0.9 and 121.4 ± 1.0 cm, respectively, for the latter. Although many of the traits analysed were not consistent with the measurements carried out in the current research, Yakubu et al. (2009), using 79 male and 125 female individuals of Nigerian White Fulani cattle with age ranging from 1.5 to 3.6 years, reported mean values for Height at Withers of 101.11 ± 2.19 cm in males and 100.48 ± 2.70 cm in females.

Local populations belonging to the long-horned West African zebu cattle group are expected to show large morphological differences. However, the current research confirms that West African zebu is a basically unstructured population at the body measurements level. Differences of the means of the traits assessed due to the breed or population are relatively poor (Tables 1 and 2) and when body traits are considered as a whole it is not possible to distinguish

Table 4. Frequencies (in percentage) of each level of the 11 qualitative traits assessed in Zebu Bororo sires and cows. Additionally, frequencies observed in four other West African zebu populations, comprising a total of 311 cows and 64 sires, for the same traits are given as a reference.

Trait	Class		Zebu Bororo		Other
	Code	Definition	Sires	Cows	West African Zebu*
N			29	357	375
Cephalic profile[ns]	1	concave		1.7	1.8
	2	convex	10.3	6.2	5.7
	3	straight	89.7	92.2	92.5
Ear shape	1	horizontal	10.3	6.2	99.8
	2	drop	89.7	92.2	0.2
	3	upright		1.6	
Muzzle pigmentation	1	pigmented	24.1	6.2	70.9
	2	not pigmented	75.9	93.8	29.1
Eyelid pigmentation	1	pigmented	100.0	95.5	70.5
	2	not pigmented		4.5	29.5
Hoof pigmentation	1	pigmented	100.0	91.6	86.5
	2	not pigmented		8.4	13.5
Horn colour	1	black	6.9	1.4	14.1
	2	grey	89.7	92.4	5.7
	3	brown	3.5	6.2	51.6
	4	two coloured			28.7
Dewlap size	1	well developed	44.8	22.1	14.0
	2	poorly developed	55.2	74.2	50.6
	3	small		3.6	35.4
Backline profile[ns]	1	straight	96.6	94.7	95.9
	2	concave		2.5	1.2
	3	convex	3.5	2.8	2.8
Horn shape	1	cup	6.9	0.8	16.0
	2	crescent		2.2	40.0
	3	lyre	93.1	95.8	30.6
	4	wheel			1.3
	5	crown		1.1	11.7
	7	en arrière			0.4
Spotting pattern	1	absence	82.8	81.2	53.8
	2	pied	17.2	13.2	15.1
	3	spotted		5.6	31.1
Coat colour pattern	1	black	34.5	3.4	4.6
	2	black-pied			20.7
	3	white		0.3	10.8
	4	red		0.8	1.5
	5	red-pied	58.6	91.0	9.5
	6	roan	6.9	1.4	1.4
	7	fawn		2.5	1.0
	8	diluted fawn			10.9
	9	grey		0.3	9.4
	10	blond			0.3
	11	fawn-blond			14.9
	12	dun-red		0.3	8.3
	13	fawn-red			6.9

* Includes individuals belonging to the Zebu Azawak, Zebu Mbororo and Zebu Peul breeds sampled in Burkina Faso and Benin. "ns" as superscript means that frequencies observed for each level of these traits among Bororo cattle and other West African zebu did not depart from random expectation at $p < 0.05$.

Table 5. Between-breeds Reynolds' distance matrices computed within sex using the individual scores of 11 qualitative type traits. The distance matrix computed on the cows' dataset are below diagonal. The distance values corresponding to the sires' dataset are above diagonal. Standard deviations of the estimates of the Reynolds' distance pairs (in brackets) were computed via bootstrapping using 1000 replicates.

	Zebu Bororo	Zebu Azawak	Zebu Mbororo	Zebu Peul (BF)	Zebu Peul (B)
Zebu Bororo		0.482 (0.062)	0.331 (0.048)	0.373 (0.04)	0.370 (0.049)
Zebu Azawak	0.239 (0.016)		0.324 (0.067)	0.121 (0.026)	0.219 (0.043)
Zebu Mbororo	0.253 (0.015)	0.152 (0.024)		0.159 (0.039)	0.198 (0.049)
Zebu Peul (BF)	0.424 (0.016)	0.035 (0.004)	0.069 (0.010)		0.103 (0.026)
Zebu Peul (B)	0.391 (0.019)	0.081 (0.010)	0.099 (0.017)	0.057 (0.006)	

West African zebu populations straightforwardly (Fig. 2). Although the two long-horned zebu analysed (Bororo and Mbororo) tended to have higher values on body traits (Tables 1 and 2) the inclusion of the effect of the type of West African zebu did not allow a better fit of the model to data. Traoré et al. (2015), using taurine, zebu and sanga individuals, identified a geographical cline in body measurements in West African cattle with a continuous decrease in size southwards. This is consistent with the increase in size with latitude assessed here using West African zebu only. Zebu Peul sampled in Benin tended to have the lower mean values in body traits (Tables 1 and 2) but also the larger 75 % confidence intervals (Fig. 2) suggesting this cattle can be a transition population between West African zebu and sanga (taurine × zebu crosses) cattle.

At the qualitative traits level, Zebu Bororo cattle showed significant differentiation from the other West African zebu populations analysed. Zebu Bororo had qualitative type features such as dropped ears, lyre-shaped horns and red-pied coat colour that are not present in high frequency in the other West African zebu populations. Furthermore, the non-Niger West African zebu analysed had larger within-populations variability for these key qualitative traits than Zebu Bororo (Traoré et al., 2015, 2016). Other West African zebu breeds, such as White Fulani (Tawah and Rege, 1996a), are known to have virtually fixed qualitative type traits such as coat colour. However, this may not be the rule in African cattle in which no selection programmes exist. In such scenario, it is usually difficult to differentiate between cattle breeds according to qualitative traits due to the existence of a very high within-breed variation due to local preferences of the stock keepers (Desta et al., 2011).

The differentiation found in Zebu Bororo may be a consequence of a directional selection due to a rough consensus of the stock-keepers. At the neutral molecular level it is well known that livestock differentiation in West Africa is more likely due to geographic distance rather than to a breed differences or to different expected origins of the livestock populations (Ibeagha-Awemu and Erhardt, 2006; Traoré et al., 2012; Álvarez et al., 2014). Further genetic analyses are advised to assess the degree of uniqueness of Zebu Bororo in the West African zebu framework.

In summary, the current research confirmed that West African zebu cattle can be considered a single large population at the zoometric level. However, there may exist between-subpopulations differentiation at the qualitative type traits level. The results presented may support the implementation of future work lines aiming at the ascertainment of possible relationships between such differences, related to breeding, and genetic or production differences.

Competing interests. The authors declare that they have no conflict of interest.

Acknowledgements. The authors are grateful for the financial support provided by West African Agricultural Productivity Programme (WAAPP/Niger). The authors are indebted to the breeders involved in this study for their kind collaboration. This study made use of data generated through CORAF/WECARD-World Bank no. 03/GRN/16. Félix Goyache, Iván Fernández and Isabel Álvarez are supported by grant FICYT GRUPIN14-113 and MICIIN-FEDER AGL2016-77813-R. CompGen.

Edited by: Steffen Maak

References

Álvarez, I., Traoré, A., Fernández, I., Lecomte, T., Soudré, A., Kaboré, A., Tamboura, H. H., and Goyache, F.: Assessing introgression of Sahelian zebu genes into native *Bos taurus* breeds in Burkina Faso, Mol. Biol. Rep., 41, 3745–3754, https://doi.org/10.1007/s11033-014-3239-x, 2014.

Desta, T. T., Ayalew, W., and Hedge, B. P.: Breed and trait preferences of Sheko cattle keeper in southern Ethiopia, Trop. Anim. Health Prod., 43, 851–856, https://doi.org/10.1007/s11250-010-9772-2, 2011.

Grema, M., Traoré, A., Issa, M., Hamani, M., Abdou, M., Fernández, I., Soudré, A., Álvarez, I., Sanou, M., Tamboura, H. H., Alhassane, Y., and Goyache, F.: Morphological assessment of Niger Kuri cattle using multivariate methods, S. Afr. J. Anim. Sci., 47, 505–515, https://doi.org/10.4314/sajas.v47i1, 2017.

Gutiérrez, J. P., Royo, L. J., Álvarez, I., and Goyache, F.: MolKin v2.0: a computer program for genetic analysis of populations using molecular coancestry information, J. Hered., 96, 718–721, https://doi.org/10.1093/jhered/esi118, 2005.

Hall, S. J. G.: Body dimensions of Nigerian cattle, sheep and goats, Anim. Prod., 53, 61–69, https://doi.org/10.1017/S0003356100005985, 1991.

Hanotte, O., Bradley, D. G., Ochieng, J. W., Verjee, Y., Hill, E. W., and Rege, J. E.: African pastoralism: genetic imprints of origins and migrations, Science, 296, 336–339, https://doi.org/10.1126/science.1069878, 2002.

Ibeagha-Awemu, E. M. and Erhardt G.: An evaluation of genetic diversity indices of the Red Bororo and White Fulani cattle breeds with different molecular markers and their implications for current and future improvement options, Trop. Anim. Health Prod., 38, 431–441, https://doi.org/10.1007/s11250-006-4347-y, 2006.

Magnavita C.: Ancient Humped Cattle in Africa: A View from the Chad Basin, Afr. Archaeol. Rev., 23, 55–84, https://doi.org/10.1007/s10437-006-9008-z, 2006.

Mwacharo, J. M., Okeyo, A. M., Kamande, G. K., and Rege, J. E. O.: The small East African shorthorn zebu cows in Kenya. I: Linear body measurements, Trop. Anim. Health Prod, 38, 65–74, https://doi.org/10.1007/s11250-006-4266-y, 2006.

Parés-Casanova, P. M. and Jordana, J.: Relaciones genéticas entre razas ibéricas de caballos utilizando caracteres morfológicos (prototipos raciales), Anim. Genet. Resour. Inf., 26, 75–94, https://doi.org/10.1017/S1014233900001218, 1999.

Payne, W. J. A.: Cattle production in the tropics, 1st Edn., Longman, London, 1970.

Pérez-Pardal, L., Royo, L. J., Beja-Pereira, A., Chen, S., Cantet, R. J. C., Traoré, A., Curik, I., Sölkner, J., Bozzi, R., Fernández, I., Álvarez, I., Gutiérrez, J. P., Gómez, E., Ponce de Leon, F. A., and Goyache, F.: Multiple paternal origins of domestic cattle revealed by Y-specific interspersed multilocus microsatellites, Heredity, 105, 511–519, https://doi.org/10.1038/hdy.2010.30, 2010.

Rege, J. E. O. and Tawah, C. L.: The state of African cattle genetic resources II. Geographical distribution, characteristics and uses of present-day breeds and strains, Anim. Genet. Resour. Inf., 26, 1–26, https://doi.org/10.1017/S1014233900001152, 1999.

Rege, J. E. O., Hanotte, O., Mamo, Y., Asrat, B., and Dessie, T.: Domestic Animal Genetic Resources Information System (DAGRIS), International Livestock Research Institute, Addis Ababa, Ethiopia, http://dagris.ilri.cgiar.org (last access: 27 February 2017), 2007.

Simon, D. L.: European approaches to conservation of farm animal genetic resources, Anim. Genet. Resour. Inf., 25, 79–99, https://doi.org/10.1017/S1014233900005794, 1999.

Tawah, C. L. and Rege, J. E. O.; White Fulani cattle of West and Central Africa, Anim. Genet. Resour. Inf., 17, 137–158, https://doi.org/10.1017/S101423390000064X, 1996a.

Tawah, C. L. and Rege, J. E. O.: Gudali cattle of West and Central Africa, Anim. Genet. Resour. Inf., 17, 159–170, https://doi.org/10.1017/S1014233900000651, 1996b.

Traoré, A., Tamboura, H. H., Kabore, A., Royo, L. J., Fernández, I., Álvarez, I., Sangare, M., Bouchel, D., Poivey, J. P., Francois, D., Toguyeni, A., Sawadogo, L., and Goyache, F.: Multivariate characterization of morphological traits in Burkina Faso sheep, Small Rum. Res., 80, 62–67, https://doi.org/10.1016/j.smallrumres.2008.09.011, 2008a.

Traoré, A., Tamboura, H. H., Kabore, A., Royo, L. J., Fernández, I., Álvarez, I., Sangare, M., Bouchel, D., Poivey, J. P., Sawadogo, L., and Goyache, F.: Multivariate analyses on morphological traits of goats in Burkina Faso, Arch. Anim. Breed., 51, 588–600, 2008b.

Traoré, A., Álvarez, I., Fernández, I., Pérez-Pardal, L., Kaboré, A., Ouédraogo-Sanou, G. M. S., Zaré, Y., Tamboura, H. H., and Goyache, F.: Ascertaining gene flow patterns in livestock populations of developing countries: a case study in Burkina Faso goat, BMC Genetics, 13, 35, https://doi.org/10.1186/1471-2156-13-35, 2012.

Traoré, A., Koudandé, D. O., Fernández, I., Soudré, A., Álvarez, I., Diarra, S., Diarra, F., Kaboré, A., Sanou, M., Tamboura, H. H., and Goyache, F.: Geographical assessment of body measurements and qualitative type traits in West African cattle, Anim. Health Prod., 47, 1505–1513, https://doi.org/10.1007/s11250-015-0891-7, 2015.

Traoré, A., Koudandé, D. O., Fernández, I., Soudré, A., Álvarez, I., Diarra, S., Diarra, F., Kaboré, A., Sanou, M., Tamboura, H. H., and Goyache, F.: Multivariate characterization of morphological traits in West African cattle sires, Arch. Anim. Breed., 59, 337–344, https://doi.org/10.5194/aab-59-337-2016, 2016.

Yakubu, A., Ogah, D. M., and Idahor, K. O.: Principal component analysis of the morphostructural indices of White Fulani cattle, Trakia J. Sci., 7, 67–73, 2009.

24

Individual and combined effects of *CAPN1*, *CAST*, *LEP* and *GHR* gene polymorphisms on carcass characteristics and meat quality in Holstein bulls

Sena Ardicli, Hale Samli, Deniz Dincel, Bahadir Soyudal, and Faruk Balci

Laboratory of Genetics, Department of Genetics, Faculty of Veterinary Medicine, Uludag University,
16059 Nilufer, Bursa, Turkey

Correspondence to: Faruk Balci (fbalci@uludag.edu.tr)

Abstract. The objective of this study was to determine the association of single nucleotide polymorphisms (SNPs) with carcass characteristics and meat quality traits in selected candidate genes in Holstein bulls. Five SNPs in four genes, i.e. calpain 1 (*CAPN1*), calpastatin (*CAST*), leptin (*LEP*) and growth hormone receptor (*GHR*), were genotyped in 400 purebred bulls using PCR-RFLP. Statistically significant associations were as follows: *CAPN1* G316A with live weight, carcass weight, back fat thickness, m. longissimus thoracis et lumborum area and carcass measurements; *CAPN1* V530I with pH and L^*; *CAST* S20T with live weight, inner chest depth and b^* value; and *GHR* with ph, a^* and h^*. In addition, significant genotypic interactions were observed for dressing percentage (*LEP* A80V × *CAST* S20T), pH (*CAPN1* V530I × *GHR* S555G and *LEP* A80V × *GHR* S555G) and rump width (*CAPN1* V530I × *CAST* S20T). There was no association between the *LEP* A80V marker and any of the traits evaluated, nor was there any association of the tested SNPs with chest width, C^* and marbling score. The present results could therefore be indicative for future studies on meat yield and quality.

1 Introduction

Selection for increased meat production and quality in the beef industry have been the primary emphasis of selection programmes. Recently, the trend of improving these programmes has gradually changed from traditional phenotypical selection methods to genotypic selection by utilizing molecular markers and a better understanding of DNA polymorphisms that have an effect on carcass and meat quality may lead to important applications through marker-assisted selection (MAS) programmes (Guo et al., 2016). However, genetic variations regarding both the quantity and quality of beef already exist among breeds and even among different populations of the same breed (Burrow et al., 2001). In 2016, 1.2 million tonnes of red meat was produced from 9.8 million animals slaughtered in Turkey. The slaughter population was made up of 3.9 million cattle, 4.1 million sheep, 1.8 million goats and 1400 water buffaloes. Of this production, 1.1 million tonnes (approximately 91 % of total) was composed of beef (Turkish Statistical Institute, 2016). In some coun-

tries, the beef industry is exclusively based on specific beef herds, but conversely, cattle farms in many countries, such as Turkey, generally consist of dairy cattle and dual-purpose breeds, and the number of beef breeds is limited. Among these, the Holstein breed comprises by far the most common cattle breed in Turkey, with 5.5 millions purebreds and 856 000 crossbreeds. Considering the approximately 14 million total cattle count, the Holstein breed has a significant impact on Turkish animal husbandry (Turkish Ministry of Food, Agriculture and Livestock Database, 2015). Therefore, evaluating the potential of Holstein meat can be considered as a strategically important point in breeding programmes and meat production (Ardicli et al., 2017a).

In recent years, genes associated with meat quality have been identified and single nucleotide polymorphisms (SNPs) of many candidate genes have been specifically determined (Li et al., 2013). In this context, one of the most investigated genes is bovine micromolar calcium-activated neutral protease 1 (*CAPN1*), which is located on chromosome 29, encoding the large subunit of the enzyme μ-calpain involved in degrading myofibrillar proteins under postmortem conditions. *CAPN1* was suggested as a genetic factor influencing the postmortem tenderization process (Koohmaraie, 1996; Page et al., 2002; Miquel et al., 2009). The SNP G316A in exon 9 (alleles C/G) of the *CAPN1* gene has been associated with meat quality traits (Gill et al., 2009; Miquel et al., 2009; Smith et al., 2009; Bonilla et al., 2010; Mazzucco et al., 2010; Pinto et al., 2010; Kaneda et al., 2011), final weight and average daily weight gain (Miquel et al., 2009; Tait et al., 2014). The SNP V530I in exon 14 of the *CAPN1* gene has been associated with beef tenderness (Corva et al., 2007; Allais et al., 2011), meat colour and drip loss (Ribeca et al., 2013). The bovine calpastatin (*CAST*) gene, mapped to chromosome 7, is considered a candidate gene for beef tenderness (Schenkel et al., 2006). The S20T polymorphism in the *CAST* gene has been shown to be associated with meat quality traits (Juszczuk-Kubiak et al., 2004). The leptin (*LEP*) gene, also known as the "obese gene", located on bovine chromosome 4, encodes leptin, which is secreted by adipose tissue. The concentration of leptin can be involved in food consumption, energy expenditure and adipose tissue development (Buchanan et al., 2002; Lagonigro et al., 2003) and plays an indicator role in marbling, intramuscular fat content, back fat depth and meat quality grade in feedlot cattle (Geary et al., 2003; Li et al., 2013). The A80V polymorphism in exon 3 has been shown to be a candidate marker for milk yield and composition traits (Liefers et al., 2003; Kulig, 2005; Kulig et al., 2010); however, its association with meat quality traits has not been fully depicted. Growth hormone (GH), also known as somatotropin, plays an important role in growth and metabolism (Di Stasio et al., 2005; Waters et al., 2011) by interacting with a specific receptor (GHR), which activates an intracellular signalling pathway (Zhou and Jiang, 2006). Variations in the *GHR* gene, mapped to chromosome 20, have been associated with performance traits in cattle (Blott et al., 2003; Ge et al., 2003; Viitala et al., 2006; Garrett et al., 2008; Waters et al., 2011). The polymorphism in exon 10 of the *GHR* gene has been associated with meat quality (Di Stasio et al., 2005; Reardon et al., 2010), growth performance and body size traits (Waters et al., 2011).

The genetic markers studied in our study and their association with meat yield/quality have also been investigated by other researchers, but the results were often inconsistent. In addition, there is very limited information about the effects of these markers on meat yield and quality of the Holstein breed, which constitutes a significant proportion in Turkey's meat industry. Therefore, the aim of this study was to investigate associations of SNPs at *CAPN1*, *CAST*, *LEP* and *GHR* genes with carcass characteristics, meat yield and quality traits.

2 Materials and methods

2.1 Animals, management and slaughter procedures

Data from 400 Holstein bulls randomly selected from a commercial herd, with a herd size 20 000 cattle, located in South Marmara region and slaughtered at 14–21 months of age, were used in the study. All animals were recorded for the Pedigree Project of the Turkish Ministry of Food, Agriculture and Livestock, and Cattle Breeders Association. Ethical approval was received from Uludag University (approval number: 2012-10/05). The farm was located in Bandırma, northern Balıkesir province (40°18′06.0″ N and 27°56′28.5″ E). The animals were maintained in a semi-open free-stall barn, with straw as bedding. Maximum and minimum ambient air temperatures (°C) in the sheds during the period of the study were 10.1 ± 1.1 and 2.1 ± 0.4 in winter, 19.1 ± 1.4 and 7.06 ± 1.4 in spring and 30.5 ± 1.9 and 16.8 ± 0.8 in summer and relative humidity percentages (%) were 66.7 ± 1.4, 58.9 ± 2.4 and 70.5 ± 0.8 in the same seasons, respectively. The fattening period were initiated after 2 weeks of training. During the fattening period, growing and finishing rations contained corn, potato and tomato pomace silage; barley straw; barley butter; pasta; corn; corn gluten meal; corn bran; sugar-beet pulp; soybean meal; sunflower meal; vitamin and mineral premix; limestone; and salt. The growing ration contained 13.8 % of crude protein and $10.2 \,\mathrm{MJ\,kg^{-1}}$ of metabolizable energy on a dry matter basis and the finishing ration contained 10.3 % of crude protein and $11.5 \,\mathrm{MJ\,kg^{-1}}$ of energy on a dry matter basis. All animals were fed ad libitum with the same diets and had full access to water throughout the experiment. At the end of the finishing period, the animals, in a non-fasted state, were transported to the nearest slaughterhouse (40°21′23.6″ N and 27°56′41.1″ E). The duration of transport from farm to slaughterhouse was approximately 1–2 h. Prior to slaughter, final live weight (LW) was recorded by precision scale (100 g sensitivity). Cattle were stunned by captive bolt before being slaughtered by means of exsanguination and dressed using standard commercial practices, after being kept for 24 h in paddocks and deprived of feed but with full access to water. After slaughter, all of the carcasses were electrically stimulated for a duration of 30 s (60 V), suspended through the Achilles tendons and chilled for 24 h at 4 °C.

2.2 Carcass characteristics

In the present study, non-carcass components were removed and then hot carcass weight (HCW) was determined. Hot carcass did not include kidneys and perinephric or pelvic fat. Chilled carcass weight (CCW) was measured after 24 h at 4 °C and the dressing percentage (DP) was calculated based on HCW. After slaughter, carcass measurements including carcass length (CL), rump length (RL), rump width (RW), chest width (CW) and inner chest depth (ICD) were measured with a caliper, cane and ruler according to following anatomic regions as described by Sagsoz et al. (2005): CL – the distance from the os pubis to the tip of the first rib; RL – the distance from the os calcaneus to the median point of the os pubis; RW – from the rump circumference starting from the point opposite the meat section to the line connecting the centre of the os pubis and the os calcaneus; CW – outer side of the half carcass section from the sixth rib tip to the sixth vertebra; ICD – measured from the sixth rib tip to the sixth vertebra on the inner side of the half carcass section. Back fat thickness (BFT) was measured from the lateral side of the m. longissimus thoracis et lumborum (LTL) at the 12th rib and the same rib surface was evaluated to calculate the LTL area by using a planimeter (Ushikata X-Plan 380d III, Tokyo, Japan).

2.3   Meat quality analyses

Meat quality characteristics investigated in the current study were marbling score (MS), carcass pH and meat colour (L^*, a^*, b^*, C^* and h^*). Marbling was subjectively evaluated corresponding to fat distribution among the muscle fibres in the m. longissimus thoracis (LT) at the 12th–13th rib interface to represent $9°$ (practically devoid, traces, slight, small, modest, moderate, slightly abundant, moderately abundant, abundant) of marbling (Hilton et al., 1998). Carcass pH was measured in the LT between the 12th and 13th ribs at 24 h postmortem using a digital pH meter (Testo 205, Lenzkirch, Germany). Meat colour parameters including L^* (lightness), a^* (redness) and b^* (yellowness) values were evaluated using a spectrocolorimeter (Konica Minolta CM508d, Minolta Co., Ltd, Osaka, Japan) with illuminant D65 as the light source. The device was set to make three measurements and take their average after the calibration corresponding to the standard white plate. Three-times-repeated colour measurements were performed from each sample of the LTL after 24 h storage at 4 °C on cut surface of fat-free area and the average of these measurements was evaluated as the final value (Ekiz et al., 2009). Chroma value (C^*) was calculated as $(a^{*2} + b^{*2})^{1/2}$ and hue angle (h^*) as arctan (b^*/a^*).

2.4   Genomic DNA isolation

DNA was isolated from 4 mL blood samples obtained from the vena jugularis of each bull and collected in K_3EDTA tubes (Vacutest Kima, SRL, Italy) by a phenol–chloroform method as described by Green and Sambrook (2012). The amount and purity of the DNA samples was measured with a spectrophotometer (NanoDrop 2000c, Thermo Scientific, Wilmington, DE, USA). DNA samples were stored at −80 °C until the genotyping was performed.

2.5   Markers used and genotyping

In the present study, the polymorphisms in four candidate genes were genotyped, which included the G316A and the V530I in the CAPN1 gene, the S20T in the CAST gene, A80V in the LEP gene and the S555G in the GHR gene. Marker G316A (GenBank accession number: AF252504) is a cytosine/guanine (C/G) polymorphism in exon 9 of the CAPN1 gene that produces an amino acid substitution (glycine/alanine) in position 316. Marker V530I (GenBank accession number: AF248054) of the same gene is an adenine/guanine (A/G) polymorphism in exon 14 that also produces an amino acid substitution (isoleucine/valine) in position 530 (Page et al., 2002; Casas et al., 2005). The SNP S20T (GenBank accession number: AF117813) in the CAST gene is a guanine/cytosine (G/C) polymorphism located in exon 1C/1D that produces serine/threonine substitution at protein position 20 (Juszczuk-Kubiak et al., 2004). Marker A80V (GenBank accession number: AF536174.1) in exon 3 of the LEP gene expresses the existence of a cytosine/thymine (C/T) substitution that causes coding of alanine instead of valine in position 80 (Haegeman et al., 2000; Lagonigro et al., 2003). Marker S555G (GenBank accession number: AF140284) is the guanine/adenine (G/A) polymorphism at position 257 in exon 10 and induces serine/glycine substitution in position 555 of the GHR gene (Di Stasio et al., 2005).

Genotyping of the SNPs in the CAPN1, CAST, LEP and GHR genes was performed by PCR-RFLP. Primer sequences and PCR conditions for amplification are shown in Table 1. The PCR amplification was performed in a total volume of 50 µL containing 33.5 µL of ddH_2O, 5 µL of 10× buffer, 5 µL of MgSO_4, 1 µL of dNTPs (2.5 mM), 2.5 U of Taq DNA polymerase (Biomatik, Cambridge, Canada, A1003-500U, 5 U µL^{-1}), 1 µL (0.025 µM) of each primer, and 3 µL of the DNA sample at a concentration of 100 ng µL^{-1}. The DNA amplification reactions were performed in a thermal cycler (Palm Cycler GC1-96, Corbett Research, Australia). After amplification, 15 µL of the PCR product with each SNP was digested in 15 U of the corresponding restriction enzyme (Table 1) by incubating at 37 °C for 16 h. Afterwards, the digestion products were electrophoresed in 3 % agarose gel (Sigma Aldrich, Steinheim, Germany) at 85–90 V for 1 h after incubation and visualized by a gel imaging system (DNr-Minilumi, DNR Bio-Imaging Systems, Israel).

2.6   Statistical analysis

The Hardy–Weinberg equilibrium (HWE) was tested for all alleles by using POPGENE software v1.32 (Yeh et al., 2000). The population genetic indexes including gene heterozygosity (He), effective allele numbers (Ne) and polymorphism information content (PIC) were estimated as described by Nei and Roychoudhury (1974) and Botstein et al. (1980). Association analysis was carried out by the least-squares method

Table 1. Primers sequences (from 5′ to 3′), PCR conditions and restriction enzymes used for genotyping the polymorphisms in the current study.

SNP name*	Allele	PCR amplicon (bp)	Primers (5′ to 3′)	PCR conditions	Restriction enzyme	Reference
CAPN1 G316A			F: 5′GACTGGGGTCTCTGGACTT3′ R: 5′GGAACCTCTGGCTCTTGA3′	95 °C 5′ (95 °C 45 s, 63 °C 45 s, 72 °C 45 s) 35 cycles, 72 °C 5′	BtgI	Lisa and Di Stasio (2009)
CAPN1 V530I	A/G	787	F: 5′AGCGCAGGGACCCAGTGA3′ F: 5′TCCCCTGCCAGTTGTCTGAAG3′	95 °C 5′ (95 °C 1′, 63 °C 1′, 72 °C 1′) 35 cycles, 72 °C 5′	AvaII	Soria et al. (2010)
CAST S20T	G/C	624	F: 5′TGGGGCCCAATGACGCCATCGATG3′ R: 5′GGTGGAGCAGCACTTCTGATCACC3′	94 °C 5′ (94 °C 30 s, 62 °C 45 s, 72 °C 45 s) 32 cycles, 72 °C 5′	AluI	Juszczuk-Kubiak et al. (2004)
LEP A80V	C/T	458	F: 5′GGGAAGGGCAGAAAGATAG3′ R: 5′CCAAGCTCTCCAAGCTCTC3′	94 °C 2′ (94 °C 30 s, 57 °C 1′, 72 °C 30 s) 35 cycles, 72 °C 15′	HphI	Oztabak et al. (2010)
GHR S555G	G/A	342	F: 5′GCTAACTTCATCGTGGACAAC3′ R: 5′CTATGGCATGATTTTGTTCAG3′	95 °C 5′ (94 °C 45 s, 53 °C 30 s, 72 °C 50 s) 35 cycles, 72 °C 5′	AluI	Di Stasio et al. (2005)

CAPN1 – micromolar calcium-activated neutral protease 1. CAST – calpastatin. LEP – leptin. GHR – growth hormone receptor. * SNP names were used according to translation.

as applied in a general linear model (GLM) procedure of Minitab (MINITAB®, Pennsylvania, USA, v17.1.0) according to the following statistical model:

$$Y_{ijklmnop} = \mu + S_i + W_j + AG_k + BG_l + CG_m + DG_n + EG_o + e_{ijklmnop},$$

where $Y_{ijklmnop}$ represents the studied traits, μ the overall mean, S_i the fixed effect of season at the slaughter (i = winter, spring and summer), W_j the fixed effect of age at slaughter (j = 14–21 months), AG_k the fixed effect of the CAPN1 genotype for the G316A (k = CC, CG, GG), BG_l the fixed effect of the CAPN1 genotype for the V530I (l = AA, AG, GG), CG_m the fixed effect of the CAST genotype for the S20T (m = CC, CG, GG), DG_n the fixed effect of the LEP genotype for the A80V (n = CC, CT, TT), EG_o the fixed effect of the GHR genotype for the S555G (o = AA, AG, GG) and $e_{ijklmnop}$ the random residual effect.

The models in the present study were selected by evaluating the adjusted R^2 to compare the explanatory power of models with different numbers of predictors. Markers were initially evaluated using the significance of genotype effects for each trait. Afterwards, the interactions between CAPN1, CAST, LEP and GHR genotypes were added to the model and tested for significance. When significant associations were identified, the mean values for each genotype were contrasted using Tukey's test.

3 Results

3.1 Allele, genotype frequencies and population genetic indices

Two alleles and three genotypes in each SNP were found in the present study. The gene frequencies, population genetic indices including heterozygosity (He), effective allele numbers (Ne), polymorphism information content (PIC) and compatibility with the Hardy–Weinberg equilibrium (HWE) are shown in Table 2. A total of 400 individuals were used in

the study. HWE was tested for all alleles and genotypic frequencies and was compatible except for the LEP A80V and GHR S555G polymorphisms in Holstein population. Results indicated that the genotypes CC and AA, at the G316A and V530I markers of the CAPN1 gene, respectively, had relatively low frequencies but an adequate number of animals had these genotypes to estimate their association with phenotypic traits. The genotype frequency of GC (52 %) was rather high compared to the other two genotypes for the CAST S20T marker. In addition, the genotypic frequencies of TT (65 %) and AA (63.5 %) at the LEP A80V and GHR S555G markers, respectively, were remarkably high in the current study. The minor allele frequencies (MAF) ranged from 0.18 to 0.43, while Ne ranged from 1.42 to 1.96. Nevertheless, the CAPN1 V530I marker showed the low frequency of the allele A (0.18), resulting in low genetic variabilities of He (0.2952) and PIC (0.2516) compared with other SNPs showing relatively high values of He (ranged from 0.3648 to 0.4902) and PIC (ranged from 0.2982 to 0.3705).

3.2 Marker associations

Levels of significance, least-squares means, and standard errors are reported in Tables 3 and 4 for the effects of CAPN1 G316A and V530I, CAST S20T, LEP A80V and GHR S555G on carcass characteristics and meat quality. The marker CAPN1 G316A was highly associated with LW, HCW, CCW and LTL area ($P < 0.001$). In addition, the same marker at CAPN1 gene affected the BFT ($P < 0.01$). Moreover, CAPN1 G316A marker was also associated with CL, RL, RW and ICD at different levels of significance (Table 4). The results indicated that the GG genotype had noteworthy effects on the mentioned carcass measurement, supporting the results of evaluating live weight and carcass weights. The SNP V530I of the CAPN1 gene had a significant effect on pH at 24 h ($P < 0.05$). Animals with the AA genotype had higher values for pH of LT compared to the other two genotypes (Table 3). In addition, AA genotype was significantly associated with L^* value. Correspondingly, meat colour eval-

Table 2. Gene frequencies, population genetic indices and HWE test of polymorphisms in the *CAPN1*, *CAST*, *LEP* and *GHR* genes in a Holstein population.

Gene	*CAPN1*						*CAST*			*LEP*			*GHR*		
SNP	G316A				V530I			S20T			A80V			S555G	
Genotypes	CC	CG	GG	AA	AG	GG	GG	GC	CC	CC	CT	TT	AA	GA	GG
N	26	169	205	12	119	269	67	208	125	59	81	260	254	100	46
%	6.50	42.25	51.25	3.00	29.75	67.25	16.75	52.00	31.25	14.75	20.25	65.00	63.50	25.00	11.50
MAF		0.28			0.18			0.43			0.25			0.24	
He		0.4032			0.2952			0.4902			0.3750			0.3648	
Ne		1.68			1.42			1.96			1.60			1.57	
PIC		0.3219			0.2516			0.3705			0.3046			0.2982	
χ^2 (HWE)		1.28			0.07			1.55			83.97*			39.61*	

χ^2 (HWE) – Hardy–Weinberg equilibrium. χ^2 value – *N* – number of experimental bulls. MAF – minor allele frequency. He – heterozygosity. Ne – number of effective alleles. PIC – polymorphism information content. * $P < 0.001$, not consistent with HWE.

uation indicated that meat from GG animals had a higher L^* value.

The *CAST* S20T marker was effective on LW, ICD and b^* value ($P < 0.05$). Animals with the GG genotype had significantly higher values of LW and ICD but a lower b^* value compared to the other two genotypes (Tables 3 and 4). No association was observed between the *LEP* A80V and any of the phenotypic traits evaluated. The marker *GHR* S555G showed associations with pH at 24 h, a^* and h^*. Higher pH and lower a^* were observed in meat from individuals with GG genotype than those carrying the AA and GA genotypes. There was no association between any of the tested SNPs with DP (individual effects of the markers) and CW traits, nor was there any association with variation in C^* and MS ($P > 0.05$).

In this study, the *CAST* S20T × *LEP* A80V interaction was associated with DP ($P < 0.05$); the *CAPN1* V530I × *GHR* S555G and the *LEP* A80V × *GHR* S555G interactions were associated with variation in pH values ($P < 0.01$) and the *CAST* S20T × *CAPN1* V530I interaction was associated with RW ($P < 0.01$), as shown in Table 5.

4 Discussion

The primary objective of the current study was to determine whether DNA markers commonly studied (*CAPN1* G316A and V530I, *CAST* S20T, *LEP* A80V and *GHR* S555G) in various beef cattle populations could be applied in Holstein bulls, which comprise by far the most important share of meat production in Turkey. The present results showed a deviation from HWE for the *LEP* A80V and *GHR* S555G polymorphisms in Holstein population. Deviations from HWE at particular markers may be associated with population characteristics. Accordingly, this disequilibrium can be a result of inbreeding or indirect selection for these loci from the selection for milk production in the Holstein breed (Lacorte et al., 2006). Menezes et al. (2006) described a polymorphic locus as the frequency of the most common allele being lower than 0.95; accordingly all markers used in the present study

were polymorphic. Further, these markers are considered as mildly informative according to the classification reported by Botstein et al. (1980). *CAPN1*, *CAST*, *LEP* and *GHR* genes were chosen because they have been shown to be involved in the regulation of appetite, growth rate, carcass traits and meat quality in many beef cattle breeds, and our results indicated that *CAPN1*, *CAST* and *GHR* markers may be associated with carcass traits and meat quality. Among them, *CAPN1* G316A was highly associated with LW and carcass weights ($P < 0.001$) given that the G is the favourable allele for these traits. Animals with the GG genotype had +52.7 kg heavier LW and +34.4 kg heavier HCW compared to the CC genotype in the present study. In the literature, *CAPN1* G316A was evaluated as an effective marker on beef tenderness in several studies indicating that the C allele is associated with more tender meat (Casas et al., 2005; Corva et al., 2007; Gill et al., 2009; Miquel et al., 2009; Smith et al., 2009; Curi et al., 2010). It has been shown that, along with loci affecting beef tenderness, other loci associated with weaning weight and carcass weights were mapped to bovine chromosome 29 (Casas et al., 2005). Miquel et al. (2009) reported that final weight and average daily gain differentiated between the *CAPN1* G316A marker genotypes and that choosing animals with the favourable marker genotype (CC) for tenderness resulted in a selection of animals with lower average daily gain and final weight in Angus and Brangus steers. In addition, Pintos and Corva (2011) found significant associations between the same marker with birth weight, weaning weight and live weight recorded at 18 months of age in Angus cattle. Ardicli et al. (2017b) reported that homozygous animals for allele G at the *CAPN1* G316A marker reached the highest final weight and total weight gain in a shorter fattening period with higher average daily gain in Simmental bulls. Among the factors considered, it is possible that Warner–Bratzler shear force (WBSF) values show an association with LW and that selection for this marker may lead to changes in both traits. Conversely, Corva et al. (2007) and Tait et al. (2014) reported that final body weight and carcass weight were not affected by the *CAPN1* G316A marker genotypes. The breed

Table 3. Levels of significance, least-squares means, and standard errors for the effect of *CAPN1*, *CAST*, *LEP* and *GHR* on carcass traits and meat quality in a Holstein population.

Genotype	N	LW (kg)	HCW (kg)	CCW (kg)	DP (%)	BFT (mm)	LTLA (cm^2)	MS (1–9)	pH
CAPN1 G316A									
CC	26	437.8 ± 12.07^b	229.6 ± 6.83^b	225.6 ± 6.75^b	53.36 ± 0.47	2.27 ± 0.22^b	90.89 ± 3.01^b	2.76 ± 0.22	5.70 ± 0.04
CG	169	481.5 ± 7.61^a	259.0 ± 4.30^a	254.7 ± 4.26^a	53.54 ± 0.32	2.81 ± 0.14^a	101.92 ± 1.92^a	2.70 ± 0.13	5.70 ± 0.03
GG	205	493.5 ± 7.64^a	264.0 ± 4.32^a	259.7 ± 4.28^a	53.48 ± 0.30	2.98 ± 0.14^a	102.47 ± 1.94^a	2.61 ± 0.13	5.71 ± 0.03
		$P < 0.001$	$P < 0.001$	$P < 0.001$	NS	$P < 0.01$	$P < 0.001$	NS	NS
CAPN1 V530I									
AA	12	464.9 ± 15.80	248.0 ± 8.91	243.8 ± 8.84	53.06 ± 0.63	2.57 ± 0.29	100.98 ± 4.08	2.59 ± 0.28	5.89 ± 0.07^a
AG	119	475.3 ± 7.15	252.7 ± 4.07	248.5 ± 4.00	53.06 ± 0.28	2.74 ± 0.13	97.92 ± 1.79	2.71 ± 0.12	5.63 ± 0.02^b
GG	269	472.6 ± 6.58	251.9 ± 3.70	247.7 ± 3.68	53.21 ± 0.26	2.74 ± 0.12	96.37 ± 1.64	2.76 ± 0.11	5.59 ± 0.02^b
		NS	NS	NS	NS	NS	NS	NS	$P < 0.05$
CAST S20T									
GG	67	482.9 ± 9.54^a	256.4 ± 5.40	252.2 ± 5.34	52.87 ± 0.38	2.77 ± 0.17	97.83 ± 2.41	2.75 ± 0.16	5.71 ± 0.03
GC	208	464.7 ± 7.97^b	247.3 ± 4.51	243.1 ± 4.46	53.09 ± 0.31	2.66 ± 0.14	98.08 ± 2.02	2.66 ± 0.13	5.70 ± 0.02
CC	125	465.2 ± 8.16^b	249.0 ± 4.62	244.7 ± 4.57	53.37 ± 0.32	2.63 ± 0.15	99.36 ± 2.06	2.66 ± 0.14	5.69 ± 0.02
		$P < 0.05$	NS	NS	NS	NS	NS	NS	NS
LEP A80V									
CC	59	471.8 ± 9.63	251.8 ± 5.45	247.6 ± 5.39	53.17 ± 0.38	2.63 ± 0.17	97.61 ± 2.42	2.66 ± 0.16	5.66 ± 0.03
CT	81	472.0 ± 8.92	251.3 ± 5.05	247.1 ± 4.99	53.06 ± 0.35	2.74 ± 0.16	98.56 ± 2.27	2.80 ± 0.15	5.73 ± 0.03
TT	260	469.0 ± 8.06	249.6 ± 4.56	245.3 ± 4.51	53.11 ± 0.32	2.68 ± 0.14	99.10 ± 2.02	2.60 ± 0.14	5.72 ± 0.02
		NS	NS	NS	NS	NS	NS	NS	NS
GHR S555G									
AA	254	473.9 ± 7.48	252.5 ± 4.23	248.3 ± 4.18	53.18 ± 0.29	2.62 ± 0.13	98.06 ± 1.89	2.68 ± 0.13	5.57 ± 0.02^b
GA	100	461.9 ± 8.59	246.4 ± 4.86	242.2 ± 4.81	53.19 ± 0.34	2.74 ± 0.15	99.94 ± 2.16	2.77 ± 0.14	5.58 ± 0.02^b
GG	46	477.0 ± 10.40	253.7 ± 5.88	249.4 ± 5.82	52.98 ± 0.41	2.70 ± 0.19	97.27 ± 2.60	2.62 ± 0.17	5.97 ± 0.03^a
		NS	NS	NS	NS	NS	NS	NS	$P < 0.01$

N – number of experimental bulls. LW – live weight. HCW – hot carcass weight. CCW – chilled carcass weight. DP – dressing percentage. BFT – back fat thickness. LTLA – m. longissimus thoracis et lumborum area. MS – marbling score. NS – not significant. [a,b] Different superscripts within a column indicate significant difference.

Table 4. Levels of significance, least-squares means, and standard errors for the effect of *CAPN1*, *CAST*, *LEP* and *GHR* on carcass measurement and meat colour in a Holstein population.

Genotype	N	CL (cm)	RL (cm)	RW (cm)	CW (cm)	ICD (cm)	L^*	a^*	b^*	C^*	h^*
CAPN1 G316A											
CC	26	138.7 ± 0.72^c	63.67 ± 0.42^b	98.16 ± 0.52^b	80.53 ± 0.75	58.58 ± 0.57^b	35.15 ± 0.94	10.55 ± 0.68	9.15 ± 0.66	14.27 ± 0.61	0.70 ± 0.04
CG	169	139.9 ± 0.43^b	63.83 ± 0.25^b	100.12 ± 0.31^b	81.26 ± 0.45	59.30 ± 0.34^a	33.69 ± 0.59	11.10 ± 0.43	9.14 ± 0.42	14.79 ± 0.38	0.67 ± 0.03
GG	205	140.9 ± 0.43^a	64.31 ± 0.25^a	100.89 ± 0.31^a	81.70 ± 0.45	59.86 ± 0.34^a	33.98 ± 0.60	10.69 ± 0.43	9.29 ± 0.42	14.6 ± 0.39	0.70 ± 0.03
		$P < 0.001$	$P < 0.05$	$P < 0.01$	NS	$P < 0.05$	NS	NS	NS	NS	NS
CAPN1 V530I											
AA	12	140.0 ± 0.92	64.11 ± 0.54	98.63 ± 0.68	81.49 ± 0.96	59.52 ± 0.73	32.47 ± 1.24^b	10.78 ± 0.89	9.18 ± 0.87	14.60 ± 0.80	0.70 ± 0.06
AG	119	139.6 ± 0.40	63.79 ± 0.23	100.33 ± 0.35	80.91 ± 0.41	59.03 ± 0.31	34.99 ± 0.56^a	10.89 ± 0.40	8.95 ± 0.39	14.46 ± 0.36	0.67 ± 0.02
GG	269	139.9 ± 0.37	63.91 ± 0.22	100.20 ± 0.29	81.09 ± 0.39	59.20 ± 0.29	35.37 ± 0.51^a	10.67 ± 0.37	9.45 ± 0.36	14.61 ± 0.33	0.70 ± 0.02
		NS	NS	NS	NS	NS	$P < 0.05$	NS	NS	NS	NS
CAST S20T											
GG	67	140.0 ± 0.53	64.10 ± 0.31	100.22 ± 0.38	81.57 ± 0.56	59.83 ± 0.42^a	34.14 ± 0.74	10.98 ± 0.54	$8.80^b \pm 0.52$	14.47 ± 0.48	0.66 ± 0.03
GC	208	139.8 ± 0.45	63.87 ± 0.26	100.02 ± 0.32	80.94 ± 0.47	58.96 ± 0.36^b	33.82 ± 0.62	10.76 ± 0.45	$9.70^a \pm 0.44$	14.88 ± 0.40	0.72 ± 0.03
CC	125	139.7 ± 0.46	63.84 ± 0.27	$99.62 0.33$	80.98 ± 0.48	58.95 ± 0.37^b	34.87 ± 0.64	10.60 ± 0.46	$9.08^a \pm 0.45$	14.32 ± 0.41	0.69 ± 0.03
		NS	NS	NS	NS	$P < 0.05$	NS	NS	$P < 0.05$	NS	NS
LEP A80V											
CC	59	140.0 ± 0.54	64.28 ± 0.31	99.95 ± 0.38	81.19 ± 0.56	59.13 ± 0.42	34.15 ± 0.75	10.17 ± 0.54	8.86 ± 0.53	14.90 ± 0.49	0.69 ± 0.03
CT	81	139.5 ± 0.50	63.82 ± 0.29	99.79 ± 0.36	81.24 ± 0.52	59.26 ± 0.39	34.97 ± 0.70	11.32 ± 0.50	9.15 ± 0.49	14.94 ± 0.45	0.68 ± 0.04
TT	260	140.0 ± 0.46	63.71 ± 0.27	100.12 ± 0.33	81.07 ± 0.48	59.35 ± 0.36	33.70 ± 0.63	10.85 ± 0.45	9.58 ± 0.44	14.83 ± 0.41	0.71 ± 0.03
		NS	NS	NS	NS	NS	NS	NS	NS	NS	NS
GHR S555G											
AA	254	139.8 ± 0.43	63.76 ± 0.25	100.06 ± 0.31	81.38 ± 0.45	59.34 ± 0.34	34.50 ± 0.58	11.47 ± 0.42^a	8.71 ± 0.41	14.80 ± 0.38	0.64 ± 0.02^b
GA	100	139.6 ± 0.48	63.84 ± 0.28	99.93 ± 0.34	81.32 ± 0.50	59.56 ± 0.38	34.30 ± 0.67	10.49 ± 0.48^b	9.42 ± 0.47	14.50 ± 0.43	0.72 ± 0.03^a
GG	46	140.1 ± 0.58	64.22 ± 0.33	99.87 ± 0.41	80.79 ± 0.60	58.85 ± 0.45	34.02 ± 0.81	10.38 ± 0.58^b	9.46 ± 0.57	14.37 ± 0.53	0.72 ± 0.04^a
		NS	NS	NS	NS	NS	NS	$P < 0.01$	NS	NS	$P < 0.01$

N – number of experimental bulls. CL – carcass length. RL – rump length. RW – rump width. CW – chest width. ICD – inner chest depth. L^* (lightness), a^* (redness), b^* (yellowness), C^* (chroma), h^* (hue angle) – meat colour parameters. NS – not significant. [a,b,c] Different superscripts within a column indicate significant difference.

Table 5. Least-squares means, and standard errors for the significant associations of genotypic interactions with related traits in a Holstein population ($N = 400$).

CAST S20T × LEP A80V[a]			CAPN1 V530I × GHR S555G[b]			LEP A80V × GHR S555G[b]			CAPN1 V530I × CAST S20T[b]		
Genotype	N	DP (%)	Genotype	N	pH	Genotype	N	pH	Genotype	N	RW (cm)
CCCC	**19**	**54.33 ± 0.53**	AAAA	6	5.57 ± 0.06	CCAA	35	5.59 ± 0.03	AACC	5	94.74 ± 1.28
CCCT	29	54.02 ± 0.48	AAGA	3	5.54 ± 0.10	CCGA	14	5.55 ± 0.05	**AAGC**	**4**	**101.13 ± 1.03**
CCTT	77	53.13 ± 0.35	**AAGG**	**3**	**6.56 ± 0.17**	CCGG	10	5.83 ± 0.08	AAGG	3	100.04 ± 1.33
GCCC	31	53.07 ± 0.46	GAAA	61	5.57 ± 0.03	CTAA	55	5.52 ± 0.03	GACC	43	100.42 ± 0.39
GCCT	44	53.08 ± 0.40	GAGA	37	5.63 ± 0.03	CTGA	16	5.60 ± 0.04	GAGC	64	100.50 ± 0.36
GCTT	133	53.34 ± 0.35	GAGG	23	5.68 ± 0.04	**CTGG**	**10**	**6.09 ± 0.09**	GAGG	12	100.09 ± 0.66
GGCC	9	52.20 ± 0.72	GGAA	187	5.56 ± 0.02	TTAA	164	5.59 ± 0.03	GGCC	77	99.83 ± 0.33
GGCT	8	51.75 ± 0.75	GGGA	60	5.55 ± 0.03	TTGA	70	5.57 ± 0.04	GGGC	140	100.23 ± 0.31
GGTT	50	53.38 ± 0.42	GGGG	22	5.67 ± 0.04	TTGG	26	5.99 ± 0.06	GGGG	52	100.53 ± 0.40

N – number of experimental bulls. DP – dressing percentage. RW – rump width. [a] $P < 0.05$, [b] $P < 0.01$.

of the animals and the production procedures determine the slaughter weight and carcass traits (Sañudo et al., 2004) and inconsistent results about the associations between the same genetic markers with these traits can be evaluated as a common circumstance. Apart from these associations, carcass measurements (CL, RL, RW and ICD), BFT and LTL area were differentiated in the CAPN1 G316A genotypes in the current study. Animals with the CC genotype had significantly lower values of the traits mentioned above. In the literature, associations of CAPN1 G316A marker with carcass and growth traits have been shown in various cattle populations (Miquel et al., 2009; Pintos and Corva, 2011). To the best of our knowledge, this is the first study indicating that a portion of the differences in live and carcass weight and growth traits associated with this marker may be dependent on the body size (according to carcass measurement) of the individual. In the current study, animals with the CC genotype had 11.58 and 11.03 cm^2 lower mean value for LTL area, compared to GG and GC genotypes, respectively. These results indicated that selecting animals with the GG genotype induced higher values of LW, carcass weights and measurements and moreover higher BFT and LTL area as well. This knowledge may be useful for marker-assisted selection programmes. The trend of beef production in many countries has gradually changed from meat yield to meat quality. However, evaluating the ways to improve meat yield may be strategically important to achieve significant economic benefits in the countries with meat production deficit. Dairy-type animals, which yield a higher percent lean and less fat meat when compared with conventional beef breeds, could be exploited more commonly for beef production (Ntunde et al., 1977).

The CC genotype at the CAPN1 G316A marker was absent or the frequency was rather low to estimate their association with phenotypic traits in several studies conducted on various cattle populations (Curi et al., 2010; Soria et al., 2010; Allais et al., 2011; Li et al., 2013). However, satisfying results were obtained for the frequencies of the CC genotype (0.06) and

the C allele (0.28) in the current study. Bovine CAPN1 has been mapped to the telomeric end of BTA29 (Smith et al., 2000; Page et al., 2002), including considerable overlap of QTLs regulating not only beef tenderness but also growth (weaning weight, carcass weight) and feed efficiency (Casas et al., 2003; Pintos and Corva, 2011). Hence, it is possible to obtain novel genetic associations among these traits by evaluating this genomic region.

The amino acid variations may cause a functional change in the μ-calpain protease. The μ-calpain isoform including V530I and G316A, or both, may be a functional protein variation in myofibrillar proteolysis and resulted in differences in meat quality (Page et al., 2002). The present results indicated that the CAPN1 V530I marker was significantly associated with meat pH and L^* values. Choosing animals with the AA genotype at CAPN1 V530I marker may have resulted in a selection of animals with lower L^* but higher pH values. Improper meat values of pH > 5.8 were observed for the AA genotype. It is worth noting that ultimate pH is one of the most important indices of meat quality and high quality-meat has ultimate pH at the range of 5.4–5.6 (Pipek et al., 2003). Moreover, the correlations between meat pH and all colour parameters proved to be significant. For example, the increase in meat pH may cause the deterioration of all colour parameters (Węglarz, 2010). Additionally, environmental conditions that, for example, cause additional stress on animals in the pre-slaughter period lead to high postmortem pH values and should be avoided (Pipek et al., 2003; Węglarz, 2010). Further experiments with larger populations should be conducted in order to evaluate the consequences of selection for the marker on optimum meat pH and quality, especially for the colour parameters.

The present results indicated that the CAST S20T polymorphism was associated with LW and ICD in Holstein bulls ($P < 0.05$). Animals with the GG genotype displayed a higher mean LW and ICD than those with the CC and heterozygous genotypes. Studies on the association of the CAST S20T polymorphism with LW and carcass measurements in cattle are insufficient and larger populations may be needed to perform an adequate evaluation. Apart from the associations mentioned above, the b^* value indicating the degree of yellow appearance of meat from GC animals was higher than that estimated in the meat from those with other genotypes ($P < 0.05$). Consistent with our results, Juszczuk-Kubiak et al. (2004) reported that the meat from GC bulls had higher b^* values and was definitely darker. One possible connection between the *CAST* marker and meat colour may be through the calpain proteolytic system. The polymorphisms in genes related to calpain/calpastatin activity might affect meat colour traits, directly or indirectly (Li et al., 2013). Moreover, assessment of epistasis, genetic linkage and pleiotropy may be useful to consider different combinations of the polymorphisms associated with economically important quantitative traits.

The polymorphism A80V of the *LEP* gene has been reported as an effective marker on weight and average daily body weight gains (Kulig and Kmieć, 2009), marbling and carcass traits in feedlot cattle (Geary et al., 2003; Silva et al., 2014). However, there was no association between the *LEP* A80V polymorphism and any of the phenotypic traits evaluated in the current study. Studies on the association of the *LEP* A80V polymorphism with meat and carcass traits were conducted mostly in beef cattle populations. The reason for the lack of corresponding result in the present study may be the breed type.

Growth hormone influences growth and metabolism by interacting with GHR. The polymorphism S555G in exon 10 of the *GHR* gene has previously been shown to be associated with meat quality (Di Stasio et al., 2005; Reardon et al., 2010), growth and body size (Waters et al., 2011). Here, this SNP was associated with meat pH and meat colour parameters (a^* and h^*). Animals with GG genotype had higher pH values compared to the alternative genotypes. Conversely, Reardon et al. (2010) and Ribeca et al. (2010) found no association between this marker and meat pH. Environmental factors and pre-slaughter conditions of the abattoir may offer an explanation for variation in pH values reported in the different studies. The statistical analysis revealed that the effect of the GG genotype was significantly greater than the AA and heterozygote genotypes and that selection of animals with GG resulted in higher pH. In the study by Reardon et al. (2010), association of this marker at the *GHR* gene with L^* values of LTL and semimembranosus muscle was observed but a^* and b^* values were not differentiated between the *GHR* genotypes. Conversely, in this study, significant associations were found between *GHR* and a^* and h^* values but not L^* value. Meat derived from animals with AA genotype seemed to have higher a^* values (darker red colour) compared to those with other genotypes. One possible explanation about this association may be the effect of meat pH on colour parameters. High postmortem pH values influence meat colour negatively (Węglarz, 2010). In this study, AA genotype exhibited low pH and higher red values. Such

genotypic information may have potential for incorporation into management systems for meat quality.

In the current study, we hypothesized that interactions between polymorphisms of the selected genes may have significant effects on meat yield and quality and, thereby, may provide novel perspectives to evaluate the availability of these polymorphisms. In this case, the interactions among *CAPN1*, *CAST*, *LEP* and *GHR* genotypes were investigated to acquire possible associations between genotypic combinations and phenotypic traits. Among these, the *LEP* A80V × *CAST* S20T was associated with DP ($P < 0.05$). Animals with the CCCC genotype exhibited the highest value of DP. Our results suggested that the individual genetic effects of these SNPs were not statistically significant for DP. Interestingly, the combined effects of these polymorphisms indicated a significant association in the interaction analysis. Evaluating non-allelic gene interactions and linkage in the corresponding genomic regions may be required before considering them in marker-assisted selection. The *CAPN1* V530I × *GHR* S555G and the *LEP* A80V × *GHR* S555G interactions were associated with variation in pH values ($P < 0.01$). Animals with the AAGG genotype of the *CAPN1* V530I × *GHR* S555G and animals with CTGG genotype of the *LEP* A80V × *GHR* S555G had very high pH values (6.56 and 6.09, respectively). High ultimate pH of meat can result in DFD (dark, firm, dry) meat occurrence. Moreover, high pH is improper for sorting, confectioning and vacuum packaging of meat (Pipek et al., 2003; Węglarz, 2010). Selecting animals with these genotypes may have an inadequate impact to commercial markets. In addition, AAGC animals of the *CAPN1* V530I × *CAST* S20T exhibited significantly higher means for RW compared to alternative genotypes ($P < 0.01$). However, investigation of a larger number of animals would be desirable, especially because of the genotypes with low frequency.

Information of SNPs in *CAPN1*, *CAST*, *LEP* and *GHR* genes may provide very important clues on how many and which polymorphisms can explain genetic variations in carcass characteristics and meat quality to produce constant meat products as well as to maintain commercial lines. In the current study, the *CAPN1*, *CAST* and *GHR* genotypes confirmed significant associations with important traits in adequate numbers of animals. Thus, G316A and V530I markers in the bovine *CAPN1* gene can be used as genetic markers in breeding programmes to improve meat quantity and quality traits, respectively. Similarly, *GHR* S555G and *CAST* S20T markers may be evaluated in conventional selection procedures, regarding improvements of meat colour and carcass traits.

Holstein cattle, which are bred mainly for dairy purposes, carry a potential for improvement of beef production due to their genetic variability for beef traits. Therefore, the dual capacity of the Holstein breed may be used in several countries to cover the shortage in milk and meat production (Calo et al., 1973). Further genetic studies should be conducted for efficient selection procedures.

Consequently, information of polymorphisms at coding regions of the mentioned genes, genotypic interactions and significant genetic associations may be used to control meat production traits, concerning improvement of the genotypic structure.

5 Conclusions

This study focused on the associations of markers in the *CAPN1*, *CAST*, *LEP* and *GHR* genes with meat yield and quality traits. The present results confirm that the variation in these traits is influenced by the corresponding genotypes, except *LEP*. *CAPN1* G316A, a candidate marker for meat tenderness, also showed association with live weight and carcass characteristics. In addition, novel associations between genotypic interactions and phenotypic traits were observed. Further studies in larger populations consisting of various breeds with different genomic backgrounds should be conducted to evaluate valuable and useful marker associations.

Author contributions. SA, FB and HS designed the study and wrote the paper. SA, HS and DD performed the experiments. SA, BS and DD collected the samples. FB and SA analysed the data and did the statistical analysis.

Competing interests. The authors declare that they have no conflict of interest.

Acknowledgements. The authors would like to thank Banvit Incorporated Company for their help in managing the animals and collecting blood samples. The language assistance of I. Taci Cangul is appreciated. This research was supported by the "Uludag University Scientific Research Projects Funds" (grant no. UAPV-2011/20).

Edited by: Steffen Maak

References

Allais, S., Journaux, L., Leveziel, H., Payet-Duprat, N., Raynaud, P., Hocquette, J. F., Lepetit, J., Rousset, S., Denoyelle, C., Bernard-Capel, C., and Renand, G.: Effects of polymorphisms in the calpastatin and μ-calpain genes on meat tenderness in 3 French beef breeds, J. Anim. Sci., 89, 1–11, 2011.

Ardicli, S., Samli, H., Alpay, F., Dincel, D., Soyudal, B., and Balci, F.: Association of single nucleotide polymorphisms in the FABP4 gene with carcass characteristics and meat quality in Holstein bulls, Ann. Anim. Sci., 17, 117–130, 2017a.

Ardicli, S., Dincel, D., Samli, H., and Balci, F.: Effects of polymorphisms at *LEP, CAST, CAPN1, GHR, FABP4* and *DGAT1* genes on fattening performance and carcass traits in Simmental bulls, Arch. Anim. Breed., 60, 61–70, https://doi.org/10.5194/aab-60-61-2017, 2017b.

Blott, S., Kim, J. J., Moisio, S., Schmidt-Küntzel, A., Cornet, A., Berzi, P., Cambisano, N., Ford, C., Grisart, B., and Johnson, D.: Molecular dissection of a quantitative trait locus: a phenylalanine-to-tyrosine substitution in the transmembrane domain of the bovine growth hormone receptor is associated with a major effect on milk yield and composition, Genetics, 163, 253–266, 2003.

Bonilla, C., Rubio, M., Sifuentes, A., Parra-Bracamonte, G., Arellano, V., Méndez, M., Berruecos, J., and Ortiz, R.: Association of CAPN1 316, CAPN1 4751 and TG5 markers with bovine meat quality traits in Mexico, Genet. Mol. Res., 9, 2395–2405, 2010.

Botstein, D., White, R. L., Skolnick, M., and Davis, R. W.: Construction of a genetic linkage map in man using restriction fragment length polymorphisms, Am. J. Hum. Genet., 32, 314–331, 1980.

Buchanan, F. C., Fitzsimmons, C. J., Van Kessel, A. G., Thue, T. D., Winkelman-Sim, D. C., and Schmutz, S. M.: Association of a missense mutation in the bovine leptin gene with carcass fat content and leptin mRNA levels, Genet. Sel. Evol., 34, 105–116, 2002.

Burrow, H., Moore, S., Johnston, D., Barendse, W., and Bindon, B.: Quantitative and molecular genetic influences on properties of beef: a review, Anim. Reprod. Sci., 41, 893–919, 2001.

Calo, L. L., McDowell, R. E., Dale Van Vleck, L., and Miller, P. D.: Genetic aspects of beef production among Holstein-Friesians pedigree selected for milk production, J. Anim. Sci., 37, 676–682, 1973.

Casas, E., Shackelford, S. D., Keele, J. W., Koohmaraie, M., Smith, T. P., and Stone, R. T.: Detection of quantitative trait loci for growth and carcass composition in cattle, J. Anim. Sci., 81, 2976–2983, 2003.

Casas, E., White, S. N., Riley, D. G., Smith, T. P., Brenneman, R. A., Olson, T. A., Johnson, D. D., Coleman, S. W., Bennett, G. L., and Chase, C. C., Jr.: Assessment of single nucleotide polymorphisms in genes residing on chromosomes 14 and 29 for association with carcass composition traits in Bos indicus cattle, J. Anim. Sci., 83, 13–19, 2005.

Corva, P., Soria, L., Schor, A., Villarreal, E., Cenci, M. P., Motter, M., Mezzadra, C., Melucci, L., Miquel, C., and Paván, E.: Association of CAPN1 and CAST gene polymorphisms with meat tenderness in Bos taurus beef cattle from Argentina, Genet. Mol. Biol., 30, 1064–1069, 2007.

Curi, R. A., Chardulo, L. A. L., Giusti, J., Silveira, A. C., Martins, C. L., and de Oliveira, H. N.: Assessment of GH1, CAPN1 and CAST polymorphisms as markers of carcass and meat traits in Bos indicus and Bos taurus–Bos indicus cross beef cattle, Meat. Sci., 86, 915–920, 2010.

Di Stasio, L., Destefanis, G., Brugiapaglia, A., Albera, A., and Rolando, A.: Polymorphism of the GHR gene in cattle and relationships with meat production and quality, Anim. Genet., 36, 138–140, 2005.

Ekiz, B., Yilmaz, A., Ozcan, M., Kaptan, C., Hanoglu, H., Erdogan, I., and Yalcintan, H.: Carcass measurements and meat quality of Turkish Merino, Ramlic, Kivircik, Chios and Imroz lambs raised under an intensive production system, Meat. Sci., 82, 64–70, 2009.

Garrett, A., Rincon, G., Medrano, J., Elzo, M., Silver, G., and Thomas, M.: Promoter region of the bovine growth hormone receptor gene: single nucleotide polymorphism discovery in cattle and association with performance in Brangus bulls, J. Anim. Sci., 86, 3315–3323, 2008.

Ge, W., Davis, M., Hines, H., Irvin, K., and Simmen, R.: Association of single nucleotide polymorphisms in the growth hormone and growth hormone receptor genes with blood serum insulin-like growth factor I concentration and growth traits in Angus cattle, J. Anim. Sci., 81, 641–648, 2003.

Geary, T., McFadin, E., MacNeil, M., Grings, E., Short, R., Funston, R., and Keisler, D.: Leptin as a predictor of carcass composition in beef cattle, J. Anim. Sci., 81, 1–8, 2003.

Gill, J. L., Bishop, S. C., McCorquodale, C., Williams, J. L., and Wiener, P.: Association of selected SNP with carcass and taste panel assessed meat quality traits in a commercial population of Aberdeen Angus-sired beef cattle, Genet. Sel. Evol., 41, 36–47, 2009.

Green, M. R. and Sambrook, J.: Isolation of high-molecular-weight DNA from mammalian cells using proteinase K and phenol, in: Molecular Cloning: A Laboratory Manual, 4, Cold Spring Harbor Laboratory Press, Cold Spring Harbor, New York, USA, 47–48, 2012.

Guo, P., Zhao, Z., Yan, S., Li, J., Xiao, H., Yang, D., Zhao, Y., Jiang, P., and Yang, R.: PSAP gene variants and haplotypes reveal significant effects on carcass and meat quality traits in Chinese Simmental-cross cattle, Arch. Anim. Breed., 59, 461–468, https://doi.org/10.5194/aab-59-461-2016, 2016.

Haegeman, A., Van Zeveren, A., and Peelman, L.: New mutation in exon 2 of the bovine leptin gene, Anim. Genet., 31, 79–79, 2000.

Hilton, G., Tatum, J., Williams, S., Belk, K., Williams, F., Wise, J., and Smith, G.: An evaluation of current and alternative systems for quality grading carcasses of mature slaughter cows, J. Anim. Sci., 76, 2094–2103, 1998.

Juszczuk-Kubiak, E., Rosochacki, S. J., Wicinska, K., and Szreder, T. S.: A novel RFLP/AluI polymorphism of the bovine calpastatin (CAST) gene and its association with selected traits of beef, Anim. Sci. Pap. Rep., 22, 195–204, 2004.

Kaneda, M., Lin, B. Z., Sasazaki, S., Oyama, K., and Mannen, H.: Allele frequencies of gene polymorphisms related to economic traits in Bos taurus and Bos indicus cattle breeds, Anim. Sci. J., 82, 717–721, 2011.

Koohmaraie, M.: Biochemical factors regulating the toughening and tenderization processes of meat, Meat. Sci., 43, 193–201, 1996.

Kulig, H.: Association between leptin combined genotypes and milk performance traits of Polish Black-and-White cows, Arch. Tierzucht., 48, 547–554, 2005.

Kulig, H. and Kmiec, M.: Association between leptin gene polymorphisms and growth traits in Limousin cattle, Russ. J. Genet.+, 45, 738–741, 2009.

Kulig, H., Kmiec, M., and Wojdak-Maksymiec, K.: Associations between leptin gene polymorphisms and somatic cell count in milk of Jersey cows, Acta Vet. Brno., 79, 237–242, 2010.

Lagonigro, R., Wiener, P., Pilla, F., Woolliams, J., and Williams, J.: A new mutation in the coding region of the bovine leptin gene associated with feed intake, Anim. Genet., 34, 371–374, 2003.

Lacorte, G., Machado, M., Martinez, M., Campos, A., Maciel, R., Verneque, R., Teodoro, R., Peixoto, M., Carvalho, M., and

Fonseca, C.: DGAT1 K232A polymorphism in Brazilian cattle breeds, Genet. Mol. Res., 5, 475–482, 2006.

Li, X., Ekerljung, M., Lundstrom, K., and Lunden, A.: Association of polymorphisms at DGAT1, leptin, SCD1, CAPN1 and CAST genes with color, marbling and water holding capacity in meat from beef cattle populations in Sweden, Meat. Sci., 94, 153–158, 2013.

Liefers, S. C., te Pas, M. F., Veerkamp, R. F., Chilliard, Y., Delavaud, C., Gerritsen, R., and van der Lende, T.: Association of leptin gene polymorphisms with serum leptin concentration in dairy cows, Mamm. Genome., 14, 657–663, 2003.

Lisa, C. and Di Stasio, L.: Variability of μ-calpain and calpastatin genes in cattle, Ital. J. Anim. Sci., 8, 99–101, 2009.

Mazzucco, J. P., Melucci, L. M., Villarreal, E. L., Mezzadra, C. A., Soria, L., Corva, P., Motter, M. M., Schor, A., and Miquel, M. C.: Effect of ageing and μ-calpain markers on meat quality from Brangus steers finished on pasture, Meat. Sci., 86, 878–882, 2010.

Menezes, M. P. C., Martinez, A. M., Ribeiro, M. N., Pimenta Filho, E. C., and Bermejo, J. V. D.: Genetic characterization of Brazilian native breeds of goats using 27 markers microsatellites, Rev. Soc. Bras. Zootecn., 35, 1336–1341, 2006.

Miquel, M. C., Villarreal, E., Mezzadra, C., Melucci, L., Soria, L., Corva, P., and Schor, A.: The association of CAPN1 316 marker genotypes with growth and meat quality traits of steers finished on pasture, Genet. Mol. Biol., 32, 491–496, 2009.

Nei, M. and Roychoudhury, A.: Sampling variances of heterozygosity and genetic distance, Genetics, 76, 379–390, 1974.

Ntunde, B. N., Usborne, W. R., and Ashton, G. C.: Responses in meat characteristics of Holstein–Friesian males to castration and diet, Can. J. Anim. Sci., 57, 449–458, 1977.

Oztabak, K., Toker, N. Y., Un, C., and Akis, I.: Leptin gene polymorphisms in native Turkish cattle breeds, Kafkas. Univ. Vet. Fak., 16, 921–924, 2010.

Page, B. T., Casas, E., Heaton, M. P., Cullen, N. G., Hyndman, D. L., Morris, C. A., Crawford, A. M., Wheeler, T. L., Koohmaraie, M., Keele, J. W., and Smith, T. P.: Evaluation of single-nucleotide polymorphisms in CAPN1 for association with meat tenderness in cattle, J. Anim. Sci., 80, 3077–3085, 2002.

Pinto, L. F., Ferraz, J. B., Meirelles, F. V., Eler, J. P., Rezende, F. M., Carvalho, M. E., Almeida, H. B., and Silva, R. C.: Association of SNPs on CAPN1 and CAST genes with tenderness in Nellore cattle, Genet. Mol. Res., 9, 1431–1442, 2010.

Pintos, D. and Corva, P. M.: Association between molecular markers for beef tenderness and growth traits in Argentinian angus cattle, Anim. Genet., 42, 329–332, 2011.

Pipek, P., Haberl, A., and Jelenikova, J.: Influence of slaughterhouse handling on the quality of beef carcasses, Czech. J. Anim. Sci., 48, 371–378, 2003.

Reardon, W., Mullen, A., Sweeney, T., and Hamill, R.: Association of polymorphisms in candidate genes with colour, water-holding capacity, and composition traits in bovine M. longissimus and M. semimembranosus, Meat. Sci., 86, 270–275, 2010.

Ribeca, C., Bittante, G., Albera, A., Bonfatti, V., Maretto, F., and Gallo, L.: Investigation on variability of candidate genes for meat quality traits in Piemontese cattle, Ital. J. Anim. Sci., 8, 132–134, 2010.

Ribeca, C., Bonfatti, V., Cecchinato, A., Albera, A., Maretto, F., Gallo, L., and Carnier, P.: Association of polymorphisms in calpain 1, (mu I^{-1}) large subunit, calpastatin, and cathepsin D genes with meat quality traits in double-muscled Piemontese cattle, Anim. Genet., 44, 193–196, 2013.

Sagsoz, Y., Coban, O., Lacin, E., Sabuncuoglu, N., and Yildiz, A.: Esmer ve Şarole x Esmer Danaların Besi Performansıve Karkas Özellikleri/Fattening Performance and Carcass Features of Brown Swiss and Charolais x Brown Swiss calves, Atatürk Univ. Ziraat Fak. Derg., 36, 163–169, 2005.

Sañudo, C., Macie, E., Olleta, J., Villarroel, M., Panea, B., and Alberti, P.: The effects of slaughter weight, breed type and ageing time on beef meat quality using two different texture devices, Meat. Sci., 66, 925–932, 2004.

Schenkel, F., Miller, S., Jiang, Z., Mandell, I., Ye, X., Li, H., and Wilton, J.: Association of a single nucleotide polymorphism in the calpastatin gene with carcass and meat quality traits of beef cattle, J. Anim. Sci., 84, 291–299, 2006.

Silva, D., Crispim, B., Silva, L., Oliveira, J., Siqueira, F., Seno, L., and Grisolia, A.: Genetic variations in the leptin gene associated with growth and carcass traits in Nellore cattle, Genet. Mol. Res., 13, 3002–3012, 2014.

Smith, T., Thomas, M., Bidner, T., Paschal, J., and Franke, D.: Single nucleotide polymorphisms in Brahman steers and their association with carcass and tenderness traits, Genet. Mol. Res., 8, 39–46, 2009.

Smith, T. P., Casas, E., Rexroad, C. E., Kappes, S. M., and Keele, J. W.: Bovine CAPN1 maps to a region of BTA29 containing a quantitative trait locus for meat tenderness, J. Anim. Sci., 78, 2589–2594, 2000.

Soria, L. A., Corva, P., Huguet, M., Mino, S., and Miquel, M.: Bovine μ-calpain (CAPN1) gene polymorphisms in Brangus and Brahman bulls, J. Basic Appl. Sci., 21, 61–69, 2010.

Tait, R. G., Shackelford, S. D., Wheeler, T. L., King, D. A., Keele, J. W., Casas, E., Smith, T. P., and Bennett, G. L.: CAPN1, CAST, and DGAT1 genetic effects on preweaning performance, carcass quality traits, and residual variance of tenderness in a beef cattle population selected for haplotype and allele equalization, J. Anim. Sci., 92, 5382–5393, 2014.

Turkish Ministry of Food: Agriculture and Livestock Database, available at: www.turkvet.gov.tr (last access: 6 June 2016), 2015.

Turkish Statistical Institute: www.turkstat.gov.tr (last access: 1 August 2017), 2016.

Viitala, S., Szyda, J., Blott, S., Schulman, N., Lidauer, M., Mäki-Tanila, A., Georges, M., and Vilkki, J.: The role of the bovine growth hormone receptor and prolactin receptor genes in milk, fat and protein production in Finnish Ayrshire dairy cattle, Genetics, 173, 2151–2164, 2006.

Waters, S., McCabe, M., Howard, D., Giblin, L., Magee, D., MacHugh, D., and Berry, D.: Associations between newly discovered polymorphisms in the Bos taurus growth hormone receptor gene and performance traits in Holstein–Friesian dairy cattle, Anim. Genet., 42, 39–49, 2011.

Węglarz, A.: Meat quality defined based on pH and colour depending on cattle category and slaughter season, Czech. J. Anim. Sci., 55, 548–556, 2010.

Yeh, F. C., Yang, R. C., Boyle, T. B., Ye, Z., and Mao, J. X.: POPGENE, the user-friendly shareware for population genetic analysis, Molecular biology and biotechnology centre, University of Alberta, Canada, 2000.

Zhou, Y. and Jiang, H.: A milk trait-associated polymorphism in the bovine growth hormone receptor gene does not affect receptor signaling, J. Dairy. Sci., 89, 1761–1764, 2006.

25

Three-step in vitro maturation culture of bovine oocytes imitating temporal changes of estradiol-17β and progesterone concentrations in preovulatory follicular fluid

Minami Matsuo, Kazuma Sumitomo, Chihiro Ogino, Yosuke Gunji, Ryo Nishimura, and Mitsugu Hishinuma

Laboratory of Theriogenology, Joint Department of Veterinary Medicine, Faculty of Agriculture, Tottori University, 4-101 Koyama-minami, Tottori 680-8553, Japan

Correspondence to: Mitsugu Hishinuma (mhishi@muses.tottori-u.ac.jp)

Abstract. The objective of the article is to evaluate the effect of three-step in vitro maturation (IVM) culture system imitating estradiol-17β (E_2) and progesterone (P_4) concentrations in preovulatory follicles on in vitro bovine embryo production. The cumulus–oocyte complexes (COCs) were collected from follicles (2 to 8 mm in diameter) of bovine ovaries obtained from a local slaughterhouse. For IVM, the COCs were cultured for 22 h in a three-step system: (1) culture in medium 199, containing $700\,\mathrm{ng\,mL^{-1}}$ E_2 and $50\,\mathrm{ng\,mL^{-1}}$ P_4, for 5 h, followed by the medium containing $150\,\mathrm{ng\,mL^{-1}}$ E_2 and $150\,\mathrm{ng\,mL^{-1}}$ P_4 for 11 h, and then the medium containing $20\,\mathrm{ng\,mL^{-1}}$ E_2 and $300\,\mathrm{ng\,mL^{-1}}$ P_4 for 6 h (EP group); (2) culture in the medium containing $700\,\mathrm{ng\,mL^{-1}}$ E_2 for 5 h, followed by the medium containing $150\,\mathrm{ng\,mL^{-1}}$ E_2 for 11 h, and then the medium containing $20\,\mathrm{ng\,mL^{-1}}$ E_2 for 6 h (E group); or (3) culture in the medium containing $50\,\mathrm{ng\,mL^{-1}}$ P_4 for 5 h, followed by the medium containing $150\,\mathrm{ng\,mL^{-1}}$ P_4 for 11 h, and then the medium containing $300\,\mathrm{ng\,mL^{-1}}$ P_4 for 6 h (P group). The COCs were cultured in the medium containing $1000\,\mathrm{ng\,mL^{-1}}$ E_2 for 22 h (control group). After IVM, the COCs were co-incubated with sperm and further cultured. At 48 h after insemination, the cleavage rate of embryos was not different among the groups. At 192 h after insemination, the blastocyst formation rate of EP group was significantly higher than that of the other groups. The total cell number of blastocysts did not differ among the groups. In conclusion, these results demonstrate that the three-step IVM culture system of bovine oocytes imitating temporal changes of E_2 and P_4 concentrations in preovulatory follicular fluid improves the developmental potential of embryos in vitro.

1 Introduction

The development of techniques for the effective production of bovine preimplantation embryos from oocytes matured and fertilized in vitro is important for embryo transfer and basic scientific research. Temperature (Lenz et al., 1983), oxygen concentration (Hashimoto et al., 2000), nutrients (Takahashi and First, 1992; Kim et al., 1993) and hormones (Fukui et al., 1982; Silva and Knight, 2000; Beker et al., 2002; Mingoti et al., 2002) during in vitro maturation (IVM) culture are considered as factors to improve in vitro development of bovine oocytes. However, developmental rate of bovine oocytes to blastocyst in vitro was still lower than that of in vivo (Leibfried-Rutledge et al., 1987; Rizos et al., 2002). Recently, simulated physiological oocyte maturation (SPOM) system using an inhibitor of phosphodiesterase and activator of adenylate cyclase (cAMP-mediated pre-IVM) was proposed to mimic some characteristics of bovine oocyte maturation in vivo (Albuz et al., 2010). However, effectiveness of the SPOM system on bovine embryo production has not been established (Guimarães et al., 2015).

Change in steroid hormone concentration in preovulatory follicles is thought to relate oocyte maturation (Moor et al., 1980; Dieleman et al., 1983; Wrenzycki and Stinshoff, 2013). Ovulation occurs at 24 ± 1.4 h after luteinizing hormone (LH) surge in cattle (Dieleman et al., 1983). Estradiol-17β (E_2) concentration of bovine follicular fluid decreased from 798–1648 ng mL^{-1} at 0–5 h before LH surge to 180–256 ng mL^{-1} at 6–15 h after LH surge, and to 80–125 ng mL^{-1} at 20–23 h after LH surge, estimated as just before ovulation (Dieleman et al., 1983; Fortune and Hansel, 1985; Hansen et al., 1988; Li et al., 2007). Simultaneously, progesterone (P_4) concentration of bovine follicular fluid increased from 50–122 ng mL^{-1} at 0–5 h before LH surge to 41–150 ng mL^{-1} at 6–15 h after LH surge, and to 280–475 ng mL^{-1} at 20–23 h after LH surge (Dieleman et al., 1983; Fortune and Hansel, 1985; Hansen et al., 1988; Li et al., 2007; Fortune et al., 2009).

In many studies, supplementation of 1000 ng mL^{-1} E_2 to the IVM medium was effective for resumption of meiosis and promotion of maturation of bovine oocytes (Fukui et al., 1982; Fukushima and Fukui, 1985; Younis et al., 1989; Beker et al., 2002). P_4 was added to the IVM medium for bovine oocytes at a concentration of 50–5000 ng mL^{-1}, and various results were reported on embryo production (Silva and Knight, 2000; Aparicio et al., 2011; Syoji et al., 2014). Concentration of P_4 in the IVM medium was fixed during the culture except with Syoji et al. (2014), and E_2 and P_4 concentrations in preovulatory follicular fluid were not considered for medium preparation in the previous studies. In porcine embryo production, three-step IVM culture system was developed to mimic hormonal changes observed in vivo (Kawashima et al., 2008). Porcine oocytes were pre-cultured with follicle-stimulating hormone (FSH) and E_2 for 10 h, after which time 10 ng mL^{-1} P_4 was added for another 10 h. The oocytes were then transferred to fresh medium containing LH, epidermal growth factor (EGF) and 100 ng mL^{-1} P_4. Similar culture system could be applied to bovine IVM, although duration of bovine IVM culture is about half compared to porcine IVM culture. Therefore, temporal changes of E_2 and P_4 concentration to mimic changes in preovulatory follicular fluid should be examined for IVM of bovine oocytes.

The objective of the present study was to evaluate the effect of three-step IVM culture system imitating temporal changes of steroid hormone concentrations in preovulatory follicular fluid on bovine embryo production in vitro. Bovine oocytes were cultured in the IVM medium containing various concentrations of E_2 and P_4 for 22 h, co-incubated with sperm and further cultured to examine developmental competence to blastocyst.

2 Materials and methods

2.1 Collection of cumulus–oocyte complexes (COCs)

Ovaries were obtained from Japanese black, Holstein and their crossbred cows at a local slaughterhouse in Tottori prefecture, Japan, and transported to laboratory in sterile physiological saline at 20 °C within 4–5 h. The COCs were collected from follicles (2 to 8 mm in diameter) by aspiration using an 18-gauge needle attached to a 10 mL syringe and washed with HEPES-buffered medium 199 (31100-035, Gibco, Grand Island, NY, USA). Intact COCs (normal oocytes with ooplasm > 120 μm in diameter and surrounded with more than three layers of unexpanded cumulus cells) were selected under a stereomicroscope (SMZ 645-3, Nikon, Japan) and used for further experiment.

2.2 IVM

Medium 199 Earle's salts (12340-030, Gibco) supplemented with 10 % (v/v) fetal bovine serum (FBS, 26140-087, Life Technologies) inactivated at 56 °C for 30 min, 0.2 mM sodium pyruvate (P5280, Sigma-Aldrich, St. Louis, MO, USA), 50 μg mL^{-1} gentamicin sulfate (G3632, Sigma-Aldrich) and 20 μg mL^{-1} FSH from porcine pituitary (F2293, Sigma-Aldrich) was used as a basal medium for IVM. Thirty COCs were cultured in 300 μL of IVM medium supplemented with steroid hormone in a well of a 48-well dish (150687, Thermo Fisher Scientific, USA) for 22 h at 39 °C in a humidified atmosphere of 5 % CO_2 in air.

2.3 In vitro fertilization (IVF) and in vitro culture (IVC)

Frozen–thawed semen from a Japanese black bull was used for IVF. The motile sperm were separated by centrifugation through a percoll gradient as described previously (Takahashi et al., 1996). Briefly, percoll (Pharmacia BioProcess, Uppsala, Sweden) was diluted to 45 and 90 % (v/v) with modified Brackett and Oliphant isotonic medium (Brackett and Oliphant, 1975) without bovine serum albumin (BSA, BO medium), and the 2 mL of 90 % percoll solution were overlaid with 2 mL of 45 % percoll solution (two-layer percoll gradient). Frozen–thawed semen was placed on top of the percoll gradient and centrifuged at $700 \times g$ for 20 min. The top layers were removed and remaining sperm pellet (at the bottom of the 90 % percoll solution) was washed using BO medium by centrifugation at $500 \times g$ for 5 min. The sperm pellet was resuspended in the same medium to yield a concentration 10×10^6 sperm mL^{-1}. A 50 μL sperm suspension was added to the 50 μL droplet of BO medium supplemented with 6 mg mL^{-1} BSA (A6003, Sigma-Aldrich) and 5 mM theophylline (T1633, Sigma-Aldrich). After IVM culture, 10 to 15 COCs were co-incubated with sperm (5×10^6 sperm mL^{-1}) in a 100 μL droplet of BO medium containing 3 mg mL^{-1} BSA and 2.5 mM theophylline covered with liquid paraffin for 18 h at 39 °C in a humidified atmosphere of 5 % CO_2 in air.

After insemination, the cumulus cells were removed from oocytes by vortexing in BO medium and washed three times in 100 µL of IVC medium that is a modified synthetic oviduct fluid with 3 mg mL^{-1} BSA instead of polyvinyl alcohol (Takahashi et al., 1996). The embryos were cultured for 192 h (8 days after start of IVF culture) in groups of 10 to 15 presumptive zygotes per 40 µL droplet of IVC medium covered with liquid paraffin at 39 °C in a humidified atmosphere of 5 % CO_2, 5 % O_2 and 90 % N_2.

2.4 Examination of embryo development

At 48 h after insemination (day 2 of IVC), cleavage rate of embryos was examined by observation of the embryos under a stereomicroscope. Developmental competence to blastocyst was evaluated at 192 h after insemination (day 8 of IVC). The embryos that formed blastocoele were classified into early blastocyst, blastocyst, expanded blastocyst and hatched blastocyst (Linder and Wight, 1983). The total cell number of blastocysts was counted using the air-drying method (Takahashi and First, 1992). Briefly, blastocysts were placed in a hypotonic solution (0.9 % sodium citrate supplemented with 0.3 % FBS) for 15 min. They were then treated with fixative (methanol : acetic acid = 3 : 1) and dried on a glass slide. After staining with 3 % Giemsa solution, the total cell number of each blastocyst was counted under a bright-field microscope (CX41, Nikon).

2.5 Experimental design

The IVM basal medium supplemented with 1000 ng mL^{-1} E_2 (E8875, Sigma-Aldrich) was used as control medium (Fukui et al., 1982; Fukushima and Fukui, 1985; Younis et al., 1989; Beker et al., 2002). The COCs were cultured in the control medium for 22 h (control group). To imitate temporal change of E_2 and P_4 concentrations in bovine follicular fluid corresponding to 3 to 25 h after LH surge, the COCs were cultured in the basal medium containing 700 ng mL^{-1} E_2 and 50 ng mL^{-1} P_4 (P8783, Sigma-Aldrich) for 5 h, then moved to fresh medium containing 150 ng mL^{-1} E_2 and 150 ng mL^{-1} P_4 and cultured for 11 h. Finally, the COCs were moved to fresh medium containing 20 ng mL^{-1} E_2 and 300 ng mL^{-1} P_4 and cultured for 6 h (EP group). To imitate temporal change of only E_2 concentrations in the follicular fluid, the COCs were cultured in the medium containing 700 ng mL^{-1} E_2 for 5 h, followed by the medium containing 150 ng mL^{-1} E_2 for 11 h, and then the medium containing 20 ng mL^{-1} E_2 for 6 h (E group). To imitate temporal change of only P_4 concentrations in the follicular fluid, the COCs were cultured in the medium containing 50 ng mL^{-1} P_4 for 5 h, followed by the medium containing 150 ng mL^{-1} P_4 for 11 h, and then the medium containing 300 ng mL^{-1} P_4 for 6 h (P group). These three-step IVM culture systems are

Table 1. Three-step IVM culture of bovine oocytes.

Experimental groups	Steroid hormones	Concentration of steroid hormones at each time of IVM culture (ng mL^{-1})		
		0–5 h	5–16 h	16–22 h
Control	E_2	1000	1000	1000
	P_4	0	0	0
EP	E_2	700	150	20
	P_4	50	150	300
E	E_2	700	150	20
	P_4	0	0	0
P	E_2	0	0	0
	P_4	50	150	300

Periods of IVM culture at 0–5, 5–16 and 16–22 h correspond to 22–17, 17–6 and 6–0 h before ovulation, respectively. IVM medium was exchanged at 5 h and 16 h of IVM culture.

shown in Table 1. In preliminary experiment using the control medium, medium exchange at 5 and 16 h of IVM culture did not affect subsequent embryo development to blastocysts. After IVM culture, the COCs were further cultured for IVF and IVC. The cleavage rate, blastocyst formation rate and total cell number of blastocysts were examined.

2.6 Statistical analysis

Experiments were replicated five times. Values are presented as means ± standard deviation (SD). Statistical analysis was performed using StatView software (Abacus Concepts, Berkeley, CA, USA). The data of cleavage rate and blastocyst formation rate were arcsine-transformed before the analysis. The data were analyzed by one-way ANOVA followed by the Bonferroni post hoc test. Difference with $P < 0.05$ was considered significant.

3 Results

The results of IVF-IVC after three-step IVM culture of bovine oocytes with various concentrations of E_2 and P_4 are shown in Table 2. At 48 h after insemination (day 2), the cleavage rate of embryos was not different among the groups. At 192 h after insemination (day 8), the blastocyst formation rate of EP group was higher than that of the other groups ($P < 0.05$). The blastocyst formation rate was not different among E, P and control groups. Percentages of early blastocyst, blastocyst, expanded blastocyst and hatched blastocyst were similar between EP and control groups. The percentage of embryos reaching the early blastocyst stage tended to be higher in E group without significant differences, while the percentage of embryos reaching the hatched blastocyst stage was highest in P group and significantly higher than E

Table 2. The effect of three-step IVM culture imitating temporal changes of E_2 and P_4 concentrations in preovulatory follicles on in vitro maturational and developmental competence of bovine oocytes.

Experimental groups	Total number of oocytes cultured for IVC	Cleavage rate[1] (%)	Blastocyst rate[2] (%)	Percentage of each blastocyst stage[3] (total number of blastocyst)				Total cell number of blastocysts[3] (number of blastocysts examined)	
				E	B	EX	H	B	EX
Control	409	81.2 ± 3.3	30.8 ± 1.8^B	28.1 ± 15.3 (34)	19.3 ± 12.7 (25)	22.6 ± 13.4 (29)	30.0 ± 19.1^{AB} (38)	109.2 ± 30.4 (25)	161.1 ± 40.5 (29)
EP	147	83.7 ± 6.4	39.4 ± 4.4^A	32.3 ± 17.6 (19)	13.5 ± 10.0 (8)	26.0 ± 6.2 (15)	27.8 ± 10.3^{AB} (16)	104.3 ± 18.2 (8)	166.9 ± 36.9 (15)
E	140	79.9 ± 2.4	30.0 ± 2.1^B	43.9 ± 22.3 (18)	23.3 ± 20.8 (10)	18.6 ± 23.1 (8)	13.9 ± 12.1^B (6)	127.8 ± 44.6 (10)	187.1 ± 27.3 (8)
P	148	75.7 ± 4.2	29.0 ± 2.7^B	14.4 ± 15.9^b (6)	11.4 ± 11.9^b (5)	22.4 ± 17.0^{ab} (10)	51.7 ± 24.4^{Aa} (22)	111.8 ± 21.4 (5)	138.5 ± 35.4 (10)

Percentage data are means ± SD of five replicates. E_2 and P_4 concentrations in the three-step IVM culture are shown in Table 1. [1] At 48 h after insemination (day 2). No. of 2–4 cell embryos / no. of oocytes cultured. [2] At 192 h after insemination (day 8). No. of blastocysts / no. of oocytes cultured. [3] At 192 h after insemination (day 8). E: early blastocyst; B: blastocyst; EX: expanded blastocyst; H: hatched blastocyst. [A,B] Values with different superscripts within a column differ significantly ($P < 0.05$). [a,b] Values with different superscripts within a row among percentages of blastocyst stages differ significantly ($P < 0.05$).

group ($P < 0.05$). The total cell number of blastocysts and expanded blastocysts did not differ among the groups.

4 Discussion

In the present study, we improved the developmental competence of bovine oocytes to blastocyst stage using a three-step IVM culture system imitating temporal changes of E_2 and P_4 concentrations in preovulatory follicular fluid. The three-step IVM culture system imitating either E_2 or P_4 concentrations in preovulatory follicular fluid (E and P groups) did not affect oocyte development to blastocyst, so that both E_2 and P_4 stimulations during IVM culture are necessary for enhancement of bovine oocyte development. In porcine embryo production, Kawashima et al. (2008) developed a three-step IVM culture system to enhance the development to blastocyst stage after IVF and IVC. Porcine oocytes were precultured with FSH and $100\,\text{ng}\,\text{mL}^{-1}$ E_2 for 10 h, after which time $10\,\text{ng}\,\text{mL}^{-1}$ P_4 was added for another 10 h. After 20 h, COCs were moved to fresh medium containing LH, EGF and $100\,\text{ng}\,\text{mL}^{-1}$ P_4. Although neither anterior pituitary hormones nor growth factors were considered in the present three-step IVM culture system of bovine oocytes, both culture systems showed effectiveness of sequential stimulation of steroid hormones during IVM.

Various concentrations of P_4 (50–$5000\,\text{ng}\,\text{mL}^{-1}$) were added to bovine IVM medium (Silva and Knight, 2000; Aparicio et al., 2011; Syoji et al., 2014). IVM culture of bovine oocytes with 50 or $100\,\text{ng}\,\text{mL}^{-1}$ P_4 did not affect development to the blastocyst (Aparicio et al., 2011), while that with $94\,\text{ng}\,\text{mL}^{-1}$ P_4 decreased the blastocyst formation rate (Silva and Knight, 2000). In the present study, IVM culture of bovine oocytes with 50–$300\,\text{ng}\,\text{mL}^{-1}$ P_4 (P group) did not affect the blastocyst formation rate. However, supplementation with 1000 or $5000\,\text{ng}\,\text{mL}^{-1}$ P_4 during the last half of IVM culture increased the blastocyst formation rate of bovine oocytes (Syoji et al., 2014). Exposure of bovine

oocytes to P_4 at concentration in the preovulatory follicles may not affect or adversely affect the developmental competence in vitro, but P_4 at extremely high concentration may enhance it. In the present study, the percentage of embryos reaching the early blastocyst stage tended to be higher in E group without significant differences, whereas the percentage of embryos reaching the hatched blastocyst stage was highest in P group and significantly higher than E group ($P < 0.05$). Exposure of bovine oocytes to P_4 at the preovulatory level might accelerate the embryo development in vitro.

In the present study, the $E_2 + P_4$ treatment during IVM (EP group) seems advantageous in comparison to the other treatments, because of the highest rate of blastocyst. However, the actual differences are rather small and are nearly in the range of normal blastocyst rate after IVM. These results suggest that the other system in addition to steroid hormone action exists for improving in vitro developmental competence of bovine oocytes, for example, pituitary hormone action as shown in porcine oocytes (Kawashima et al., 2008). The present results showed the highest rate of most advanced (i.e., hatched) blastocysts in P group. This result suggests a possibility that the additional treatment rather than steroid hormones, such as pituitary hormones, may improve the developmental rate of P_4 treatment, even though the normal blastocyst rate was higher in EP group than P group.

Bovine COCs secreted E_2 and P_4 during IVM (Mingoti et al., 2002; Schoenfelder et al., 2003; Salhab et al., 2011; Blaschka et al., 2015). During IVM of bovine COCs ($30\,\text{COCs}/300\,\mu\text{L}$ IVM medium) for 24 h, P_4 concentration in the medium significantly increased (3.3 to $10.4\,\text{ng}\,\text{mL}^{-1}$), but E_2 concentration did not change (52.8 to $74.7\,\text{pg}\,\text{mL}^{-1}$; Blaschka et al., 2015). Therefore, the influence of P_4 secreted by COCs should be considered when P_4 was added to the IVM medium. However, in the present three-step IVM culture system using 20–$700\,\text{ng}\,\text{mL}^{-1}$ E_2 and 50–$300\,\text{ng}\,\text{mL}^{-1}$ P_4, E_2 and P_4 secretion by bovine COCs can be ignored due to their small amount and removal by medium exchange twice during the culture.

Supplementation with E_2 and P_4 to the IVM medium affected gene expressions of proteins by combining with each receptor in the nucleus passing through a cumulus and oocyte plasma membrane (Bain et al., 2007). Beker-van Woudenberg et al. (2004) supplemented with E_2-BSA conjugate, which is a non-cell-permeable E_2, to the IVM medium and demonstrated that $1000\,\mathrm{ng\,mL^{-1}}$ (3.67 µM) E_2 had detrimental effects on maturation of bovine "denuded" oocyte. Aparicio et al. (2011) reported the presence of genomic and nongenomic P_4 receptors (PRs) in bovine COCs both before and after IVM. The protein expression of genomic nPR-A and nPR-B and nongenomic mPRα and mPRβ increased in cumulus cells after IVM, whereas genomic nPR-A and nongenomic mPRα and mPRβ decreased in oocytes after IVM, indicating a different role for each receptor in bovine oocyte maturation. In the present three-step IVM culture system of bovine oocytes, E_2 and P_4 added to the medium may act through various receptors in cumulus cells and oocytes.

In conclusion, the present study demonstrates that three-step IVM culture system of bovine oocytes imitating temporal changes of E_2 and P_4 concentrations in preovulatory follicular fluid improves the developmental potential of embryos in vitro. Further experiments are needed to evaluate the effect of temporal changes of E_2 and P_4 concentrations on maturational mechanism in bovine oocytes.

Author contributions. All authors contributed to the work described in the manuscript, and all take responsibility for it. MM, KS, CO, YG and RN as co-authors made a significant contribution to the conception and design of the experiments, as well as the analysis and interpretation of the data. Moreover, MH participated in drafting the article as well as reviewing and revising it for contents.

Competing interests. The authors declare that they have no conflict of interest.

Acknowledgements. We would like to thank staff of the meat inspection office in Tottori prefecture for collection of bovine ovaries. We thank staff of the livestock experimental station in Tottori prefecture for provision of bovine frozen semen. We also thank Shambhu Shah for English proofreading of the manuscript.

Edited by: Manfred Mielenz

References

Albuz, F. K., Sasseville, M., Lane, M., Armstrong, D. T., Thompson, J. G., and Gilchrist, R. B.: Simulated physiological oocyte maturation (SPOM): a novel in vitro maturation system that substantially improves embryo yield and pregnancy outcomes, Hum. Reprod., 25, 2999–3011, https://doi.org/10.1093/humrep/deq246, 2010.

Aparicio, I. M., Garcia-Herreros, M., O'Shea, L. C., Hensey, C., Lonergan, P., and Fair, T.: Expression, regulation, and function of progesterone receptors in bovine cumulus oocyte complexes during in vitro maturation, Biol. Reprod., 84, 910–921, https://doi.org/10.1095/biolreprod.110.087411, 2011.

Bain, D. L., Heneghan, A. F., Connaghan-Jones, K. D., and Miura, M. T.: Nuclear receptor structure: implications for function, Annu. Rev. Physiol., 69, 201–220, https://doi.org/10.1146/annurev.physiol.69.031905.160308, 2007.

Beker, A. R., Colenbrander, B., and Bevers, M. M.: Effect of 17β -estradiol on the in vitro maturation of bovine oocytes, Theriogenology, 58, 1663–1673, https://doi.org/10.1016/S0093-691X(02)01082-8, 2002.

Beker-van Woudenberg, A. R., van Tol, H. T., Roelen, B. A., Colenbrander, B., and Bevers, M. M.: Estradiol and its membrane-impermeable conjugate (estradiol-bovine serum albumin) during in vitro maturation of bovine oocytes: effects on nuclear and cytoplasmic maturation, cytoskeleton, and embryo quality, Biol. Reprod., 70, 1465–1474, https://doi.org/10.1095/biolreprod.103.025684, 2004.

Blaschka, C., Stinshoff, H., Poppicht, F., and Wrenzycki, C.: Temporal pattern of steroid hormone concentrations in medium used for IVM of bovine oocytes, Reprod. Dom. Anim., 50, 23–24 (abstr.), https://doi.org/10.1071/RDv27n1Ab275, 2015.

Brackett, B. G. and Oliphant, G.: Capacitation of rabbit spermatozoa in vitro, Biol. Reprod., 12, 260–274, 1975.

Dieleman, S. J., Kruip, T. A., Fontijne, P., Jong, W. H., and Weyden, G. C.: Changes in oestradiol, progesterone and testosterone concentrations in follicular fluid and in the micromorphology of preovulatory bovine follicles relative to the peak of luteinizing hormone, J. Endocrinol., 97, 31–42, https://doi.org/10.1677/joe.0.0970031, 1983.

Fortune, J. E. and Hansel, W.: Concentrations of steroids and gonadotropins in follicular fluid from normal heifers and heifers primed for superovulation, Biol. Reprod., 32, 1069–1079, 1985.

Fortune, J. E., Willis, E. L., Bridges, P. J., and Yang, C. S.: The periovulatory period in cattle: progesterone, prostaglandins, oxytocin and ADAMTS proteases, Anim. Reprod., 6, 60–71, 2009.

Fukui, Y., Fukushima, M., Terawaki, Y., and Ono, H.: Effect of gonadotropins, steroids and culture media on bovine oocyte maturation in vitro, Theriogenology, 18, 161–175, https://doi.org/10.1016/0093-691X(82)90100-5, 1982.

Fukushima, M. and Fukui, Y.: Effects of gonadotropins and steroids on the subsequent fertilizability of extrafollicular bovine oocytes cultured in vitro, Anim. Reprod. Sci., 9, 323–332, https://doi.org/10.1016/0378-4320(85)90061-2, 1985.

Guimarães, A. L., Pereira, S. A., Leme, L. O., and Dode, M. A.: Evaluation of the simulated physiological oocyte maturation system for improving bovine in vitro embryo production, Theriogenology, 83, 52–57, https://doi.org/10.1016/j.theriogenology.2014.07.042, 2015.

Hansen, T. R., Randel, R. D., and Welsh, T. H.: Granulosa cell steroidogenesis and follicular fluid steroid concentrations after the onset of oestrus in cows, J. Reprod. Fertil., 84, 409–416, https://doi.org/10.1530/jrf.0.0840409, 1988.

Hashimoto, S., Minami, N., Takakura, R., Yamada, M., Imai, H., and Kashima, N.: Low oxygen tension during in vitro maturation is beneficial for supporting the subsequent development of bovine cumulus-oocyte complexes, Mol. Reprod. Dev., 57, 353–360, 2000.

Kawashima, I., Okazaki, T., Noma, N., Nishibori, M., Yamashita, Y., and Shimada, M.: Sequential exposure of porcine cumulus cells to FSH and/or LH is critical for appropriate expression of steroidogenic and ovulation-related genes that impact oocyte maturation in vivo and in vitro, Reproduction, 136, 9–21, https://doi.org/10.1530/REP-08-0074, 2008.

Kim, J. H., Niwa, K., Lim, J. M., and Okuda, K.: Effects of phosphate, energy substrates, and amino-acids on development of in vitro-matured, in vitro-fertilized bovine oocytes in a chemically defined, protein-free culture-medium, Biol. Reprod., 48, 1320–1325, https://doi.org/10.1095/biolreprod48.6.1320, 1993.

Leibfried-Rutledge, M. L., Critser, E. S., Eyestone, W. H., Northey, D. L., and First, N. L.: Development potential of bovine oocytes matured in vitro or in vivo, Biol. Reprod., 36, 376–383, https://doi.org/10.1095/biolreprod36.2.376, 1987.

Lenz, R. W., Ball, G. D., Leibfried, M. L., Ax, R. L., and First, N. L.: In vitro maturation and fertilization of bovine oocytes are temperature-dependent processes, Biol. Reprod., 29, 173–179, https://doi.org/10.1095/biolreprod29.1.173, 1983.

Li, Q., Jimenez-Krassel, F., Bettegowda, A., Ireland, J. J., and Smith, G. W.: Evidence that the preovulatory rise in intrafollicular progesterone may not be required for ovulation in cattle, J. Endocrinol., 192, 473–483, https://doi.org/10.1677/JOE-06-0020, 2007.

Linder, G.M. and Wight, R.W. Jr.: Bovine embryo morphology and evaluation, Theriogenology, 20, 407–416, https://doi.org/10.1016/0093-691X(83)90201-7, 1983.

Mingoti, G. Z., Garcia, J. M., and Rosa-e-Silva, A. A.: Steroidogenesis in cumulus cells of bovine cumulus-oocyte-complexes matured in vitro with BSA and different concentrations of steroids, Anim. Reprod. Sci., 69, 175–186, https://doi.org/10.1016/S0378-4320(01)00187-7, 2002.

Moor, R. M., Polge, C., and Willadsen, S. M.: Effect of follicular steroids on the maturation and fertilization of mammalian oocytes, J. Embryol. Exp. Morphol., 56, 319–335, 1980.

Rizos, D., Ward, F., Duffy, P., Boland, M. P., and Lonergan P.: Consequences of bovine oocyte maturation, fertilization or early embryo development in vitro versus in vivo: Implications for blastocyst yield and blastocyst quality, Mol. Reprod. Dev., 61, 234–248, https://doi.org/10.1002/mrd.1153, 2002.

Salhab, M., Tosca, L., Cabau, C., Papillier, P., Perreau, C., Dupont, J., Mermillod, P., and Uzbekova, S.: Kinetics of gene expression and signaling in bovine cumulus cells throughout IVM in different mediums in relation to oocyte developmental competence, cumulus apoptosis and progesterone secretion, Theriogenology, 75, 90–104, https://doi.org/10.1016/j.theriogenology.2010.07.014, 2011.

Schoenfelder, M., Schams, D., and Einspanier, R.: Steroidogenesis during in vitro maturation of bovine cumulus oocyte complexes and possible effects of tri-butyltin on granulosa cells, J. Steroid Biochem. Mol. Biol., 84, 291–300, https://doi.org/10.1016/S0960-0760(03)00042-6, 2003.

Silva, C. C. and Knight, P. G.: Effects of androgens, progesterone and their antagonists on the developmental competence of in vitro matured bovine oocytes, J. Reprod. Fertil., 119, 261–269, https://doi.org/10.1530/jrf.0.1190261, 2000.

Syoji, K., Imai, K., Koyama, H., and Dochi, O.: Effect of progesterone supplementation of maturation medium on the development of in vitro-matured-in vitro-fertilized-in vitro-cultured bovine embryos, Reprod. Fertil. Dev., 26, 180 (abstr.), https://doi.org/10.1071/RDv26n1Ab133, 2014.

Takahashi, Y. and First, N. L.: In vitro development of bovine one-cell embryos: Influence of glucose, lactate, pyruvate, amino acids and vitamins, Theriogenology, 37, 963–978, https://doi.org/10.1016/0093-691X(92)90096-A, 1992.

Takahashi, Y., Hishinuma, M., Matsui, M., Tanaka, H., and Kanagawa, H.: Development of in vitro matured/fertilized bovine embryos in a chemically defined medium: influence of oxygen concentration in the gas atmosphere, J. Vet. Med. Sci., 58, 897–902, https://doi.org/10.1292/jvms.58.897, 1996.

Wrenzycki, C. and Stinshoff, H.: Maturation environment and impact on subsequent developmental competence of bovine oocytes, Reprod. Dom. Anim., 48, 38–43, https://doi.org/10.1111/rda.12204, 2013.

Younis, A. I., Brackett, B. G., and Fayrer-Hosken, R. A.: Influence of serum and hormones on bovine oocyte maturation and fertilization in vitro, Gamete Res., 23, 189–201, https://doi.org/10.1002/mrd.1120230206, 1989.

Permissions

All chapters in this book were first published in AAB, by Copernicus Publications; hereby published with permission under the Creative Commons Attribution License or equivalent. Every chapter published in this book has been scrutinized by our experts. Their significance has been extensively debated. The topics covered herein carry significant findings which will fuel the growth of the discipline. They may even be implemented as practical applications or may be referred to as a beginning point for another development.

The contributors of this book come from diverse backgrounds, making this book a truly international effort. This book will bring forth new frontiers with its revolutionizing research information and detailed analysis of the nascent developments around the world.

We would like to thank all the contributing authors for lending their expertise to make the book truly unique. They have played a crucial role in the development of this book. Without their invaluable contributions this book wouldn't have been possible. They have made vital efforts to compile up to date information on the varied aspects of this subject to make this book a valuable addition to the collection of many professionals and students.

This book was conceptualized with the vision of imparting up-to-date information and advanced data in this field. To ensure the same, a matchless editorial board was set up. Every individual on the board went through rigorous rounds of assessment to prove their worth. After which they invested a large part of their time researching and compiling the most relevant data for our readers.

The editorial board has been involved in producing this book since its inception. They have spent rigorous hours researching and exploring the diverse topics which have resulted in the successful publishing of this book. They have passed on their knowledge of decades through this book. To expedite this challenging task, the publisher supported the team at every step. A small team of assistant editors was also appointed to further simplify the editing procedure and attain best results for the readers.

Apart from the editorial board, the designing team has also invested a significant amount of their time in understanding the subject and creating the most relevant covers. They scrutinized every image to scout for the most suitable representation of the subject and create an appropriate cover for the book.

The publishing team has been an ardent support to the editorial, designing and production team. Their endless efforts to recruit the best for this project, has resulted in the accomplishment of this book. They are a veteran in the field of academics and their pool of knowledge is as vast as their experience in printing. Their expertise and guidance has proved useful at every step. Their uncompromising quality standards have made this book an exceptional effort. Their encouragement from time to time has been an inspiration for everyone.

The publisher and the editorial board hope that this book will prove to be a valuable piece of knowledge for researchers, students, practitioners and scholars across the globe.

List of Contributors

Agnieszka J. Rozbicka-Wieczorek, Katarzyna A. Krajewska-Bienias and Marian Czauderna
The Kielanowski Institute of Animal Physiology and Nutrition, Polish Academy of Sciences, 05-110 Jabłonna, Poland

Mohammad Taghi Vajed Ebrahimi, Mohammadreza Mohammadabadi, and Ali Esmailizadeh
Department of Animal Science, Faculty of Agriculture, Shahid Bahonar University of Kerman, Kerman, Iran

Buhari Habibu, Mohammed Kawu, Tagang Aluwong, Lukman Yaqub and Tavershima Dzenda
Department of Veterinary Physiology, Ahmadu Bello University, Zaria, Nigeria

Hussaina Makun
Small Ruminant Research Programme, National Animal Production Research Institute, Zaria, Nigeria

Hajarah Buhari
Samaru College of Agriculture, Division of Agricultural Colleges, Ahmadu Bello University, Zaria, Nigeria

José Carlos Ferreira-Silva, Paulo Francisco Maciel Póvoas Souto, Paulo Castelo Branco Gouveia Filho, Marcelo Tigre Moura and Marcos Antonio Lemos Oliveira
Laboratório de Biotécnicas Aplicadas à Reprodução, Universidade Federal Rural de Pernambuco, Recife, PE, Brasil

Tracy Anne Burnett
Faculty of Land and Food Systems. University of British Columbia, Vancouver, BC, V6T 1Z4, Canada

Lucas Carvalho Pereira and Mariana Vieira Araujo
Laboratório Nordeste In Vitro, Maceió, AL, Brasil

Tomasz Stankiewicz
West Pomeranian University of Technology, Szczecin, Faculty of Biotechnology and Animal Husbandry, Department of Animal Reproduction Biotechnology and Environmental Hygiene, 29 Klemensa Janickiego Street, 71-270 Szczecin, Poland

Amir Meimandipour and Ali Soleimani
Department of Animal Biotechnology, National Institute of Genetic Engineering and Biotechnology (NIGEB), Tehran, Iran

Ali Nouri Emamzadeh
Department of Animal Science, Garmsar Branch, Islamic Azad University, Garmsar, Iran

Ling Chen, Mengxing Zhai, Si Chen, Na Li and Xiaolin Liu
College of Animal Science and Technology, Northwest A&F University, Shaanxi Key Laboratory of Molecular Biology for Agriculture, Yangling, Shaanxi 712100, P.R. China

HongliangWang
State Key Laboratory for Molecular Biology of Special Economic Animals, Institute of Special Animal and Plant Sciences, Chinese Academy of Agricultural Sciences, Changchun 130112, P.R. China

Zhixiong Li
College of Animal Science and Technology, Northwest A&F University, Shaanxi Key Laboratory of Molecular Biology for Agriculture, Yangling, Shaanxi 712100, P.R. China
College of Life Science and Technology, Southwest University for Nationalities, Chengdu 610041, P.R. China

RalfWassmuth
University of Applied Sciences, Am Kruempel 31, 49090 Osnabrück, Germany

Christoph Biestmann and Heiko Janssen
Chamber of Agriculture Lower Saxony, Mars-la-Tour-Str. 6, 26121 Oldenburg, Germany

Lana Vranković, Jasna Aladrović and Zvonko Stojević
Department of Physiology and Radiobiology, University of Zagreb Faculty of Veterinary Medicine, Heinzelova 55, 10000 Zagreb, Croatia

Daria Octenjak, Dušanka Bijelić and Luka Cvetnić
Students of Faculty of Veterinary Medicine, University of Zagreb, Heinzelova 55, 10000 Zagreb, Croatia

Sophie Rothammer and Ivica Medugorac
Chair of Animal Genetics and Husbandry, LMU Munich, 80539 Munich, Germany

Maren Bernau and Armin M. Scholz
Livestock Center of the Faculty of Veterinary Medicine, LMU Munich, 85764 Oberschleissheim, Germany

Prisca V. Kremer-Rücker
Livestock Center of the Faculty of Veterinary Medicine, LMU Munich, 85764 Oberschleissheim, Germany
University of Applied Sciences Weihenstephan-Triesdorf, 91746 Weidenbach, Germany

Mehdi Salmanzadeh, Yahya Ebrahimnezhad, Habib Aghdam Shahryar and Jamshid Ghiasi Ghaleh-Kandi
Departments of Animal Science, Shabestar branch, Islamic Azad University, Shabestar, Iran

Ana González, Francisco Peña and Andrés L. Martínez
Department of Animal Production, University of Córdoba, Córdoba, Spain

Dolores Ayuso and Mercedes Izquierdo
Departament of Animal Production, CICYTEX, Badajoz, Spain

Justyna Batkowska and Antoni Brodacki
Department of Biological Basis of Animal Production, University of Life Sciences in Lublin,13 Akademicka St., 20-950 Lublin, Poland

StanisławWinnicki and Jerzy Jugowar
Institute of Technology and Life Sciences, Poznań Branch, ul. Biskupińska 67, 60-463 Poznań, Poland

Zbigniew Sobek, Anna Nienartowicz-Zdrojewska and Jolanta Ró˙zańska-Zawieja
Department of Genetics and Animal Breeding, Poznan University of Life Sciences, ul. Wołyńska 33, 60-637 Poznań, Poland

Ryszard Kujawiak
Sano Agrar Institut, ul. Lipowa 10, 64-541 S_ekowo, Poland

Shifeng Pan, Hua Xing and Tangjie Zhang
College of Veterinary Medicine, Yangzhou University, Yangzhou, 225009, China
Jiangsu Co-Innovation Center for the Prevention and Control of Important Animal Infectious Disease and Zoonoses, Yangzhou, Jiangsu, 225009, China

Cong Wang, Xuan Dong and Mingliang Chen
College of Veterinary Medicine, Yangzhou University, Yangzhou, 225009, China

Guifen Liu, Hongbo Zhao, Xiuwen Tan, Haijian Cheng, Wei You, Fachun Wan, Yifan Liu, Enliang Song and Xiaomu Liu
Shandong Key Lab of Animal Disease Control and Breeding, Sangyuan Road 8 Number, Ji'nan City, Shandong Province, 250100, China

Institute of Animal Science and Veterinary Medicine, Shandong Academy of Agricultural Sciences, Sangyuan Road 8 Number, Ji'nan City, Shandong Province, 250100, China

Katarzyna Ropka-Molik, Tomasz Szmatoła and Katarzyna Piórkowska
Department of Genomics and Animal Molecular Biology, National Research Institute of Animal Production, 32-083 Balice, Poland

Jan Knapik
Department of Animal Genetics and Breeding, National Research Institute of Animal Production, 32-083 Balice, Poland

Marek Pieszka
Department of Animal Nutrition and Feed Science, National Research Institute of Animal Production, 32-083 Balice, Poland

Janez Jeretina and Drago Babnik
Agricultural Institute of Slovenia, Hacquetova ulica 17, 1000 Ljubljana, Slovenia

Dejan Škorjanc
University of Maribor, Faculty of Agriculture and Life Sciences, Pivola 10, 2311 Hoče, Slovenia

Frank Liebert
Chair of Animal Nutrition, University of Göttingen, Kellnerweg 6, 37077 Göttingen, Germany

Han Xu, Sihuan Zhang, Xiaoyan Zhang, Ruihua Dang, Chuzhao Lei, Hong Chen, and Xianyong Lan
College of Animal Science and Technology, Northwest A&F University, Shaanxi Key Laboratory of Molecular Biology for Agriculture, Yangling, Shaanxi 712100, China

Jieping Huang, Lijun Jiang, Fen Li and Yun Ma
College of Life Sciences, Xinyang Normal University, Xinyang, Henan, China
Institute for Conservation and Utilization of Agro-Bioresources in Dabie Mountains, Xinyang, Henan, China

Qiuzhi Zheng, Shuzhe Wang, Ruijie Hao and Qiongqiong Zhang
College of Life Sciences, Xinyang Normal University, Xinyang, Henan, China

Hulya Yalcintan, Bulent Ekiz, Omur Kocak, Nursen Dogan, P. Dilara Akin and Alper Yilmaz
Istanbul University Veterinary Faculty, Department of Animal Breeding and Husbandry, Avcilar, 34320 Istanbul, Turkey

M. Maaouia A. Moussa, Moumouni Issa, Moustapha Grema, Marichatou Hamani and Yenikoye Alhassane
Université Abdou Moumouni de Niamey, Faculté des sciences et techniques, BP 10960 Niamey, Niger

Amadou Traoré, Moumouni Sanou and Hamidou H. Tamboura
INERA, Laboratoire de Biologie et santé animals, 04 BP 8645 Ouagadougou 04, Burkina Faso

Iván Fernández, Isabel Álvarez and Félix Goyache
SERIDA-Deva, Camino de Rioseco 1225, 33394 Deva-Gijón (Asturias), Spain

Albert Soudré
Université de Koudougou, BP 376 Koudougou, Burkina Faso

Sena Ardicli, Hale Samli, Deniz Dincel, Bahadir Soyudal and Faruk Balci
Laboratory of Genetics, Department of Genetics, Faculty of Veterinary Medicine, Uludag University, 16059 Nilufer, Bursa, Turkey

Minami Matsuo, Kazuma Sumitomo, Chihiro Ogino, Yosuke Gunji, Ryo Nishimura and Mitsugu Hishinuma
Laboratory of Theriogenology, Joint Department of Veterinary Medicine, Faculty of Agriculture, Tottori University, 4-101 Koyama-minami, Tottori 680-8553, Japan

Index